Periodic Table of the Elements with the Gmelin System Numbers

1 H 2																1 H 2	2 He 1
3 Li 20	4 Be 26											5 B 13	6 C 14	7 N 4	8 O 3	9 F 5	10 Ne 1
11 Na 21	12 Mg 27											13 Al 35	14 Si 15	15 P 16	16 S 9	17 Cl 6	18 Ar 1
19 * K 22	20 Ca 28	21 Sc 39	22 Ti 41	23 V 48	24 Cr 52	25 Mn 56	26 Fe 59	27 Co 58	28 Ni 57	29 Cu 60	30 Zn 32	31 Ga 36	32 Ge 45	33 As 17	34 Se 10	35 Br 7	36 Kr 1
37 Rb 24	38 Sr 29	39 Y 39	40 Zr 42	41 Nb 49	42 Mo 53	43 Tc 69	44 Ru 63	45 Rh 64	46 Pd 65	47 Ag 61	48 Cd 33	49 In 37	50 Sn 46	51 Sb 18	52 Te 11	53 I 8	54 Xe 1
55 Cs 25	56 Ba 30	57** La 39	72 Hf 43	73 Ta 50	74 W 54	75 Re 70	76 Os 66	77 Ir 67	78 Pt 68	79 Au 62	80 Hg 34	81 Tl 38	82 Pb 47	83 Bi 19	84 Po 12	85 At 8a	86 Rn 1
87 Fr 25a	88 Ra 31	89*** Ac 40	104 71	105 71													* NH_4 23

**Lanthanides 39	58 Ce	59 Pr	60 Nd	61 Pm	62 Sm	63 Eu	64 Gd	65 Tb	66 Dy	67 Ho	68 Er	69 Tm	70 Yb	71 Lu
***Actinides	90 Th 44	91 Pa 51	92 U 55	93 Np 71	94 Pu 71	95 Am 71	96 Cm 71	97 Bk 71	98 Cf 71	99 Es 71	100 Fm 71	101 Md 71	102 No 71	103 Lr 71

A Key to the Gmelin System is given on the Inside Back Cover

Gmelin Handbook of Inorganic and Organometallic Chemistry

8th Edition

Gmelin Handbook of Inorganic and Organometallic Chemistry

8th Edition

Gmelin Handbuch der Anorganischen Chemie

Achte, völlig neu bearbeitete Auflage

PREPARED AND ISSUED BY

Gmelin-Institut für Anorganische Chemie
der Max-Planck-Gesellschaft
zur Förderung der Wissenschaften

Director: Ekkehard Fluck

FOUNDED BY Leopold Gmelin

8TH EDITION 8th Edition begun under the auspices of the Deutsche Chemische Gesellschaft by R. J. Meyer

CONTINUED BY E. H. E. Pietsch and A. Kotowski, and by Margot Becke-Goehring

Springer-Verlag Berlin Heidelberg GmbH

GMELIN HANDBOOK

Dr. J. von Jouanne

Dr. L. Berg, Dr. H. Bergmann, Dr. J. Faust, J. Füssel, Dr. H. Katscher, Dr. R. Keim, Dipl.-Phys. D. Koschel, Dr. A. Kubny, Dr. P. Merlet, Dr. M. Mirbach, Prof. Dr. W. Petz, Dr. F. A. Schröder, Dr. A. Slawisch, Dr. W. Töpper

Dr. R. Albrecht, Dr. G. Bär, D. Barthel, Dr. N. Baumann, Dr. K. Behrends, Dr. W. Behrendt, D. Benzaid, Dr. R. Bohrer, K. D. Bonn, Dipl.-Chem. U. Boßlet, Dr. U. Busch, A.-K. Castro, Dipl.-Ing. V. A. Chavizon, E. Cloos, A. Dittmar, Dipl.-Geol. R. Ditz, R. Dowideit, Dr. H.-J. Fachmann, B. Fischer, Dr. D. Fischer, Dipl.-Ing. N. Gagel, Dr. K. Greiner, Dipl.-Bibl. W. Grieser, Dr. R. Haubold, Dipl.-Min. H. Hein, H.-P. Hente, H. W. Herold, U. Hettwer, G. Hoell, Dr. G. Hönes, Dr. W. Hoffmann, G. Horndasch, Dr. W. Huisl, Dr. M. Irmler, B. Jaeger, Dr. R. Jotter, Dipl.-Chem. P. Kämpf, Dr. B. Kalbskopf, H.-G. Karrenberg, Dipl.-Phys. H. Keller-Rudek, Dipl.-Chem. C. Koeppel, Dr. M. Körfer, R. Kolb, Dr. M. Kotowski, E. Kranz, E. Krawczyk, Dipl.-Chem. I. Kreuzbichler, Dr. V. Kruppa, Dr. W. Kurtz, M. Langer, Dr. B. Ledüc, H. Mathis, E. Meinhard, M. Meßer, C. Metz, K. Meyer, E. Mlitzke, Dipl.-Chem. B. Mohsin, Dr. U. Neu-Becker, K. Nöring, Dipl.-Min. U. Nohl, Dr. U. Ohms-Bredemann, Dr. H. Pscheidl, Dipl.-Phys. H.-J. Richter-Ditten, E. Rudolph, G. Rudolph, Dr. B. Sarbas, Dr. H. Schäfer, Dr. R. Schemm, Dr. D. Schiöberg, V. Schlicht, Dipl.-Chem. D. Schneider, E. Schneider, A. Schwärzel, Dr. B. Schwager, R. Simeone, Dr. F. Stein, Dr. C. Strametz, Dr. G. Swoboda, Dr. D. Tille, A. Tuttas, Dipl.-Phys. J. Wagner, R. Wagner, M. Walter, Dr. E. Warkentin, Dr. C. Weber, Dr. A. Wietelmann, Dr. M. Winter, Dr. B. Wöbke, K. Wolff

GMELIN ONLINE

Dr. R. Deplanque

Dr. P. Kuhn, Dr. G. Olbrich

Dr. R. Baier, Dr. B. Becker, Dipl.-Chem. E. Best, Dr. H.-U. Böhmer, Dipl.-Phys. R. Bost, Dr. A. Brandl, Dr. R. Braun, Dr. T. Buck, Dipl.-Chem. R. Durban, R. Hanz, Dipl.-Phys. C. Heinrich-Sterzel, Dr. S. Kalwellis-Mohn, Dr. A. Kirchhoff, Dipl.-Chem. H. Köttelwesch, Dr. M. Kunz, Dipl.-Chem. R. Maass, Dr. K. Müller, Dr. A. Nebel, Dipl.-Chem. R. Nohl, Dr. M. Nohlen, H. Reinhardt, Dr. B. Rempfer, Dipl.-Ing. H. Vanecek

Gmelin Handbook of Inorganic and Organometallic Chemistry

8th Edition

B

Boron Compounds

4th Supplement Volume 2

Boron and Oxygen

With 23 illustrations

AUTHOR Gert Heller, Freie Universität Berlin, Institut für Anorganische und Analytische Chemie, Berlin

EDITORIAL ASSISTANCE Rainer Bohrer, Gmelin-Institut, Frankfurt/Main

EDITORS Jürgen Faust, Gmelin-Institut, Frankfurt/Main

Kurt Niedenzu, Department of Chemistry, University of Kentucky, Lexington, Kentucky, USA

System Number 13

Springer-Verlag Berlin Heidelberg GmbH

LITERATURE CLOSING DATE: END OF 1988
IN SOME CASES MORE RECENT DATA HAVE BEEN CONSIDERED

Library of Congress Catalog Card Number: Agr 25-1383

ISBN 978-3-662-06152-7 ISBN 978-3-662-06150-3 (eBook)
DOI 10.1007/978-3-662-06150-3

Originally published by Springer-Verlag Berlin in 1993.
Softcover reprint of the hardcover 8th edition 1993

Preface

The present issue, Volume 2 of "Boron Compounds" 4th Supplement of the Gmelin Handbook, updates the previous issues by reporting the literature on boron–oxygen systems published up to 1988. For some important recent developments literature is covered through mid-1992; this concerns, for example, the compounds β-$Ba_3[B_3O_6]_2$ and $Li[B_3O_5]$ which became of interest as materials with nonlinear optical properties. The volume directly complements the earlier "Boron Compounds" 3rd Supplement Volume 2.

In the original literature, alternative formulations are frequently used for the same compound. This is especially true for many borates. Often, these species are neither completely heteropolar nor covalent, and an experimentally based decision has not been made. Hence, the use of brackets does not necessarily reflect a truly salt-like character.

Volume 1 (systems with hydrogen and noble gases) of this particular supplement will be published subsequently, whereas Volume 3a (boron and nitrogen), Volume 3b (boron and nitrogen, boron and fluorine), and Volume 4 (boron compounds containing Cl, Br, I, S, Se, and Te, as well as a section containing carboranes) have already been published. All volumes of the 4th supplement will be augmented by a formula index.

The IUPAC nomenclature is generally adhered to; thf means tetrahydrofuran; and occasionally additional abbreviations for compounds are explained in the text. Positive signs for chemical shifts of the NMR signals indicates downfield shifts from the references, usually internal $(CH_3)_4Si$ for δ^1H and $\delta^{13}C$ with others being specified.

Lexington, Kentucky (USA) — Kurt Niedenzu
Frankfurt am Main — Jürgen Faust
October 1993

Boron and Boron Compounds in the Gmelin Handbook (Syst. No. 13)

"Bor" (Main Volume)	Historical Occurrence. The Element. Compounds of B with H, O, N, the Halogens, S, Se, and Te. Literature closing date: end of 1925.
"Bor" (Supplement Volume 1)	Occurrence. The Element. Compounds of B with H, O, N, the Halogens, S, and C. Literature closing date: end of 1949.
"Borverbindungen" 1	Boron Nitride. B–N–C Heterocycles. Polymeric B–N Compounds. Literature coverage from 1950 up to 1972.
"Borverbindungen" 2	Carboranes. Part 1. Nomenclature and Types of Carboranes. Carboranes (without Hetero- and Metallocarboranes, and Higher Carboranes). Literature coverage from 1950 up to 1973 or 1970, respectively.
"Borverbindungen" 3	Compounds of B Containing Bonds to S, Se, Te, P, As, Sb, Si, and Metals. Literature coverage from 1950 to the end of 1973.
"Borverbindungen" 4	Compounds with Isolated Trigonal Boron Atoms and Covalent Boron-Nitrogen Bonding (Aminoboranes and B–N Heterocycles). Literature coverage from 1950 to the end of 1973.
"Borverbindungen" 5	Boron-Pyrazole. Derivatives and Spectroscopic Studies on Trigonal B–N Compounds. Literature coverage from 1950 to the end of 1973.
"Borverbindungen" 6	Carboranes, Part 2. Hetero- and Metallocarboranes. Polymeric Carborane Derivatives. Electronic Properties. Literature coverage from 1950 up to 1974 or 1971, respectively.
"Borverbindungen" 7	Boron Oxides. Boric Acids. Borates. Literature coverage from 1950 to the end of 1973.
"Borverbindungen" 8	The Tetrahydroborate Ion and Its Derivatives. Literature coverage from 1950 to the end of 1974.
"Borverbindungen" 9	Boron–Halogen Compounds, Part 1. Literature coverage from 1950 to the end of 1974.
"Borverbindungen" 10	Boron Compounds with Coordination Number 4. Literature coverage from 1950 to the end of 1975.
"Borverbindungen" 11	Carboranes. Part 3. Dicarba-*closo*-dodecaboranes. Literature coverage from 1950 to the end of 1975.
"Borverbindungen" 12	Carboranes. Part 4. Dicarba-*closo*-dodecaboranes. Literature coverage from 1950 to the end of 1975.
"Borverbindungen" 13	Boron-Oxygen Compounds, Part 1. Literature coverage from 1950 to the end of 1975.
"Borverbindungen" 14	Boron-Hydrogen Compounds, Part 1. Literature coverage from 1950 to the end of 1975.
"Borverbindungen" 15	Amine-boranes. Literature coverage from 1950 to the end of 1975.
"Borverbindungen" 16	Boron-Oxygen Compounds, Part 2. Literature coverage from 1950 to the end of 1975.

"Borverbindungen" 17	Borazine and its Derivatives. Literature coverage from 1950 to the end of 1976.
"Borverbindungen" 18	Boron-Hydrogen Compounds, Part 2. Literature coverage from 1950 to the end of 1976.
"Borverbindungen" 19	Boron-Halogen Compounds, Part 2. Literature coverage from 1950 to the end of 1976.
"Borverbindungen" 20	Boron-Hydrogen Compounds, Part 3. Literature coverage from 1950 to the end of 1976.
"Boron Compounds"	Formula Index (for the volumes "Borverbindungen" 1 to 20).
"Boron Compounds" 1st Suppl. Vol. 1	Boron and Rare Gases. Boron and Hydrogen Boron and Oxygen. Literature coverage through 1977.
"Boron Compounds" 1st Suppl. Vol. 2	Boron and Nitrogen. Boron and Halogens. Literature coverage through 1977.
"Boron Compounds" 1st Suppl. Vol. 3	Boron and Chalcogens. Carboranes. Formula Index for 1st Suppl. Vol. 1 to 3. Literature coverage through 1977.
"Boron Compounds" 2nd Suppl. Vol. 1	Boron and Nobles Gases. Boron and Hydrogen. Boron and Oxygen. Boron and Nitrogen. Formula Index. Literature coverage through 1980.
"Boron Compounds" 2nd Suppl. Vol. 2	Boron and Chalcogens. Boron and Chalcogens. Carboranes. Formula Index. Literature coverage through 1980.
"Boron Compounds" 3rd Suppl. Vol. 1	Boron and Hydrogen. Literature coverage through 1984.
"Boron Compounds" 3rd Suppl. Vol. 2	Boron and Oxygen. Literature coverage through 1984.
"Boron Compounds" 3rd Suppl. Vol. 3	Boron and Nitrogen. Boron and Fluorine – 1988. Literature coverage through 1984.
"Boron Compounds" 3rd Suppl. Vol. 4	Boron and Cl, Br, I, S, Se, Te. Carboranes – 1988. Literature coverage through 1984.
"Boron Compounds"	Formula Index – 1988 (for the volumes "Boron Compounds" 3rd Suppl. Vol. 1 to 4).
"Boron Compounds" 4th Suppl. Vol. 2	Boron and Oxygen – 1993. Literature coverage through 1988 (**present volume**)
"Boron Compounds" 4th Suppl. Vol. 3a	Boron and Nitrogen – 1991. Literature coverage through 1988.
"Boron Compounds" 4th Suppl. Vol. 3b	Boron and Nitrogen, Boron and Fluorine – 1992. Literature coverage through 1988.
"Boron Compounds" 4th Suppl. Vol. 4	Boron and Cl, Br, I, S, Se, Te. Carboranes – 1991. Literature coverage through 1988.

Table of Contents

3 The System Boron–Oxygen

Gert Heller
Institut für Anorganische und Analytische Chemie
Freie Universität Berlin
Federal Republic of Germany

3.1 General Remarks

This presentation covers the years 1985 to 1988 and, in part, to 1992. It continues the previous discussion of the system boron–oxygen in "Boron Compounds" 3rd Suppl. Vol. 2, 1987, pp. 1/184, and earlier literature cited therein. All formulas including radicals are written according to IUPAC nomenclature [1].

3.2 Boron Oxides

3.2.1 The Boron Monoxide Radical, $[BO]^{\bullet}$

The **$[BO]^{\bullet}$** radical is one of the fundamental species appearing as intermediates in oxidation reactions. Hot-pressed B_4C was oxidized in wet nitrogen and/or wet air atmosphere at a water vapor pressure of 1.5 to 20 kPa and in dry air of 900 up to 1400°C. The oxidation of B_4C by water vapor and dry air at 900°C results in a weight loss; hence, it is assumed that the formation of $[BO]^{\bullet}$ (and/or HBO_2) seems to be the major reaction. The oxidation rate of B_4C by water vapor at 900°C could be expressed by surface chemical reaction-controlled kinetics with an apparent activation energy of 200 kJ/mol [2].

In the reaction of molecular oxygen with crystalline α-boron powder, oxygen atoms in concentrations exceeding the thermodynamical equilibrium value (above an inert surface) by four to seven orders of magnitude were observed. By the use of the resonance fluorescence of oxygen atoms between 350 and 1000°C, the proposed mechanism of the reaction is: $(B)+O_2 \rightarrow [BO]^{\bullet}+O^{\bullet}$; $O^{\bullet}+(B) \rightarrow [BO]^{\bullet}$; and $[BO]^{\bullet}+O_2 \rightarrow [BO_2]^{\bullet}+O^{\bullet}$; the detection of boron-containing radicals in the reaction zone should be an important subsequent step in the study of this reaction [3].

Boron ignition and combustion with its surrounding atmosphere was studied by time-resolved emission spectroscopy in the range of 0.2 to 5.5 μm. In one experiment, boron powder was burned in an oxygen atmosphere with initiation by a pyrotechnic device. The energy transfer by the hot gases led to a glowing phase of the boron particles, which then changed to a high-temperature combustion and ended in a second glowing phase. The two glowing phases emitted continuous emission spectra, and the burning phase emitted the bands of $[BO]^{\bullet}$ and $[BO_2]^{\bullet}$. Another experimental setup was used to feed boron particles into a hot oxidizing atmosphere, which was provided by a propane-air flame. The reaction was observed by high-speed cinematography and time-resolved emission spectroscopy. The flame contained a small amount of background radiation, and the emitted bands of $[BO]^{\bullet}$, $[BO_2]^{\bullet}$, HBO_2, CO, and CO_2 were identified [4].

Chemical excitation processes, including the formation of $[BO]^{\bullet}$, have been classified; the expansion of the spectral band width of the radiation from electronic-transition chemical lasers was analyzed [5].

In studying the temperature dependence of the kinetics of gas-phase boron oxidation reactions, experimental measurements were made between 440 and 1830 K. The kinetic behavior of the boron atom and boron monoxide radical oxidation reactions as affected by the temperature is of interest. The measurements were made using the high-temperature fast-flow reactor technique for isolated elementary reactions in a heat bath. Laser-induced fluorescence was used to monitor the boron atom or radical reactant concentrations as a function of time, concentration of the (excess) oxidant, temperature, and pressure [6].

The cross-contamination during silicon implantation in GaAs with **$[^{11}B^{18}O]^{\bullet}$**, up to 40% of the dose for the ^{29}Si implantation, was determined by secondary ion mass spectrometry (SIMS) and by electrical measurements [7]. The rate constant for the reaction of boron, instantaneously produced by photolysis of BCl_3 with an excimer laser, and oxygen in argon at 298 K and 25 Torr total pressure giving $[BO]^{\bullet}$ was measured as 7×10^{-11} mL·molecule^{-1}·s^{-1}. This value was determined using the laser photolysis/laser-induced fluorescence technique; the value increases slightly with temperature up to 1180 K [8].

The microwave spectrum of the $[BO]^{\bullet}$ radical generated in an absorption cell by a glow discharge in a gaseous mixture of BCl_3, nitric oxide, and nitrogen was observed [9]. A source configuration which lies intermediate between a low-pressure effusing molecular beam and a high-pressure flow device was used to generate large concentrations of small clusters in a highly oxidizing environment, for example, boron in N_2O or NO_2 at $\leq$2520 K and 10^{-1} Torr. From these studies the first quantal information on the energy levels and optical signatures of $[BO]^{\bullet}$ and $[BO_2]^{\bullet}$ clusters was obtained. The study outlines the potential for chemiluminescent probes of the cluster quantum levels, not only within themselves but also as a means of suggesting future laser fluorescent probes. Chemiluminescent spectra are depicted that result from the multiple collision oxidation of boron with NO_2 in argon showing the $[BO]^{\bullet}$ A $^2\Pi \rightarrow$ X $^2\Sigma$ emission system emanating from v′ = 0, 1, 2, A $^2\Pi$, and what appears to be a new system at over 5900 Å (when the boron flux is increased substantially) [10].

The microwave spectrum of the $[BO]^{\bullet}$ radical in the electronic ground state X $^2\Sigma^+$ was observed by using a glow discharge cell spectrometer. The observed transitions were N = 1-0 and 3-2 in the 100 and 300 GHz regions, respectively. The latter transition was observed by using a frequency multiplier working in the 300 GHz region as a microwave source. The spectra of the two isotopic species $[^{11}BO]^{\bullet}$ and $[^{10}BO]^{\bullet}$ were analyzed to obtain molecular parameters including the rotational constants B, the centrifugal distortion constants D, the spin-rotation interaction constants γ, the magnetic hyperfine interaction constants b and c, the nuclear quadrupole interaction constants eQq, the nuclear spin-rotation interaction constants C_I, and the hyperfine constants $A_{iso} = (b+c)/3$ and $A_{dip} = c/3$; see Table 3/1. The spin density of the unpaired σ electron was calculated from the magnetic hyperfine coupling constants and was found to be in good accordance with the result of ab initio calculations: $|\Psi(O)|^2 = 0.7164(2)$ and $\langle(3\cos^2\Theta - 1)/r^3\rangle = 0.3169(1)$ [9].

For both $[^{10}BO]^{\bullet}$ and $[^{11}BO]^{\bullet}$, emission spectra between 2000 and 7500 Å were obtained. A full rotational analysis with 29 vibrational bands of the A–X band system and 33 bands of the B–X system was performed, and molecular parameters were obtained. The emission spectra of these two isotopic species were excited by a microwave discharge. The A $^2\Pi_i \rightarrow$ X $^2\Sigma^+$ and B $^2\Sigma^+ \rightarrow$ X $^2\Sigma^+$ transitions were analyzed, and rotational and vibrational constants were determined with a good accuracy for the three electronic states. From rotational constants and band origin, the following equilibrium constants (see Table 3/2) have been determined: for the

[^{11}BO]• ground state X $^2\Sigma^+$ (constants in cm^{-1}) B_0=1.772861(79), D_0=6.236(35)×10^{-6}, and γ_0= 0.359(40)×10^{-2}. A previously unreported perturbation in the v=8 level of the A $^2\Pi_i$ state was interpreted as due to the v″=20 level of the X $^2\Sigma^+$ ground state [11].

Table 3/1
Molecular Parameters of [BO]• (in MHz)*) [9].

constant	for [^{11}BO]•	for [^{10}BO]•
B	53165.5156(46)	56291.2073(67)
D	0.19148(26)	0.21581(37)
γ	177.604(10)	188.033(14)
b	1000.27(31)	344.890(56)
c	81.382(32)	27.146(39)
eQq	−2.486(38)	−5.14(12)
C_I	0.0190(28)	0.0050(20)
A_{iso}	1027.40(31)	343.939(58)
A_{dip}	27.127(11)	9.049(13)

*) Values in parentheses are one standard error in units of the last significant figure of the corresponding constant.

Table 3/2
Spectroscopic Characteristics for the [^{10}BO]• and [^{11}BO]• States [11].

[^{11}BO]•	X $^2\Sigma^+$	A $^2\Pi_i$	B $^2\Pi_i$	
T_e	0	23897.153(32)	43173.192(68)	(in eV)
ω_e	1885.286(41)	1260.782(28)	1283.318(60)	(in cm^{-1})
$\omega_e\chi_e$	11.694(11)	11.1914(11)	11.562(21)	(in cm^{-1})
$\omega_e y_e$	−0.00952(83)	0.05411(65)	0.1655(21)	(in cm^{-1})
B_e	1.718110(31)	1.411199(67)	1.516751(89)	(in cm^{-1})
$D_e\cdot10^6$	6.247(14)*)	6.935(18)*)	8.392(23)*)	(in eV)
A_e	—	−122.459(28)	—	

[^{10}BO]•	X $^2\Sigma^+$	A $^2\Pi_i$	B $^2\Pi_i$	
T_e	0	23897.461(43)	43173.84(11)	(in eV)
ω_e	1940.308(23)	1297.466(36)	1320.54(17)	(in cm^{-1})
$\omega_e\chi_e$	12.4873(42)	11.861(10)	12.230(84)	(in cm^{-1})
$\omega_e y_e$	—	0.05873(82)	0.179(11)	(in cm^{-1})
B_e	1.886156(46)	1.494275(27)	1.606260(83)	(in cm^{-1})
$D_e\cdot10^6$	6.928(20)*)	7.724(34)*)	9.222(50)*)	(in eV)
A_e	—	−122.485(18)	—	

*) Mean values.

Configuration-interaction studies were performed on low-lying $^2\Sigma^+$, $^2\Sigma^-$, $^2\Pi$, $^2\Delta$, $^4\Sigma^+$, $^4\Sigma^-$, $^4\Pi$, and $^4\Delta$ states of [BO]•; potential energy curves for all these states are shown in

 References on pp. 10/1

Fig. 3-1 and **Fig. 3-2**. Sixteen stable states have been found, among them four $3s_B$ Rydberg character states ($4\,^2\Sigma^+$, $2\,^4\Pi$, $3\,^2\Pi$, $4\,^2\Pi$). Spectroscopic constants were compared with those of the four experimental states (see Table 3/3). A comparison with the isoelectronic molecules CN and $[CO]^+$ reveals many similarities, especially with $[CO]^+$ [12].

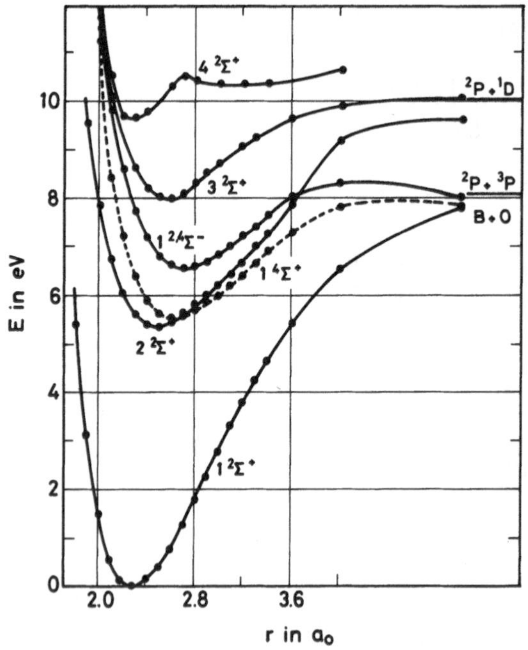

Fig. 3-1. Potential curves for Σ^+ and Σ^- states of $[BO]^\bullet$ [12].

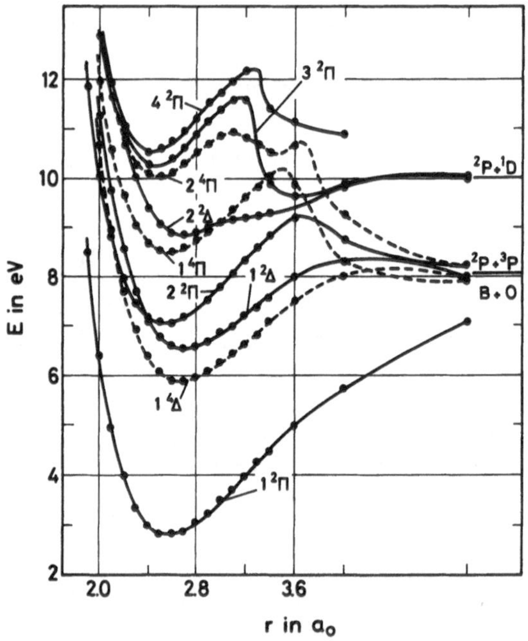

Fig. 3-2. Potential curves for Π and Δ states of $[BO]^\bullet$ [12].

Table 3/3
Calculated Spectroscopic Constants of [BO]• [12] and Comparison with Experimental Results (exp) [11].

molecular state	T_e (eV)	D_e (eV)	r_e (pm)	ω_e (cm^{-1})	$\omega_e x_e$ (cm^{-1})	B_e (cm^{-1})	α_e (cm^{-1})
1X $^2\Sigma^+$	0	8.14	121.1	1870.1	15.5	1.763	0.0064
exp	0	8.32	120.5	1885.3	11.7	1.781	0.0165
1A $^2\Pi$	2.81	5.33	135.7	1284.0	14.7	1.405	0.0264
exp	2.96	—	135.3	1260.8	11.2	1.411	0.0185
2B $^2\Sigma^+$	5.35	4.69	130.8	1289.6	3.3	1.511	0.0267
exp	5.35	—	130.5	1283.3	11.6	1.516	0.0220
1 $^4\Sigma^+$	5.49	2.64	139.7	1237.6	3.6	1.324	0.0132
1 $^4\Delta$	5.90	2.24	141.3	1246.0	15.5	1.294	0.0152
1 $^2\Delta$	6.55	1.20	143.1	1185.3	14.5	1.262	0.0148
1 $^2\Sigma^-$	6.55	1.59	143.4	1179.5	6.8	1.257	0.0130
1 $^4\Sigma^-$	6.57	1.57	142.8	1184.7	20.2	1.268	0.0219
2C $^2\Pi$	7.06	1.08	132.6	1306.4	14.0	1.442	0.0200
exp	6.86	—	132.0	1315.3	11.1	1.483	0.018
3 $^2\Sigma^+$	7.92	2.12	135.6	1514.9	20.0	1.406	0.0109
1 $^4\Pi$	8.53	—	136.0	1266.0	14.8	1.397	0.0185
2 $^2\Delta$	8.85	1.20	143.9	1140.0	6.3	1.249	0.0052
4 $^2\Sigma^+$	9.62	—	120.4	≈2126	≈200.2	1.786	—
2 $^4\Pi$	10.06	—	131.5	1345.6	−20.1	1.498	0.0027
3 $^2\Pi$	—	—	130.5	1428.0	−36.5	1.525	−0.0012
4 $^2\Pi$	—	—	127.5	1437.0	41.7	1.588	0.0465

The efficacy of the recently proposed Morse-Korwar-Navati oscillator model has been tested by taking the X $^2\Sigma^+$ and A $^2\Pi$ states; for comparison, the Morse function was also processed. The authors approximation method was used for testing by applying it to the A $^2\Pi \rightarrow$ X $^2\Sigma^+$ transition of the [BO]• molecule. The eigenfunctions obtained by yet another method known as Langer's approximation method have been evaluated in order to compare the performance of the two methods. Franck-Condon factors $q_{0,0}$ to $q_{2,4}$ and r-centroids, obtained by using the wave functions, have been computed by both the methods; the new method is more reliable than Langer's method [13].

A study of the efficacy of Linnett's potential constants a and b, representing the potential energy of a diatomic molecule, was undertaken using experimental values of the equilibrium internuclear distance r_e (in Å), μ_A (in atomic weight units), and $\omega_e x_e$ (in cm^{-1}). Taking the values for the X $^2\Sigma^+$ state of [BO]• from Herzberg (1950) with r_e=1.20498 a.u., D_e=9.1 eV, m=3, and

References on pp. 10/1

$n = 1.72$ Å^{-1}, there were determined $a = 286946$ Å^3/cm and $b = 1885710$ cm^{-1}. Linnett's potential for the B $^2\Sigma^+$ state of $[BO]^•$ was constructed; the results were reduced to the scale of the true potential curves as obtained by Singh and Rai [14], and were compared with the true potential with only small Δr_e values between 0.085 and 0.393 cm^{-1} [15].

The variation of the electronic transition moment with the equilibrium internuclear distance r_e in the cases of the A $^2\Pi_i \rightarrow X\ ^2\Sigma^+$ and B $^2\Sigma^+ \rightarrow X\ ^2\Sigma^+$ band transitions of $[BO]^•$ has been reinvestigated elsewhere. The conventional Morse potential does not yield meaningful information about the electronic transition moment; rather, the Rydberg-Klein-Rees-Vanderslice potential is preferred [16].

Among 66 diatomic molecules found in cosmic objects spectra, the radical $[BO]^•$ is found with an ionization potential of $I = 7.0 \pm 0.5$ eV, a dissociation energy of $D_0 = 8.28$ eV, the total atomic number $Z_1 + Z_2 = 13$, and the difference of atomic numbers $Z_1 - Z_2 = 3$, which may be of astrophysical significance, because $D_0 \geq 4.19$ eV and $Z_1 \pm Z_2 < 25$ are theoretical reasons for the existence in space [17].

Least-squares polynomial fits are presented for polyatomic partition functions mostly derived from the recently revised JANAF Thermochemical Tables. The determination of the equation of state calculations on a grid of temperatures between 1000 and 6000 K and pressures of 0.1 to 106 Pa was used to restrict the current partition function tabulation at the 10^{-5} level, for instance for $[BO]^•$ [18].

In 1970/71 Anderson and Parr calculated vibrational force constants from electronic densities, e.g., for $[BO]^•$ to be $k_e = 13.1 \times 10^5$ cgs units (experimentally $k_e = 13.5 \times 10^5$ cgs units) [19]. The dissociation energy was calculated by the same authors to be 9.2 eV (experimentally 9.2 eV) and the parameter $m_e = 516 \times 10^{21}$ cgs units (experimentally 525×10^{21} cgs units) [20]. The perfectly following density (PFD) model of Anderson and Parr [19, 20] was used to formulate a semiempirical model relating the electron density at the saddle point ρ_M of the harmonic force constant, i.e., of $[BO]^•$ to be $(\bar{c}f/\pi)^{1/2} = 0.375$ a.u. for $\rho_M = 0.325$ a.u. ($\bar{c}$ = mean constant, characteristic for the atoms; f = assumed density distribution) [21].

Molecular energy components were analyzed on the basis of the vibrational potential function in the effective nuclear charge (ENC) model. The analytical formulas for the electronic kinetic energy and the potential energies of $[BO]^•$ were derived from such an ENC model potential. The effective nuclear charge is $Z_{ij}^* = 5.1611$, the r_e-independent energy constant $W_0 = -97.2869$, and the total energy $E_0 = -60.7120$ hartree [22]. Two homogeneity hypotheses for the total molecular and electronic energies, proposed by Parr and Gadre [23], were examined through derivation of the harmonic and anharmonic force constants by making use of the intramolecular potential function in the ENC model. The homogeneity postulate for the total molecular energy can be compared favorably with that of the electronic energy. An approximate Hartree-Fock method, with a homogeneity constraint for the total molecular energy, was developed. The approximate value of a Lagrange multiplier, fixing the internuclear distance in equilibrium and homogeneity parameter $k = 7/3$ for $[BO]^•$, has been predicted to be $\lambda = -0.0668$ [24]. From detailed considerations of two homogeneity postulates of the total molecular and electronic energies proposed by Parr and Gadre [23], a new homogeneity hypothesis of the total molecular energy W is presented: $\Sigma_i Z_i(\delta W/\delta Z_i)_N = k_0 \cdot W_0 + k_r \cdot W_r$, where Z_i is the atomic number, W_0 and W_r are the r_e-independent and r_e-dependent molecular energies, and k_0 and k_r are the local and nonlocal homogeneity parameters; r_e (indexed as r) is the equilibrium internuclear distance. For $[BO]^•$, the local homogeneity parameter $k_0 = 2.3759$, and the approximate local and nonlocal Lagrange multipliers are $\lambda_0 = -0.0194$ and $\lambda_r = 0$ at the fixed homogeneity parameters $k_0 = 2.3688$ and $k_r = 2$. Approximate Hartree-Fock methods with the new homogeneity constraint were developed [25].

An annihilated-unrestricted Hartree Fock (AUHF) wave function in which the first spin contaminant has been annihilated self-consistently was used as the starting point for subsequent Møller-Plesset (MP) perturbation calculations. If the original unrestricted Hartree Fock (UHF) wave function is moderately to heavily spin contaminated, MP energies using the annihilated-unrestricted wave function (AUMP) are lower than normal even though the self-consistent field (SCF) energy is higher; see Table 3/4 [26].

Table 3/4
SCF and MP Energies (in hartree) for the [BO]• Radical [26].
For r=118.7 pm; (A)UHF means (annihilated-) unrestricted Hartree-Fock and ROHF means restricted open-shell Hartree-Fock.

basis	method	EHF	$\langle S^2 \rangle$	EMP2	EMP3	EMP4
STO-3G	UHF	−98.17437	0.9178	−98.27252	−98.27383	−98.29002
	AUHF	−98.16517	0.7556	−98.27720	−98.27329	−98.29692
	ROHF	−98.16497	0.7500	—	—	—
3-21G	UHF	−98.95936	0.8331	−99.12869	−99.11679	−99.13786
	AUHF	−98.95440	0.7544	−99.12922	−99.11527	−99.13916
	ROHF	−98.95423	0.7500	—	—	—
6-31G*	UHF	−99.51995	0.8004	−99.75341	−99.74896	−99.76735
	AUHF	−99.51658	0.7536	−99.75301	−99.74778	−99.76738
	ROHF	−99.51640	0.7500	—	—	—
6-31G**	UHF	−99.52494	0.7954	−99.76149	−99.75596	−99.77543
	AUHF	−99.52183	0.7534	−99.76104	−99.75481	−99.77534
	ROHF	−99.52166	0.7500	—	—	—
6-311G**	UHF	−99.54846	0.7969	−99.80433	−99.79660	−99.81959
	AUHF	−99.54529	0.7535	−99.80396	−99.79545	−99.81956
	ROHF	−99.54511	0.7500	—	—	—

Approximate relationships between density power integrals I_k, moments of the momentum density $\langle p^k \rangle$, and interelectronic repulsion potential values (with median errors of 5%) have been calculated for [BO]•, assuming a bond length r=227.5 pm. Hartree-Fock quality values are according to the equation $\langle p^k \rangle = c_k \cdot I_k$ for k=−2 to +4, in which $c_k = 3(3\pi^2)^{k/3}(k+3)^{-1}$: $\langle p^{-2} \rangle = 15.73$, $\langle p^{-1} \rangle = 9.955$, $\langle p \rangle = 35.07$, $\langle p^2 \rangle = 199.0$, $\langle p^3 \rangle = 2118$, and $\langle p^4 \rangle = 44480$ (with median errors of 2%). The ratios $R_k = c_k \cdot I_k / \langle p^k \rangle$ are: $R_{-2} = 2.37$, $R_{-1} = 1.1529$, $R_1 = 0.9762$, $R_2 = 0.9073$, $R_3 = 0.6901$, and $R_4 = 0.333$ [27].

The dimensionless kinetic energy anisotropy α has been calculated for the valence orbitals of [BO]•, taking the experimental value $r_e = 227.5$ pm, to be for 1σ=0.0000, 2σ=−0.0003, 3σ=0.0119, 1π=−0.2513, 4σ=0.3068, and 5σ=0.1891; the total kinetic energy anisotropy is −0.0013. It was shown that the orbital values of α in free atoms depend only on the angular

References on pp. 10/1

momentum quantum numbers, and can be expressed in terms of the Condon-Shortley coefficient c^k. Comparison of α values for individual molecular orbitals with their free counterparts enables one to classify molecular orbitals in terms of their parent atomic orbitals. In favorable cases, α values can also be used to estimate the extent and nature of the reorganization of each orbital upon bond formation. The total molecular value of α correlates roughly with the difference between the numbers of p_σ and p_π electrons [28].

A spin population analysis of $[BO]^{\bullet}$ by configuration interaction calculations on SCF-MO basis is given. The Mulliken-Gross populations for estimating the percent s or p character are for B: 57% s, 39% p_z, 4% p_x, 4% p_y; and for O: 0% s, 3% p_z, −3% p_x, −3% p_y. The apparent experimental population is for B: 40% s, 42% p_z, total 82%; and for O: 0% s, 9% p_z, total 9% [29].

Atomic charges are derived from two dissimilar methods of partitioning the electron density of $[BO]^{\bullet}$; the dimensionless charge transfer Δ_q from the boron to the oxygen atom, obtained from integration of the electron density of the promolecule with surfaces, is 1.216, Δ_q from integration of the molecular electron density inside surfaces is 1.552 (from Bader's atomic charge values) and 1.592 (from Hirschfeld's atomic charge values), respectively. The charges derived are correlated closely with electronegativity differences and with dipole moments. They follow chemically sensible trends and have reasonable magnitudes. The partitioning methods used can be applied to the analysis of diffraction data for crystalline solids [30].

For the $[BO]^{\bullet}$ radical, the bond length r=120.4 pm and the stretching force constant 14.849 mdyn/Å were computed on the 4-31G level, reproducing the empirical a_{ij} and b_{ij} parameters of "Badger's rule" for a series of diatomic molecules [31].

AM1 calculations (a "third generation" treatment) with optimized parameters for boron result in better values for $[BO]^{\bullet}$ than those given by modified neglect of diatomic overlap (MNDO) calculations: the dipole moment is 1.57 D; the ionization potential is 12.34 eV; the heat of formation is $\Delta H_f = 0.8$ kcal/mol (observed 6.0 kcal/mol); and the B–O distance is for the $C_{\infty v}$ geometry r=116.8 pm (observed r=120.4 pm) [32].

The dipole moment of diatomic molecules has also been calculated by the neglect of diatomic differential overlap (NDDO) method. The obtained value for $[BO]^{\bullet}$ of 1.56 D is in good agreement with the intermediate neglect of differential overlap (INDO) value of 1.63 D [33].

The electron affinity (EA) of the $[BO]^{\bullet}$ radical has been measured to be EA(XY)= 3.1±0.1 eV; the calculated values are EA(X)=1.5 eV, EA_1(XY)=3.2 eV, and EA_2(XY)=4.2 eV (for the correction coefficient α=1.1) [34].

The calculated values for the electron affinities of $[BO]^{\bullet}$ with the annihilated-unrestricted Møller-Plesset wave function (AUMP) are with the 6-31++G** basis set for self-consistent field (SCF) the electron affinity EA=1.49 eV, for MP2 EA=2.24 eV, for MP3 EA=2.30 eV, and for MP4 EA=2.34 eV. With the 6-311++G** basis set, for SCF the electron affinity EA= 1.50 eV, for MP2 EA=2.24 eV, for MP3 EA=2.36 eV, and for MP4 EA=2.40 eV [26].

The electron affinity of $[BO]^{\bullet}$ was also calculated by using ab initio methods which included the electronic correlation at the MP2 level. The configuration interaction by perturbation selected iterations (CIPSI) value for $[BO]^{\bullet}$ is EA=2.17 eV, as compared with the experimental value of EA=3.1 eV found in 1971 [37]. For total energies, see Table 3/5 [35].

Table 3/5

Total Energy (in hartree) for the [BO]• Radical at the MP2/6-31+G* Optimized Geometry in $^2\Sigma$ Symmetry and Electron Affinity Values (in eV) [35].

UHF means unrestricted Hartree-Fock, UMP unrestricted Møller-Plesset, ROHF restricted open-shell Hartree-Fock, and CIPSI means configuration interaction by perturbation selected iterations. For Nesbet and ROMP2, cf. [35].

basis set	method	total energy	electron affinity
3-21+G	UHF	−98.978583	1.67
	UMP2	−99.157130*)	2.15
	UMP3	−99.141606*)	2.23
	Nesbet	−98.964577	2.05
	Nesbet-MP2	−99.168331	1.93
	Nesbet-CIPSI	−99.163578	2.02
	ROHF	−98.973881	1.80
	ROMP2	−99.164791	2.03
	ROHF-CIPSI	−99.160784	2.09
6-31+G*	UHF	−99.522368	1.42
	UMP2	−99.759896*)	2.16
	UMP3	−99.753134*)	2.26
	Nesbet	−99.508645	1.79
	Nesbet-MP2	−99.771630	1.96
	Nesbet-CIPSI	−99.767917	2.03
6-311+G*	UHF	−99.545834	1.43
	UMP2	−99.805216*)	2.21
	UMP3	−99.795694*)	2.33
	Nesbet	−99.532017	1.81
	Nesbet-MP2	−99.848693	2.01
	Nesbet-CIPSI	−99.844767	2.08
	ROHF	−99.542265	1.53
	ROMP2	−99.845542	2.10
	ROHF-CIPSI	−99.842904	2.13
6-311+G(2d)	UHF	−99.550160	1.43
	UMP2	−99.826397*)	2.30
	UMP3	−99.816853*)	2.41
	Nesbet	−99.536330	1.81
	Nesbet-MP2	−99.872962	2.09
	Nesbet-CIPSI	−99.869812	2.17

*) The 1s core orbitals are frozen in the MPn computation.

References on pp. 10/1

The rate constant for the reaction of $[BO]^{\bullet}$ with oxygen in argon was measured between 298 and 1180 K by using the laser photolysis/laser-induced fluorescence technique. This reaction, giving $[BO_2]^{\bullet}$ radicals, has a negative temperature dependence in a non-Arrhenius fashion and probably proceeds via a stable BO_3 intermediate complex, beginning with 2×10^{-11} mL $\cdot$ molecule$^{-1}\cdot s^{-1}$ at 298 K [8].

The reactant, saddle point, and product for the collinear abstraction reaction $[BO]^{\bullet}+H_2\rightarrow H^{\bullet}+HBO$ have been studied using ab initio multiconfiguration self-consistent field and multireference configuration interaction techniques with basis sets up to triple zeta plus double polarization quality. At the best level of theory, the reaction is calculated to be exothermic, in contrast to previous estimates. The 0 Kelvin exothermicity is 6.4 kcal/mol and the zero point-corrected barrier height is 9.5 kcal/mol. The transition state is centrally located with the H–H bond stretched by about 20% and the B–H bond stretched by about 28%. The frequency of the doubly degenerate bending vibration at the saddle point is found to be very low. As a consequence, this mode becomes active at relatively low energies. This is documented by a substantial upward curvature in the transition-state-theory Arrhenius plot at high temperatures. The Arrhenius plot is also found to be noticeably curved at low temperatures. This behavior is attributed to quantum mechanical tunneling. The computed rate coefficient for the mentioned reaction is from 300 to 3000 K well represented by the three parameter expression $k(T)=2.96\times10^{-22}\cdot T^{3.29}\cdot e^{-2200/T}$. Using a value of 0.0 kcal/mol for the $[BO]^{\bullet}$ radical leads to a calculated standard heat of formation of HBO of $\Delta H_f^{298}=-60$ kcal/mol [36].

References for 3.1 and 3.2.1:

[1] IUPAC, International Union of Pure and Applied Chemistry (Nomencl. Inorg. Chem., Chapter I/4, Formulae Final Draft **1983**).
[2] Sato, T.; Haryu, K.; Endo, T.; Shimada, M. (Zairyo **37** No. 412 [1988] 77/82 from C.A. **108** [1988] No. 208946).
[3] Aleksandrov, E. N.; Vedeneev, V. I.; Dubrovina, I. V.; Kalekin, O. Yu.; Kozlov, S. N.; Prakh, V. V.; Shcherbina, K. G. (Izv. Akad. Nauk SSSR Ser. Khim. **1988** 2185; Bull. Acad. Sci. USSR Div. Chem. Sci. **37** [1988/89] 1964).
[4] Eisenreich, N.; Liehmann, W. (Propellants Explos. Pyrotech. **12** No. 3 [1987] 88/91 from C.A. **107** [1987] No. 80591).
[5] Basov, N. G.; Gavrikov, V. F.; Shcheglov, V. A. (Kvantovaya Electron. [Moscow] **14** [1985] 1787/806 from C.A. **107** [1987] No. 225516).
[6] Fontijn, A. (AD-A204047 [1989] 1/62 from C.A. **111** [1989] No. 157003).
[7] Meuris, M.; Vandervorst, W.; Maes, H. E. (SIA Surf. Interface Anal. **12** [1987/88] 339/43 from C.A. **110** [1989] No. 16763).
[8] Oldenborg, R. C.; Baughcum, S. L. (AIP Conf. Proc. No. 146 [1986] 562/3).
[9] Tanimoto, M.; Saito, Sh.; Hirota, E. (J. Chem. Phys. **84** [1986] 1210/4).
[10] Woodward, R. W.; Le, P. N.; Temmen, M.; Gole, J. L. (J. Phys. Chem. **91** [1987] 2637/45).

[11] Mélen, F.; Dubois, I.; Bredohl, H. (J. Phys. B **18** [1985] 2423/32).
[12] Karna, S. P.; Grein, F. (J. Mol. Spectrosc. **122** [1987] 356/64).
[13] Navati, B. S.; Korwar, V. M. (Acta Phys. Hung. **60** [1986] 127/34).
[14] Singh, R. B.; Rai, D. K. (J. Quant. Spectrosc. Radiat. Transfer **5** [1965] 723/7).
[15] Walvekar, A. P.; Rama, M. A. (Indian J. Phys. B **60** [1986] 235/7).
[16] Mummigatti, V. M. (Acta Phys. Hung. **61** [1987] 343/8).
[17] Singh, M. (Astrophys. Space Sci. **140** [1988] 421/7).
[18] Irwin, A. W. (Astron. Astrophys. Suppl. Ser. **74** [1988] 145/60 from C.A. **109** [1988] No. 100938).

[19] Anderson, A. B.; Parr, R. G. (J. Chem. Phys. **53** [1970] 3375/6).
[20] Anderson, A. B.; Parr, R. G. (J. Chem. Phys. **55** [1971] 5490/3).

[21] Ptak, W. S.; Giemza, J.; Tkacz, K. (J. Mol. Struct. **142** [1986] 299/302).
[22] Ohwada, K. (J. Chem. Phys. **82** [1985] 860/7).
[23] Parr, R. G.; Gadre, S. R. (J. Chem. Phys. **72** [1980] 3669/73).
[24] Ohwada, K. (J. Chem. Phys. **84** [1986] 1670/6).
[25] Ohwada, K. (J. Chem. Phys. **85** [1986] 5882/9).
[26] Baker, J. (Chem. Phys. Lett. **152** [1988] 227/32).
[27] Pathak, R. K.; Sharma, B. S.; Thakkar, A. J. (J. Chem. Phys. **85** [1986] 958/62).
[28] Thakkar, A. J.; Sharma, B. S.; Koga, T. (J. Chem. Phys. **85** [1986] 2845/9).
[29] Knight, L. B.; Ligon, A.; Woodward, R. W.; Feller, D.; Davidson, E. R. (J. Am. Chem. Soc. **107** [1985] 2857/64).
[30] Maslen, E. N.; Spackman, M. A. (Austral. J. Phys. **38** [1985] 273/87).

[31] Lee, Ikchoon; Cho, Jeoungki; Song, Chaung Hyun (J. Chem. Soc. Faraday Trans. II **84** [1988] 1177/84).
[32] Dewar, M. J. S.; Jie, Caoxian; Zoebisch, E. G. (Organometallics **7** [1988] 513/21).
[33] Dixit, A. N. (Indian J. Pure Appl. Phys. **26** [1988] 431/2).
[34] Karachevtsev, G. V. (Zh. Fiz. Khim. **61** [1987] 2070/4; Russ. J. Phys. Chem. **61** [1987/88] 1078/81).
[35] Mota, F.; Novoa, J. J.; Ramirez, A. C. (J. Mol. Struct. **166** [1988] 153/8 [THEOCHEM **43**]).
[36] Page, M. (J. Phys. Chem. **93** [1989] 3639/43; AD-A 201 375 [1989] 1/22).
[37] Srivastava, R. D.; Uy, O. M.; Farber, M. (Trans. Faraday Soc. **67** [1971] 2941/4).

3.2.2 The Boron Dioxide Radical, $[BO_2]^•$

Vaporization of B_2O_3 occurs primarily by evolution of gaseous B_2O_3, but gaseous $[BO_2]^•$ becomes important at lower temperatures under oxidizing conditions [1].

When an intensive boron beam is agglomerated in dry ice-cooled argon and subsequently reacts with $[NO_2]^•$, the spectrum is found to consist largely of a modified $[BO_2]^•$ emission system which overlaps $[BO]^•$ emission features and appears to be a new system at $\lambda \geq 5900$ Å, which begins to dominate the total chemiluminescent spectrum as the boron flux is increased substantially. At least two $\Delta\nu = 40$ cm^{-1} sequence groupings separated by ca. 440 cm^{-1} and a second long progression or sequence grouping with $\Delta\nu \approx 142$ cm^{-1} were observed and tentatively correlated with emission from the asymmetric BBO molecule. Under both the single and multiple collision conditions ($B + [NO_2]^• + Ar$), the reaction between boron and $[NO_2]^•$ leads to the formation of $[BO_2]^•$ (A $^2\Pi$, B $^2\Sigma^+$) and $[BO]^•$ (A $^2\Pi$) excited states [2].

Laser-induced fluorescence (LIF) was applied to the detection of a variety of combustion species including $[BO_2]^•$ in low-pressure reacting flows, for example, in an atmospheric pressure boron-seeded flame. $[BO_2]^•$ is believed to be one of the key species in boron combustion chemistry. A CH_4/O_2/air premixed flame seeded with BCl_3 was probed with a pulsed dye laser and the resulting fluorescence was observed with an optical multichannel analyzer. Nonresonant LIF was produced in the A–X system of $[BO_2]^•$ principally by pumping the (00°0→00°0) transition and observing the (00°0→10°0) transition, although other combinations were also used. The spectra showed a considerable redistribution of energy due to collisions in the flame. A two-dimensional relative fluorescence profile of the $[BO_2]^•$ in the flame was also obtained [3]. $[BO_2]^•$ was generated directly by a 60 Hz discharge in a mixture of BCl_3 and O_2.

References on pp. 15/6

The (010) κ $^2\Sigma \rightarrow$(000) $^2\Pi_{3/2}$ vibration-rotation transition of $[BO_2]^{\bullet}$ was observed by infrared diode laser spectroscopy [4].

The vapor phase flash photolysis of BI_3 in the presence of OCS leads to a mixture of products, among which $[BO_2]^{\bullet}$ was detected by the ^{11}B bands at 406.55 and 409.05 nm for the transition B $^2\Sigma_u^+ \leftarrow$ X $^2\Pi_g$, and at 490.66, 495.57, 516.85, and 518.05 nm for the transition A $^2\Pi_u \leftarrow$ X $^2\Pi_g$ [5]. The $[BO_2]^{\bullet}$ radical is also formed by flash photolysis of an aqueous alkaline solution of $Na[B(OH)_4]$ and $K_2[S_2O_8]$; its absorption spectrum shows two maxima. The self-decay of this radical is second order. Rate constants for the reactions of this radical with some organic substrates were determined [6]. The $[BO]^{\bullet} + O_2 \rightleftharpoons [BO_3]^{\bullet} \rightarrow [BO_2]^{\bullet} + O^{\bullet}$ reaction rate constant at 298 K is 2×10^{-11} mL·molecule^{-1}·s^{-1}; it decreases with increasing temperature in a non-Arrhenius fashion [7].

The absorption spectrum of $[BO_2]^{\bullet}$ in the ground state X $^2\Pi_g$ was measured by high-resolution Fourier transform infrared (FTIR) spectroscopy. The (00°1→00°0) and (10°1→00°0) vibrational transitions were measured for both the $^2\Pi_{1/2}$ and $^2\Pi_{3/2}$ states. Because of the large number of vibronic levels arising from the Renner-Teller effect, there is a high likelihood of observing level crossings in this molecule, and three examples of perturbations in these states are described. Improved ground state constants are also reported [8].

The radiative lifetime of the A $^2\Pi_u$ state of the $[^{11}\mathbf{BO_2}]^{\bullet}$ radical increases slightly from 91± 4 ns for the (100) $^2\Pi_{3/2}$ level to 102±6 ns for the (002) $^2\Pi_{3/2}$ level (there is no sudden increase for levels involving the asymmetric stretching mode ν_3). Several rotational levels of the (100) and (002) A $^2\Pi_u$ vibronic states have anomalously long lifetimes but no significant shift for a laser band width of 0.03 cm^{-1}; this is attributed to coupling with nearby levels of the X $^2\Pi_g$ state [9]. The (010) κ $^2\Sigma \leftarrow$(000) $^2\Pi_{3/2}$ vibration-rotation transition of $[BO_2]^{\bullet}$ was observed using a tunable infrared diode laser. The observed transition wavenumbers were combined with the spectral data on the (011)←(010) band (cf. [10]) and were analyzed by the least-squares method to yield the ν_2-band origin at 633.8049(4) cm^{-1} and the spin uncoupling constants γ^{κ} = 0.16738(19) and 0.15348(17) cm^{-1}, respectively, for the (011) κ $^2\Sigma$ and (010) κ $^2\Sigma$ states (standard deviations in parentheses). The effects of rovibronic interaction on the spin uncoupling constant and its centrifugal distortion correction term are discussed: the (011) κ $^2\Sigma$ state is perturbed by the (120) μ $^2\Pi_{3/2}$ state through higher order rovibronic interaction [4]. Bending potential curves are reported for the X $^2\Pi_g$ ground state of symmetrical $[BO_2]^{\bullet}$ which were computed at the ab initio MRD-CI level treatment of the Renner-Teller effect, where the equilibrium internuclear distance r_e was kept fixed at 126.5 pm. Employing these data along with spin-orbit splitting results obtained to the first order in perturbation theory, the vibronic energy levels were computed, and agreement with available experimental data was noted. The following structural parameters for $[^{11}BO_2]^{\bullet}$ were calculated: r_e=126.5 pm; A^{SO}=137.8 cm^{-1}; ω_1= 1065 cm^{-1}; $\omega_2(^2B_2)$=435 cm^{-1}; $\omega_2(^2A_2)$=527 cm^{-1}; $\overline{\omega}_2$=481 cm^{-1}; $|\varepsilon\overline{\omega}_2|$=87.6 cm^{-1}; and ω_3= 1270 cm^{-1} [10].

The potential curves for the ground and low-lying excited states of $[BO_2]^{\bullet}$, i.e., X $^2\Pi_g$, A $^2\Pi_u$, B $^2\Sigma_u^+$, and C $^2\Sigma_g^+$, were calculated at the ab initio self-consistent field restricted Hartree-Fock (SCF-RHF) and configuration interaction (CI) levels. The results are consistent with a linear molecular model for all states considered. The calculated structural parameters and transition energies are in good agreement with relevant experimental data. For α=180°, the electronic energies (in hartree) have the following local minima. From self-consistent field (SCF) calculations for X $^2\Pi_g$ results, E=−174.44144 (at the equilibrium internuclear distance r_e= 123.5 pm); from configuration interaction (CI) calculations for X $^2\Pi_g$ results, E=−174.56338 (at r_e=126.5 pm). From configuration interaction (CI) calculations results for A $^2\Pi_u$ at r_e= 129.5 pm, the value E=−174.47832; for B $^2\Sigma_u^+$ at r_e=126.5 pm, E=−174.42279; for C $^2\Sigma_g^+$ at r_e=124.5 pm, E=−174.39505; and for D $^2\Delta_u$ at r_e=135.0 pm, E=−174.23921 [11].

The calculated structural parameters by CI are: r_e=126.2 pm for X $^2\Pi_g$, r_e=130.0 pm for A $^2\Pi_u$, r_e=126.9 pm for B $^2\Sigma_u^+$, r_e=124.2 pm for C $^2\Sigma_g^+$, and r_e=135 pm for D $^2\Delta_u$ [11].

The relative molecular energies are: 0.0851 a.u. or 2.3144 eV for A $^2\Pi_u$, 0.1355 a.u. or 3.6872 eV for B $^2\Sigma_u^+$, 0.1684 a.u. or 4.5825 eV for C $^2\Sigma_g^+$, and 0.3242 a.u. or 8.8221 eV for D $^2\Delta_u$ [11].

In order to obtain the value of the antisymmetric stretching fundamental frequency, ν_3, ab initio restricted self-consistent field (SCF), complete active space self-consistent field (CASSCF), and configuration interaction (CI) calculations were carried out in C_s and $C_{\infty v}$ symmetries using the SPUSH program system. The basis sets were the standard STO-3G (basis I) and 9s5p/4s2p (basis II) sets. From the reference set, 200 singly and doubly excited configurations were selected [12]. As a starting point the symmetric linear configurations O–B–O at equilibrium geometries for the basis I set with r_e(BO) = 2.47 a.u. and the basis II set with r_e(BO) = 2.38 a.u. were chosen. The CASSCF solution appears to be stable for basis I, and yields a low value of ν_3=1281 cm^{-1}, but it becomes unstable with basis II [13]. In order to investigate the O–B–O↔B–O–O rearrangement, ab initio restricted self-consistent field calculations of fragments of potential surfaces in C_s point group in the STO-3G basis set for X $^2\Pi$ ($^2A'$) and ($^2A''$) states using the SPUSH program system were carried out; r_e(BO) and r_e(OO) here varied between 2.00 and 3.20 a.u., bond angles between 180° and 60°. OBO was found to be the more stable species; for the energy of the rearrangement a value of 320 kJ/mol or 0.136 a.u. was found; see also **Fig. 3-3** [14].

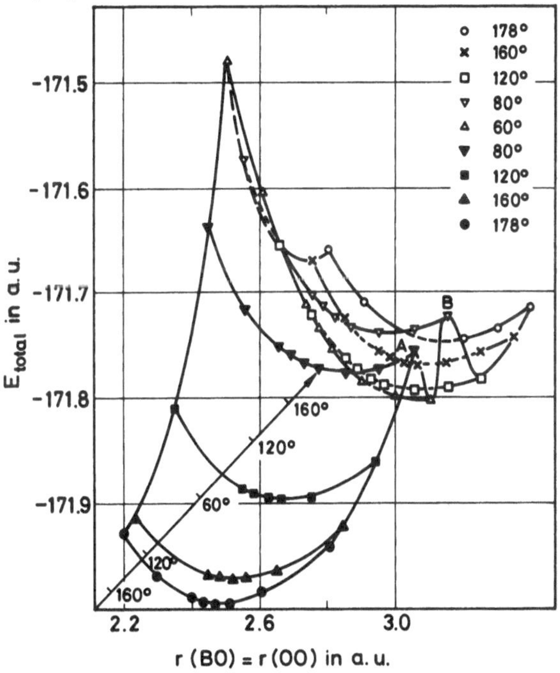

Fig. 3-3. Potential surface of O–B–O↔B–O–O rearrangement. SCF calculations in C_s point group in STO-3G basis set [14].

Nonempirical calculations of the portions of the potential surfaces (ground and low-lying exited doublet and quartet states) of the $[BO_2]^{\bullet}$ molecule, both for the linear $D_{\infty h}$ and bent C_{2v} geometrical configurations, were carried out in the minimal OST-3GF set (basis I) and then for configurations close to equilibrium in the 9s5p/4s2p set (basis II). The geometry of the states had been optimized: the vibration ν_1 for X $^2\Pi_g$ lies at 1079 cm^{-1} (basis I) or 1099 cm^{-1}

References on pp. 15/6

(basis II), and ν_2 at 464 cm^{-1} (basis I); for A $^2\Pi_u$, ν_1=1123 cm^{-1} (basis I) or 982 cm^{-1} (basis II), and ν_2=541 cm^{-1} (basis I); for B $^2\Sigma_u^+$, ν_1=906 cm^{-1} (basis II); for C $^2\Sigma_g^+$, ν_1=1149 cm^{-1} (basis II); the equilibrium internuclear distances of B–O are for X $^2\Pi_g$, r_e=134.2 pm (basis I) or 125.8 pm (basis II), and for A $^2\Pi_u$, r_e=139.6 pm (basis I) or 130.2 pm (basis II); for B $^2\Sigma_u^+$, r_e=128.9 pm (basis II); and for C $^2\Sigma_g^+$, r_e=124.6 pm (basis II). The total energy for dissociation for X $^2\Pi_g$ is E=−171.996 a.u. (basis I) or −174.360 a.u. (basis II); for A $^2\Pi_u$, E=−171.824 a.u. (basis I) or −174.246 a.u. (basis II); for B $^2\Sigma_u^+$, E=−174.229 a.u. (basis II); and for C $^2\Sigma_g^+$, E=−174.457 a.u. (basis II). The following values are all for basis II: for X $^2\Pi_g$, D_e=2.085 eV; for A $^2\Pi_u$, T_e=3.115 eV and D_e=1.158 eV; for B $^2\Sigma_u^+$, T_e=3.554 eV and D_e=0.719 eV; and for C $^2\Sigma_g^+$, T_e=4.546 eV [15].

The potential surface for the isomerization O–B–O↔B–O–O has also been calculated by Hartree-Fock methods using 4-31G and 6-31G** basis sets. The equilibrium internuclear distances r_e(BO) and r_e(OO) are always 131 pm for B–O–O and r_e(BO)=127 pm for O–B–O, the angle is always 180°. The force constant for B–O–O is f(BO)=6.0 mdyn/Å and f(OO)= 9.2 mdyn/Å; for O–B–O was found f(BO)=14.3 mdyn/Å; the total energy is E=−179.2029 a.u. for B–O–O, and E=−174.4306 a.u. for O–B–O. Using the basis set 6-31G**, the dipole moment for B–O–O is −1.59 D, the ionization potential IP=10.3 eV for B–O–O and 15.5 eV for O–B–O. The charge on the atoms for B–O–O is Z(B)=+0.15 and Z(O)=−0.23 and +0.07; for O–B–O, Z(B)=+0.56 and Z(O)=−0.28 and −0.03 [16].

The X $^2\Pi_g$→A $^2\Pi_u$ excitation energy of the radical $[BO_2]^•$ was summarized for r_e=126.5 pm; see Table 3/6 [17].

Table 3/6
Excitation and Total Energies of $[BO_2]^•$ [17].
HF means Hartree-Fock, MP means Møller-Plesset, and CI/DZ+Ryd means configuration interaction with a double zeta plus sp Rydberg basis set.

method	total energies (in hartree) X $^2\Pi_g$	A $^2\Pi_u$	vertical excitation energies $^2\Pi_u$←$^2\Pi_g$ (in eV)
HF	−174.43068	−174.29308	3.74
MP2	−174.81443	−174.74640	1.85
MP3	−174.82236	−174.72943	2.53
MP4SDQ	−174.83670	−174.74985	2.35
3R-CI/DZ+Ryd			2.685 [18]

The harmonic vibrational frequencies of $[BO_2]^•$ ($^2\Pi_g$ state) are from HF/6-31G* calculations: $\nu_1(\sigma_g^+)$=1129 cm^{-1}, $\nu_2(\pi_u)$=509 cm^{-1}, and $\nu_3(\sigma_u^+)$=1382 cm^{-1}. The first ionization potential of $[BO_2]^•$ is IP=12.84 eV at the MP4SDQ/6-31G* level [17]. AM1 calculations on $[BO_2]^•$ with $D_{\infty h}$ geometry and the equilibrium internuclear distance r_e=122.5 pm (experimental 126.3 pm) gave the ionization potential IP=14.23 eV, and the standard enthalpy of formation ΔH_f°= −70.7 kcal/mol (observed value −71.3 kcal/mol) [19].

The effusion method of Knudsen with mass-spectrometric analysis was employed to investigate the gas phase over the systems B_2O_3–PbO, B_2O_3–ZnO, and B_2O_3–Bi_2O_3 between 1200 and 1400 K; from the data, the standard enthalpy of formation of gaseous $[BO_2]^•$ was determined to be ΔH_f°=−292.1±8.4 kJ/mol. The equilibrium constants of the gas-phase reaction $2B_2O_3+O_2 \rightarrow 4[BO_2]^•$, varying for the different systems, lead to a standard enthalpy of reaction of ΔH°=250.0±2.2 kJ/mol. Using the method of ion-molecular equilibria with the ion $[AlF_4]^-$, the standard enthalpy of formation of gaseous $[BO_2]^-$ is determined to be −709±

16 kJ/mol. (For $[BO_2]^-$, see Section 3.4.1.2, pp. 72/3.) The electron affinity of the $[BO_2]^{\bullet}$ radical EA = 417 ± 18 kJ/mol = 4.33 eV [20].

By configuration interaction by perturbation selected iterations (CIPSI) calculations on the MP2 level, the electron affinity of $[BO_2]^{\bullet}$ was calculated to be EA = 4.20 eV. Total energies (in hartree) for the $[BO_2]^{\bullet}$ and $[BO_2]^-$ (cf. Section 3.4.1.2, pp. 72/3) systems at their MP2/6-31+G* optimized geometry in their fundamental states are given in Table 3/7 [21].

Table 3/7
Total Energy (in hartree) for the $[BO_2]^{\bullet}$ and $[BO_2]^-$ Systems at Their MP2/6-31+G* Optimized Geometry in Their Fundamental States and Electron Affinity Values (in eV) for $[BO_2]^{\bullet}$ [21]. UHF means unrestricted Hartree-Fock, UMP unrestricted Møller-Plesset, ROHF restricted open-shell Hartree-Fock, and CIPSI means configuration interaction by perturbation selected iterations. For Nesbet and ROMP, cf. [21].

basis set	method	total energy $[BO_2]^{\bullet}$	total energy $[BO_2]^-$	electron affinity $[BO_2]^{\bullet}$
3-21+G	UHF	−173.504217 a)	−173.627673 b)	3.36
	UMP2	−173.776540 c)	−173.953945 c)	4.83
	Nesbet	−173.462239	−173.627603	4.50
	Nesbet-MP2	−173.837976	−173.959947	3.32
	Nesbet-CIPSI	−173.802650	−173.956349	4.18
	ROHF	−173.463716	−173.627603	4.46
	ROMP2	−173.841741	−173.959947	3.22
	ROHF-CIPSI	−173.805972	−173.956349	4.09
6-31+G*	UHF	−174.436337 a)	−174.554941 b)	3.23
	UMP2	−174.825671 c)	−175.009565 c)	5.00
	Nesbet	−174.394718	−174.554167	4.34
	Nesbet-MP2	−174.884292	−175.009943	3.42
	Nesbet-CIPSI	−174.852885	−175.010575	4.29
6-311+G*	Nesbet	−174.438504	−174.597643	4.33
	Nesbet-MP2	−175.018677	−175.143106	3.39
	Nesbet-CIPSI	−174.986224	−175.140733	4.20

a) $\langle S^2\rangle$ = 0.96 (3-21+G), 0.93 (6-31+G*). – b) $\langle S^2\rangle$ = 0.00 (3-21+G and 6-31+G*). – c) The 1s core orbitals are frozen in the MPn computation.

References for 3.2.2:

[1] Lamoreaux, R. H.; Hildenbrand, D. H.; Brewer, L. (J. Phys. Chem. Ref. Data **16** [1987] 419/43).
[2] Woodward, R. M.; Le, P. N.; Temmen, M.; Gole, J. L. (J. Phys. Chem. **91** [1987] 2637/45).
[3] Schneider, G. R.; Roh, W. B. (AIP Conf. Proc. No. 172 [1987/88] 753/5 from C.A. **109** [1988] No. 218681).
[4] Kawaguchi, K.; Hirota, E. (J. Mol. Spectrosc. **116** [1986] 450/7).
[5] Briggs, A. G.; Simmons, R. E. (Spectrosc. Lett. **19** [1986] 953/61).

[6] Padmaja, S.; Ramakrishnan, V.; Rajaram, J.; Kuriacose, J. C. (Proc. Indian Acad. Sci. Chem. Sci. **100** [1988] 279/303 from C.A. **109** [1988] No. 240481).
[7] Oldenborg, P. C.; Baughcum, S. L. (AIP Conf. Proc. No. 146 [1988] 562/3).
[8] Maki, A. G.; Burkholder, J. B.; Sinha, A.; Howard, C. J. (J. Mol. Spectrosc. **130** [1988] 238/48).
[9] Hodgson, A. (J. Chem. Soc. Faraday Trans. II **81** [1985] 1445/61).
[10] Periç, M.; Bhanuprakash, K.; Buenker, R. J. (Mol. Phys. **65** [1988] 403/12).

[11] Császár, P.; Kosmus, W.; Panchenko, Yu. N. (Chem. Phys. Lett. **129** [1986] 282/6).
[12] Punicher, V. I.; Simkin, V. Ya.; Safanov, A. A.; Dement'ev, A. I. (Vestn. Mosk. Univ. Khim. **25** [1984] 161/4).
[13] Lomp, P.-E.; Stepanov, N. F.; Simkin, V. Ya.; Dement'ev, A. I. (Eesti NSV Tead. Akad. Toim. Keem. **34** [1985] 231/4).
[14] Lomp, P.-E.; Simkin, V. Ya.; Dement'ev, A. I.; Stepanov, N. F. (Eesti NSV Tead. Akad. Toim. Keem. **34** [1985] 235/6).
[15] Lomp, P.-E.; Simkin, V. Ya.; Stepanov, N. F. (Zh. Fiz. Khim. **60** [1986] 1162/5; Russ. J. Phys. Chem. **60** [1986/87] 694/6).
[16] Zyubina, T. S. (Zh. Neorg. Khim. **30** [1985] 1121/4; Russ. J. Inorg. Chem. **30** [1985] 633/5).
[17] Nguyen, M. T. (Mol. Phys. **58** [1986] 655/8).
[18] Saraswathy, V.; Diamond, J. J.; Segal, G. A. (J. Phys. Chem. **87** [1983] 718/9).
[19] Dewar, M. J. S.; Jie, Caoxian; Zoebisch, E. G. (Organometallics **7** [1988] 513/21).
[20] Semnikhin, V. I.; Minaeva, I. I.; Sorokin, I. D.; Nikitin, M. I.; Rudnyi, E. B.; Sidorov, L. N. (Teplofiz. Vys. Temp. **25** [1987] 660/70; High Temp. [USSR] **25** [1987] 497/501).

[21] Mota, F.; Novoa, J. J.; Ramirez, A. C. (J. Mol. Struct. **166** [1988] 153/8 [THEOCHEM **43**]).

3.2.3 Diboron Monoxide, B_2O

B_2O may exist as the isomers **BBO** or **BOB**. The BBO arrangement is also found in B_6O [1, 2] and $B_{12}O_2$ [3]; see Section 3.2.6, p. 68.

Approximation calculations on the basis Hartree-Fock/double-zeta Huzinaga-Dunning (HF/DZHD) gave for the transition from the ground-state BOB to the excited-state BBO an energy of isomerization of 23 kcal/mol and a barrier energy of 39 kcal/mol [4].

When an intensive boron beam is agglomerated in dry ice-cooled argon and reacts with NO_2, at least two $\Delta v = 40$ cm^{-1} sequence groupings separated by ca. 440 cm^{-1} and $\Delta v \approx$ 142 cm^{-1} are observed which have been tentatively correlated with emission from the asymmetric BBO molecule [5].

A source configuration which lies intermediate to a low-pressure effusing molecular beam and a high-pressure flow device has been used to generate boron cluster molecules in a highly oxidizing environment. Using this source operating in an N_2O or NO_2 oxidative environment, a chemiluminescent emission spectrum was generated which was attributed to the asymmetric BBO molecule. The observed spectrum is characterized by a strong $\Delta v = 0$, $\Delta v' = 40$ cm^{-1} sequence grouping and a weaker $\Delta v = +1$ sequence ($\Delta v = 40$ cm^{-1}), 440 cm^{-1} to higher energy. A second sequence with $\Delta v \approx 142$ cm^{-1} is also observed. Combining the 440 cm^{-1} upper-state frequency with the 142 cm^{-1} sequence structure implies a lower-state frequency of ca. 582 cm^{-1} for the B–B stretch, consistent with ab initio calculations [6].

B_2O was prepared by oxidizing BP with oxygen (derived from CrO_3) at 4 Pa and 1150 °C [7], 3.5 to 5.5 Pa and 1000 to 1200 °C [8], or 2 to 6 Pa and 800 to 1350 °C [9] for 12 h in a gold capsule, crushing, and washing with water. The species crystallizes as black-brown opaque triclinic crystals in the trigonal space group P3$-C_3^1$ (No. 143) with a = 297.9 pm and c = 705.2 pm; Z = 2; D_m = 2.48 g/cm³. The structure of B_2O is that of diamond-type consisting of six-membered boron rings and double six-membered alternate B–O rings, as X-ray diffraction data show (analysis by the Rietveld method). The Vicker's microhardness of sintered B_2O with a relative density of 92.5% is 40.5 GPa [8]. It has a Rockwell hardness of ca. 3700 kg/mm² (as compared to 2800 kg/mm² for B_4C), and has a specific electrical resistivity of 100 Ω·cm [9]. The semiconductive characteristics between 150 and 300 K were studied and the electrical resistivity at 300 K was found to be 500 Ω·cm, and the activation energy was 0.25 eV [8]. No thermal decomposition was found to occur at 1200 °C in air, and high resistance to corrosion was observed in molten KOH at 400 °C for one hour as well as in hot concentrated H_2SO_4 solution [8, 9].

On heating a mixture of B_2O_3 and coke powder in an Ar atmosphere in order to form a B_4C layer, B_2O gas was evolved [10].

References for 3.2.3:

[1] Makarov, V. S.; Ugai, Ya. A. (J. Less-Common Met. **117** [1986] 277/81).
[2] Werheit, H.; Haupt, H. (Z. Naturforsch. **42a** [1987] 925/34).
[3] Brodhag, C.; Thevenot, F. (J. Less-Common Met. **117** [1987] 1/6).
[4] Zyubina, T. S.; Charkin, O. P.; Zyubin, A. S.; Zakzhevskii, V. G. (Z. Neorg. Khim. **27** [1982] 558/64; Russ. J. Inorg. Chem. **27** [1982] 315/8).
[5] Woodward, R. W.; Le, P. N.; Temmen, M.; Gole, J. L. (J. Phys. Chem. **91** [1987] 2637/45).
[6] Devore, T. C.; Woodward, R. W.; Gole, J. L. (J. Phys. Chem. **92** [1988] 6919/23).
[7] National Institute of Research of Inorganic Materials (Jpn. Kokai Tokkyo Koho 85-21812 [1982/85] 1/2 from C.A. **103** [1985] No. 10458).
[8] Endo, T.; Sato, T.; Shimada, M. (Mater. Sci. Monogr. C **38** [1987] 2623/8 from C.A. **107** [1987] No. 167562).
[9] Endo, T.; Sato, T.; Shimada, M. (J. Mater. Sci. Lett. **6** [1987] 683/5).
[10] Hitachi Chemical Co., Ltd. (Jpn. Kokai Tokkyo Koho 85-131884 [1983/85] 1/16 from C.A. **104** [1986] No. 23375).

3.2.4 Diboron Dioxide, B_2O_2

Vaporization of B_2O_3 takes place primarily by evolution of gaseous B_2O_3, but gaseous $\mathbf{B_2O_2}$ becomes important at high temperatures under reducing conditions [1].

AM1 calculations for B_2O_2 gave for the isomer **O=B–B=O** of $D_{\infty h}$ geometry with r(BO) = 116.8 (120) pm and r(BB) = 150.7 (170) pm distances (values in parentheses are observed) the heat of formation ΔH_f = –133.1 (–109.0) kcal/mol and for the ionization potential IP = 14.57 eV. For the isomer **O(1)=B(2)–O(3)–B(4)** of C_s geometry with the bond distances r(B(2)–O(1)) = 117.7 pm, r(B(2)–O(3)) = 130.6 pm, and r(O(3)–B(4)) = 134.7 pm and the bond angles ∢(OBO) = 177.2° and ∢(BOB) = 120.9°, the heat of formation ΔH_f = –111.9 (–108.6) kcal/mol, the dipole moment μ = 2.31 D, and the ionization potential IP = 12.02 eV [2].

Scanning electron microscopy (SEM) and X-ray diffraction analysis were used to study the structure and mechanism in reaction-sintering of a mixture of B_4C and boron at 1600 °C. The

References on p. 18

formation of secondary phases was attributed to the presence of B_2O_2 in the gas phase. On increasing the boron concentration in the reaction mixture, the B_2O_2 concentration in the gas phase increased. The sintered material was characterized by a uniformly distributed skeletal structure of B_4C and a fine-grained binder [3].

References for 3.2.4:

[1] Lamoreaux, R. H.; Hildenbrand, D. L.; Brewer, L. (J. Phys. Chem. Ref. Data **16** [1987] 419/43, 431).
[2] Dewar, M. J. S.; Jie, Caoxian; Zoebisch, E. G. (Organometallics **7** [1988] 513/21).
[3] Markholiya, T. P.; Kozelkova, I. I.; Kazakov, S. V. (Ogneupory **1986** No. 10, pp. 24/5).

3.2.5 Diboron Trioxide, B_2O_3

3.2.5.1 Preparation, Formation, and General Properties

The overall combustion process of boron particles is separated into an ignition phase (during which the particle is covered by a layer of B_2O_3) and a clean-particle burning phase. During the ignition phase, $\mathbf{B_2O_3}$ formation results from boron diffusion across the oxide layer, rather than the generally accepted view of oxygen diffusion. During the clean-particle burning phase, boron oxidation is similar to that of carbon. The chemistry of B_2O_3 formation during this phase and the regions of kinetic and diffusion control of the burning rate were defined [1].

The behavior of B_2O_3 drops [2] as well as agglomerates of boron particles [3], formed in the combustion of boron slurry fuels, was studied in the post-flame region of a flat-flame burner. Individual particles (150 to 800 μ) were supported on a probe and could be placed rapidly into the gas environment. Measurements were made on the variation of particle size and temperature with time. Quenched agglomerate surface morphology was studied using scanning electron microscopy (SEM). Burner operating conditions were varied in order to give fuel equivalence ratios of 0.3 to 0.7 and gas temperatures between 1500 and 1975 K at atmospheric pressure. Flame environments both with and without water vapor present were considered. B_2O_3 gasification is a relatively slow process, with 1000 μm initial diameter drop lifetimes being two minutes to two hours. The presence of water vapor speeds gasification and probably accounts for the shorter ignitation delays observed for the oxidation of boron particles in wet environments. Application of a diffusion-limited equilibrium model provided excellent agreement with oxide drop-life histories at all conditions except for the low-temperature ($T<1300$ K) wet environment. In this case, results of a partial-equilibrium analysis suggest that chemical kinetics limits gasification rates [2, 3].

Observations of the surface morphology of partially reacted slurry agglomerates suggest that the ignition and combustion process consists of the heat up of a relatively porous agglomerate composed of individual boron particles covered with a solid B_2O_3 coating, followed by B_2O_3 melting to form a liquid coating of low porosity. The B_2O_3 layer then gasifies, leaving the open, porous structure of reacting boron particles. Finally, the boron melts transforming the agglomerate into a drop. Measurements of ignition times and burning rates were interpreted in this framework for agglomerates formed from three different slurries. Burning rates calculated from a diffusion-limited equilibrium analysis agreed with experimental rates [3].

An optical polarization technique was developed to measure rapidly growing and evaporating transparent liquid condensate films on a solid surface exposed to flowing combustion product gases at film thicknesses well below the onset of complications due to run off. In order to demonstrate the validity of optical techniques, both ellipsometric and optical interference techniques were first used to measure the previous unknown evaporation rates of liquid B_2O_3. As compared with the interference method, the polarization technique places less stringent requirements on the surface quality. The complementary real-time optical methods of polarization and interference hold considerable promise for application to rapid remote measurements of condensation and evaporation rates in high-temperature flow environments [4].

B_2O_3 particles are produced in asymmetric, laminar diffusion flames by burning CO/$B(OCH_3)_3$ or $CH_4/B(OCH_3)_3$ mixtures in air, as was examined by laser-light-scattering techniques. The $CH_4/B(OCH_3)_3$ flames were different: the presence of gaseous H_2O contributed to the formation of gaseous HBO_2 in favor of condensed-phase B_2O_3 [5].

The reaction of B_4C with steam in order to produce B_2O_3 was examined in an Inconel 600 system with CsOH and CsI at 1270 K; the rates of reaction are functions of time, temperature, B_4C geometry, and partial pressure of the steam and the produced boric acid. Significant reactions were observed between the B_2O_3, formed on the B_4C, and both the CsOH and the CsI vapor. The reaction product in both these cases was probably $CsBO_2$ with the iodine forming HI [6]. The oxidation of hot-pressed pure B_4C in air at 1200 K follows a near-parabolic law, and its activation energy is 108 kJ/mol; the oxidation products are B_2O_3, HBO_2, and H_3BO_3. The oxidation rate increases with temperature and is limited by the diffusion of boron and carbon from the bulk to the surface [7].

Exhaust gases such as B_2O_3 from etching or from impurity diffusion in semiconductor manufacturing plants were reviewed [11]. Nucleation and crystallization of B_2O_3/PbO glass (72.5:27.5 mol%) is catalyzed by the addition of 2 to 8.5 mol% CeO_2 which enhances phase separation; fine crystals of B_2O_3 were precipitated when the glass was heat-treated at 450°C [12].

Oxidation of sintered LaB_6, studied in air between 0 and 1500°C, establishes the formation of $La(BO_2)_3$ and B_2O_3 on the surface and principally of LaB_4 at 670°C; the process mechanism and composition of the interacting products vary depending on the temperature [8].

A comparison was performed between the laser and (classical) thermal dehydration of H_3BO_3 to yield B_2O_3. Different reaction routes and kinetics, as well as different intermediate and final products, were evidenced in a direct correlation with laser absorptivity by the reaction products [9].

The recovery of B_2O_3 from a ground borate ore (Kazakhstan) was maximal 80.3%, when it was processed in a three-phase fluidized bed for 1.5 h with the liquid phase (water at 80°C) being fed from below and the gas phase (a 2:1 air/CO_2 mixture) introduced through a central coaxial tube with radial branches perforated with round holes [10].

The kinetics of crystallization of B_2O_3 under pressure were investigated and the crystal growth mechanism was identified. A theory of motion of the crystal-melt interface at wide variations from equilibrium was given. A model explains why crystal growth has never been observed at any temperature at atmospheric pressure [13].

The crystal growth rate of B_2O_3-I in the amorphous phase was measured as it varied over five orders of magnitude with changes in temperature and pressure. The crystal nucleation barrier was eliminated by seeding the surface of boron oxide glass with crystals. The growth rate became measurable only when the pressure exceeded a threshold level near 10 kbar. By using the published thermodynamic information about the B_2O_3 system and a crude free-energy model for the crystal and glass phases, the results were interpreted qualitatively by the theory of crystal growth as limited by the rate of two-dimensional nucleation of monolayers. The constants for the prefactor, activation energy, activation volume, and ledge tension were

References on pp. 24/5

determined. By adjusting the thermodynamic parameters to a set of values that are well within the ranges delineated by the experimental uncertainties, a relation is given for the measured growth rates between 300 and 500 °C and 0 to 30 kbar. The "B_2O_3 crystallization anomaly", i.e., that crystals have never been observed to grow at atmospheric pressure, is explained, since according to the model, the frequency of two-dimensional nucleation is negligible at all temperatures at pressures less than 10 kbar [14].

A review containing 44 references shows that glass formation in melt quenching is possible only when the resistance to homogeneous crystal nucleation remains high, since the melt is deeply undercooled through its labile region to the glass temperature. This implies that the crystallization process in glass-forming materials must be one in which short-range order is reconstructed. The constraints on glass structural models imposed by this reconstructive requirement are discussed. If heterophase nucleants operate, crystallization may be suppressed during melt quenching if the activation energy for crystal growth is sufficiently high. The mechanisms of crystal growth in covalent network materials are discussed. Molten or glassy B_2O_3 under atmospheric pressure does not crystallize at a measurable rate even when seeded, but does rapidly crystallize at pressures in excess of 7 to 10 kbar. An explanation for this "B_2O_3 crystallization anomaly" is also offered and seems to have general applicability [15].

Heating of controlled porous glass (CPG) between 400 and 700 °C not only leads to a dehydroxylation process on the surface but also brings about a diffusion of the boron atoms remaining in the silica network of CPG towards the glass surface. Some authors assume that a long thermal exposure can cause the formation of B_2O_3 crystals in the pores, but small angle X-ray scattering (SAXS) data led to pore size distribution functions that show the crystallites being $x\,Na_2O \cdot y\,B_2O_3$ ($y > x$) rather than B_2O_3 [16]. Some of the materials used as chromatographic packing, adsorbents, or catalyst supports are CPG's. The chemical structure of these materials can easily be changed by heating the porous glass. Such thermal modification leads to the enrichment of boron atoms in the CPG surface. The long exposure of CPG's to thermal treatment can even cause the formation of B_2O_3 crystals in the pores of the glasses [17].

A review with 110 references on the preparation and properties of boron compounds such as low- and high-molecular weight forms of B_2O_3 was compiled. An attempt was made to classify the high-molecular forms of B_2O_3 as inorganic representatives of polyboroxane; r(B=O)=119 pm and r(B–O)=133.7 pm; ∢(BOB)=135.8° and ∢(OBO)=187.6° [18].

In the B–O system, the stable solid phase is B_2O_3 with a melting point of 723±1 K, a standard molar enthalpy of formation for solid B_2O_3, divided by the molar gas constant of $\Delta H_f^\circ/R = -153170 \pm 170$ K; for gaseous B_2O_3, $\Delta H_f^\circ/R = -100500 \pm 500$ K, and the standard molar enthalpy of fusion $\Delta H_{fus}/R = 2950 \pm 20$ K. Vaporization takes place primarily by evolution of gaseous B_2O_3. However, gaseous B_2O_2 becomes important at high temperatures under reducing conditions, and gaseous BO_2 is important at low temperatures under oxidizing conditions. The calculated vaporization equilibria and vaporization rates are shown in **Fig. 3-4**, **Fig. 3-5**, **Fig. 3-6**, and **Fig. 3-7**, pp. 21/2. Partial pressures under reducing conditions were calculated for an oxygen partial pressure of 10^{-15} bar up to 1923 K, where the calculated O_2 pressure of the B/B_2O_3 equilibrium reaches this value, and for the O_2 pressure of the B/B_2O_3 equilibrium at higher temperatures. If liquid B_2O_3 is the only condensed phase present, its composition will change during vaporization unless evolution of B_2O_3 gas dominates; if the liquid composition is changing, the partial pressures indicated are thus strictly applicable only at the start of the vaporization [19].

AM1 calculations on B_2O_3 gave a heat of formation $\Delta H_f = -206.9$ kcal/mol (experimental −201.5 kcal/mol), a dipole moment of 1.06 D, and an ionization potential of 13.56 eV, assuming point group C_{2v} with a B–O distance of 134.0 pm (experimental 136 pm), a B=O distance of 117.1 pm (experimental 120 pm), and a B–O–B angle of 105.0° [20].

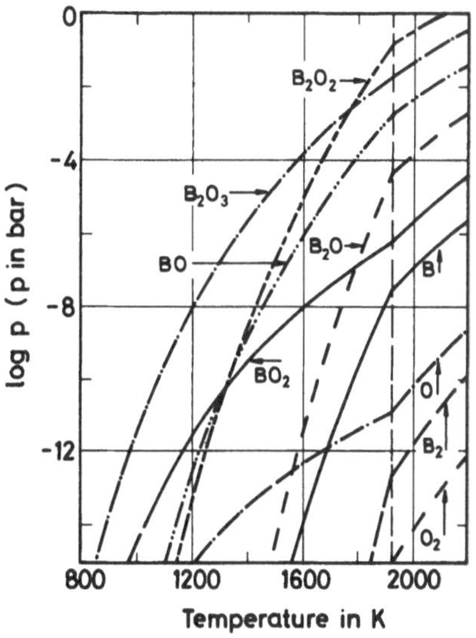

Fig. 3-4. B_2O_3 vaporization in 10^{-15} bar O_2 below 1923 K and vaporization of a B/B_2O_3 mixture above 1923 K [19].

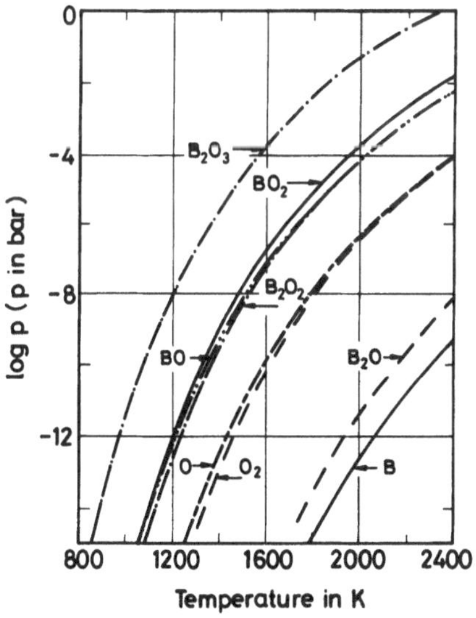

Fig. 3-5. B_2O_3 congruent vaporization [19].

References on pp. 24/5

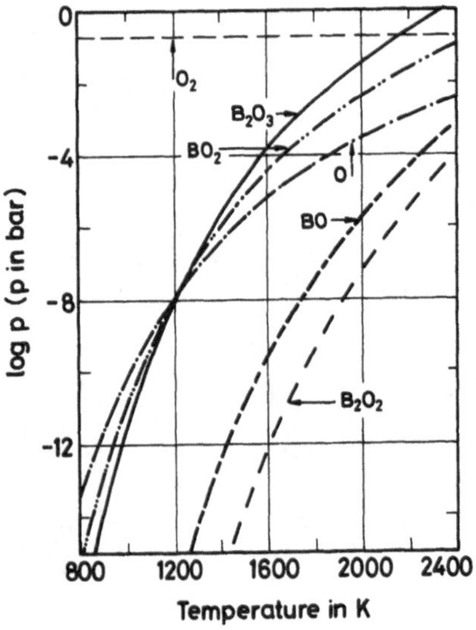

Fig. 3-6. B_2O_3 vaporization in 0.2 bar O_2 [19].

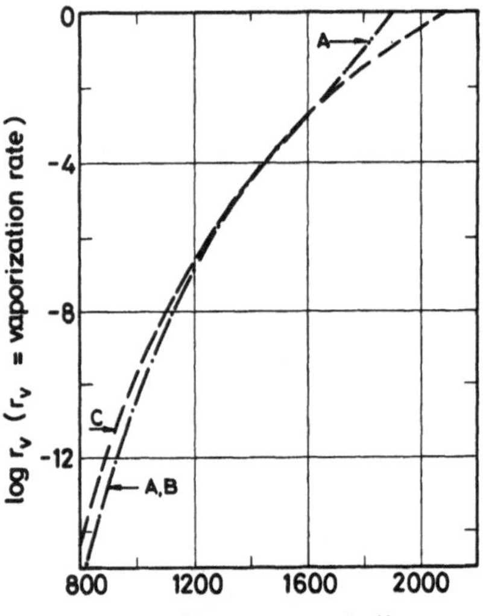

Fig. 3-7. B_2O_3 maximum vaporization rates. A: 10^{-15} bar O_2; B: congruent vaporization; C: 0.2 bar O_2 [19].

Ab initio Hartree-Fock-Roothaan calculations with large polarized basis sets gave ^{17}O nuclear quadrupole coupling constants and a molecular model for bridging oxygens in B_2O_3 which vary strongly with the angle. The calculated e^2qQ/h values of 4.8 and 6.1 MHz for the B–O–B angle of 120° and 132°, respectively, are compared with the experimental values (obtained by X-ray and neutron diffraction analysis) of 4.7 and 5.8 MHz for the two inequivalent oxygen atoms in vitreous B_2O_3, for which the B–O–B angle is 120° (boroxin rings) or 128° to 132° (BO_3 triangles, linked to boroxin rings) [21].

Molecular dynamic (MD) calculations were carried out for B_2O_3 glass using the interionic potentials derived by the MD calculations for crystals. The results are in good agreement with the data obtained by NMR spectroscopy [22].

A review with 29 references on the dielectric properties, thermal properties, and viscosity of glass-crystalline materials containing B_2O_3 for use in microelectronics was compiled [23].

For liquid B_2O_3, the high-frequency shear and compressibility moduli between 747 and 839 K, and the bulk and shear viscosities and diffusion coefficients between 1273 and 1673 K were obtained by calculations with a statistical-mechanical method and Lennard-Jones potential functions [26].

Viscosities of pure molten B_2O_3 and borate-base binary oxides, in which network structures would be formed, were measured by using an oscillating-plate viscometer. Characteristic features of the viscosity of molten B_2O_3 are expressed in a non-Arrhenius equation. As compared with the viscosities of molten B_2O_3, the viscosities of molten B_2O_3 with 2 mol% SiO_2 are high, while the viscosities of molten B_2O_3 with 2 mol% ZnO, BaO, PbO, or Na_2O are lower and decrease in that order. This order corresponds approximately to that of the relative strength of acidity or basicity for the oxides. Since network-forming melts have higher viscosity values, positive deviations from the viscosities as calculated by using an equation for simple melts would provide a quantitative approach to the network structures of melts [24].

The Tool-Narayanaswamy model of the relaxation theory of glass formation was modified for its application in nonisobaric and nonisothermal conditions. Good agreement between calculated and experimental relaxation coefficients of compressibility of B_2O_3 was obtained [25].

The thermal conductivity γ of vitreous B_2O_3 was measured under pressures up to 1.7 GPa, using a hot-wire procedure between 170 and 570 K. At room temperature and zero pressure, $\gamma = 0.52\ W \cdot m^{-1} \cdot K^{-2}$. The values of the logarithmic pressure, $g = d(\ln \lambda)/d(\ln \rho)$, where ρ is the density, were found to be 1.1 for uncompacted glass and 0.7 for glass compacted to 1.2 GPa; the variation of λ with the temperature at constant density was approximately linear, with a positive slope of $1.38 \times 10^{-3}\ W \cdot m^{-1} \cdot K^{-2}$ [27].

Auger spectra of B_2O_3 were compared with those of BN, BeO, Be_3N_2, Li_2O, and Li_3N. The low energy Auger transition of boron is generally split into a multiplet of several lines. The effect is controlled by the L_1 level of oxygen which opens with its local density of state extra decay channels at the site of the excited atom [28].

The boron K_α spectrum of B_2O_3 has peaks at 167, 181, and 193 eV, showing that the local environment of boron involves trigonal and tetrahedral arrangements of the oxygen atoms. Soft X-rays with $\lambda \approx 1$ nm have escape depths of a few µm or less. Changes in peak profiles, satellite peaks, etc., which can be correlated with chemical effects, can thus be used to infer chemical changes in sample surfaces. Hydrolysis and radiation decomposition are both indicated by changes in the boron K_α X-ray emission peak from a variety of boron compounds such as B_2O_3. The relation between X-ray emission and X-ray photoelectron spectra is discussed [29].

References on pp. 24/5

The kinetics of X-ray-induced defect accumulation in B_2O_3 at 77 K were studied by ESR. The g-factors of the defects are given. Localized vacancies and dislocations are observed. Recombination of defect pairs is observed at high radiation intensities. The activation energies for the different clustering were determined [30].

A large T^3 specific heat anomaly, associated with a vibrational mode near 40 cm^{-1}, is found in vitreous silica, but not in vitreous B_2O_3. A material-specific structural model is proposed for this anomaly [31].

Pure crystalline B_2O_3 has an enthalpy of solution in 2 N HNO_3 at 298 K of 4 kcal/mol, whereas glassy B_2O_3 has a value of 8 kcal/mol [32].

References for 3.2.5.1:

[1] Glassman, I.; Williams, F. A.; Antaki, P. (Proc. 20th Int. Symp. Combust., Princeton, N. J., 1984 [1985], pp. 2057/64 from C.A. **103** [1985] No. 162832).
[2] Turns, S. R.; Holl, J. T.; Solomon, A. S. P.; Faeth, G. M. (Combust. Sci. Technol. **43** [1985] 287/300 from C.A. **103** [1985] No. 110326).
[3] Turns, S. R.; Holl, J. T.; Solomon, A. S. P.; Faeth, G. M. (CPIA Publ. No. 412 [1984] 31/44 from C.A. **104** [1986] No. 71276).
[4] Seshadri, K.; Rosner, D. E. (Combust. Flame **61** [1985] 251/60 from C.A. **103** [1985] No. 127610).
[5] Turns, S. R.; Funari, M. J.; Khan, A. (Combust. Flame **75** [1989] 183/95 from C.A. **110** [1989] No. 98254).
[6] Elrick, R. M.; Sallach, R. A.; Ouellette, A. L.; Douglas, S. C. (SAND-87-1491 [1987] 1/77 from C.A. **109** [1988] No. 44776).
[7] Efimenko, L. N.; Lifshits, E. V.; Ostapenko, I. T.; Snezhko, I. A.; Shevyakova, E. P. (Poroshk. Metall. [Kiev] **1987** No. 4, pp. 56/60; Sov. Powder Metall. Met. Ceram. **27** [1987] 318/21 from C.A. **107** [1987] No. 82608).
[8] Gogotsi, Yu. G.; Kotlyar, D. A.; Kresanov, V. S.; Morozov, V. V. (Poroshk. Metall. [Kiev] **1987** No. 11, pp. 56/9; Sov. Powder Metall. Met. Ceram. **28** [1988] 914/7 from C.A. **108** [1988] No. 67846).
[9] Popescu, C.; Jianu, V.; Alexandrescu, R.; Mihailescu, I. N.; Morjan, I.; Pascu, M. L. (Thermochim. Acta **129** [1988] 269/76).
[10] Savinykh, Yu. G.; Polozov, A. P.; Vasanova, L. K.; Korotke, V. V. (Khim. Promst. [Moscow] **1988** No. 2, pp. 111/2 from C.A. **108** [1988] No. 134319).

[11] Abe, Y.; Sugiyama, H. (PPM **16** No. 6 [1985] 40/52 from C.A. **103** [1985] No. 75487).
[12] El-Bayoumi, O. H.; Subramanian, K. N. (Phys. Chem. Glasses **26** [1985] 64/7).
[13] Aziz, M. J. (Diss. Harvard Univ. 1984, pp. 1/118; Diss. Abstr. Intern. B **45** [1984] 638; C.A. **101** [1984] No. 161410).
[14] Aziz, M. J.; Nygren, E.; Hays, J. F.; Turnbull, D. (J. Appl. Phys. **57** [1985] 2233/42).
[15] Turnbull, D. (J. Non-Cryst. Solids **75** [1985] 197/207).
[16] Pikus, S.; Dawidowicz, A. L. (Appl. Surf. Sci. **24** [1985] 274/82 from C.A. **103** [1985] No. 184171).
[17] Dawidowicz, A. L.; Staszczuk, P. (J. Therm. Anal. **30** [1985] 793/801).
[18] Tarasevich, B. P.; Sirotkina, N. Z.; Il'in, A. S.; Kuznetsov, Ye. V. (Vysokomol. Soedin. A **27** [1985] 451/63).
[19] Lamoreaux, R. H.; Hildenbrand, D. H.; Brewer, L. (J. Phys. Chem. Ref. Data **16** [1987] 419/43).
[20] Dewar, M. J. S.; Jie, Caoxian; Zoebisch, E. G. (Organometallics **7** [1988] 513/21).

[21] Tossell, J. A.; Lazzeretti, P. (J. Non-Cryst. Solids **99** [1988] 267/75).
[22] Xu, Qiang; Kawamura, K.; Yokokawa, T. (J. Non-Cryst. Solids **104** [1988] 261/72).
[23] Kirsch, M.; Hübert, T.; Banach, U.; Kleinke, H. (Z. Chem. **25** [1985] 280/6).
[24] Iida, T.; Kawamoto, M.; Okuda, H.; Morita, Z. (Tetsu to Hagane **73** [1987] 469/75 from C.A. **106** [1987] No. 141540).
[25] Mazurin, O. V.; Leko, V. K. (Fiz. Khim. Stekla **11** [1985] 250/4 from C.A. **102** [1985] No. 224829).
[26] Gopala Rao, R. V.; Das Gupta, B. (Acustica **57** [1985] 44/7 from C.A. **102** [1985] No. 155218).
[27] Nilsson, O.; Sandberg, O.; Baeckstroem, G. (Int. J. Thermophys. **6** [1985] 267/73 from C.A. **103** [1985] No. 93915).
[28] Hanke, G.; Müller, K. (Surf. Sci. **152/153** [1985] 902/10 from C.A. **103** [1985] No. 14020).
[29] Arber, J. M.; Urch, D. S. (Spectrochim. Acta B **40** [1985] 757/62).
[30] Garibov, A. A.; Gasanov, A. M.; Gezalov, Kh. B.; Kerimov, M. K. (Sovrem. Metody YaMR EPR Khim. Tverd. Tela Mater. 4th Vses. Koord. Soveshch., Nogiwsk, USSR, 1985, pp. 228/30 from C.A. **103** [1985] No. 224735).

[31] Phillips, J. C. (Phys. Rev. B **32** [1985] 5356/61).
[32] Shul'ts, M. M.; Vedishcheva, N. M.; Shakhmatkin, B. A. (Thermochim. Acta **92** [1985] 345/8).

3.2.5.2 The Structure of B_2O_3 Glasses

This section not only presents the structure of pure B_2O_3 glassy systems but, based on chemical reasons, includes structural results on multicomponent compounds such as $B_2O_3 \cdot Na_2O$, etc., which, according to the Gmelin System of Last Position, normally would not be discussed in a "Boron" volume. However, the chemistry of these compounds is largely determined by the B–O structural systems.

The structure of glassy B_2O_3 is subject of debate since it is not yet established if BO_3 triangles and/or B_3O_6 boroxin rings form the structural unit of a continuous random network describing the structure. The macrostructure of B_2O_3 glass was investigated by the hand-built method, in which the glass was assumed to be comprised of BO_3 units and boroxin groups. The radial distribution functions (RDF), calculated by using the pair function method, were compared with the experimental RDF obtained by X-ray measurements. The results indicated that the major part of the B_2O_3 glass consisted of boroxin groups. The microstructure of B_2O_3 glass was studied by intermediate neglect of differential overlap (INDO) methods which can take all valence electrons into consideration and describe the structure of glasses. The stability of six-membered rings by varying the B–O bond length and the stability of eight-membered rings by varying the interbond angle are discussed. A bond length between 135 and 145 pm, calculated by INDO or by ab initio methods, agrees well with the experimental data as estimated by the X-ray study. The bond angle of two BO_3 units is ca. 120°. This suggests the existence of boroxin groups in the pure B_2O_3 glass [1].

The structure of B_2O_3 glass was also simulated by molecular dynamics (MD). A new type of potential calculated by the INDO method and a three-body potential were introduced into the MD simulation. Boroxin rings and diborate groups were successfully reproduced in the structural model obtained from the MD simulation. Calculated RDF data agreed well with those obtained from X-ray diffraction analysis. The distribution of O–B–O and B–O–B bond angles was discussed [2].

References on pp. 37/40

The structures of borate glasses containing heavy metal ions, i.e., $Tl_2O \cdot 2B_2O_3$ and BaO $\cdot 2B_2O_3$, were investigated using both X-ray and neutron diffraction analysis (see Table 3/8). Structural models were built, and radial distribution functions (RDF) were calculated by the pair function method. In the thallium glass, the presence of three-coordinate oxygen atoms (proposed on the basis of NMR study) was not confirmed by this analysis or by Raman spectral data. In the case of the barium glass, the calculated pair distribution function (PDF) for the crystals of the same composition showed good agreement with the observed PDF [3].

Table 3/8
Borate Groups Specific to Borate Glasses with Various Metal Oxides [3].

composition	radius of the metal ion	structural borate group
$Li_2O \cdot 2B_2O_3$	68 pm	diborate
$Na_2O \cdot B_2O_3$	97 pm	triborate di-pentaborate
$K_2O \cdot 2B_2O_3$	133 pm	orthoborate diborate di-triborate
$Tl_2O \cdot 2B_2O_3$	147 pm	orthoborate diborate di-triborate
$BaO \cdot 2B_2O_3$	135 pm	di-triborate di-pentaborate

For two different densities of pure B_2O_3 glass with D_1 =1.85 g/mL (250°C for 24 h, 1000°C for 24 h) and D_2 =1.81 g/mL (as before, but in addition at 312°C in vacuum for 24 h), the Raman spectra were recorded between 4.2 K and room temperature. The polarization behavior of the 806 cm^{-1} line, assigned to the symmetric breathing mode of the oxygen atom in the boroxin ring, was reported. The strong increase of the depolarized scattering of this line with decreasing temperature was discussed in terms of medium-range order and angular corrections [6].

IR spectral data have shown that glassy B_2O_3 forms a network with a structure which is analogous to the structure of B_2O_3 crystals [4].

Inelastic scattering of cold neutrons was used to determine the density of the vibrational states (DVS) of pure glassy B_2O_3. In the low-energy range of the spectra, the DVS has a universal, although not a Debye form. Normalized intensities of Raman scattering spectra are shown [5].

Oxygen-17 NMR spectra of vitreous B_2O_3 show two different oxygen sites: one for the boroxin ring and one for oxygens that connect the ring [7].

Modified neglect of diatomic overlap (MNDO) calculations were used to study the structure of B_2O_3 glass. The clusters $B(OH)_3$, $B_3O_3(OH)_3$, $[B_3O_3(-O)_3]^{3-}$, $(HO)_2B-O-B(OH)_2$, $(HO)_2B_3O_3-O-B_3O_3(OH)_2$, $B[O-B(OH)_2]_3$, and $B[O-B_3O_3(OH)_2]_3$ were considered and their geometries, the heats of reactions, π-electron systems and electronic structures are discussed. The geometry and the electronic structures show a good correspondence with the experimental data. The resonance stabilization effect of the π-electron system is not so large as to control the geometries and reactions [8]. The structures and basicity of binary borate glasses were studied by applying these cluster approximations and by the MNDO method. Optimized geometries of the characteristic structural unit in borate glass, e.g., pentaborate, triborate and diborate clusters, showed good agreement with the experimental determinations. The bond length variations accompanying the formation of four-coordinate borón (BO_4) or nonbridging oxygen reflect the intramolecular charge rearrangement and conform well to the prediction of Gutmann's bond length variation rule. Ten isomers of the molecular formula $(H_8B_{12}O_{23})^{2-}$ were constructed in order to represent the structures in various compositions of binary borate glass [9].

A new quality "sparkle affinity" was defined as a measure of the hard basicity of borate anion clusters. The "sparkle" is stabilized in the electrostatic field well formed by three oxygen atoms connected to the cluster containing BO_4. The "sparkle" is a virtual chemical species whose behavior is expected to be similar to the alkali or alkaline earth metal cation. The "sparkle affinity" is defined as the energetic gain when one "sparkle" is set near a borate anion cluster. The comparison is made among the delocalization energy, the proton affinity, and the "sparkle affinity" [11].

On the basis of the hard and soft acid and base (HSAB) theory, the BO_4 unit is classified rather into a hard base and nonbridging oxygen is classified rather into a soft one. A hard dopant cation is stabilized in the electrostatic field around BO_4 and a soft one coordinated by nonbridging oxygen and taken into the borate network. In the case where H_2O is the counter oxide, the proton forms the terminal OH group accompanying the BO_4 formation in the low H_2O content range, and BO_4 appears in the medium H_2O content range. Thus, the proton behaves differently than the alkali metal cations [9].

The geometries and basicity of the isomers of $(H_8B_9O_{18})^-$ and $(H_8B_{11}O_{21})^-$ were also studied by this method. The optimized geometries of the pentaborate, triborate, and diborate structures show good agreement with experimental data. The basicity in the borate glass was defined as the ability of the oxygen atoms in the glass network to donate electrons. Molecular orbital interaction between the occupied orbitals of the cluster and the LUMO of the acidic site in $[H_2B_9O_{14}]_2O$ and the proton affinity of the clusters were estimated. The basicities of the clusters which contained BO_4 and that of the clusters which contained nonbridging oxygen were compared. For low alkali metal oxide content, the cluster neutralizes itself by forming BO_4; in high alkali metal oxide content borate glass, the self-neutralization becomes less and less favorable and the formation of nonbridging oxygen becomes energetically more preferable [10].

A hard cation such as Li^+ is stabilized around the hard BO_4 unit, while a soft cation such as Pb^{2+} is donated to the lone pair of electrons from a soft nonbridging oxygen. A proton is found to be combined with one oxygen atom with high covalency to form a terminal OH group, in contrast with alkali metal ions which coordinate plural oxygen atoms via Coulombic forces [11].

Molecular dynamics (MD) calculations for sodium borate glasses using the interionic potentials derived by MD calculations of crystals show that the composition dependence of the fraction of BO_4 is in good agreement with NMR measurements. The pair distributions and bond angle distributions around BO_3 and BO_4 showed remarkable differences from each other and resulted in the particular local structures. The network is developed initially with an increase of Na_2O content and the maximum development is shown to occur near 33 mol% Na_2O. Above 33 mol% Na_2O, the network of the sodium borate glass starts to disintegrate, and a significant increase of nonbridging oxygen is found [12].

References on pp. 37/40

The optical absorption spectra of binary thallium borate glasses (irradiated by X-rays and high-energy electrons) show trapping of an electron on the Tl^+ center and generation of a Tl^0 center in glasses with a Tl_2O content below 20 mol%. The conditions of formation and diminishing of this Tl^0 center are given. The effect of doping with foreign ions such as Cd^{2+} or Cl^- was also studied. The Cd^{2+} ion reduced the formation of Tl^0 centers, whereas the Cl^- ion did not show any marked effect. The Cl_2^- center is formed only in glasses with low Tl^+ content; the same was observed in alkali borate glasses [13].

The structure of vitreous B_2O_3 consists of randomly linked planar BO_3 triangles [14 to 16].

A structural correlation length equal to the size of a pair of BO_3 triangles has been found. The addition of alkali metal oxide leads to a decrease of this correlation length. The extra oxygen atoms are mostly incorporated into BO_4 units and change the BO_3 triangles into tetrahedra, but larger ions such as Rb^+ will enhance the number of nonbridging oxygen atoms [14]. The structures of such borate glasses have been investigated by means of X-ray diffraction studies and have been simulated by use of molecular dynamics (MD) [15, 17]. Calculated radial distribution functions are in agreement. The structure of lithium borate glasses appears to consist only of randomly connected BO_3 triangles and BO_4 tetrahedra. A comparison between the slowly quenched glasses (studied by X-ray diffraction analysis) and fast-quenched glasses (studied by MD simulations) leads to the conclusion that a low quench rate leads to a preponderance for the B–O–B angles of adjacent BO_3 triangles to 120°. The frequency spectra of B–O vibrations in the MD simulations agree qualitatively with infrared transmission spectra [17]. The ionic conductivities of these glasses were investigated also. In order to understand their superionic properties, several experimental techniques were employed. Experimental results are given for density measurements, differential scanning calorimetry, ionic thermocurrents (ITC), and dielectric-loss (DEL) measurements in the frequency range between 100 and 30000 Hz. The activation energy of the ionic conductivity decreased with increasing Li_2O content [18].

The ionic conductivity of glasses of the system B_2O_3–Li_2O–Cs_2O was investigated by ITC and DEL measurements. The conductivity shows a pronounced mixed alkali metal (MA) effect, where the conductivity of pure lithium borate glass is higher than that of pure caesium borate glass with the same alkali metal content. The MA effect can be explained very well with the weak electrolyte theory, provided that a short range dipole-induced dipole interaction between alkali metal ions is assumed [16]. Raman spectra of glasses of this type were studied between 20 and 500°C. The most important features of the spectra, i.e., sharp and intense bands at 770 cm^{-1} (BO_4) and 805 cm^{-1} (nonbridging oxygen), are attributed to symmetrical boroxin rings in the B–O network structure. The intensity ratios of these bands, I(770)/I(805), provide information on the glass structure. These results were compared with the glass transition temperature data. The replacement of Li^+ by Cs^+ in the $LiBO_2$ glass leads to an enhanced number of nonbridging oxygen atoms in the glass network; in addition it leads to preference to form borate rings with one instead of two BO_4 units [19]. Addition of LiCl leads to a more open oxygen glass network and consequently to a decrease of the boron coordination number. Velocity autocorrelation functions, Raman and IR spectra have been calculated and were compared with the experimental spectra. The frequency spectra of the lithium and caesium atoms are in agreement with far-infrared (FIR) measurements. The dominant parts of the overall vibrational density of states (DOS) spectra are due to nearest neighbor B–O stretching modes. The calculated IR spectra show a main peak at ca. 1100 cm^{-1}; this maximum shifts toward lower frequencies upon the addition of an alkali metal oxide, but does not change significantly upon addition of LiCl. These results agree with the experimental observations. For the calculations of Raman spectra, only autocorrelated B–O bands were taken into consideration. These calculations fail to reproduce the dominant experimental bands at 780 and 800 cm^{-1}. For a proper calculation of the Raman spectra of borate glasses, apparently cross-correlated B–O bond stretching vibrations should be considered [15].

Low-frequency Raman spectra of glasses of the systems B_2O_3–Li_2O, B_2O_3–Rb_2O, and B_2O_3–Li_2O–LiCl differ significantly from the spectra of their crystalline counterpart. The model of Martin and Brenig (1974) [20] provides a useful tool for understanding these differences. The temperature-reduced spectra show peaks at 50 and ca. 130 cm^{-1}. The peak at 50 cm^{-1} appears to be a common property of oxide glasses and arises because the limited structural correlation length of the glass network causes a nonzero maximum of the frequency-dependent Raman coupling coefficient. The 130 cm^{-1} band can be attributed to liberational modes of BO_3 and BO_4 units. Sharp and intense bands at 770 cm^{-1} (BO_4) and 805 cm^{-1} (nonbridging oxygen) are attributed to boroxin rings. The addition of LiCl does not produce major changes in the Raman spectra and consequently does not change the $BO_3:BO_4$ ratio. The Cl^- ions are incorporated into interstitial vacancies of the network leading to an expansion of the B–O network structure [16, 19]. The ionic conductivities of these glasses were also investigated by several techniques. The presence of Cl^- ions, however, causes a drastic increase in the ionic conductivity, which could not be accounted for by only the increased number of charge carriers. The activation energy associated with the conductivity of Li^+ ions was reduced by the presence of these ions [18].

IR spectral data showed that $2B_2O_3 \cdot M_2O$ glasses with M = Li, Na, and K formed networks, the structure of which is analogous to the structure of $2B_2O_3 \cdot Li_2O$ crystals. In the glass with 33 mol% Li_2O, a partial breakdown of diborate rings was observed; the amount of tetrahedral BO_4 units was not affected. In contrast to spectra of other alkali metal borate glasses, IR spectra of glasses with M = Li showed the presence of a band at 460 cm^{-1}, which was attributed to Li–O valence vibrations in LiO_4 tetrahedra [4].

The Zachariasen random network model is not appropriate for borate glasses, but the model of Krogh-Moe – who suggested that borate glasses should contain larger structural groupings such as boroxin rings ($B_3O_{4.5}$), pentaborate ($[B_5O_8]^-$), tetraborate ($[B_8O_{13}]^{2-}$), triborate ($[B_3O_5]^-$), diborate ($[B_4O_7]^{2-}$), pyroborate ($[B_2O_5]^{4-}$), di-triborate ($(B_3O_{5.5})^{2-}$), di-pentaborate ($(B_5O_{8.5})^{2-}$), tri-metaborate ($[B_3O_6]^{3-}$), metaborate ($(BO_2)^-$), and orthoborate ($[BO_3]^{3-}$) anion units that exist in the crystalline compounds of the particular borate system – seems to be valid, since extensive ^{10}B NMR and ^{11}B NMR studies showed the amount of each structural grouping in each glass. The model can be viewed as a modified Warren-Zachariasen network, in which the structural groupings (rather than individual BO_4 or BO_3 units) are connected randomly to each other [21]. A review on the NMR data of boron-oxide glasses was given [22].

A statistical-mechanical model for the thermodynamic properties of alkali metal borate glasses was developed, alkali metal borate building blocks were defined, and the model was used to describe the short-range order and bonding arrangement of boron in the glasses. The effects were discussed with respect to changes in the fictive temperature, associated with the final metastable state of the glass, on the predictions of the model [23]. This model for systems in which the structural units need not all be bonded to the same number of neighbors has been applied. Comparison of the calculated miscibility dome with that observed for alkali metal borates suggests that there is a strong association between two nonbridging oxygen atoms and two alkali metal ions at temperatures where phase separation is observed. The effect of temperature on the fraction of tetrahedral BO_4 units (N_4) in alkali metal borate glasses is predicted [24]. Calculated values of N_4 are in good agreement with quenching experiments, that show the correctness of the model of Jellison and Bray [25], describing the effect that the diborate group is energetically stabilized by the BO_4 tetrahedra [26].

NMR data of diffusion-induced spin-lattice relaxation in lithium borate glasses cannot generally be interpreted in the framework of the classical Bloembergen-Purcell-Pound theory.

References on pp. 37/40

A combination of standard relaxation theory with a hopping model for diffusion in glasses has been presented. The observed anomalies in the NMR data can be explained as a result of anomalous diffusion [27]. In the $Li_2O \cdot 3B_2O_3$ glass the nuclear spin-lattice relaxation behavior of ^8Li and ^{12}B was studied by use of the method of β radiation detected NMR relaxation. The value obtained for the apparent activation energy indicates partially localized lithium diffusion [28].

Raman and mid-infrared spectroscopic studies were used to probe the B–O network of glasses with high Li_2O content in the system $(B_2O_3)_{1-x}(Li_2O)_x$ with $0.2 \leq x \leq 0.65$. Increasing x induced progressive formation of pentaborate, diborate, metaborate, pyroborate, and orthoborate groups. Far-infrared (FIR) spectroscopy was used to show the interactions between the Li^+ cations and the network. The square of the frequency of the Li^+ motion band ν_2 varied linearly with the composition until $x \approx 0.5$. A deviation from linearity, observed for glasses of higher Li_2O content, was attributed to the participation of lithium in the network formation. The integrated intensity of the Li^+ motion band was proportional to ν_2. These observations were discussed in view of the spectroscopic information concerning the nature of the borate network at various boron contents [29].

Far-infrared (FIR) spectra (60 to 400 cm^{-1}) were also measured for all alkali metal borate glasses $(B_2O_3)_{1-x}(M_2O)_x$ with $x \leq 0.65$ for M = Li, $x \leq 0.40$ for M = Na, and $x \leq 0.35$ for M = K, Rb, or Cs and analyzed in order to study the cation-network interactions and their compositional dependence [30, 31]. Band deconvolution of the measured spectra showed the presence of two distinct distributions of M^+ sites in lithium, sodium, and potassium glasses; similar results were obtained for rubidium and caesium borate glasses of compositions with $x > 0.25$. One distribution of cation sites was observed for the lower alkali metal content rubidium and caesium glasses. The fractions of cations in the two different network sites were also evaluated. The squares of the frequencies of the cation-motion bands were found to vary linearly with the composition, and exhibit kinks at $x \approx 0.2$ for all but caesium glasses. This behavior was explained on the basis of the network structural changes known to occur at this composition [30]. The distributions of the network anionic sites were also obtained, and the results were discussed in view of the theoretical models proposed for ionic conduction in glasses. Activation energies for ion transport were calculated and compared with experimental values [31].

The effect of the interactions between alkali metal cations and the glass network in binary alkali metal borate glasses on the glass transition temperature (T_g) was studied. T_g depends linearly on the effective force constants of these interactions, and this effective force constant was defined so as to express the effect of cations vibrating in different network sites. The relation between N_4 (fraction of tetrahedral BO_4 units) and the effective force constant was also studied. N_4 depends strongly on the nature of the alkali metal cation. Results are discussed in terms of the borate glass structure and recent NMR studies of these glasses [32].

The cation vibrational frequencies and their compositional dependence, which were obtained from far-infrared (FIR) measurements, were used to elucidate the role of the alkali metal cation on the borate glass structure. The dependence of the cation motion frequency on the symmetry and size of the anionic network site was studied by using a simplified type of the Born-Mayer potential in order to describe the cation-network interactions. Thus, comparison of calculated and experimental FIR data revealed that the anionic site charge density shows a strong cation dependence, i.e., it decreases systematically upon increasing the alkali metal cation size. This is a manifestation of the dependence of the borate glass structure, not only on the alkali metal oxide content, but also on the nature of the alkali metal cation. For crystallographic data of alkali metal borates, see Table 3/9 [33].

Table 3/9
Crystallographic Data for Alkali Metal Borates [33].

crystals	coordination number	number of M^+ ions	r(M–O) in pm
$Li_2O \cdot B_2O_3$	4	1	205
$Li_2O \cdot 2B_2O_3$	5	1	207
$Na_2O \cdot B_2O_3$	7	1	251
$Na_2O \cdot 2B_2O_3$	6	3	248
	7	1	259
α-$Na_2O \cdot 3B_2O_3$	5	1	251
	6	2	244
β-$Na_2O \cdot 3B_2O_3$	6	1	256
	7	1	257
	8	1	263
$Na_2O \cdot 4B_2O_3$	7	1	259
	8	1	256
$K_2O \cdot B_2O_3$	6	1	278
	8	1	281 to 291
$Cs_2O \cdot B_2O_3$	10	1	330 to 339

The polarization behavior of the Raman lines observed at 803 cm^{-1} and assigned to the symmetric breathing mode of the oxygen atom in the boroxin rings, and at 772 cm^{-1} assigned to the breathing mode of the six-membered ring containing both BO_3 and BO_4 units, are reported for the $0.84B_2O_3 \cdot 0.16Na_2O$ glass between 4.2 K and room temperature [6].

Glasses of high Na_2O content in the system $(B_2O_3)_{1-x}(Na_2O)_x$ with $0.35 \leq x \leq 0.75$ were studied by Raman and FIR spectroscopy. Small additions of Al_2O_3 around $x = 0.50$ were required in order to induce glass formation. At Na_2O contents around the metaborate composition, Raman spectra showed the presence of quite large borate units, which constituted a fairly well connected network through BO_4 bridges. Very high sodium contents ($x \geq 0.60$) caused the depolymerization of the network, which was finally disrupted and consisted primarily of pyroborate and orthoborate units. Such units create network sites of high charge density, as was evidenced from the FIR Na motion bands. Implications for the activation energy for ionic conduction are discussed [34].

Far-infrared (FIR) spectra of $B_2O_3(Na_2O)_x(MgO)_y$ glasses were also analyzed. The frequency of the Mg^{2+} motion band exhibits a departure from linearity in its compositional dependence. Implications of this behavior to the nature of the interaction between Mg^{2+} and its oxygen site are discussed [35]. The Mg^{2+} and Na^+ motions give characteristic absorptions at 300 to 450 cm^{-1} and 170 to 230 cm^{-1}, respectively. The cation dependence of the peak frequency of the cation-motion band was modeled through a Born-Mayer-type potential, indicating that the effective ionic charge and the integrated intensity constitute new sensitive probes of the degree of ionic character of the cation site interactions [36]. As Raman and

References on pp. 37/40

infrared spectral data show, the presence of Mg^{2+} cations for compositions with $x+y=0.33$, 0.53, and 0.67 causes mainly the destruction of diborate groups (1120 cm^{-1} in Raman) in favor of boroxin rings (805 cm^{-1}), tetraborate groups (780 cm^{-1}), pyroborate (1285 cm^{-1}, 850 cm^{-1}), and metaborate units (690 cm^{-1}). Similar borate groups were found in glass compositions $x+y=1.0$, originating from the destruction of orthoborate (945 cm^{-1}) and di-triborate groups (760 cm^{-1}). The Raman and infrared results were expressed in terms of melt equilibria between the various borate groups and proved useful in providing a structural interpretation for the compositional dependence of the Mg^{2+} motion band, observed in the far infrared [37].

The Raman spectra of cadmium borate glasses were also analyzed. Increasing the CdO amounts, the formation of nonbridging oxygen containing groups was detected. A band at 1385 cm^{-1} is attributed to a triangular B–O stretching, a band at 805 cm^{-1} to a breathing vibration of the boroxin ring, a band at 775 cm^{-1} to other six-membered borate rings containing one tetrahedral boron, and its shift to 760 cm^{-1} has been associated with the formation of six-membered borate rings containing two tetrahedral boron atoms [38]. A parallel mid- and far-infrared study shows CdO-induced changes of the borate network units from the intensity of the band envelope associated with the presence of B–O tetrahedra between 800 and 1100 cm^{-1}, relative to a band attributed to B–O triangles between 1150 and 1400 cm^{-1} [39]. As ^{11}B and ^{113}Cd NMR studies show, the N_4 (fraction of tetrahedral BO_4 units) in these glasses decreases with an increasing mole fraction of CdO along diagonals of the phase diagram [40].

Vibrational and dielectric measurements were also made on model fast ionic-conducting borate glasses. The addition of alkali metal halide causes significant structural changes in the glass network and an enhanced ionic conductivity is measured [41].

The role of the F^- ion in B_2O_3–Na_2O–NaF glasses was studied by various spectroscopic techniques. Raman, mid-infrared, and UV spectral data provided evidence that F^- participates in the borate network by forming covalent B–F bonds. The far-infrared spectra demonstrated that F^- induces the creation of high potential energy network sites with strong cation-network interactions. The implication of these phenomena on the electric conduction properties is discussed [42].

The influence of $[SO_4]^{2-}$ ions on the structure in the glass system $(B_2O_3)_{1-x-y}(Li_2O)_x(Li_2[SO_4])_y$ with x between 0.20 and 0.56 and y between 0 and 0.5 has been studied by Raman and Fourier transform infrared (FTIR) spectroscopy. Difference spectra and combined Raman and infrared results show that $Li_2[SO_4]$ addition induces the formation of six-membered rings with BO_4 tetrahedra in the $x=0.20$ series. However, for high Li_2O content ternary glasses ($x=0.56$) it was shown that sulfate anions are completely dispersed in the B–O network which is affected by their presence. For the $y=0.5$ glass, an increase of nonbridging oxygen atoms was shown [43].

IR spectroscopy was used to study the structure of basic $(B_2O_3)_3(Li_2O)_2(Li_2[SO_4])_x$ glasses with $x \leq 0.23$. Introduction of $Li_2[SO_4]$ resulted in the accumulation of sulfate in the glass network without changing the structure of the B–O matrix, whereas introduction of $[NH_4]_2$-$[SO_4]$ resulted in the depletion of Li_2O in the matrix and enrichment of Li_2O in the sulfate component of the glass [44].

Raman scattering results associated with ionic conductivity measurements on (B_2O_3)-$(Li_2O)_{0.5}(Li_2[SO_4])_{0.1}$ during the crystallization process show that in the glassy state the Li^+ and $[SO_4]^{2-}$ ions dilute the vitreous matrix without detectable interaction with the surroundings. The presence of new Li^+ ions improves the conduction. During the annealing process, the increase of this ionic conductivity drops for the first time at the glass crystal transition temperature where the material becomes an insulator. Simultaneously, Raman scattering reveals different stages for the structural change of the glass, in which the crystallization of the borate matrix is observed before that of the sulfate. The correlation of the two experiments allows one to expect that the structural changes of the material play an important role in the variation of the ionic conductivity up to $7\times10^{-6}\ \Omega^{-1}\cdot cm^{-1}$ [45].

Lithium borate glasses are fast ionic conductors in which the Li^+ conductivity is all the more important as the content of Li_2O in the Li salt is increased. Mid-infrared spectroscopy was used in order to study the glass-modifying properties of Li_2O [46, 47]. The mid-reflectivity spectra show clearly the disappearance of boroxin rings and the formation, in a first step, of tetraborate groups and later of diborate groups with increasing Li_2O content. One band at ca. 390 cm^{-1} is attributed to Li^+ vibrations. Modifications of the B–O network with the addition of $Li_2[SO_4]$ as a doping salt were also investigated in the B_2O_3–Li_2O–$Li_2[SO_4]$ system. At the same time, by far-infrared absorption on the binary glass, the vibrational motion of the Li^+ ions was confirmed [47]. Doping with halogenide ions modifies the B–O network differently: F^- participates directly to the B–O network, while Cl^- and Br^- are in interstitial positions in the glass matrix [46].

The IR spectra of B_2O_3–Li_2O–$Li_2[SO_4]$ glasses can be viewed as consisting of two parts: the low-frequency side of the charge carrier dynamics, and the higher-frequency region involving the host dynamics. The IR reflectivity increases considerably at frequencies near 500 cm^{-1}. This high reflectivity corresponds to the motion of the Li^+ ions and allows the frequency-dependent ionic conductivity to be deduced. An analysis, based on a simple double-well potential model, leads to a picture in which the characteristic vibrational frequencies of the Li^+ ion are distributed over a large frequency range corresponding to different site configurations. The extent of the frequency distribution is a function of the free-ion concentration. A drastic decrease in the bandwidth is observed when the concentration increases. The maximum in the distribution of the characteristic Li^+ vibrational frequency remains constant with an increasing Li^+ concentration, but the width of the distribution narrows, which indicates an increase in the relaxation time. With increasing Li^+ concentration, the occupation of sites by a given preferential configuration becomes more frequent, and the residence time in such sites increases [48].

The IR reflectivity spectra of B_2O_3–Li_2O–$Li_2[SO_4]$ glasses show that in the B–O matrix, the boron coordination changes as a function of the modifier content. BO_4 units are formed which constitute the negative sites for the conduction of the Li^+ ions. Such a mechanism is strongly dependent on the distribution of the sites inside the glass network. The effects of annealing at temperatures near the crystallization temperature on reflectivity were studied using an Fourier transform infrared (FTIR) microscope. The structural modification is observed by a change of the BO_4 distribution, which is responsible for the increase of the ionic conductivity [49].

In a further development of the statistical mechanical theory of the ion conductivity in B_2O_3–Li_2O–$Li_2[SO_4]$ glasses, based on the assumption that Li_2O and $Li_2[SO_4]$ behave as weak electrolytes which dissociate to provide free Li^+ ions, the effect of Coulombic interactions between free Li^+ ions was taken into account [50].

B_2O_3–Li_2O–LiX glasses with X = F, Cl, or I are fast ionic conductors. Investigations of their structure were performed by using Raman scattering and Fourier transform infrared (FTIR) spectroscopy. Binary and doped ternary glasses with the same O : B ratio do not exhibit an identical network. The modifications are attributed to interactions between the network and the anions of the doping salt. In addition, the type of the anion plays an important role in the observed effects [51].

The effects of halide substitution on the ionic conductivity in potassium borate glasses are different from those observed in their lithium and sodium analogues. Physical property and Raman spectroscopy results suggest that this behavior is associated with differences in the glass network structure induced by the presence of different interstitial alkali metal cations [52].

References on pp. 37/40

Thin micro-solid-state lithium batteries using lithium borate glass electrolytes and InSe or In_2Se_3 layered cathodes were fabricated by flash evaporation or molecular beam deposition techniques [53]. Thin films of titanium oxide sulfides are used as intercalation cathodes in solid-state microbatteries with B_2O_3–Li_2O–$Li_2[SO_4]$ glass as an electrolyte and evaporated lithium as an anode [54]. A review containing 55 references on borate-based lithium conducting glasses has been assembled [55].

Raman spectra of $(B_2O_3)_{1-x}(Li_2O)_x$ quenched glasses with $x=0$ to 0.70 were measured. Glasses with $x>0.50$ show marked spectral changes whereas the spectra of the glasses containing smaller Li_2O contents show a close resemblance to each other. The spectra revealed that the main structural units are the six-membered borate rings with BO_4 units for the glasses with small values of x, and pyroborate and orthoborate groups for those with large x values. The six-membered borate rings with BO_4 units have a maximal proportion at $x=0.33$ and the pyroborate group a maximal proportion at $x=0.67$. A large amount of orthoborate groups (completely discrete anions) was present even at $x=0.25$. The fraction of tetrahedral BO_4 units present in the glasses was determined as a function of the glass composition and compared with the results of an NMR study. It cannot be assumed that every structural unit has the same Raman scattering efficiency [56]. The Raman spectra of an $Li_4[B_2O_5]$ melt and a rapidly quenched glass are similar, and consisted of orthoborate groups, pyroborate groups, and six-membered borate ring groups with BO_4 units, while the same crystals consisted only of pyroborate groups [57, 58].

Raman spectra were also recorded for silicate- and phosphate-containing lithium borate glasses. The results suggested that nonbridging oxygen atoms were most preferentially formed in the phosphate groups, followed by the borate and the silicate groups [59]. Lithium borate glasses with 10 mol% SiO_2 show (by impedance and ^{11}B NMR measurements) a dilution of the network. The tendencies of ionic conductivity lead to the conclusion of a possible sequence of bonding strengths because of different options for Li^+ positions [60].

The existence of tetraborate groups in lithium borate glasses was proven by IR spectroscopy. The IR bands in $(B_2O_3)_{1-x}(Li_2O)_x$ with $0.15 \le x \le 0.25$ and $(B_2O_3)_{1-y}(Na_2O)_y$ with $0.15 \le y \le 0.36$ were related to vibrations predominantly localized on fragments of polyborate groups [61]. Relative intensity changes of bands in the IR reflection spectra of lithium and sodium borate glasses and melts with x or y ca. 0.25, occurring with an increase of temperature, are caused by an increase of groups corresponding in composition to B_2O_3 and metaborate groups due to a decrease of tetraborate groups. In the lithium borate system, the changes are also due to an increase of diborate groups. These transformations are accompanied by a decrease of N_4 (fraction of tetrahedral BO_4 units) and are not observed in spectra recorded below the glass transition temperature T_g [62]. The B_2O_3–M_2O–MX glass systems with M = Li or Na and X = F or Cl (B : M = 3 : 1, variable amounts of MX) form BO_3F groups with a B–F bond, but not BO_3Cl groups, as IR spectra show; the formation of BO_3F in the lithium system was not conclusive [63].

The $(B_2O_3)_{1-x-y}(Li_2O)_x(LiF)_y$ glasses with $0.2125 \le x \le 0.25$ and $0 \le y \le 0.15$ were also studied by optical spectroscopy. The B–O chain structure is not affected by the presence of LiF. The type of chain structure depends only on the B : O ratio. The fluorine atoms do not replace any of the oxygen atoms in the chains [64].

The glass structure of B_2O_3–CaO–Al_2O_3 glasses was determined using IR spectroscopy. The aluminium atoms are in tetrahedral coordination with oxygen atoms and closely approached by boron atoms. The short-range order is different from those in crystals [65]. Boron-11 NMR spectral data have been used to determine N_4 (fraction of tetrahedral BO_4 units) in CABAL (B_2O_3–CaO–Al_2O_3) and SiO_2–CABAL glasses subjected to various thermal histories. Aluminium competes with boron for bonding with oxygen. In the SiO_2–CABAL glass system, the boron coordination depends on the quench rate [7]. The IR spectra were also obtained of thin glass films prepared by pouring melts of B_2O_3–M_1O–Al_2O_3 with M_1 = Sr or Ba at 1200°C. As the B_2O_3 content decreased and the M_1O content increased, the intensity of absorption bands between 800 and 1150 cm^{-1}, relative to the intensity of 1150 to 1600 cm^{-1} bands, decreased. This indicates that N_4 decreases as M_1 increases. In some glasses, the four-coordinate boron atoms are not observed [66].

In $(B_2O_3)_{1-x}(PbO)_x$ glasses with $0.2 \leq x \leq 0.3$, the predominant borate groups are those corresponding in composition to B_2O_3, diborate groups, and, apparently, groups with the same composition as diborate groups but differing from them structurally [67].

Densities of lithium borate glasses were determined for a wide range of compositions (Q is the molar ratio of lithium to boron). A semiempirical model for the density was developed which makes use of the structural ideas of Krogh-Moe. The results supported the assumption that the densities reflect the underlying atom arrangements in a simple manner. CO_2 retention occurs for glasses prepared from $Li_2[CO_3]$ and having extremely high Li_2O contents but has no effect on the densities. There are present at Q = 1.00 metaborate ions, at Q = 2.00 pyroborate ions (calculated density D_c = 2.01 g/cm³; experimentally, D_m = 2.04 g/cm³), and at Q = 3.00 orthoborate ions (D_c = 2.13 g/cm³; D_m = 2.16 g/cm³) [68]. The model was also applied to the densities of sodium borate glasses. The analysis indicated the volumes of the structural units present in these glasses were larger than in the corresponding lithium borates [69].

The model was further applied to the densities of potassium borate glasses in order to determine the volumes of the structural units present. These calculated values, in conjunction with volumes found in the lithium and sodium cases, were used in a general discussion of the packing fractions of the units. The fractions indicated that the size increase of the structural groupings, as one goes from lithium to sodium and to potassium, was primarily due to the alkali metal being used, and that the same volumes are present for each of the boron-oxygen configurations [70].

The model was also applied to the mixed Na + Li and Li + Rb borate glasses over an extremely wide range of alkali metal content from B_2O_3 through the orthoborate composition. The results indicate that the size increases of the structural groupings in going from small to large alkali metal are primarily due to the alkali metal being used [71].

Reported densities from rubidium, caesium, silver, and thallium borate glasses were used in a quantitative model involving atomic arrangements. The packing fractions of the borate units (four-coordinate boron) are Li 0.67, Na 0.57, Ag 0.65, K 0.52, Rb 0.49, Tl 0.63, and Cs 0.52. The packing fractions of the borate units (three-coordinate boron) are 0.34 for Li, Na, Ag, Rb, Cs; 0.35 for K; the average of the packing fraction for Tl is also 0.35. The packing fraction is defined as the ratio of the volume of a unit calculated from ionic radii to the same volume found from the density measurements [72].

In $(B_2O_3)_{1-x}(Na_2O)_x$ glasses with $0 \leq x \leq 0.36$, the boron coordination was determined by X-ray diffraction techniques with Ag K_α radiation on quenched and normally cooled samples. The coordination number of the quenched glass is a little smaller than that of a normally cooled one. BO_4 tetrahedra seem to be stable at high temperatures [73]. In the same glasses with $0.36 \leq x \leq 0.64$, ^{11}B NMR spectroscopy was used to study the temperature-dependent coordination of boron. Satisfactory agreement between NMR and X-ray diffraction data was observed. The mean coordination numbers of boron in conventionally cooled and quenched glasses were determined by NMR spectroscopy as 3.37 ± 0.02 and 3.32 ± 0.02, respectively [74]. For the same glasses with x = 0.33, a structural model is proposed in which half of the boron is tetrahedrally and the other half is triangularly coordinated [75].

References on pp. 37/40

According to the concepts of Krogh-Moe, the structural units of borate glasses are not BO_3 triangles and BO_4 tetrahedra, but rather groups of more complex arrangement. The network of the sodium borate glass $(B_2O_3)_{1-x}(Na_2O)_x$ with $x=0.33$ is supposed to consist exclusively of diborate groups with a small percentage of tetraborate groups. Since pyroborax, a noncrystalline material prepared from borax by dehydration, essentially consists of diborate groups, it can be regarded as a model for the short-range order of this glass. Accurate measurements of the scattering curves of pyroborax of a glass melted from borax at 800°C, and of normally prepared sodium borate glasses lend new support to the Krogh-Moe model of borate glass structure [76].

$(B_2O_3)_{1-x-y}(Ag_2O)_x(AgI)_y$ superionic glasses are becoming a model vitreous electrolyte system. They have been investigated by ^{109}Ag NMR studies [77] and ^{11}B NMR spin-lattice relaxation experiments [78]. These glasses are considered to consist of the network former B_2O_3, the network modifier Ag_2O, and the doping salt AgI. The covalently bonded glass network consists of BO_3 and BO_4 units which are linked by bridging oxygen atoms; there are also nonbridging oxygen atoms. The ^{109}Ag NMR chemical shift range is covered by the different forms of AgI and the dependence of the isotropic chemical shift upon the BO_4 units and nonbridging oxygen atoms concentrations [77]. The spin-lattice relaxation time was discussed, in terms of BO_4 units, above the glass transition temperature T_g; the rates increase rapidly with the temperature and become independent of the frequency. This behavior is anomalous [78].

Molecular dynamics (MD) simulation was performed on glasses $(B_2O_3)_{1-x}(Ag_2O)_x$ using the pair potential of the modified Born-Mayer-Huggins form without the dispersion terms. This simulation confirms the previous hypothesis concerning the structure of these glasses in which the coordination of boron changes from three to four with a relative transformation of planar BO_3 groups to tetrahedral BO_4 groups, and indicates a localization of Ag^+ ions in the interstices of the network near the oxygen atoms [79]. Size effects of the Ag^+ radius on the MD simulation were studied also. The results with $r=63$ pm are compared with those of $r=110$ pm. The glassy network is quite well reproduced by both of the radii. The main differences that arise by using a smaller Ag^+ radius are a higher degree of distortion, which characterizes the structural units building the glassy network, and a clustering effect of the Ag^+ ions [80].

A study of the ultrasound velocity in these glasses at 5 MHz and between 77 and 450 K gives further evidence for the structural hypothesis of a glassy network formed from a matrix of silver borate in which weakly bonded microdomains of AgI are dispersed [81]. There are dispersive effects, the contribution of which increases with the AgI content. These effects arise from the thermally activated jumps of Ag^+ ions between nearly equivalent positions available in the glassy network [82]. An ultrasonic study between 10 and 77 K leads to calculations of the tunneling frequency for the Ag^+ ions [83]. Measurements of the acoustic attenuation and relative acoustic velocity between 1 and 150 K in $(AgI)_x(Ag_2O\cdot nB_2O_3)_{1-x}$ glasses with $0.1\leq x\leq 0.6$ and $2\leq n\leq 10$ show that the composition-dependent acoustic loss is correlated with the formation of BO_4 tetrahedra and their rotation [84]. The effects of hydrostatic pressure on the elastic behavior were studied by the changes of the velocities of ultrasonic waves. The increase of the AgI concentration in glasses of a fixed $Ag_2O:B_2O_3$ ratio of 0.33 produces an increase in the pressure gradients. The implication that the Ag^+ ions introduced via Ag_2O and AgI occupy different sites is in accord with the microdomain structure model [85]. Specific heat and thermal conductivity measurements of glasses between 1 and 10 K show an anomaly which is interpreted in terms of phonon localization [86].

$(AgI)_x(Ag_2O\cdot 2B_2O_3)_{1-x}$ glasses with $x=0.1$ and $x=0.6$ were investigated by comparative neutron-diffraction measurements. For $x=0.6$, a strong and sharp diffraction peak is attributed either to the formation of AgI clusters or to density deficits in the B–O network; this peak is not present in the $x=0.1$ glass [87]. There was also observed Brillouin scattering, supporting the model in which AgI is introduced into the glass network in the form of α-AgI-like microdomains [88, 89]. Both the observations are compared with current models for fast ion diffusion in glasses [90]. NMR relaxation and conductivity measurements were also correlated for B_2O_3–Ag_2O–AgI glasses [91].

Low-frequency Raman scattering of these glasses was measured as a function of the concentration of AgI [92, 93, 94]. The X-ray absorption spectroscopy (XAS) and the extended X-ray absorption fine structure spectroscopy (EXAFS) measurements with $x \leq 0.5$ glasses show that Ag^+ ions are bonded to BO_4 tetrahedra. Because of the lack of nonbridging oxygen atoms for $x \approx 0.33$, the oxygen nearest neighbors of an Ag^+ ion have to belong to the same BO_4 tetrahedron [95, 96].

The coordination of iodine in these glasses was also studied by EXAFS measurements. The mean local environment of iodine changes with the AgI content. The two coordinations of Ag^+ ions were shown also by EXAFS measurements; the coordination number is about 2. The results are consistent with the hypothesis that AgI tends to reproduce distorted tetrahedral units without affecting the host matrix. This structural model is consistent also with the conductivity properties [97].

An $(Ag_2O)_x(AgX)_y(B_2O_3)_{1-x-y}$ glass with X=I and y=0.2, doped with 0.01 mol $^{57}Fe_2O_3$, was studied by Mößbauer spectroscopy. The Fe^{3+} ions are present at the substitutional sites of the tetrahedral boron atoms constituting BO_4 units as a network former. Increasing x exceeding 0.16 leads to formation of nonbridging oxygen atoms [98]; with X=Br and y=0.2 the Mößbauer spectroscopy results are in good agreement with results obtained by differential thermal analysis (DTA) in which the glass transition temperature T_g shows a distinct dependence on the composition with a maximum at $x \approx 0.12$. The increase at lower Ag_2O contents is ascribed to the well-known structural change from BO_3 to BO_4 units observed in ordinary borate glasses [99].

Conductometric, spectroscopic, dynamical, and diffractometric studies on these glasses show that AgX (X=I, Br, or Cl) crystalline cluster models do not seem able to account for all these experimental observations. It is suggested that an adequate model could be based on AgX amorphous clusters, i.e., nearly structureless agglomerates of Ag and X atoms [100].

References for 3.2.5.2:

[1] Aoki, N.; Suzuki, K.; Hasegawa, H.; Yasui, I. (Yogyo Kyokaishi **93** [1985] 327/33 from C.A. **103** [1985] No. 75075).
[2] Inoue, H.; Aoki, N.; Yasui, I. (J. Am. Ceram. Soc. **70** [1987] 622/7).
[3] Yasui, I.; Hasegawa, H.; Saito, Y. (J. Non-Cryst. Solids **106** [1988] 30/3).
[4] Kolesova, V. A. (Fiz. Khim. Stekla **12** [1986] 4/13 from C.A. **104** [1986] No. 233144).
[5] Zemlyanov, M. G.; Malinovskii, V. K.; Novikov, V. N.; Parshin, P. P.; Sokolov, A. P. (Pis'ma Zh. Eksp. Teor. Fiz. **49** [1989] 521/3 from C.A. **111** [1989] No. 86404).
[6] Ramos, M. A.; Vieira, S.; Calleja, J. M. (Solid State Commun. **64** [1987] 455/7).
[7] Bray, P. J. (J. Non-Cryst. Solids **71** [1985] 411/28).
[8] Uchida, N.; Maekawa, T.; Yokokawa, T. (J. Non-Cryst. Solids **74** [1985] 25/36).
[9] Uchida, N.; Maekawa, T.; Yokokawa, T. (Nippon Kagaku Kaishi **1986** 1414/24 from C.A. **106** [1987] No. 73199).
[10] Uchida, N.; Maekawa, T.; Yokokawa, T. (J. Non-Cryst. Solids **85** [1986] 290/308).

[11] Uchida, N.; Maekawa, T.; Yokokawa, T. (J. Non-Cryst. Solids **88** [1986] 1/10).
[12] Xu, Qiang; Kawamura, K.; Yokokawa, T. (J. Non-Cryst. Solids **104** [1988] 261/72).
[13] Maekawa, T.; Kubota, Y.; Yokokawa, T. (Phys. Chem. Glasses **30** No. 4 [1989] 142/8).

[14] Soppe, W.; Ebens, W.; Den Hartog, H. W. (J. Non-Cryst. Solids **105** [1988] 251/7).
[15] Soppe, W.; Den Hartog, H. W. (J. Non-Cryst. Solids **108** [1989] 260/8).
[16] Soppe, W.; Althof, V.; Den Hartog, H. W. (J. Non-Cryst. Solids **104** [1988] 22/30).
[17] Soppe, W.; Van der Marel, C.; Den Hartog, H. W. (J. Non-Cryst. Solids **101** [1988] 101/10).
[18] Soppe, W.; Aldenkamp, F.; Den Hartog, H. W. (J. Non-Cryst. Solids **91** [1987] 351/74).
[19] Soppe, W.; Kleerebezem, J.; Den Hartog, H. W. (J. Non-Cryst. Solids **93** [1987] 142/54).
[20] Martin, A. J.; Brenig, W. (Phys. Status Solidi B **64** [1974] 163/72).

[21] Bray, P. J. (J. Non-Cryst. Solids **75** [1985] 29/36).
[22] Bray, P. J.; Holupka, E. J.; Zong, J.; Mulkern, R. V.; Brinker, C. J. (Wiss. Z. Friedrich-Schiller-Univ. Jena Math. Naturwiss. Reihe **36** [1987] 735/43).
[23] Bray, P. J.; Mulkern, R. V.; Holupka, E. J. (J. Non-Cryst. Solids **75** [1985] 37/43).
[24] Araujo, R. J. (Struct. Bonding Noncryst. Solids Int. Symp., Reston, Va., 1983 [1986], pp. 13/28 from C.A. **106** [1987] No. 71756).
[25] Jellison, G. E.; Bray, P. J. (J. Non-Cryst. Solids **29** [1978] 187/206).
[26] Araujo, R. J. (J. Non-Cryst. Solids **81** [1986] 251/4).
[27] Schirmacher, W.; Schirmer, A. (Solid State Ionics **28/30** [1987/88] 134/7).
[28] Schirmer, A.; Heitjans, P.; Ackermann, H.; Bader, B.; Freiländer, P.; Stöckmann, H. J. (Solid State Ionics **28/30** [1987/88] 717/21).
[29] Kamitsos, E. I.; Karakassides, M. A.; Chryssikos, G. D. (Phys. Chem. Glasses **28** [1987] 203/9).
[30] Kamitsos, E. I.; Karakassides, M. A.; Chryssikos, G. D. (J. Phys. Chem. **91** [1987] 5807/13).

[31] Kamitsos, E. I.; Karakassides, M. A.; Chryssikos, G. D. (Solid State Ionics **28/30** [1987/88] 687/92).
[32] Kamitsos, E. I.; Chryssikos, G. D.; Karakassides, M. A. (Phys. Chem. Glasses **29** [1988] 121/6).
[33] Kamitsos, E. I. (J. Phys. Chem. **93** [1989] 1604/11).
[34] Kamitsos, E. I.; Karakassides, M. A. (Phys. Chem. Glasses **30** [1989] 19/26).
[35] Kamitsos, E. I.; Karakassides, M. A.; Chryssikos, G. D. (Solid State Commun. **60** [1986] 885/8).
[36] Kamitsos, E. I.; Chryssikos, G. D.; Karakassides, M. A. (J. Phys. Chem. **91** [1987] 1067/73).
[37] Kamitsos, E. I.; Karakassides, M. A.; Chryssikos, G. D. (J. Phys. Chem. **91** [1987] 1073/9).
[38] Chryssikos, G. D.; Kamitsos, E. I.; Risen, W. M. (J. Non-Cryst. Solids **93** [1987] 155/68).
[39] Chryssikos, G. D.; Turcotte, D. E.; Mulkern, R. V.; Bray, P. J.; Risen, W. M. (J. Non-Cryst. Solids **85** [1986] 54/8).
[40] Mulkern, R. V.; Chung, S. J.; Bray, P. J.; Chryssikos, G. D.; Turcotte, D. E.; Risen, W. M. (J. Non-Cryst. Solids **85** [1986] 69/73).

[41] Turcotte, D. E. (Diss. Brown Univ., Providence, R.I., 1987, pp. 1/229 from C.A. **107** [1987] No. 226840).
[42] Kamitsos, E. I.; Karakassides, M. A. (Solid State Ionics **28/30** [1987/88] 783/7).
[43] Kamitsos, E. I.; Karakassides, M. A.; Chryssikos, G. D. (J. Phys. Chem. **90** [1986] 4528/33).
[44] Kolesova, V. A. (Fiz. Khim. Stekla **12** [1986] 106/8 from C.A. **105** [1986] No. 10593).
[45] Balkanski, M.; Ayyadi, A.; Cadet, P.; Jouanne, M.; Julien, C.; Massot, M.; Scagliotti, M.; Levasseur, A. (Solid State Commun. **57** [1986] 41/6).

[46] Massot, M.; Julien, C.; Balkanski, M. (Infrared Phys. **29** [1989] 775/9).
[47] Julien, C.; Massot, M.; Balkanski, M.; Krol, A.; Nazarewicz, W. (Mater. Sci. Eng. B **3** [1989] 307/12 from C.A. **111** [1989] No. 139055).
[48] Balkanski, M.; Darianian, I.; Burret, P. A.; Massot, M.; Julien, C. (Mater. Sci. Eng. B **3** [1989] 177/63 from C.A. **111** [1989] No. 101704).
[49] Julien, C.; Massot, M. (Solid State Ionics **34** [1989] 269/73).
[50] Balkanski, M.; Wallis, R. F.; Deppe, J. (Mater. Sci. Eng. B **3** [1989] 65/8 from C.A. **111** [1989] No. 206424).

[51] Massot, M.; Haro, E.; Oueslati, M.; Balkanski, M. (Mater. Res. Soc. Symp. Proc. **135** [1989] 207/17 from C.A. **111** [1989] No. 104702).
[52] Fusco, F. A.; Massot, M.; Oueslati, M.; Haro, E.; Tuller, H. L.; Balkanski, M. (Mater. Res. Soc. Symp. Proc. **135** [1989] 189/97 from C.A. **111** [1989] No. 124617).
[53] Balkanski, M.; Julien, C.; Emery, J. Y. (J. Power Sources **26** [1989] 615/22 from C.A. **111** [1989] No. 137482).
[54] Meunier, G.; Dormoy, R.; Levasseur, A. (Mater. Sci. Eng. B **3** [1989] 19/23 from C.A. **111** [1989] No. 137484).
[55] Levasseur, A.; Menetrier, M. (Mater. Chem. Phys. **23** [1989] 1/12 from C.A. **111** [1989] No. 206291).
[56] Tatsumisago, M.; Takahashi, M.; Minami, T.; Tanaka, M.; Umesaki, N.; Iwamoto, N. (Yogyo Kyokaishi **94** [1986] 464/9 from C.A. **104** [1986] No. 229134).
[57] Tatsumisago, M.; Kowada, Y.; Minami, T. (Chem. Express **2** [1987] 149/52 from C.A. **106** [1987] No. 181251).
[58] Tatsumisago, M.; Takahashi, M.; Minami, T.; Umesaki, N.; Iwamoto, N. (Phys. Chem. Glasses **28** [1987] 95/6).
[59] Kowada, Y.; Tatsumisago, M.; Minami, T. (J. Phys. Chem. **93** [1989] 2147/71).
[60] Müller, W.; Kruschke, D.; Torge, M.; Grimmer, A. R.; Schütt, H. J. (Solid State Ionics **23** [1987] 53/8).

[61] Chekhovskii, V. G. (Fiz. Khim. Stekla **11** [1985] 24/33 from C.A. **102** [1985] No. 171260).
[62] Chekhovskii, V. G.; Sizonenko, A. P. (Fiz. Khim. Stekla **14** [1988] 194/9 from C.A. **109** [1988] No. 78295).
[63] Eliseev, S. Yu.; Chekhovskii, V. G. (Fiz. Khim. Stekla **11** [1985] 118/20 from C.A. **102** [1985] No. 194241).
[64] Petrakov, V. N.; Gorbachev, V. V.; Chekhovskii, V. G. (Fiz. Khim. Stekla **13** [1987] 475/8 from C.A. **107** [1987] No. 141912).
[65] Keiss, J.; Chekhovskii, V. G.; Pauks, P. (Fiz. Khim. Stekla **13** [1987] 22/8 from C.A. **106** [1987] No. 220429).
[66] Chekhovskii, V. G.; Keiss, J.; Petrov, Yu. A.; Yurkova, S. N.; Pauks, P. (Fiz. Khim. Stekla **14** [1988] 150/3 from C.A. **108** [1988] No. 191376).
[67] Chekhovskii, V. G.; Yurkova, S. N.; Egorov, F. K.; Ushakov, D. F. (Fiz. Khim. Stekla **14** [1988] 673/9 from C.A. **110** [1989] No. 80965).
[68] Shibada, M.; Sanchez, C.; Patel, H.; Feller, S.; Stark, J.; Sumcad, G.; Kasper, J. (J. Non-Cryst. Solids **85** [1986] 29/41).
[69] Karki, A.; Feller, S.; Lim, H. P.; Stark, J.; Sanchez, C.; Shibada, M. (J. Non-Cryst. Solids **92** [1987] 11/9).
[70] Lim, H. P.; Karki, A.; Feller, S.; Kasper, J. E.; Sumcad, G. (J. Non-Cryst. Solids **91** [1987] 324/32).

[71] Chong, B. C. L.; Choo, S. H.; Feller, S.; Teoh, B.; et al. (J. Non-Cryst. Solids **109** [1989] 105/13).

[72] Lim, Hun C.; Feller, S. (J. Non-Cryst. Solids **94** [1987] 36/44).
[73] Derno, M.; Herms, G.; Steil, H. (Phys. Status Solidi A **93** [1986] K5/K8).
[74] Grimmer, A. R.; Herms, G. (Z. Chem. **26** [1986] 452/3).
[75] Herms, G.; Weigt, N. (Rostocker Phys. Manuskr. No. 11 [1987] 68/71 from C.A. **108** [1988] No. 119312).
[76] Herms, G.; Derno, M.; Steil, H. (J. Non-Cryst. Solids **88** [1986] 381/7).
[77] Villa, M.; Chiodelli, G.; Magistris, A.; Licheri, G. (J. Chem. Phys. **85** [1986] 2392/400).
[78] Villa, M.; Farrington, G. C. (Philos. Mag. [8] B **56** [1987] 147/53 from C.A. **108** [1988] No. 47938).
[79] Abramo, M. C.; Pizzimenti, G.; Carini, G. (J. Non-Cryst. Solids **85** [1986] 233/8).
[80] Abramo, M. C.; Pizzimenti, G.; Carini, G. (Solid State Ionics **28/30** [1987/88] 148/51).

[81] Carini, G.; Cutroni, M.; Federico, M.; Tripodo, G. (Phys. Rev. [3] B **32** [1985] 8264/7).
[82] Carini, G.; Cutroni, M.; Federico, M.; Tripodo, G. (Solid State Ionics **18/19** [1986] 415/20).
[83] Carini, G.; Cutroni, M.; Federico, M.; Tripodo, G. (Phys. Rev. [3] B **37** [1988] 7021/6).
[84] Carini, G.; Cutroni, M.; Federico, M.; Galli, G.; Tripodo, G. (Philos. Mag. [8] B **59** [1989] 43/8 from C.A. **110** [1989] No. 199579).
[85] Saunders, G. A.; Sidek, H. A. A.; Comins, J. D.; Carini, G.; Federico, M. (Philos. Mag. [8] B **56** [1987] 1/13 from C.A. **107** [1988] No. 159917).
[86] Avogadro, A.; Aldrovandi, S.; Carini, G.; Siri, A. (Philos. Mag. [8] B **59** [1989] 32/42 from C.A. **110** [1989] No. 220038).
[87] Boerjesson, L.; Torell, L. M.; Dahlborg, U.; Howells, W. S. (Phys. Rev. [3] B **39** [1989] 3404/7).
[88] Boerjesson, L.; Torell, L. M.; Martin, S. W.; Liu, C.; Angell, C. A. (Phys. Lett. A **125** [1987] 330/4).
[89] Boerjesson, L. (Phys. Rev. [3] B **36** [1987] 4600/12).
[90] Boerjesson, L.; Torell, L. M.; Howells, W. S. (Philos. Mag. [8] B **59** [1989] 105/23 from C.A. **110** [1989] No. 199580).

[91] Angell, C. A.; Martin, S. W. (Mater. Res. Soc. Symp. Proc. **135** [1989] 73/81 from C.A. **111** [1989] No. 124615).
[92] Fontana, A.; Rocca, F.; Fontana, M. P. (Philos. Mag. [8] B **56** [1987] 251/5 from C.A. **107** [1988] No. 225143).
[93] Fontana, A.; Rocca, F. (Phys. Rev. [3] B **36** [1987] 9279/82).
[94] Fontana, A.; Rocca, F.; Tomasi, A. (Solid State Ionics **28/30** [1987/88] 722/5).
[95] Dalba, G.; Fornasini, P.; Rocca, F.; Bernieri, E.; Burattini, E.; Mobilo, S. (J. Non-Cryst. Solids **91** [1987] 153/64).
[96] Rocca, F.; Dalba, G.; Fornasini, P. (Mater. Chem. Phys. **23** [1989] 85/98 from C.A. **111** [1989] No. 219392).
[97] Dalba, G.; Fornasini, P.; Fontana, A.; Rocca, F.; Burattini, E. (Solid State Ionics **28/30** [1987/88] 713/6).
[98] Nishida, T.; Ogata, M.; Takashima, Y. (Bull. Chem. Soc. Jpn. **60** [1987] 2381/6).
[99] Nishida, T.; Ogata, M.; Takashima, Y. (Phys. Chem. Glasses **29** [1988] 22/5).
[100] Musinu, A.; Paschina, G.; Piccaluga, G.; Pinna, G. (Solid State Ionics **34** [1989] 187/93).

3.2.5.3 Chemical and Physical Data

Helium solubility in sodium borate melts and glasses was determined between 400 and 900 °C and in Na/Ca borate melts between 500 and 900 °C. The results show that the replacement of B_2O_3 by Na_2O and of CaO by Na_2O decreases the solubility. Although the solubility increases with increasing temperature in all cases, the temperature dependence of the solubility decreases with increasing modifier oxide concentration. The results are explained in terms of the free volume variations with composition and temperature [1].

Diffusion of tritium in lithium borate glass was studied by a radiometric method during thermal evolution. The rate of tritium evolution is determined by the bulk diffusion. The diffusion parameters do not depend on the presence of interfaces; the contribution of surface processes is appreciable only at the initial stage of the kinetic curves. The diffusion coefficient of 8.5×10^{-9} cm^2/s at 650 °C is smaller than for glasses containing alkali metal ions. Tritium evolves only in an oxidized form as HTO or T_2O [2].

Oxygen incorporation in epitaxial silicon produced by molecular-beam epitaxy and boron doped by coevaporation of B_2O_3 was investigated [3].

Anisotropic borate glasses, obtained by gas-phase deposition of B_2O_3/H_2O mixtures, show copolymerization. In the vapor phase between 700 and 1200 °C, B_2O_3 monomers were separated [4].

The difficulty of completely dehydrating B_2O_3, even at ca. 1000 °C, was illustrated by a study of the reactions between B_2O_3 and CsI. At ca. 600 °C, this system generates HI (accompanied by a partial decomposition to H_2 and I_2) in the absence of oxygen and produces much larger quantities of I_2 in the presence of oxygen [5].

The effects of aqueous acid and alkaline media on alkali metal borate glass were examined in order to identify factors affecting glass-media equilibria. The glass weight loss decreased with increasing glass soda content, increased with immersion time, and increased with increasing concentration of the immersion solution [6].

The electrolytic behavior of molten sodium borate glasses and the effect of water was studied by determining current-potential curves on Pt electrodes. In the $(B_2O_3)_{10-x}(Na_2O)_x$ system (x=1 to 5) containing small amounts of water, experimental decomposition voltages were 1.3 to 1.6 V at 1000 °C, increasing with increasing x from 1 to 5. In this case, elemental boron was produced at the cathode by the discharge of borate anions, and no hydrogen evolution was observed [7].

Pyrex glass tubes were autoclaved for various times with distilled water between 200 and 260 °C. Above 220 °C, the amount of B_2O_3 extracted from the glass surface continues to increase linearly with time. The energy of the B_2O_3 extraction was calculated to be 27.0 kcal/mol. The rates of Na_2O and SiO_2 extractions approach zero once the aqueous phase is saturated with the extracted SiO_2. IR spectra measured on the autoclaved thin films of Pyrex glass show the existence of both molecular water and hydroxyl groups [8].

The kinetics of selective dissolution of the metastable sodium borate phase in phase-separated sodium borosilicate glass by 0.1N HNO_3 was studied for the preparation of porous glasses. The depth of the acid leaching zone was unaffected by the channel size or volume fraction of the metastable phase that resulted from phase-separation heat-treatments within the range 550 to 700 °C [9].

The transpiration method was used to determine the vapor density of molten borax between 900 and 1115 °C in an N_2 atmosphere and in N_2/H_2O vapor mixtures. The results imply a relatively small effect of H_2O vapor on the density of borate vapors and are indicative of quasi-congruent vaporization resulting from the reaction of the melt with water vapor [10].

References on pp. 46/9

The B_2O_3–Li_2O system becomes metastable at a critical lithium concentration of 4.2 mol% and a critical temperature of 430 °C, based on scattering data of visible light [11].

According to the acid-base theorie of Lux [89], B_2O_3 behaves as a "Lux acid" in B_2O_3–Na_2O melts or glasses, but it behaves as a "Lux base" in the B_2O_3–Na_2O–P_2O_5 system. The basicity of a $B_2O_3 \cdot 2\,Na_2O \cdot 2\,P_2O_5$ melt was determined at 900 °C from the electromotive force values of a galvanic concentration cell [12].

The enthalpies of mixing of the liquid mixtures of $B_2O_3 + Na_3[B_3O_6]$, $B_2O_3 + (NaPO_3)_2$, and $Na_3[B_3O_6] + Na_4[P_2O_7]$ were measured by liquid-liquid calorimetry. The results are discussed in terms of tentative structural models of the melts, and in terms of tentatively structural models of the O^{2-} ion transfer from phosphorus groups to boron groups in the mixture $B_2O_3 + (NaPO_3)_2$, and from boron groups to phosphorus groups in $Na_3[B_3O_6] + Na_4[P_2O_7]$ [13]. The enthalpies of mixing in the quasi-binary liquid system $NaBO_2$–SiO_2 were measured by solid-liquid calorimetry at 1394 K. After correcting for the known enthalpy of fusion of SiO_2 at this temperature, the corresponding liquid-liquid enthalpies of mixing, H_{mix}, and the enthalpy interaction parameters (N = mole fraction) were also derived; the latter quantity, $\Delta H_{mix}/N(NaBO_2) \cdot N(SiO_2)$, changes from ca. −30 kJ/mol in pure $NaBO_2$ to ca. −20 kJ/mol in $0.33\,B_2O_3 \cdot 0.67\,SiO_2$. From the experimentally determined curve for H_{mix}, the partial enthalpies of the two components were obtained by the method of intercepts [14].

The changes in volume accompanying the crystallization of $Li_2[B_4O_7]$, $Na_2[B_8O_{13}]$, $Cs_2[B_6O_{10}]$, $Ba[B_4O_7]$, and $Pb[B_4O_7]$ from binary borate melts were determined throughout the crystal growth range. The degree of similarity between the structure of the crystals and that of its corresponding melt may be inferred from the change in volume associated with the crystallization [15].

The relationship between the acid-base properties of oxides and the atomic characteristics of their components has been studied. The increase in the basicity from Li_2O to Cs_2O is due to an increase in the ionic character of metal–oxygen bonds during the formation of alkali metal borates from oxides [16].

The heat capacities were measured for $(B_2O_3)_{1-x}(Cs_2O)_x$ with $0.05 \le x \le 0.4$ for glasses and $0.1 \le x \le 0.5$ for crystals between 200 and 1100 K. The data for the crystalline and glassy compound of the same composition are compared. Values were correlated as functions of x at a constant temperature. Below the glass transition temperature, T_g, the values for the crystalline and glassy oxides are very close. Near T_g, the characters of the temperature- and composition-dependence change and the heat capacity values show sharp jumps. The effect of the Cs_2O concentration varies as x changes; this is related to the changing Cs–O and B–O bonds [17]. Applying differential scanning calorimetry to these caesium borates leads to the determination of heats, entropies, and free energies of the phase transitions as functions of the temperature as well as of the composition [18].

A series of crystalline rubidium borates (see Table 3/10) was prepared. The indexing of the diffractograms for β-$Rb_2O \cdot B_2O_3$ was completed. Under given conditions, the crystallization product of the glass with the composition $0.286\,Rb_2O \cdot 0.714\,B_2O_3$ is not the individual compound $2\,Rb_2O \cdot 5\,B_2O_3$, but a mixture of β-$Rb_2O \cdot B_2O_3$ and $Rb_2O \cdot 3\,B_2O_3$. The product of the glass $0.1\,Rb_2O \cdot 0.9\,B_2O_3$ is β-$Rb_2O \cdot 5\,B_2O_3$ instead of $Rb_2O \cdot 9\,B_2O_3$ [19, 20]. In the same manner, caesium borates (see also Table 3/10) were synthesized. The values obtained for the interplanar distances of compounds $Cs_2O \cdot B_2O_3$ and $Cs_2O \cdot 4\,B_2O_3$ differ from those available in the literature. The following parameters of the elementary cell of the tetraborate were determined: a = 1434(2) pm and c = 1238(2) pm [20, 21].

Table 3/10
Thermodynamic Properties of Rubidium and Caesium Borates at 298 K (uncertainty last figure) [20, 22, 27].

compound	coordination No. of $[M]^+$	$-\Delta H_f^\circ$ (kcal/mol)	$-\Delta H_{cr}^\circ$ (kcal/mol)
β-$Rb_2O \cdot 5B_2O_3$	8	−0.7 (1)	22.3(2)
$Rb_2O \cdot 4B_2O_3$	9	−0.40(5)	25.5(2)
$Rb_2O \cdot 3B_2O_3$	8	0.5 (1)	29.6(3)
$Rb_2O \cdot 2B_2O_3$	7	3.6 (1)	35.3(4)
β-$Rb_2O \cdot B_2O_3$	7	15.3 (2)	41.0(4)
β-$Cs_2O \cdot 9B_2O_3$	8	0.72(8)	14.3(2)
β-$Cs_2O \cdot 5B_2O_3$	9	0.05(5)	22.1(2)
$Cs_2O \cdot 4B_2O_3$	8	0.19(5)	25.6(2)
$Cs_2O \cdot 3B_2O_3$	8	0.51(9)	30.6(3)
$Cs_2O \cdot 2B_2O_3$	7	5.1 (1)	35.0(4)
$Cs_2O \cdot B_2O_3$	7	17.7 (2)	40.3(4)

Differential microcalorimetry was used in order to determine the heats of formation $-\Delta H_f^\circ$ of potassium, rubidium, and caesium borate glasses containing $K_2O \leq 35.9\%$, $Rb_2O \leq 34.7\%$, and $Cs_2O \leq 42.5\%$, respectively, from the heat of dissolution in 2 N HNO_3. The exothermal effect of the B_2O_3–M_2O reaction (M = alkali metal) was due to the formation of additional M–O bonds in the alkali metal borate in comparison with pure M_2O [22].

The thermodynamic properties of $(B_2O_3)_{1-x}(Cs_2O)_x$ glasses and crystals with $0 \leq x \leq 0.4$ were studied by differential scanning calorimetry (DSC) between 200 and 1100 K. The coincidence of the heat capacity of stable and metastable melts of the same composition was observed. The variation of the heat capacity per mol with the composition is related to the nonlinear changes of the glass and crystal structure with increasing x. The increase of x is followed by the appearance of nonbridging oxygen atoms and destruction of the framework [23, 24]. The thermodynamic properties and chemical structure of alkali metal borates in crystals and glass-like states are also discussed [25].

The heats of formation of glassy and crystalline rubidium and caesium borates from oxides and the heat of crystallization were reported. Heats of solution in aqueous HNO_3 at 298 K were also determined. The systems have considerable negative deviations from ideal behavior due to the strong acid-base interactions of the components [26].

Solution calorimetry was used to determine the enthalpies of solution for $Rb_2O \cdot nB_2O_3$ and $Cs_2O \cdot nB_2O_3$ crystals (cf. Table 3/10) in 2 N HNO_3 at 298 K; the minimum is at n = 5, the maximum at n = 1 with ΔH_{sol} = 16 kcal/mol or 18 kcal/mol, respectively. The heats of crystallization $-\Delta H_{cr}^\circ$ have been determined according to the equation of the reaction: xM_2O (cryst.) + yB_2O_3 (cryst.) → $xM_2O \cdot yB_2O_3$ (cryst.) with M = Rb or Cs and x + y = 1 at 298 K [27].

The stability relative to decomposition to oxides was established for alkali metal borates $M_2O \cdot nB_2O_3$ with n = 1 to 4 and M = Li, Na, K, Rb, and Cs. The correlation between the degrees of dissociation and the enthalpies of formation is discussed [28]. The enthalpies of solution in 2 N HNO_3 at 298 K were measured for alkali metal borate glasses and crystals [29].

References on pp. 46/9

Films of Li_2O–B_2O_3 were prepared in the amorphous state by conventional vacuum deposition methods. As long as the substrate temperature is low, the lithium content in films is very close to that of $Li_2O \cdot B_2O_3$ for evaporation sources in the range between 33 and 50 mol% of Li_2O [30].

Differential thermal analysis (DTA) and thermodynamical calculations were used to study the phase diagram B_2O_3–Na_2O–$2SiO_2$–TiO_2 [31, 32].

The crystallinity of gel powders and gel monoliths in the system B_2O_3–Na_2O–SiO_2 was investigated above the transition point [33].

The volatilization rate of B_2O_3 melts between 1000 and 1200°C was seen to be independent of time as was expected for congruent evaporation [34].

The vapor composition in B_2O_3-PbO melts between 1100 and 1482 K was established by mass spectrometry and component activities to be a mixture of Pb_2BO_3, Pb_3BO_4, Pb_4BO_5, Pb_5BO_6, PbB_2O_4, $Pb_2B_2O_4$, $Pb_3B_2O_6$, $Pb_4B_2O_7$, and $Pb_5B_2O_8$. Standard heats of formation were calculated for these species in the vapor phase over B_2O_3–PbO [35]. In studying the gas phase of the systems B_2O_3–ZnO and B_2O_3–Bi_2O_3, the enthalpy of formation of gaseous $[BO_2]^-$ was determined [36].

The reaction of B_2O_3 with CoO was studied by diffuse reflection electronic spectroscopy between 20 and 900°C. At 15 wt% CoO the system exhibits two phases, one of which is Co_2O_3 [37]. The reaction of B_2O_3 with NiO was studied using infrared and electronic spectra, differential thermal analysis, and thermogravimetry (TG). At low NiO concentrations, Ni^{2+} ions are very sensitive to the B_2O_3 structure and are localized at defects. At 10% NiO content, the system exhibits two phases, and the excess Ni is aggregated as nonstoichiometric NiO_{1+x} [38].

Immiscibility in the system B_2O_3–BaO–TeO_2 was studied. The network former TeO_2 acts as a network modifier oxide [39]. Anomalies in the surface tension curves for the B_2O_3–PbO–TeO_2 system between 1173 and 1273 K are explained in terms of changes in coordination numbers of B (3→4), Te (4→3), and Pb (3→4) [40].

The following compounds were identified by phase diagram studies:

$3Li_2O \cdot B_2O_3$ [41], $LiCaBO_3$ (congruent melting point 915±5°C) [42], $Li_2B_2O_4 \cdot 4BaB_2O_4$ [43], $LiBO_2 \cdot 3Li_2MoO_4$ (melts incongruently) [44], $3Li_2O \cdot 2ZnO \cdot 3B_2O_3$, $2Li_2O \cdot 3ZnO \cdot 2B_2O_3$, $Li_2O \cdot 2ZnO \cdot B_2O_3$, $Li_3CdB_3O_7$, α-$LiCdBO_3$, β-$LiCdBO_3$ [45], $Li_2Y(BO_2)_5$, $Li_3Y_2(BO_3)_3$, $Li_6Y(BO_3)_3$ [46], $LiGeBO_4$ [47];

$NaCa[B_5O_9]$, $Na_3Ca[B_5O_{10}]$, $Ca_2Na[BO_4]$ (all incongruently melting compounds) [48], $Na[(VO_3)_2(BO)]$, $Na_3[(VO_3)_2(B_3O_5)]$ [49];

$3K_2O \cdot 5BaB_2O_4$, $K_6Ba_4B_8O_{19}$ [50];

$Mg_2[B_2O_5]$, $MgFe[BO_4]$, and $Mg(Fe_{0.9}Ga_{0.1})[BO_4]$ [51], $Mg_xMn_y[BO_3]$ (x, y not defined) [52];

boron-containing apatite in the system CaO–P_2O_5–B_2O_3 [53];

$Sr_x \cdot 0.5La_{1-x}(MnO_3) \cdot 0.5B_2O_3$ ($0.2 \leq x \leq 1$) [54];

$BaO \cdot Al_2O_3 \cdot Nd_2O_3 \cdot 6B_2O_3$, possibly space group P6mm–C_{6v}^1 (No. 183) with a = 456.6(2) pm and c = 2493(1) pm; Z = 1; D_m = 3.77 g/cm³ and D_c = 3.72 g/cm³ [55], $Ba_3Ge_2(B_6O_{16})$ [56], $Ba_2Ti_2(B_2O_9)$ [57];

$PbO \cdot xP_2O_5 \cdot yB_2O_3$ [58], $8PbO \cdot 3Bi_2O_3 \cdot 7B_2O_3$ [59, 60], $4PbO \cdot B_2O_3$ [61];

$Co_{17}Fe_{37}B_{15}O_{32}$ (amorphous) [62];

$Cu_3B_4O_6$ [63], CuB_2O_4 [63, 64], $Cu_2[B_2O_5]$, $Cu_3[BO_3]_2$ [64];

$3ZnO \cdot B_2O_3$ (melting point 610°C), $ZnO \cdot B_2O_3$ (melting point 710°C) [65], α-$5ZnO \cdot B_2O_3$, β-$ZnO \cdot B_2O_3$, α-$ZnO \cdot B_2O_3$ [66], $4ZnO \cdot B_2O_3$, $3ZnO \cdot B_2O_3$, $2ZnO \cdot B_2O_3$, $ZnO \cdot B_2O_3$ [67], $ZnO \cdot 2Bi_2O_3 \cdot B_2O_3$ [68];

$9Al_2O_3 \cdot 2B_2O_3$ [69, 70], $Al_6[BO_3]_5F_3$, space group $P6_3/m-C^2_{6h}$ (No. 176) with a=855.2(4) pm and c=818.7(3) pm; D_c=3.28 g/cm³ and R=3.88% [71];

$3Bi_2O_3 \cdot B_2O_3$ (melting point 500°C), $2Bi_2O_3 \cdot B_2O_3$ (melting point 570°C), $Bi_2O_3 \cdot B_2O_3$ (melting point 600°C), $3Bi_2O_3 \cdot 5B_2O_3$ (melting point 640°C), $Bi_2O_3 \cdot 3B_2O_3$ (melting point 660°C) [65], $12Bi_2O_3 \cdot B_2O_3$, $3Bi_2O_3 \cdot B_2O_3$, $2Bi_2O_3 \cdot B_2O_3$, $Bi_2O_3 \cdot B_2O_3$, $3Bi_2O_3 \cdot 5B_2O_3$, $Bi_2O_3 \cdot 3B_2O_3$, and $Bi_2O_3 \cdot 4B_2O_3$ [72];

$Fe^{II}_2Fe^{III}[O_2|(BO_3)]$ [73], $Fe^{III}Tb(BO_2)_6$, monoclinic with a=785 pm, b=983 pm, and c=524 pm, β=114°; melting point 1083°C; D_c=3.23 g/cm³, and hardness 7 to 7.5 (on Mohs' scale) [74], $Fe^{III}Nd(BO_2)_6$ (melting point 1092°C) [75];

$NdBO_3$ (high-temperature form) [76]; $Ho_2O_3 \cdot GeO_2 \cdot 2B_2O_3$ (melting point 1124°C), $Ho_2O_3 \cdot GeO_2 \cdot 10B_2O_3$ (melting point 1184°C) [77], $NdGeBO_5$ (melting point 1170°C), $Nd_{14}Ge_2B_8O_{37}$ (incongruently melting at ca. 1320°C) [78]; $LnGeBO_5$ with Ln=La, Pr, and Nd (low-temperature form) in a hexagonal stillwellite-like phase, with Ln=Nd to Er and Y in a monoclinic phase, also with tetrahedral boron, and no boron-containing phase is obtained for Ln=Tm to Lu [79].

B_4C was prepared by reacting B_2O_3 with carbon black in a plasma furnace in an argon atmosphere. The content of B_4C in the reaction product increases to maximal 94.2% with the increase of the content of B_2O_3 (to 80%) in the reaction mixture. The structure of the product was characterized by X-ray diffraction analysis, revealing ca. 5% of graphite. The product can be used for the manufacture of high-temperature-resistant construction materials, and as an abrasive for the chemical, metallurgical, aeronautic, and nuclear industries [80]. The binary systems B_2O_3–MgO, B_2O_3–CaO, and B_2O_3–BaO and the formation of boron nitrides by treating the borates with a gaseous stream of NH_3/N_2 were studied. The formation of boron nitride decreases with a decrease in the proportion of B_2O_3 in the system with the exception of the CaO system [81].

By treating pyrocatechol (1,2-$(HO)_2C_6H_4$) with B_2O_3 and aminophenols, organyloxoborates and boroxins containing three-coordinate boron have been prepared [82]. Arylborates, prepared from B_2O_3, 1,2-$(HO)_2C_6H_4$, and aromatic hydroxy compounds, have been used as heat stabilizers in poly(ethylene terephthalate) fibers. Best results were attained with 4-hydroxyphenyl *o*-phenylene borate and 1-naphthyl *o*-phenyleneborate heat stabilizers [83]. Mono- and diesters of boric acid were prepared from B_2O_3 by esterification with 1,2-$(HO)_2C_6H_4$ and bisphenol A for stabilization of low- and high-density polyethylene. The structure of the esters was determined by infrared and mass spectrometry and chemical analysis. The esters are effective antioxidants; the monoborate has also a photostabilizing effect. The effectiveness of the monoborate as an antioxidant and light stabilizer was related to the internal synergism due to the presence of boron and C_6H_4OH fragments in its molecule [84]. The reaction of B_2O_3 and 1,2-$(HO)_2C_6H_4$ leads to *p*-[2-(4-hydroxyphenyl)prop-2-yl]phenyl-*o*-phenyleneborate; B_2O_3 and bisphenol A give (*p*-phenylene-2-propane)[bis(*o*-phenylene)borate]. The thermal stability and photostability of plasticized PVC increases in the presence of these borates, especially in the presence of the latter [85].

The effect of B_2O_3 on the thermal oxidative stabilization of carborane polymers was tested employing a resin which was prepared by condensation of B_2O_3, phenol, and paraformaldehyde [86].

B_2O_3 is an effective reagent for the preparation of esters of formic acid by the direct esterification of HCOOH with alcohols [87].

$Pd[P(C_6H_5)_3]_4$ catalyzes the allylic alkylation of carbon nucleophiles with B_2O_3 and allyl alcohols under neutral conditions in tetrahydrofuran at 65°C [88].

References on pp. 46/9

References for 3.2.5.3:

[1] Donohoe, L. M.; Kohli, J. T.; Shelby, J. E. (Adv. Fusion Glass Proc. 1st Int. Conf., Alfred, N. Y., 1988, pp. 37.1/37.6 from C.A. **111** [1989] No. 200206).

[2] Khodyakov, A. A.; Bogdanov, V. L.; Gromov, V. V.; Dubrovin, A. L.; Saunin, E. I.; Fedorushkova, E. B. (Zh. Fiz. Khim. **59** [1985] 2579/81; Russ. J. Phys. Chem. **59** [1985] 1536/8).

[3] Tuppen, C. G.; Prior, K. A.; Gibbings, C. J.; Houghton, D. C.; Jackman, T. E. (J. Appl. Phys. **64** [1988] 2751/4).

[4] Tarasevich, B. P.; Sirotkina, N. Z.; Il'in, A. S.; Kuznetsov, E. V. (Vysokomol. Soedin. A **27** [1985] 451/63).

[5] Beattie, I. R.; Bowsher, B. R.; Gilson, T. R.; Jones, P. J. (AERE-R-11974 [1986] 253/74 from C.A. **106** [1987] No. 109613).

[6] El-Hadi, Z. A.; Gammal, M.; Ezz-el-Din, F. M.; Moustaffa, F. A. (Cent. Glass Ceram. Res. Inst. Bull. **32** No. 1/2 [1985] 15/9 from C.A. **105** [1986] No. 10616).

[7] Miura, Y.; Akiyama, Y.; Takahashi, K. (Yogyo Kyokaishi **94** [1986] 425/31 from C.A. **104** [1986] No. 191422).

[8] Yamanaka, S.; Akagi, J.; Hattori, M. (J. Non-Cryst. Solids **70** [1985] 279/90).

[9] Roskova, G. P.; Antropova, T. V.; Tsekhomskaya, T. S.; Anfimova, I. N. (Fiz. Khim. Stekla **11** [1985] 578/86 from C.A. **104** [1987] No. 38603).

[10] Hlavac, J.; Plasil, I. (Silikaty [Prague] **31** [1987] 309/15 from C.A. **107** [1987] No. 241339).

[11] Bokov, N. A. (Fiz. Khim. Stekla **14** [1988] 135/7 from C.A. **108** [1988] No. 191372).

[12] Kubicek, P.; Lesko, J.; Wozniakova, B. (Chem. Listy **79** [1985] 534/8 from C.A. **103** [1985] No. 134015).

[13] Julsrud, S.; Kleppa, O. J. (Z. Naturforsch. **42a** [1987] 463/70).

[14] Julsrud, S.; Kleppa, O. J. (Geochim. Cosmochim. Acta **50** [1986] 1201/4).

[15] Richards, E. A.; Bergeron, C. G. (Phys. Chem. Glasses **26** [1985] 177/81 from C.A. **103** [1985] No. 187066).

[16] Shul'ts, M. M.; Vedishcheva, N. M.; Shakhmatkin, B. A. (Thermochim. Acta **110** [1987] 443/7).

[17] Shul'ts, M. M.; Ushakov, V. M.; Borisova, N. V.; Starodubtsev, A. M. (Fiz. Khim. Stekla **12** [1986] 504/7 from C.A. **105** [1986] No. 233412).

[18] Shul'ts, M. M.; Borisova, N. V.; Ushakov, V. M. (Fiz. Khim. Stekla **12** [1986] 713/5 from C.A. **106** [1987] No. 91199).

[19] Shul'ts, M. M.; Vedishcheva, N. M.; Shakhmatkin, B. A.; Polyakova, I. G.; Fokin, V. M. (Fiz. Khim. Stekla **12** [1986] 651/9 from C.A. **106** [1987] No. 108862).

[20] Shul'ts, M. M.; Vedishcheva, N. M.; Shakhmatkin, B. A. (Dokl. Akad. Nauk SSSR **291** [1986] 166/8; Dokl. Phys. Chem. Proc. Acad. Sci. USSR **289/291** [1986] 999/1001).

[21] Shul'ts, M. M.; Vedishcheva, N. M.; Shakhmatkin, B. A. (Fiz. Khim. Stekla **12** [1986] 536/43 from C.A. **106** [1987] No. 24036).

[22] Shul'ts, M. M.; Vedishcheva, N. M.; Shakhmatkin, B. A.; Starodubtsev, A. M. (Fiz. Khim. Stekla **11** [1985] 472/9 from C.A. **103** [1985] No. 199892).

[23] Ushakov, V. M.; Borisova, N. V.; Shul'ts, M. M. (Thermochim. Acta **93** [1985] 235/7).

[24] Ushakov, V. M.; Borisova, N. V.; Shul'ts, M. M. (Therm. Anal. Proc. 8th Int. Conf., Bratislava 1985, Vol. 2, pp. 235/7 from C.A. **107** [1987] No. 44862).

[25] Shul'ts, M. M.; Vedishcheva, N. M.; Shakhmatkin, B. A. (Fiz. Khim. Silikatov L. **1987** 5/28 from C.A. **109** [1988] No. 80829 or No. 157561).

[26] Shul'ts, M. M.; Vedishcheva, N. M.; Shakhmatkin, B. A. (Therm. Anal. Proc. 8th Int. Conf., Bratislava 1985, Vol. 1, pp. 345/8 from C.A. **107** [1987] No. 103866).

[27] Shul'ts, M. M.; Vedishcheva, N. M.; Shakhmatkin, B. A. (Thermochim. Acta **92** [1985] 345/8).
[28] Vedishcheva, N. M.; Shakhmatkin, B. A. (J. Therm. Anal. **32** [1987] 1693/6).
[29] Vedishcheva, N. M.; Shakhmatkin, B. A.; Shul'ts, M. M. (J. Therm. Anal. **33** [1987/88] 923/7).
[30] Zhang, L. W.; Kobayashi, M.; Goto, K. S. (Solid State Ionics **18/19** [1986] 741/6).

[31] Zargarova, M. I.; Shalumov, B. Z.; Mamedov, A. N.; Ganf, K. L.; Dzhakhandarov, Sh. D.; D'yakonov, S. S. (Zh. Neorg. Khim. **30** [1985] 2985/7; Russ. J. Inorg. Chem. **30** [1985/86] 1701/2).
[32] Salmanova, Sh. Kh.; Mamedov, A. N. (Izv. Akad. Nauk SSSR Neorg. Mater. **23** [1987] 699/700; Inorg. Mater. [USSR] **23** [1987] 628/9 from C.A. **106** [1987] No. 221083).
[33] Mukherjee, S. P. (Mater. Sci. Res. **17** [1984] 95/109 from C.A. **102** [1984] No. 189542).
[34] Cable, M.; Fernandes, M. H. V. (Glass Technol. **28** No. 3 [1987] 135/40 from C.A. **107** [1987] No. 82452).
[35] Seminikhin, V. I.; Sorokin, I. D.; Yurkov, L. F.; Sidorov, L. N. (Fiz. Khim. Stekla **13** [1987] 542/7 from C.A. **107** [1987] No. 244022).
[36] Seminikhin, V. I.; Minaeva, I. I.; Sorokin, I. D.; Nikitin, M. I.; Rudnyi, E. B.; Sidorov, L. N. (Teplofiz. Vys. Temp. **25** [1987] 660/70 from C.A. **107** [1987] No. 208107).
[37] Nivarov, V. A.; Agzamkhodzhaeva, D. R.; Vorob'ev, V. N.; Abidova, M. F.; Razikov, K. Kh. (Zh. Neorg. Khim. **30** [1985] 2483/6; Russ. J. Inorg. Chem. **30** [1985/86] 1414/6).
[38] Vorob'ev, V. N.; Agzamkhodzhaeva, D. R.; Razikov, K. Kh.; Abidova, M. F. (Zh. Obshch. Khim. **55** [1985] 85/91; J. Gen. Chem. [USSR] **55** [1985] 73/7).
[39] Reiss, H.; Nass, H.; Vogel, W. (Silikattechnik **39** [1988] 141/2).
[40] Buler, P. I.; Spiridonova, E. V.; Frolov, A. A. (Rasplavy **1989** No. 1, pp. 98/100 from C.A. **111** [1989] No. 64518).

[41] Barsoum, M. W.; Tuller, H. L. (Solid State Ionics **18/19** [1986] 388/92).
[42] Yi, Xiande; Fan, Xinyong; Huang, Qingzhen (Wuji Huaxue **4** [1988] 119/23 from C.A. **110** [1989] No. 32851).
[43] Huang, Qingzhen; Liang, J.-K.; Gon, W. (Kexue Tongbao [Foreign Lang. Ed.] **31** [1986] 610/4 from C.A. **105** [1986] No. 67451).
[44] Khakulov, Z. L.; Shurdumov, G. K.; Mokhosoev, M. V. (Zh. Neorg. Khim. **31** [1986] 1265/70; Russ. J. Inorg. Chem. **31** [1986] 719/22).
[45] Buludov, N. T.; Karaev, Z. Sh.; Abdullaev, G. K. (Zh. Neorg. Khim. **30** [1985] 1523/6; Russ. J. Inorg. Chem. **30** [1985] 868/70 and Zh. Neorg. Khim. **31** [1986] 2108/11; Russ. J. Inorg. Chem. **31** [1986/87] 1215/7).
[46] Zargarova, M. I.; Kuli-Zade, E. S.; Shuster, N. S. (Izv. Akad. Nauk SSSR Neorg. Mater. **24** [1988] 427/9; Inorg. Mater. [USSR] **24** [1988] 346/8 from C.A. **109** [1988] No. 46835).
[47] Rulmont, A.; Tarte, P.; Winand, J. M. (J. Mater. Sci. Lett. **6** [1987] 659/62 from C.A. **107** [1987] No. 105248).
[48] Lawson, R. P.; Glasser, F. P. (Sci. Ceram. **14** [1988] 531/7 from C.A. **110** [1989] No. 28085).
[49] Buludov, N. T.; Mamedaliev, F. D.; Karaev, Z. Sh.; Abdullaev, G. K. (Zh. Neorg. Khim. **30** [1985] 1816/8; Russ. J. Inorg. Chem. **30** [1985/86] 1032/4).
[50] Wang, G.-F.; Huang, Qingzhen (Wuli Xuebao **34** [1985] 562/6 from C.A. **103** [1985] No. 28068).

[51] Wanklyn, B. M.; Watts, B. E.; Wondre, F. R.; Davison, W. (J. Mater. Sci. Lett. **5** [1986] 499/502).
[52] Cooper, J. J.; Tilley, R. J. D. (J. Solid State Chem. **63** [1986] 129/38).

[53] Ito, A.; Aoki, H.; Akao, M.; Miura, N.; Otsuka, R.; Tsutsumi, S. (Nippon Seramikkusu Kyokai Gakujutsu Ronbunshi **96** [1988] 707/9 from C.A. **109** [1988] No. 97678).
[54] Inomata, K.; Hashimoto, S.; Nakamura, Sh. (Jpn. J. Appl. Phys. **27** Pt. 2 [1988] L883/L885 from C.A. **109** [1988] No. 162015).
[55] Kong, Huashuang; He, Chongfan; Wang, Peiling; Li, Deyu (Yingyong Kexue Xuebao **4** [1986] 79/83).
[56] Liang, Jingkui; Chai, Zhang; Zhang, Jinping (Wuli Huaxue Xuebao **3** [1987] 508/13 from C.A. **108** [1988] No. 101823).
[57] Millet, J. M.; Roth, R. S.; Parker, H. S. (J. Am. Ceram. Soc. **69** [1986] 811/4).
[58] Mal'tsev, V. T.; Bondarev, Yu. M.; Akhtyrskii, V. G.; Andrushchenko, E. S. (Zh. Neorg. Khim. **31** [1986] 2112/5; Russ. J. Inorg. Chem. **31** [1986/87] 1217/9).
[59] Zargarova, M. I.; Shuster, N. S.; Afuzova, A. O. (Zh. Neorg. Khim. **32** [1987] 2014/7; Russ. J. Inorg. Chem. **32** [1987/88] 1185/8).
[60] Zargarova, M. I.; Shuster, N. S. (Izv. Akad. Nauk SSSR Neorg. Mater. **21** [1985] 273/7; Inorg. Mater. [USSR] **21** [1985] 223/7 from C.A. **102** [1985] No. 171250).

[61] Gelyasin, A. E.; Lutsko, L. A.; Mikhnevich, V. V. (Vestsi Akad. Navuk BSSR Ser. Khim. Navuk **1987** No. 5, pp. 104/7 from C.A. **108** [1988] No. 80449).
[62] Corrias, A.; Ennas, G.; Licheri, G.; Marongiu, G.; Musinu, A.; Paschina, G.; Piccaluga, G.; Pinna, G.; Magini, M. (J. Mater. Sci. Lett. **7** [1988] 407/9 from C.A. **109** [1988] No. 77356).
[63] Prasser, C.; Müller, F. (High-Temp. High Pressures **17** [1985] 551/7 from C.A. **104** [1986] No. 156844).
[64] Fomina, A. E.; Sedmale, G. P.; Sedmalis, Yu. Ya. (Neorg. Stekla Pokrytiya Mater. No. 8 [1987] 67/74; Latv. PSR Zinat. Akad. Vestis Khim. Ser. **1987** No. 1, pp. 117/8 from C.A. **106** [1987] No. 223279).
[65] Zargarova, M. I.; Korbanov, T. Ch.; Kasumova, M. F. (Thermochim. Acta **93** [1985] 449/52).
[66] Smolyar, A. S.; Kryuchkova, A. R. (Adgez. Rasplavov Paika Mater. No. 17 [1986] 76/9 from C.A. **107** [1987] No. 44846).
[67] Dafinova, R. (J. Mater. Sci. Lett. **7** [1988] 69/70 from C.A. **109** [1988] No. 82487; Bulg. J. Phys. **15** [1988] from C.A. **109** [1988] No. 118673).
[68] Zargarova, M. I.; Kasumova, M. F.; Abdullaev, G. K. (Zh. Neorg. Khim. **32** [1987] 1211/4; Russ. J. Inorg. Chem. **32** [1987] 737/9).
[69] Potgieter, E.; Res, M. A.; Sigalas, I.; Schonberger, H. (S. Afr. J. Phys. **9** [1986] 142/6 from C.A. **106** [1987] No. 71743).
[70] Potgieter, E.; Res, M. A.; Sigalas, I.; Schonberger, H.; Bednarik, J. (J. Mater. Sci. **22** [1987] 752/6 from C.A. **106** [1987] No. 142594).

[71] Sokolova, E. V.; Egorov-Tismenko, Yu. K.; Kargal'tsev, S. V.; Klyakhin, V. A.; Urnsov, V. S. (Vestn. Mosk. Univ. Geol. **1987** No. 3, pp. 82/4 from C.A. **108** [1988] No. 67644).
[72] Zargarova, M. I.; Kasumova, M. F. (Azerb. Khim. Zh. **1984** No. 6, pp. 98/101 from C.A. **103** [1985] No. 114902).
[73] Barrese, E.; Burragato, F.; Flamini, A. (Neues Jahrb. Mineral. Monatsh. **1984** 483/9).
[74] Aliev, O. A.; Rustamov, P. G.; Allakhverdiev, Kh. M. (Zh. Neorg. Khim. **32** [1987] 813/5; Russ. J. Inorg. Chem. **32** [1987] 455/6).
[75] Aliev, O. A.; Rustamov, P. G.; Allakhverdiev, Kh. M. (Dokl. Akad. Nauk Az. SSR **43** No. 8 [1987] 61/2 from C.A. **109** [1988] No. 12743).
[76] De Villiers, D. R.; Res, M. A.; Richter, P. W. (S. Afr. J. Phys. **9** [1986] 139/41 from C.A. **106** [1987] No. 71742).

[77] Aliev, O. A.; Rustamov, P. G.; Allakhverdiev, Kh. M. (Dokl. Akad. Nauk Az. SSR **44** No. 5 [1988] 64/6 from C.A. **110** [1989] No. 102647).
[78] Lysanova, G. V.; Dzhurinskii, B. F.; Komova, M. G.; Sorokina, O. V.; Chistova, V. I. (Izv. Akad. Nauk SSSR Neorg. Mater. **24** [1988] 255/8; Inorg. Mater. [USSR] **24** [1988] 190/3 from C.A. **108** [1988] No. 138678).
[79] Rulmont, A.; Tarte, P. (J. Solid State Chem. **75** [1988] 244/50).
[80] Becherescu, D.; Marx, F.; Lazau, I. (Mater. Constr. [Bucharest] **15** [1985] 92/4 from C.A. **103** [1985] No. 180302).

[81] Hervas Moreno, C. (Quim. Ind. [Madrid] **32** [1986] 555/62 from C.A. **107** [1987] No. 50698).
[82] Grachek, V. I.; Motol'ko, G. R.; Naumova, S. F.; Kozlov, N. S. (Dokl. Akad. Nauk BSSR **32** [1988] 235/8 from C.A. **109** [1988] No. 231110).
[83] Bondareva, O. M.; Grachek, V. I.; Lopatik, D. V.; Motol'ko, G. R.; Naumova, S. F.; Osipenko, I. F.; Prokopchuk, N. R. (Khim. Volokna **1986** No. 6, pp. 24/6 from C.A. **106** [1987] No. 103642).
[84] Naumova, S. F.; Motol'ko, G. R.; Isakovich, V. N.; Shingel, I. A. (Dokl. Akad. Nauk BSSR **30** [1986] 742/4 from C.A. **105** [1986] No. 227836).
[85] Naumova, S. F.; Motol'ko, G. R.; Isakovich, V. N.; Strelkova, L. D. (Plast. Massy **1987** No. 1, pp. 52/3 from C.A. **106** [1987] No. 139203).
[86] Rodionov, Yu. M.; Danilov, S. I. (Dokl. Akad. Nauk SSSR **288** [1986] 143/6; Dokl. Chem. Proc. Acad. Sci. USSR **286/291** [1986] 132/5).
[87] Carlson, C. G.; Hall, J. E.; Huang, Y. Y.; Kotila, S.; Rauk, A.; Tavares, D. F. (Can. J. Chem. **65** [1987] 2461/3).
[88] Lu, Xiyan; Jiang, Xiaohui; Tao, Xiaochan (J. Organomet. Chem. **344** [1988] 109/18).
[89] Lux, H. (Z. Elektrochem. **45** [1939] 303/9).

3.2.5.4 Catalysis and Other Applications

The oxidation of *n*-hexadecane with oxygen in the presence of B_2O_3 shows that ionol (2,6-bis(1,1-dimethylethyl)-4-methyl-phenol) has the strongest inhibitor ability among a number of benzene derivatives [1]. Air oxidation of paraffins in $(CH_3CO)_2O$ in the presence of B_2O_3 gave 75 to 80% higher molecular aliphatic alcohols (40% hydrocarbon conversion) [2]. The catalytic efficiency of supported B_2O_3 as a solid acid is enhanced by a new method of chemical vapor deposition (CVD) [3]. Cyclohexanone oxime has been converted to ε-caprolactam in high yield (93%) at 250°C over an SiO_2-supported B_2O_3 catalyst which was prepared from contacting $B(OC_2H_5)_3$ vapor with silica gel at 350°C in the presence of air. This catalyst is much more efficient than SiO_2–B_2O_3 or Al_2O_3–B_2O_3, which are obtained by the usual impregnation method [4]. The catalytic behavior of B_2O_3–Al_2O_3 prepared from $B(OC_2H_5)_3$ vapor and γ-Al_2O_3 greatly changes with the B_2O_3 content. At higher B_2O_3 contents (ca. 15 wt%), the samples exhibit higher catalytic efficiencies, e.g., for the vapor-phase Beckmann rearrangement of cyclohexanone oxime or for the *m*-xylene isomerization [5].

High conversions and selectivities in CH_3OH hydrogenation were obtained with catalysts containing B_2O_3 on 13X zeolite and also B_2O_3 on γ-Al_2O_3. The highest selectivity to form C_2 to C_4 hydrocarbons (85% at 41.5% CH_3OH conversion) was obtained with a B_2O_3-13X zeolite catalyst at 300°C and 1.2 h^{-1} space velocity. The catalysts were fully characterized by chemical analysis, X-ray diffractograms, differential thermal analysis (DTA), IR spectroscopy, and porosimetry [6].

References on pp. 61/4

The relationship between structure and acidity in nonstoichiometric B_2O_3–Al_2O_3 catalysts has been studied. These oxides are amorphous, refractory materials and exhibit properties which are dependent on the stoichiometry of the catalyst. Distinct crystalline phases were detected after calcining at 1000 °C. Both ^{11}B and ^{27}Al NMR spectroscopy were used to determine the local structure. The surface acidity was measured by 2-propanol dehydration and was shown to be closely related to the relative concentration of tetrahedral BO_4 groups [7].

The method of molding a B_2O_3–Al_2O_3–Cr_2O_3 support for a low-temperature catalyst for conversion of CO with H_2O vapor has been studied [8]. An activity study of B- and Al-containing catalysts confirm the proposed mechanism of the ultrasound homogenization of the catalyst system [11]. A B_2O_3–Cr_2O_3–V_2O_5 catalyst is used for the synthesis of aromatic nitriles [9]. B_2O_3–Al_2O_3–NiO (with and without Cr_2O_3) is a mechanically stable catalyst for the conversion of natural gas with water vapor [10].

Melts of B_2O_3 and metal oxides have been used for the growth of single crystals of GaAs [12, 13], InP [14, 15], CdTe [16], YIG microcrystals and a $BaFe_{12}O_{19}$ ferrite phase [17], $Gd_3Fe_5O_{12}$ garnet crystals [18], $Na[NbO_3]$ and $Na[Nb_3O_8]$ [19], orthorhombic $Na[TaO_3]$ and orthorhombic $Na_2O \cdot 4Ta_2O_5$ [20], $PbTiO_3$ crystals [21], $DyFeO_3$ crystals [22], $CoFe_2O_4$ microcrystals [23], and superconducting $YBa_2Cu_3O_{7-x}$ in a borate glass-ceramic matrix [24].

Superionic conductivity has been found and applied in lithium borate thin-film solid state batteries [25, 26], B_2O_3–Li_2O–$Li_2[SO_4]$ glasses [27], and B_2O_3–Ag_2O–AgX glasses; see Section 3.2.5.2 (pp. 28 and 36).

If boron is included in a chemical equilibrium estimate of aerosols that would be produced in an overheated light-water reactor core, it will significantly alter or dominate the components of the aerosol in the form of B_2O_3 [28].

The use of $B_2O_3 \cdot ZnO$ as flame retardant, smoke- and afterglow-suppressant additive in PVC, epoxy resins, polyolefins, polyesters, polyurethans, and other polymers was reviewed [29]. The fire endurance of a borate polyethylene shielding material was investigated [30].

Barium metaborate pigments were incorporated into anticorrosive primers free of lead and chromate [31].

$2B_2O_3 \cdot 9Al_2O_3$ can be used as a refractory material in thick masonry subjected to high pressure, e.g., in coal gasification reactors [32].

A survey of luminescent phosphors prepared from B_2O_3 is given in Table 3/11 and Table 3/12, p. 53. For practical reasons, all borate units are written in parentheses as (B_xO_y) without regard to the ionic or homopolar character of the structure. The first column lists the contained elements in alphabetical order except for boron, oxygen, and halogens. In regard to the often arbitrary decisions between solid solutions and doped compounds, doped compounds are listed twice: in Table 3/11 without and in Table 3/12 with all dopants. Double doping of compounds takes place, if after the colon two ions are separated with a comma.

Table 3/11
Undoped Luminescent Phosphors Prepared from B_2O_3.

element	formula	Ref.
BaCeTb	$BaCe_{1-x}Tb_x(B_9O_{16})$	[35]
BaGd	$BaGd(B_9O_{16})$	[35]
BaLa	$BaLa(B_9O_{16})$	[33, 35]

Table 3/11 (continued)

element	formula	Ref.
BaY	$BaY(B_9O_{16})$	[35]
Bi	$Bi(BO_3)$	[65]
Bi	$Bi_3(B_5O_{12})$	[43]
CeGd	$Gd_{1-x}Ce_x(B_3O_6)$	[64]
CeGdMgTb	$Gd_{1-x-y}Ce_xTb_yMg(B_5O_{10})$	[54]
CeGdSm	$Gd_{1-x-y}Ce_xSm_y(B_3O_6)$	[63]
CeGdTb	$Gd_{1-x-y}Ce_xTb_y(B_3O_6)$	[62]
CeLaMgTb	$La_{1-x-y}Ce_xTb_yMg(B_5O_{10})$	[54]
CeMg	$CeMg(B_5O_{10})$	[47]
CeMgTbY	$Y_{1-x-y}Ce_xTb_yMg(B_5O_{10})$	[54]
DyGd	$Gd_{1-x}Dy_x(B_3O_6)$	[64]
DyMg	$DyMg(B_5O_{10})$	[47]
DyY	$Y_{1-x}Dy_x(BO_3)$	[71]
ErMg	$ErMg(B_5O_{10})$	[47]
ErY	$Y_{1-x}Er_x(BO_3)$	[71]
EuGd	$Gd_{0.9}Eu_{0.1}(BO_3)$	[73]
EuGdLi	$Li_6Gd_{1-x}Eu_x(BO_3)_3$	[59, 86, 87]
EuGdMg	$Gd_{1-x}Eu_xMg(B_5O_{10})$	[48, 49]
EuGdNa	$Na_3Gd_{1-x}Eu_x(BO_3)_2$	[59]
EuGdW	$Gd_{3-x}Eu_x(BWO_9)$	[88]
EuLa	$La_{0.8}Eu_{0.2}(BO_3)$	[73]
EuLaMo	$La_{1-x}Eu_x(BMoO_6)$	[88]
EuLaW	$La_{1-x}Eu_x(BWO_6)$	[88]
EuLaW	$La_{3-x}Eu_x(BWO_9)$	[88]
EuLi	$Li_6Eu(BO_3)_3$	[86, 87]
EuMg	$EuMg(B_5O_{10})$	[47, 49]
EuW	$Eu_3(BWO_9)$	[88]
EuWY	$Y_{3-x}Eu_x(BWO_9)$	[88]

References on pp. 61/4

Table 3/11 (continued)

element	formula	Ref.
EuY	$Y_{0.9}Eu_{0.1}(BO_3)$	[73]
Gd	$Gd(B_3O_6)$	[44]
GdLi	$Li_6Gd(BO_3)_3$	[84, 85]
GdMg	$GdMg(B_5O_{10})$	[44, 45, 47]
GdNaTb	$Na_3Gd_{1-x}Tb_x(BO_3)_2$	[59]
GdSr	$SrGd(B_9O_{16})$	[35]
GdTb	$Gd_{0.9}Tb_{0.1}(BO_3)$	[73]
HoMg	$HoMg(B_5O_{10})$	[47]
HoY	$Y_{1-x}Ho_x(BO_3)$	[71]
LaSr	$SrLa(B_9O_{16})$	[35]
LaTb	$La_{0.9}Tb_{0.1}(BO_3)$	[73]
MgLa	$LaMg(B_5O_{10})$	[47]
MgNd	$NdMg(B_5O_{10})$	[47]
MgPr	$PrMg(B_5O_{10})$	[47]
MgSm	$SmMg(B_5O_{10})$	[47]
MgTb	$TbMg(B_5O_{10})$	[47, 51]
MnPbSr	$Sr_{1-x-y}Mn_xPb_y(B_6O_{10})$	[41]
MnSnSr	$Sr_{1-x-y}Mn_xSn_y(B_6O_{10})$	[41]
PbSr	$Sr_{2-x}Pb_x(B_5O_9Br)$	[55]
PbSr	$Sr_{2-x}Pb_x(B_5O_9Cl)$	[55]
PrY	$Y_{1-x}Pr_x(BO_3)$	[71]
SmY	$Y_{1-x}Sm_x(BO_3)$	[71]
SrY	$SrY(B_9O_{16})$	[35]
TbY	$Y_{0.9}Tb_{0.1}(BO_3)$	[73]
TmY	$Y_{1-x}Tm_x(BO_3)$	[71]
Zn	$Zn_4O(B_6O_{12})$	[38, 39]

Table 3/12

Doped Luminescent Phosphors Prepared from B_2O_3.
"Ln" means not specified rare earth metals, "M" stands for metals in general.

element	formula	Ref.
Ba	$Ba_2(B_5O_9Br):Eu^{2+}$	[56]
Ba	$Ba(B_8O_{13}):Ce^{3+}$	[37]
Ba	$Ba(B_8O_{13}):Ce^{3+},Tb^{3+}$	[37]
Ba	$Ba(B_8O_{13}):Tb^{3+}$	[37]
BaBiGd	$BaGd(B_9O_{16}):Bi^{3+}$	[34, 35]
BaBiLa	$BaLa(B_9O_{16}):Bi^{3+}$	[35]
BaBiY	$BaY(B_9O_{16}):Bi^{3+}$	[35]
BaCe	$Ba(B_8O_{13}):Ce^{3+}$	[37]
BaCeGd	$BaGd(B_9O_{16}):Ce^{3+}$	[35]
BaCeLa	$BaLa(B_9O_{16}):Ce^{3+}$	[35]
BaCeLn	$BaLn(B_9O_{16}):Ce^{3+}$	[36]
BaCeTb	$Ba(B_8O_{13}):Ce^{3+},Tb^{3+}$	[37]
BaCeY	$BaY(B_9O_{16}):Ce^{3+}$	[35]
BaDyGd	$BaGd(B_9O_{16}):Dy^{3+}$	[34]
BaEu	$Ba_2(B_5O_9Br):Eu^{2+}$	[56]
BaEuGd	$BaGd(B_9O_{16}):Eu^{3+}$	[34]
BaGd	$BaGd(B_9O_{16}):Bi^{3+}$	[34, 35]
BaGd	$BaGd(B_9O_{16}):Ce^{3+}$	[35]
BaGd	$BaGd(B_9O_{16}):Dy^{3+}$	[34]
BaGd	$BaGd(B_9O_{16}):Eu^{3+}$	[34]
BaGd	$BaGd(B_9O_{16}):Sm^{3+}$	[34]
BaGd	$BaGd(B_9O_{16}):Tb^{3+}$	[34]
BaGdSm	$BaGd(B_9O_{16}):Sm^{3+}$	[34]
BaGdTb	$BaGd(B_9O_{16}):Tb^{3+}$	[34]
BaLa	$BaLa(B_9O_{16}):Bi^{3+}$	[35]
BaLa	$BaLa(B_9O_{16}):Ce^{3+}$	[35]
BaLn	$BaLn(B_9O_{16}):Ce^{3+}$	[36]

References on pp. 61/4

Table 3/12 (continued)

element	formula	Ref.
BaLn	$BaLn(B_9O_{16}):Tb^{3+}$	[36]
BaLnTb	$BaLn(B_9O_{16}):Tb^{3+}$	[36]
BaMnPbSr	$Sr_{1-x-y}Mn_xPb_y(B_6O_{10}):Ba^{2+}$	[41]
BaPbZr	$BaZr(BO_3)_2:Pb^{2+}$	[82]
BaTb	$Ba(B_8O_{13}):Tb^{3+}$	[37]
BaTbZr	$BaZr(BO_3)_2:Tb^{3+}$	[82]
BaTiZr	$BaZr(BO_3)_2:Ti^{4+}$	[82]
BaY	$BaY(B_9O_{16}):Bi^{3+}$	[35]
BaY	$BaY(B_9O_{16}):Ce^{3+}$	[35]
BaZr	$BaZr(BO_3)_2:Pb^{2+}$	[82]
BaZr	$BaZr(BO_3)_2:Tb^{3+}$	[82]
BaZr	$BaZr(BO_3)_2:Ti^{4+}$	[82]
BiGd	$Gd(BO_3):Bi^{3+}$	[76]
BiGd	$Gd(B_3O_6):Bi^{3+}$	[61]
BiGdLa	$La_{0.8}Gd_{0.2}(B_3O_6):Bi^{3+}$	[60]
BiGdSm	$Gd_{1-x}Sm_x(BO_3):Bi^{3+}$	[79]
BiGdTb	$Gd(B_3O_6):Bi^{3+},Tb^{3+}$	[60]
BiLa	$La(BO_3):Bi^{3+}$	[76]
BiLu	$Lu(BO_3):Bi^{3+}$	[76]
BiSc	$Sc(BO_3):Bi^{3+}$	[76]
BiY	$Y(BO_3):Bi^{3+}$	[76]
CaGdSr	$Sr_{1-x}Ca_x(B_6O_{10}):Gd^{3+}$	[40]
CaMnPbSr	$Sr_{1-x-y}Mn_xPb_y(B_6O_{10}):Ca^{2+}$	[41]
CaPbSr	$Sr_{1-x}Ca_x(B_6O_{10}):Pb^{2+}$	[40]
CaPbZr	$CaZr(BO_3)_2:Pb^{2+}$	[82]
CaSr	$Sr_{1-x}Ca_x(B_6O_{10}):Gd^{3+}$	[40]
CaSr	$Sr_{1-x}Ca_x(B_6O_{10}):Pb^{2+}$	[40]
CaTbZr	$CaZr(BO_3)_2:Tb^{3+}$	[82]
CaZr	$CaZr(BO_3)_2:Pb^{2+}$	[82]

Table 3/12 (continued)

element	formula	Ref.
CaZr	$CaZr(BO_3)_2:Tb^{3+}$	[82]
CaZr	$CaZr(BO_3)_2:Ti^{4+}$	[82]
CdMnPbSr	$Sr_{1-x-y}Mn_xPb_y(B_6O_{10}):Cd^{2+}$	[41]
CeDyGd	$Gd_{1-x}Ce_x(B_3O_6):Dy^{3+}$	[64]
CeGd	$Gd_{1-x}Ce_x(B_3O_6):Dy^{3+}$	[64]
CeGd	$Gd_{1-x}Ce_x(B_3O_6):Tb^{3+}$	[62]
CeGd	$Gd(B_3O_6):Ce^{3+}$	[61]
CeGdLaMg	$La_{1-x}Gd_xMg(B_5O_{10}):Ce^{3+}$	[53]
CeGdLaMgTb	$La_{0.75}Gd_{0.15}Mg(B_5O_{10}):Ce^{3+},Tb^{3+}$	[54]
CeGdMg	$GdMg(B_5O_{10}):Ce^{3+}$	[46, 54]
CeGdMgTb	$GdMg(B_5O_{10}):Ce^{3+},Tb^{3+}$	[54]
CeGdSm	$Gd_{1-x}Sm_x(BO_3):Ce^{3+}$	[79]
CeGdTb	$Gd_{1-x}Ce_x(B_3O_6):Tb^{3+}$	[62]
CeLaMgTb	$LaMg(B_5O_{10}):Ce^{3+},Tb^{3+}$	[54]
CeMgLa	$LaMg(B_5O_{10}):Ce^{3+}$	[54]
CeMgTbY	$YMg(B_5O_{10}):Ce^{3+},Tb^{3+}$	[54]
CeMgY	$YMg(B_5O_{10}):Ce^{3+}$	[54]
CeSc	$Sc_{1-x}Ce_x(BO_3):Tb^{3+}$	[68]
CeSc	$Sc(BO_3):Ce^{3+}$	[67]
CeScTb	$Sc_{1-x}Ce_x(BO_3):Tb^{3+}$	[68]
CeSr	$Sr(B_8O_{13}):Ce^{3+}$	[37]
CeSrTb	$Sr(B_8O_{13}):Ce^{3+},Tb^{3+}$	[37]
CrSc	$Sc(BO_3):Cr^{3+}$	[67]
DyGdPr	$Gd(BO_3):Pr^{3+},Dy^{3+}$	[77]
DyM	$M(BO_3):Dy^{3+}$	[74]
DyY	$Y(BO_3):Dy^{3+}$	[69]
ErIn	$In(BO_3):Er^{3+}$	[80]
ErLu	$Lu(BO_3):Er^{3+}$	[80]
ErM	$M(BO_3):Er^{3+}$	[74]

References on pp. 61/4

Table 3/12 (continued)

element	formula	Ref.
ErY	$Y(BO_3):Er^{3+}$	[70, 72, 80]
EuGd	$Gd(B_3O_6):Eu^{3+}$	[57 to 59, 61]
EuGdLi	$Li_6Gd_{1-x}Eu_x(BO_3)_3:Nd^{3+}$	[87]
EuGdLiNd	$Li_6Gd_{1-x}Eu_x(BO_3)_3:Nd^{3+}$	[87]
EuGdPr	$Gd(BO_3):Pr^{3+},Eu^{3+}$	[77]
EuGdPrY	$Y_{1-x}Gd_x(BO_3):Pr^{3+},Eu^{3+}$	[78]
EuLaMg	$LaMg(B_5O_{10}):Eu^{3+}$	[52]
EuLi	$Li_6Eu(BO_3)_3:Nd^{3+}$	[86, 87]
EuLu	$Lu(BO_3):Eu^{3+}$	[81]
EuM	$M(BO_3):Eu^{3+}$	[74]
EuMg	$EuMg(B_5O_{10}):Nd^{3+}$	[48 to 50]
EuMg	$EuMg(B_5O_{10}):Ni^{2+}$	[48, 50]
EuMgNd	$EuMg(B_5O_{10}):Nd^{3+}$	[48 to 50]
EuMgNi	$EuMg(B_5O_{10}):Ni^{2+}$	[48, 50]
EuMgTb	$TbMg(B_5O_{10}):Eu^{3+}$	[51]
EuMnSr	$Sr(B_6O_{10}):Mn^{2+},Eu^{2+}$	[42]
EuPrY	$Y(BO_3):Pr^{3+},Eu^{3+}$	[78]
EuSc	$Sc(BO_3):Eu^{3+}$	[67]
EuSr	$Sr(B_6O_{10}):Eu^{2+}$	[42]
EuY	$Y(BO_3):Eu^{3+}$	[78]
EuZn	$n ZnO \cdot B_2O_3:Eu^{3+}$ (n=1 to 4)	[90]
Gd	$Gd(BO_3):Bi^{3+}$	[76]
Gd	$Gd(BO_3):Pr^{3+}$	[78]
Gd	$Gd(BO_3):Pr^{3+},Dy^{3+}$	[77]
Gd	$Gd(BO_3):Pr^{3+},Eu^{3+}$	[77]
Gd	$Gd(BO_3):Pr^{3+},Sm^{3+}$	[77]
Gd	$Gd(BO_3):Pr^{3+},Tb^{3+}$	[77]
Gd	$Gd(BO_3):Sb^{3+}$	[75]
Gd	$Gd(B_3O_6):Bi^{3+}$	[61]

Table 3/12 (continued)

element	formula	Ref.
Gd	$Gd(B_3O_6):Bi^{3+},Tb^{3+}$	[60]
Gd	$Gd(B_3O_6):Ce^{3+}$	[61]
Gd	$Gd(B_3O_6):Eu^{3+}$	[57 to 59, 61]
Gd	$Gd(B_3O_6):Sb^{3+}$	[61]
Gd	$Gd(B_3O_6):Tb^{3+}$	[61]
GdLa	$La_{0.8}Gd_{0.2}(B_3O_6):Bi^{3+}$	[60]
GdLaMg	$La_{0.75}Gd_{0.15}Mg(B_5O_{10}):Ce^{3+},Tb^{3+}$	[54]
GdLaMg	$La_{1-x}Gd_xMg(B_5O_{10}):Ce^{3+}$	[53]
GdLaMg	$La_{1-x}Gd_xMg(B_5O_{10}):Tb^{3+}$	[53]
GdLaMgTb	$La_{1-x}Gd_xMg(B_5O_{10}):Tb^{3+}$	[53]
GdMg	$GdMg(B_5O_{10}):Ce^{3+}$	[46, 54]
GdMg	$GdMg(B_5O_{10}):Ce^{3+},Tb^{3+}$	[54]
GdMg	$GdMg(B_5O_{10}):Tb^{3+}$	[54]
GdMgTb	$GdMg(B_5O_{10}):Tb^{3+}$	[54]
GdPr	$Gd(BO_3):Pr^{3+}$	[78]
GdPrSm	$Gd_{1-x}Sm_x(BO_3):Pr^{3+}$	[79]
GdPrSm	$Gd(BO_3):Pr^{3+},Sm^{3+}$	[77]
GdPrTb	$Gd(BO_3):Pr^{3+},Tb^{3+}$	[77]
GdPrTbY	$Y_{1-x}Gd_x(BO_3):Pr^{3+},Tb^{3+}$	[78]
GdPrY	$Y(BO_3):Pr^{3+},Gd^{3+}$	[77]
GdSb	$Gd(B_3O_6):Sb^{3+}$	[61]
GdSb	$Gd(BO_3):Sb^{3+}$	[75]
GdSbSm	$Gd_{1-x}Sm_x(BO_3):Sb^{3+}$	[79]
GdSc	$Sc(BO_3):Gd^{3+}$	[67]
GdSm	$Gd_{1-x}Sm_x(BO_3):Bi^{3+}$	[79]
GdSm	$Gd_{1-x}Sm_x(BO_3):Ce^{3+}$	[79]
GdSm	$Gd_{1-x}Sm_x(BO_3):Pr^{3+}$	[79]
GdSm	$Gd_{1-x}Sm_x(BO_3):Sb^{3+}$	[79]
GdTb	$Gd(B_3O_6):Tb^{3+}$	[61]

References on pp. 61/4

Table 3/12 (continued)

element	formula	Ref.
GdY	$Y_{1-x}Gd_x(BO_3):Pr^{3+},Tb^{3+}$	[78]
HoM	$M(BO_3):Ho^{3+}$	[74]
HoY	$Y(BO_3):Ho^{3+}$	[70]
In	$In(BO_3):Er^{3+}$	[80]
La	$La(BO_3):Bi^{3+}$	[76]
La	$La(BO_3):Pr^{3+}$	[66]
La	$La(BO_3):Sb^{3+}$	[75]
LaMg	$LaMg(B_5O_{10}):Ce^{3+},Tb^{3+}$	[54]
LaMg	$LaMg(B_5O_{10}):Eu^{3+}$	[52]
LaPr	$La(BO_3):Pr^{3+}$	[66]
LaSb	$La(BO_3):Sb^{3+}$	[75]
Lu	$Lu(BO_3):Bi^{3+}$	[76]
Lu	$Lu(BO_3):Er^{3+}$	[80]
Lu	$Lu(BO_3):Eu^{3+}$	[81]
Lu	$Lu(BO_3):Sb^{3+}$	[75]
LuSb	$Lu(BO_3):Sb^{3+}$	[75]
M	$M(BO_3):Dy^{3+}$	[74]
M	$M(BO_3):Er^{3+}$	[74]
M	$M(BO_3):Eu^{3+}$	[74]
M	$M(BO_3):Ho^{3+}$	[74]
M	$M(BO_3):Sm^{3+}$	[74]
M	$M(BO_3):Tb^{3+}$	[74]
MSm	$M(BO_3):Sm^{3+}$	[74]
MTb	$M(BO_3):Tb^{3+}$	[74]
MgLa	$LaMg(B_5O_{10}):Ce^{3+}$	[54]
MgLa	$LaMg(B_5O_{10}):Tb^{3+}$	[54]
MgLaTb	$LaMg(B_5O_{10}):Tb^{3+}$	[54]
MgMnTb	$TbMg(B_5O_{10}):Mn^{2+}$	[51]
MgTb	$TbMg(B_5O_{10}):Eu^{3+}$	[51]

Table 3/12 (continued)

element	formula	Ref.
MgTb	$TbMg(B_5O_{10}):Mn^{2+}$	[51]
MgTbY	$YMg(B_5O_{10}):Tb^{3+}$	[54]
MgY	$YMg(B_5O_{10}):Ce^{3+}$	[54]
MgY	$YMg(B_5O_{10}):Ce^{3+},Tb^{3+}$	[54]
MgY	$YMg(B_5O_{10}):Tb^{3+}$	[54]
MnPbSr	$Sr_{1-x-y}Mn_xPb_y(B_6O_{10}):Ba^{2+}$	[41]
MnPbSr	$Sr_{1-x-y}Mn_xPb_y(B_6O_{10}):Ca^{2+}$	[41]
MnPbSr	$Sr_{1-x-y}Mn_xPb_y(B_6O_{10}):Cd^{2+}$	[41]
MnPbSr	$Sr(B_6O_{10}):Mn^{2+},Pb^{2+}$	[42]
MnSnSr	$Sr(B_6O_{10}):Mn^{2+},Sn^{2+}$	[42]
MnSr	$Sr(B_6O_{10}):Mn^{2+}$	[41]
MnZn	$nZnO \cdot B_2O_3:Mn^{2+}$ (n=1 to 4)	[89, 90]
MnZn	$Zn_4O(B_6O_{12}):Mn^{2+}$	[38]
PbSr	$Sr(B_6O_{10}):Pb^{2+}$	[42]
PbSrZr	$SrZr(BO_3)_2:Pb^{2+}$	[82]
PrSc	$Sc(BO_3):Pr^{3+}$	[66]
PrY	$Y(BO_3):Pr^{3+}$	[66, 77, 78]
SbY	$Y(BO_3):Sb^{3+}$	[75]
Sc	$Sc(BO_3):Bi^{3+}$	[76]
Sc	$Sc(BO_3):Ce^{3+}$	[67]
Sc	$Sc(BO_3):Cr^{3+}$	[67]
Sc	$Sc(BO_3):Eu^{3+}$	[67]
Sc	$Sc(BO_3):Gd^{3+}$	[67]
Sc	$Sc(BO_3):Pr^{3+}$	[66]
Sc	$Sc(BO_3):Sb^{3+}$	[75]
SnSr	$Sr(B_6O_{10}):Sn^{2+}$	[42]
Sr	$Sr(B_6O_{10}):Eu^{2+}$	[42]
Sr	$Sr(B_6O_{10}):Mn^{2+}$	[41]
Sr	$Sr(B_6O_{10}):Mn^{2+},Eu^{2+}$	[42]

References on pp. 61/4

Table 3/12 (continued)

element	formula	Ref.
Sr	$Sr(B_6O_{10})$: Mn^{2+},Pb^{2+}	[42]
Sr	$Sr(B_6O_{10})$: Mn^{2+},Sn^{2+}	[42]
Sr	$Sr(B_6O_{10})$: Pb^{2+}	[42]
Sr	$Sr(B_6O_{10})$: Sn^{2+}	[42]
Sr	$Sr(B_8O_{13})$: Ce^{3+}	[37]
Sr	$Sr(B_8O_{13})$: Ce^{3+},Tb^{3+}	[37]
Sr	$Sr(B_8O_{13})$: Tb^{3+}	[37]
SrTb	$Sr(B_8O_{13})$: Tb^{3+}	[37]
SrTbZr	$SrZr(BO_3)_2$: Tb^{3+}	[82]
SrTiZr	$SrZr(BO_3)_2$: Ti^{4+}	[82]
SrZr	$SrZr(BO_3)_2$: Pb^{2+}	[82]
SrZr	$SrZr(BO_3)_2$: Tb^{3+}	[82]
SrZr	$SrZr(BO_3)_2$: Ti^{4+}	[82]
Y	$Y(BO_3)$: Bi^{3+}	[76]
Y	$Y(BO_3)$: Dy^{3+}	[69]
Y	$Y(BO_3)$: Er^{3+}	[70, 72, 80]
Y	$Y(BO_3)$: Eu^{3+}	[78]
Y	$Y(BO_3)$: Ho^{3+}	[70]
Y	$Y(BO_3)$: Pr^{3+}	[66, 77, 78]
Y	$Y(BO_3)$: Pr^{3+},Eu^{3+}	[78]
Y	$Y(BO_3)$: Pr^{3+},Gd^{3+}	[77]
Y	$Y(BO_3)$: Sb^{3+}	[75]
Y	$Y(BO_3)$: Yb^{3+}	[72]
YYb	$Y(BO_3)$: Yb^{3+}	[72]
Zn	$nZnO \cdot B_2O_3$: Eu^{3+} (n=1 to 4)	[90]
Zn	$nZnO \cdot B_2O_3$: Mn^{2+} (n=1 to 4)	[89, 90]
Zn	$Zn_4O(B_6O_{12})$: Mn^{2+}	[38]

For additional luminescent properties, see also $Li_2[B_4O_7]$ (Section 3.4.4.1, p. 123) and compounds of the types $MAl_3[BO_3]_4$ (Section 3.4.1.3, pp. 85/6).

The first room-temperature tunable laser (787 to 892 nm) in a borate single crystal is $Sc(BO_3)$:Cr [91 to 93].

For the possible application of B_2O_3 in a nonlinear optical material, see β-$Ba_3[B_3O_6]_2$ (Section 3.4.3.2.1, p. 104) and $Li[B_3O_5]$ (Section 3.4.3.1, p. 99).

References for 3.2.5.4:

[1] Bavika, V. I.; Gridneva, N. V. (Neftekhimiya **27** [1987] 806/12 from C.A. **108** [1988] No. 111600).
[2] Grozhan, M. M.; Reguchadze, G. R.; Eremeev, A. P.; Kruglov, A. I.; Lebedeva, G. P.; Fedorova, M. N. (Zh. Prikl. Khim. [Leningrad] **61** [1988] 1127/31; J. Appl. Chem. [USSR] **61** [1988] 1026/30 from C.A. **110** [1989] No. 212223).
[3] Izumi, Y. (Shokubai **30** [1988] 108/11 from C.A. **109** [1988] No. 24445).
[4] Sato, S.; Sakurai, H.; Urabe, K.; Izumi, Y. (Chem. Lett. **1985** 277/8).
[5] Sakurai, H.; Sato, S.; Urabe, K.; Izumi, Y. (Chem. Lett. **1985** 1783/4).
[6] Zaman Khan, M. K. (Adv. Catal. Proc. 7th Natl. Symp. Catal., Baroda, India, 1985, pp. 355/64 from C.A. **104** [1986] No. 7997).
[7] Peil, K. P.; Galya, L. G.; Marcelin, G. (J. Catal. **115** [1989] 441/51 from C.A. **110** [1989] No. 102511; Proc. 9th Int. Congr. Catal., Calgery, Alberta, 1988, Vol. 4, pp. 1712/8 from C.A. **111** [1989] No. 121677).
[8] Vdovets, B. S.; Tarasov, L. A.; Sharkov, A. V.; Lebedeva, G. N. (Katal. Protsessy Katalizatory, L. **1987** 67/70 from C.A. **110** [1989] No. 117089).
[9] Ray, S. C.; Singh, B.; Choudhury, A.; Roy, S. K.; Mukherjee, P. N. (Adv. Catal. Proc. 7th Natl. Symp. Catal., Baroda, India, 1985, pp. 413/22 from C.A. **105** [1986] No. 42390).
[10] Kochetkova, N. V.; Anufrieva, T. A.; Anokhin, V. N.; Peregudov, V. A.; Titova, E. A. (Zh. Prikl. Khim. [Leningrad] **59** [1986] 1469/72; J. Appl. Chem. [USSR] **59** [1986] 1365/8 from C.A. **106** [1987] No. 87316).

[11] Romenskii, A. V.; Popik, I. V.; Loboiko, A. Ya.; Atroshchenko, V. I. (Khim. Tekhnol. [Kiev] **1985** No. 6, pp. 23/5 from C.A. **104** [1986] No. 40495).
[12] Zhang, M.-Q.; Zou, Y.-X.; Fang, D.-F. (Jinshu Xuebao **21** [1985] A81/A85 from C.A. **103** [1985] No. 113540).
[13] Emori, H.; Kikuta, T.; Inada, T.; Obokata, T.; Fukuda, T. (Jpn. J. Appl. Phys. **24** Pt. 2 [1985] L291/L293 from C.A. **103** [1985] No. 30962).
[14] Kubota, E.; Katsui, A. (Jpn. J. Appl. Phys. **24** Pt. 2 [1985] L344/L346 from C.A. **103** [1985] No. 46622).
[15] Zakharenkov, L. F.; Zykov, A. M.; Makarenko, V. G.; Samorukov, B. E. (Legir. Poluprovodn. Mater. **1985** 52/6 from C.A. **104** [1986] No. 43836).
[16] Imhoff, D.; Gelsdorf, F.; Pellissier, B.; Castaing, J. (Phys. Status Solidi A **90** [1985] 537/43).
[17] Ram, S.; Datta, S.; Chakravorthy, D.; Bahadur, D. (J. Non-Cryst. Solids **91** [1987] 165/9).
[18] Liu, C.-X.; Jia, W.-Y.; Zhang, X.-X. (Guisuanyan Xuebao **13** [1985] 124/6 from C.A. **103** [1985] No. 14726).
[19] Potgieter, E.; Res, M. A.; Sigalas, I.; Schonberger, H. (S. Afr. J. Phys. **9** [1986] 142/6 from C.A. **106** [1987] No. 71743).
[20] Potgieter, E.; Res, M. A.; Sigalas, I.; Schonberger, H.; Bednarik, J. (J. Mater. Sci. **22** [1987] 752/6 from C.A. **106** [1987] No. 142594).

[21] Gelyasin, A. E.; Lutsko, L. A.; Mikhnevich, V. V. (Vestsi Akad. Navuk BSSR Ser. Khim. Navuk **1987** No. 5, pp. 104/7 from C.A. **108** [1988] No. 80449).

[22] Kotru, P. N.; Kachroo, S. K.; Wanklyn, B. M.; Watts, B. E. (J. Mater. Sci. **22** [1987] 4484/98 from C.A. **108** [1988] No. 214142).
[23] Edel'man, I. S.; Zarubina, T. V.; Kim, T. A.; Arkhipov, A. K.; Gorelova, A. V.; Smyk, A. A. (Fiz. Khim. Stekla **13** [1987] 848/53 from C.A. **108** [1988] No. 103273).
[24] Bhargava, A.; Varshneya, A. K.; Snyder, R. L. (Mater. Lett. **8** [1989] 41/5 from C.A. **111** [1989] No. 124784; Supercond. Its Appl. Proc. 2nd Ann. Conf. Supercond. Appl., Buffalo 1988, pp. 124/8 from C.A. **111** [1989] No. 139011).
[25] Balkanski, M. (Appl. Surf. Sci. **33/34** [1987/88] 1260/8 from C.A. **110** [1989] No. 26558).
[26] Dzwonkowski, P.; Julien, C.; Balkanski, M. (Appl. Surf. Sci. **33/34** [1987/88] 838/43 from C.A. **110** [1989] No. 16659).
[27] Ayyadi, A.; Massot, M.; Balkanski, M. (Phys. Scr. **38** [1988] 75/7 from C.A. **109** [1988] No. 160807).
[28] Wichner, R. P.; Spence, R. D. (Nucl. Technol. **70** [1985] 376/93 from C.A. **103** [1985] No. 130713).
[29] Shen, K. K. (Plast. Compd. **8** No. 5 [1985] 66/7, 69/72, 74, 76, 78, 80 from C.A. **104** [1986] No. 169293).
[30] Foote, K. L. (UCID-21338 [1988] 1/18 from C.A. **111** [1989] No. 24418).

[31] Gomaa, A. Z.; Gad, H. A. (J. Oil Colour Chem. Assoc. **71** [1988] 50/5 from C.A. **108** [1988] No. 223130).
[32] Rymon-Lipinski, T.; Hennicke, H. W.; Lingenberg, W. (Keram. Z. **37** [1985] 450/3 from C.A. **104** [1986] No. 93995).
[33] Shen, C. F.; Ximen, L. L.; Zong, X. F. (Mater. Res. Bull. **24** [1989] 1223/30 from C.A. **112** [1990] No. 14553).
[34] Srivastava, A. M.; Sobieraj, M. T.; Gieger, R.; Banks, E. (Mater. Chem. Phys. **21** [1989] 327/33).
[35] Fu, W. T.; Fouassier, C.; Hagenmuller, P. (Mater. Res. Bull. **22** [1987] 899/909 from C.A. **107** [1987] No. 123648).
[36] Fu, W. T.; Fouassier, C.; Hagenmuller, P. (Mater. Res. Bull. **22** [1987] 389/97 from C.A. **107** [1987] No. 16619).
[37] Koskentalo, T.; Niinistö, L.; Leskelä, M. (J. Less-Common Met. **112** [1985] 67/70).
[38] Otero de la Gandara, M. J.; Conesa, J. C.; Soria, J. (Phys. Status Solidi A **87** [1987] K81/K84).
[39] Otero de la Gandara, M. J.; Molina, M.; Carmona, P. (Opt. Pura Appl. **21** [1988] 255/62 from C.A. **111** [1989] No. 67075).
[40] Tews, W.; Becker, P.; Herzog, G.; Künzler, G. (Z. Phys. Chem. [Leipzig] **268** [1987] 985/92).

[41] Koskentalo, T.; Leskelä, M.; Niinistö, L. (Mater. Res. Bull. **20** [1985] 265/74 from C.A. **102** [1985] No. 212048).
[42] Leskelä, M.; Koskentalo, T.; Blasse, G. (J. Solid State Chem. **59** [1985] 272/9).
[43] Blasse, G.; Oomen, E. W. J. L.; Liebertz, J. (Phys. Status Solidi B **137** [1986] K77/K81).
[44] Blasse, G. (J. Less-Common. Met. **112** [1985] 79/82).
[45] Blasse, G.; Kiliaan, H. S.; De Vries, A. J. (J. Less-Common Met. **126** [1986] 139/46).
[46] Blasse, G. (Chem. Mag. [Rijswijk, Neth.] **1986** 717/9).
[47] Vlasse, M.; Fouassier, C. (J. Solid State Chem. **34** [1980] 271/3).
[48] Buijs, M.; Blasse, G. (Chem. Phys. Lett. **113** [1985] 384/6).
[49] Berdowski, P. A. M.; Buijs, M.; Blasse, G. (J. Phys. Colloq. [Paris] **46** [1985] C7-31/C7-34).
[50] Buijs, M.; Blasse, G. (J. Lumin. **34** [1986] 263/78 from C.A. **104** [1986] No. 98687).

[51] Buijs, M.; Van Vliet, J. P. M.; Blasse, G. (J. Lumin. **35** [1986] 213/22 from C.A. **105** [1986] No. 161335).
[52] Holsa, J.; Leskelä, M. (Mol. Phys. **54** [1985] 657/67).
[53] Ratner, I. M.; Martsokha, V. I.; Mirokhina, L. L. (Sb. Nauchn. Tr. Vses. Nauchn. Issled. Inst. Lyuminoforov Osobo Chist. Veshchestv No. 30 [1986] 51/4 from C.A. **108** [1988] No. 46195).
[54] Ding, Xiyi (J. Less-Common. Met. **148** [1988/89] 393/7).
[55] Meijerink, A.; Jetten, H.; Blasse, G. (J. Solid State Chem. **76** [1988] 115/23).
[56] Meijerink, A.; Blasse, G. (Proc. Electrochem. Soc. **88**-24 [1988] 279/89 from C.A. **110** [1989] No. 84631).
[57] Blasse, G. (Inorg. Chim. Acta **142** [1988] 153/4).
[58] Verwey, J. W. M.; Dirksen, G. J.; Blasse, G. (J. Non-Cryst. Solids **107** [1988] 49/54).
[59] Fu, Wen Tian; Garcia, A.; Fouassier, C. (Rev. Chim. Miner. **24** [1988] 509/24).
[60] De Vries, A. J.; Blasse, G. (J. Phys. Colloq. [Paris] **46** [1985] C7-109/C7-112).

[61] Hao, Z.-R.; Blasse, G. (Mater. Chem. Phys. **12** [1985] 257/74 from C.A. **102** [1985] No. 157372).
[62] Guo, Fengyu; Peng, Yian (J. Lumin. **40/41** [1988] 175/6 from C.A. **109** [1988] No. 82488; Zhongguo Xitu Xuebao **6** No. 1 [1988] 23/9 from C.A. **109** [1988] No. 200336).
[63] Peng, Yian; Guo, Fengyu (Zhongguo Xitu Xuebao **7** No. 2 [1989] 35/40 from C.A. **112** [1990] No. 13647).
[64] Guo, Fengyu; Peng, Yian (Beijing Daxue Xuebao Ziran Kexueban **24** [1988] 672/9 from C.A. **111** [1989] No. 67045).
[65] Sviridov, V. V.; Loginova, N. V.; Shevchenko, G. P. (J. Inf. Rec. Mater. **16** [1988] 285/94 from C.A. **109** [1988] No. 201264).
[66] Blasse, G.; Van Vliet, J. P. M.; Verwey, J. W. M.; Hoogendam, R.; Wiegel, M. (J. Phys. Chem. Solids **50** [1989] 583/5).
[67] Blasse, G.; Dirksen, G. J. (Inorg. Chim. Acta **145** [1988] 303/8).
[68] Batezat, O. Yu.; Gubanova, E. R.; Efryushina, N. P.; Zhikhareva, E. A. (Vysokochist. Veshchestva **1988** 222/5 from C.A. **108** [1988] No. 231054).
[69] Dotsenko, V. P.; Ermakova, S. V.; Efryushina, N. P.; Zhikhareva, E. A. (Dopov. Akad. Nauk Ukr. RSR B **1985** No. 4, pp. 47/50 from C.A. **103** [1985] No. 29568).
[70] Dotsenko, V. P.; Ermakova, S. V.; Efryushina, N. P.; Zhikhareva, E. A. (Ukr. Fiz. Zh. [Russ. Ed.] **31** [1986] 1513/6 from C.A. **106** [1987] No. 25191).

[71] Dotsenko, V. P.; Efryushina, N. P.; Bol'shukhin, V. A.; Seifullina, T. Z.; Chopovskii, V. A. (Ukr. Khim. Zh. [Russ. Ed.] **54** [1988] 1130/4 from C.A. **110** [1989] No. 201933).
[72] Efryushina, N. P.; Dotsenko, V. P.; Zhikhareva, E. A.; Ermakova, S. V. (Rare Earths Spectrosc. Proc. Int. Symp., Wroclaw 1984 [1985], pp. 427/32 from C.A. **104** [1986] No. 138591).
[73] Radzki, S. (Przemysl Chem. **64** [1985] 316/8 from C.A. **104** [1986] No. 98563).
[74] Dotsenko, V. P.; Berezovskaya, I. V.; Efryushina, N. P.; Ermakova, S. V.; Zhikhareva, E. A. (Izv. Akad. Nauk SSSR Ser. Fiz. **50** [1986] 596/8; Bull. Acad. Sci. USSR Phys. Ser. **50** No. 3 [1986] 167/70 from C.A. **104** [1986] No. 196114).
[75] Oomen, E. W. J. L.; Van Gorkom, L. C. G.; Smit, W. M. A.; Blasse, G. (J. Solid State Chem. **65** [1986] 156/67).
[76] Wolfert, A.; Oomen, E. W. J. L.; Blasse, G. (J. Solid State Chem. **59** [1985] 280/90).
[77] Srivastava, A. M.; Sobieraj, M. T.; Ruan, S. K.; Banks, E. (Mater. Res. Bull. **21** [1986] 1455/63).
[78] De Vries, A. J.; Blasse, G.; Pet, R. J. (Mater. Res. Bull. **22** [1987] 1141/50).

[79] Jagannathan, R.; Rao, R. P.; Kutty, T. R. N. (Mater. Chem. Phys. **23** [1989] 329/33 from C.A. **111** [1989] No. 204490).
[80] Dotsenko, V. P.; Zhikhareva, E. A.; Ermakova, S. V.; Berezovskaya, I. V. (Ukr. Khim. Zh. [Russ. Ed.] **51** [1985] 641/3 from C.A. **103** [1985] No. 169078).

[81] Holsa, J. (Inorg. Chim. Acta **139** [1987] 257/9).
[82] Blasse, G.; Sas, S. J. M.; Smit, W. M. A.; Konijnendijk, W. L. (Mater. Chem. Phys. **14** [1986] 253/8 from C.A. **104** [1986] No. 119087).
[83] Kiliaan, H. S.; Blasse, G. (Mater. Chem. Phys. **18** [1987] 155/70 from C.A. **107** [1987] No. 246958).
[84] Garapon, C. T.; Jacquier, B.; Chaminade, J. P.; Fouassier, C. (J. Lumin. **34** [1985] 211/22 from C.A. **104** [1986] No. 26404).
[85] Garapon, C. T.; Jacquier, B.; Salem, Y.; Moncorge, R. (J. Phys. Colloq. [Paris] **46** [1985] C7-141/C7-145).
[86] Buijs, M.; Vree, J. I.; Blasse, G. (Chem. Phys. Lett. **137** [1987] 381/5).
[87] Fu, W. T.; Fouassier, C.; Hagenmuller, P. (J. Phys. Chem. Solids **48** [1987] 245/8).
[88] Dzhurinskii, B. F.; Zolin, V. F.; Tsaryuk, V. I.; Lysanova, G. V.; Komova, M. G.; Markushev, V. M. (Izv. Akad. Nauk SSSR Neorg. Mater. **23** [1987] 1525/30; Inorg. Mater. [USSR] **23** [1987] 1346/50 from C.A. **108** [1988] No. 13316).
[89] Dafinova, R. (Bulg. J. Phys. **15** [1988] 60/5 from C.A. **109** [1988] No. 118673).
[90] Dafinova, R. (J. Mater. Sci. Lett. **7** [1988] 69/70 from C.A. **109** [1988] No. 82487).

[91] Lai, S. T.; Chai, B. H. T.; Long, M.; Morris, R. C. (IEEE J. Quantum Electron. **22** [1986] 1931/3 from C.A. **105** [1986] No. 161811).
[92] Chai, B. H. T.; Long, M.; Morris, R. C.; Lai, S. T. (Springer Ser. Opt. Sci. **52** [1986] 76/81 from C.A. **106** [1987] No. 224708).
[93] Lai, S. T.; Chai, B. H. T.; Long, M.; Shinn, M. D.; Caird, J. A.; Marion, J. E.; Staver, P. R. (Springer Ser. Opt. Sci. **52** [1986] 145/50 from C.A. **106** [1987] No. 204766).

3.2.5.5 Analytical Aspects

Wet chemical methods have been described for the determination of B_2O_3 in glass batches; problems of sampling are discussed with respect to the homogeneity and average composition of batches [1]. Other methods are described for the qualitative and semiquantitative determination of B_2O_3 in pigments used in paintings [2].

Micro amounts of boron in iron or steels have been determined spectrophotometrically with 1-(salicylideneamino)-8-hydroxynaphthalene-3,6-disulfuric acid (Azomethine H) after distilling boron as $B(OCH_3)_3$ into a 4% NaOH solution and drying of this absorbent [3]. The time for this determination can be considerably shortened by the simultaneous distillation of $B(OCH_3)_3$ and evaporation of the distillate for analysis by inductively coupled plasma atomic emission spectrometry (ICPAES). After distilling with a large quantity of methanol, the $B(OCH_3)_3$ is collected in 4% NaOH solution heated in a water-bath at 90°C. As the boiling range of the mixture of absorption solution and methanol is lower than 90°C, the solvent mixture is easily evaporated. The limit of detection corresponds to 35 ng boron/0.5 g steel [4]. The same technique was used with a 1% NaOH solution, dissolving the dried salt in HCl/acetic acid, developing the color by the addition of a curcumin/acetic acid solution and sulfuric acid/acetic acid mixture. After 20 minutes, the solution was diluted to 100 mL with 75 mL of methanol and ca. 15 mL of distilled water, and the absorbance was measured at 548 nm after 10 minutes.

The absorbance of the blank could be reduced from 9.4 to 4.8%, and the detection limit was reduced from 0.04 to 0.02 ng of boron per gram steel. The molar absorption coefficient was 1.62×10^5 $L \cdot mol^{-1} \cdot cm^{-3}$, and Beer's law was obeyed at concentrations up to 5 µg boron/100 mL [5]. The same technique was used with a 0.8% NaOH solution. After coloration of the dried salt by the addition of curcumin, the boron-curcumin complex was extracted with 4-methyl-2-pentanone. The absorbance of the extracted phase was measured at 555 nm. The molar absorption coefficient was 1.65×10^5 $L \cdot mol^{-1} \cdot cm^{-3}$, and Beer's law was obeyed at concentrations up to 6.7 µg boron/100 mL. The detection limit was 0.048 ppm boron for a sample weight of 0.5 g [6].

Boron was determined in Ni base alloys and low-alloy steel by extraction from HCl into 15% 2,4-dimethyl-4,6-octanediol in toluene and molecular emission spectrometry at 546 nm with diffusion H_2/N_2O aerosol flames. The sensitivity with these flames is higher than with premixed flames. The detection limits were 0.05 and 0.1 µg boron/mL, respectively [7]. A method has been developed for the quantitative determination of boron in auxiliary neutron absorbers of critical assemblies in a reactor by neutron activation analysis. In borated cellulose with a boron content of 0.3 to 2.5 wt%, unfiltered beams of neutrons were used, but in a homogeneous mixture of Al and B_4C with a boron content of ca. 10%, beams of neutrons filtered with Cd were used. The operation was conducted in an automated facility for activation analysis in a beam of neutrons using a Be converter at the IR-8 reactor facility [8].

A $LiBO_2$ flux (from $Li_2[CO_3]$: H_3BO_3, 1:5; between 950 and 1050°C) was used to dissolve the following materials and determine the elements after solution in 4% HNO_3, dilute $HCl/SrCl_2$, or dilute H_2SO_4: soil samples by wet-chemical methods [9], rock samples by atomic absorption spectrometry (AAS) [10], mineral samples by AAS or spectrophotometry [11], silicates and slags by X-ray fluorescence (XRF) [12], silicates (after addition of I_2O_5) by XRF [13], rock melts by XRF [14], silicate rock samples by XRF [15], CeO_2/Y_2O_3 mixtures by XRF [16], silicate rocks (after addition of $SrCl_2$) by AAS (HC≡CH/air or $HC{\equiv}CH/N_2O$ flames) [17], geological samples by AAS or inductively coupled plasma atomic emission spectrometry (ICPAES) [18], granite, amphibolite, and biotite samples (Mn) by AAS [19], fly ash (Tl) by AAS [20], in the standard reference material BCR 176 (Be) by Zeeman AAS [21], in soils and rocks by instrumental analysis [22], in soil clay samples by comparing AAS with XRF and ICPAES [23], in silicate rocks (Mn) by a spectrometric method [24] or by ion-exchange chromatography [25], in coal (with lithium peroxide) [26] and (without lithium peroxide) [27], and rocks (rare earth elements) by atomic-emission determinations [28], K, Ca, Ti, V, Cr, Mn, Fe, Co, Ni, Zn, and Zr in $LiBO_2$ fusion pellets by energy-dispersive XRF spectrometry [29], in chromite materials (platinum-group elements) by a fire-assay procedure with NiS as the collector [30], and in ores (Cr) after proton irradiation [31]. Suitable additives to minimize the structure effect in spectral analysis are $LiBO_2$ or $Li_2[B_4O_7]$ [32].

The well-known sample preparation method for the preparation of alkali metal-borate glass pellets can also be used in energy-dispersive XRF analysis. The detection limit in the element range of $Z \approx 14$ (Si) is ca. 0.01 to 0.1% for a sample to glass ratio of 1:10 [33].

The isotopic distribution of boron in minerals has been determined by thermal ionization mass spectrometry (TIMS) [34]. A method for the high-precision determination of the $^{10}B/^{11}B$ isotopic distribution (which varies between 0.24663 and 0.24737) in nuclear reactor materials by TIMS using the $[Cs_2(BO_2)]^+$ ion, is capable of producing an isotope ratio analysis with a standard deviation of 0.12 ppm. This is a distinct improvement over results obtained with the normally used $[Na_2(BO_2)]^+$ ion [35]. The absolute amount of ^{10}B (0.1 µg/mL) and, therefore, the concentration and isotopic composition of boron has been analyzed in heavy water moderators as well as in B_4C samples, used in nuclear reactors, by application of $[Cs_2(BO_2)]^+$ in TIMS; the certified atom ratio is 18.80 ± 0.02 [36].

References on pp. 66/7

The reaction of guanidinium halides, $[(NH_2)_3C]X$ (X = Cl or Br), with oxygen-containing substances was used as a method of oxygen separation for isotopic analysis in borates [37]. A procedure is described whereby rapid and accurate isotopic measurements can be performed in H_3BO_3 and B_4C after fusion of these compounds with $Ca[CO_3]$ to give $Ca(BO_2)_2$ (better than the use of $Ba[CO_3]$). It allows the determination of $^{10}B/^{11}B$ in H_3BO_3 with an isotopic fractionation correction factor $K = 0.99470 \pm 0.0028$ by isotope-dilution mass spectrometry [38].

A gas flow proportional detection system for the determination of boron in selected materials such as borosilicate glass or borated polyethylene is described that uses detection of the charged particles produced in the $^{10}B(n,\alpha) \rightarrow ^{7}Li$ reaction induced by thermal neutrons [39].

Analytical methods for the determination of the oxidation state of copper in borate glasses have been developed; ESR spectroscopy was used for Cu^{2+} [40]. A spectrocolorimetric method for the determination of Sb^{3+} in borate glass was developed using ICl [41]. 10^{-5} to 10^{-1}% Tb^{3+} was determined by measuring the luminescence of $Tb_xSc_{1-x}[BO_3]$ crystallophosphors activated at 1100°C. The preferred wavelength for the intensity measurement was 551 nm. The relative standard deviation was ca. 0.16 [42].

Hydrogen and oxygen have been determined in powdered boron, H_3BO_3, and B_2O_3 by using reducing fusion involving pulsed electrothermal heating in a graphite capsule in an inert gas atmosphere between 2100 and 2400°C and chromatography; the evolution of gas in two stages is associated with the reduction of H_3BO_3 and B_2O_3, respectively [43].

References for 3.2.5.5:

[1] Lange, J. (Silikattechnik **36** [1985] 35/7 from C.A. **102** [1985] No. 208162).
[2] Pokorny, J. (Sb. Vys. Sk. Chem. Technol. Praze Polym. Chem. Vlastnosti Zprac. S **13** [1985] 149/69 from C.A. **105** [1986] No. 5775).
[3] Hosoya, M.; Takeuchi, M. (Bunseki Kagaku **35** [1986] 854/7 from C.A. **106** [1986] No. 187998).
[4] Hosoya, M.; Tozawa, K.; Takada, K. (Talanta **33** [1986] 691/3).
[5] Takeuchi, M.; Hosoya, M. (Bunseki Kagaku **37** [1988] T92/T95 from C.A. **109** [1988] No. 221544).
[6] Ishikuro, M.; Kimura, J. (Bunseki Kagaku **37** [1988] 498/502 from C.A. **109** [1986] No. 221460).
[7] Panov, V. A.; Semenenko, K. A.; Kuzyakov, Yu. Ya. (Zh. Anal. Khim. **43** [1988] 832/8 from C.A. **109** [1988] No. 243146).
[8] Alekseev, I. Ya.; Bogomolov, L. M.; Borisov, G. I.; Komkov, M. M.; Kuz'michev, V. A.; Loboda, S. V. (At. Energ. **65** [1988] 28/32 from C.A. **109** [1988] No. 117774).
[9] Chen, Sh.-Q.; Lu, Q.-G. (Turang Tongbao **16** No. 3 [1985] 126/8 from C.A. **103** [1985] No. 226593).
[10] Gao, C.-S.; Zhang, B.-C.; Hu, Sh.-F. (Fenxi Huaxue **13** [1985] 139/42 from C.A. **102** [1985] No. 230995).

[11] Xu, Jiaquan (Fenxi Shiyanshi **7** No. 2 [1988] 62 from C.A. **111** [1989] No. 224339).
[12] Eddy, B. T.; Balaes, A. M. E. (X-Ray Spectrom. **17** [1988] 17/8 from C.A. **108** [1988] No. 215411).
[13] Xu, Jirong; Wu, Zhihong; Le, Chun; Li, Qihung (Yanshi Kuangwu Ji Ceshi **5** [1986] 201/6 from C.A. **109** [1988] No. 103789).
[14] Gunicheva, T. N. (Metody Rentgenospektr. Anal. **1986** 46/57 from C.A. **107** [1987] No. 50871).

[15] Ye, Y.-W.; Nishido, H.; Sakamoto, T.; Doi, A. (Okayama Rika Daigaku Kiyo A No. 20 [1984/85] 77/86 from C.A. **103** [1985] No. 226608).
[16] Pella, P. A.; Tao, G. Y.; Dragoo, A. L.; Epp, J. M. (Anal. Chem. **57** [1985] 1752/4).
[17] Zaikin, I. D.; Zubova, O. A.; Borodavkina, O. N. (Zh. Anal. Khim. **43** [1988] 809/13 from C.A. **109** [1988] No. 203978).
[18] Deliiska, A.; Blazheva, T.; Petkova, L.; Dimov, L. (Fresenius' Z. Anal. Chem. **332** [1988] 362/5).
[19] Angeles Guerrero, M.; Milagros Tobias, M.; Garcia Vargas, M. (Afinidad **42** No. 397 [1985] 300/4 from C.A. **103** [1985] No. 188614).
[20] Grognard, M.; Piolon, M. (At. Spectrosc. **6** No. 5 [1985] 142/3 from C.A. **104** [1986] No. 135453).

[21] Grognard, M. (At. Spectrosc. **8** No. 5 [1987] 153/4 from C.A. **108** [1988] No. 30971).
[22] Brueckner, H. P.; Drews, G.; Kritsotakis, K.; Tobschall, H. J. (Chem. Erde **45** [1986] 53/6 from C.A. **104** [1986] No. 161197).
[23] Bartenfelder, D. C.; Karathanasis, A. D. (Commun. Soil Sci. Plant Anal. **19** [1988] 471/92 from C.A. **109** [1988] No. 31268).
[24] Kuroda, R.; Matsuzawa, Y.; Oguma, K. (Fresenius' Z. Anal. Chem. **326** [1987] 156/7).
[25] Kuroda, R.; Suzuki, N.; Oguma, K. (Fresenius' Z. Anal. Chem. **324** [1986] 43/6).
[26] Rigin, V. I.; Laktionova, N. V. (Khim. Tverd. Topl. [Moscow] **1986** No. 5, pp. 109/13 from C.A. **106** [1987] No. 105018).
[27] Rigin, V. I. (Zh. Anal. Khim. **40** [1985] 253/7 from C.A. **102** [1985] No. 151815).
[28] Vocke, R. D.; Hanson, G. N.; Gruenenfelder, M. (Contrib. Mineral. Petrol. **95** [1985] 145/54).
[29] Wegscheider, W.; Ellis, A. T.; Goldbach, K.; Leyden, D. E.; Mahan, K. I. (Anal. Chim. Acta **188** [1986] 59/66).
[30] Robert, R. V. D. (MINTEK-324 [1987] 1/13 from C.A. **108** [1988] No. 48379 or C.A. **109** [1988] No. 103690).

[31] Olivier, C.; De Wet, B. S.; Morland, H. J. (Radiochim. Acta **42** [1987] 143/9).
[32] Kardos, J.; Zimmer, K.; Florian, K. (Spectrochim. Acta B **42** [1987] 765/71).
[33] Beckmann, W. (GIT Fachz. Lab. **32** [1988] 799/801 from C.A. **109** [1988] No. 214845).
[34] Spivack, A. J.; Edmond, J. M. (Anal. Chem. **58** [1986] 31/5).
[35] Ramakumar, K. L.; Parab, A. R.; Khodade, P. S.; Almaula, A. I.; Chitambar, S. A.; Jain, H. C. (J. Radioanal. Nucl. Chem. Lett. **94** [1985] 53/62).
[36] Ramakumar, K. L.; Khodade, P. S.; Parab, A. R.; Chitambar, S. A.; Jain, H. C. (J. Radioanal. Nucl. Chem. Lett. **107** [1986] 215/23).
[37] Berezovskii, F. I.; Korostyshevskii, I. Z.; Demikhov, Yu. N.; Lyuta, N. N. (Zh. Anal. Khim. **40** [1985] 1848/53).
[38] Duchaleau, N. L.; Verbruggen, A.; Hendrickx, F.; De Bièvre, P. (Talanta **33** [1986] 291/4).
[39] Chabot, G. E. (J. Radioanal. Nucl. Chem. Articles **123** [1988] 491/515).
[40] Duran, A.; Valle, F. J. (Glass Technol. **26** [1985] 179/85 from C.A. **103** [1985] No. 127838).

[41] Singh, S. P.; Pyare, R.; Nath, P. (Glass Technol. **26** [1985] 176/8 from C.A. **103** [1985] No. 127837).
[42] Gubanova, E. R.; Efryushina, N. P.; Zhikhareva, E. A.; Bazetat, O. Yu. (Zavod. Lab. **54** No. 5 [1988] 20/1 from C.A. **109** [1988] No. 85324).
[43] Oganezov, K. A., Bairamashvilii, I. A.; Tsagareishvili, G. V.; Tabutsidze, M. L.; Andriasova, I. A. (Poroshk. Metall. [Kiev] **1986** No. 2, pp. 75/7; Sov. Powder Metall. Met. Ceram. **25** [1986] 141/3 from C.A. **105** [1986] No. 237533).

3.2.6 Higher Boron Oxides

Thermal effects of the interaction between $B_{12}O_2$ (formulated as B_6O) and strong nitric acid were investigated by calorimetric dissolution. The standard enthalpy of formation was determined as $\Delta H_f^\circ = 527 \pm 32$ kJ/mol [1]. Reflectivity spectra of B_6O were measured in the medium range (MIR) and far infrared (FIR) spectral range. Plasma resonance spectra yield qualitative information on the dependence of electronic transport on material properties [2].

$B_{12}O_2$ powder has been prepared by the reduction of metal oxides, preferentially ZnO, but also MgO, SnO, Ga_2O_3, or CdO, with boron. Massive $B_{12}O_2$ can be prepared by hot pressing at 1800°C. The reaction of boron with B_2O_3 to give crystallized $B_{12}O_2$ containing an amorphous phase was studied with graphical data processing. It appears that the amorphous phase could be boron clusters bound with oxygen based on the variation in the O/B ratio observed without significant change in the rhombohedral cell dimensions [3].

References for 3.2.6:

[1] Makarov, V. S.; Ugai, Ya. A. (J. Less-Common Metals **117** [1986] 277/81).
[2] Werheit, H.; Haupt, H. (Z. Naturforsch. **42a** [1987] 925/34).
[3] Brodhag, C.; Thevenot, F. (J. Less-Common Metals **117** [1986] 1/6).

3.3 Cationic Species

For radical cations, see p. 241.

$[BO]^+$. The molecular cation $[BO]^+$ has been produced by passing B^+ ions (obtained by electron bombardment of BCl_3) through NO_2. The low-lying singlet, triplet, and quintet states of $[BO]^+$ were studied by configuration interaction (CI) methods. The ground state is $1\,^1\Sigma^+$ resulting from the closed shell configuration $(1\sigma \rightarrow 4\sigma)^2\, 1\pi^4$. Many stable states are predicted, similar in structure and ordering to those of BeO and BN. Three states, namely $2\,^1\Pi$, $2\,^3\Pi$, and $1\,^5\Pi$, have unusually flat minima (at large values of r_e). Dipole moments (in Debye) for the lowest states are: $1\,^1\Sigma^+$, −4.89; $1\,^3\Pi$, −2.57; and $1\,^1\Pi$, −2.60 [1].

The adiabatic ionization potentials (vertical ionization potentials in parentheses, all in eV), based on the equilibrium internuclear distance r_e=121.1 pm and E=−99.8038 hartree, are for $[BO]^\bullet$: X $^2\Sigma^+ \rightarrow [BO]^+$: X $^1\Sigma^+$ $(5\sigma \rightarrow \infty)$ 12.81 (12.81); for $[BO]^\bullet$: X $^2\Sigma^+ \rightarrow [BO]^+$: 1 $^1\Pi$ $(1\pi \rightarrow \infty)$ 13.40 (13.76); and for $[BO]^\bullet$: X $^2\Sigma^+ \rightarrow [BO]^+$: 1 $^3\Pi$ $(1\pi \rightarrow \infty)$ 13.12 (13.49) [1].

The ion $[BO]^+$ with its dissociation energy of ≤9.7 eV, its total atomic number $Z_1 + Z_2 = 12$, and the difference of atomic numbers $Z_1 - Z_2 = 2$ is doubtful among 31 diatomic molecules found in cosmic objects spectra, but may be of astrophysical significance since $D_0 \geq 4.19$ eV and $Z_1 \pm Z_2 < 25$ are theoretical reasons for the existence in space [2].

The existence of 3 to 4 MeV beams of the double ionized $[^{10}B^{16}O]^{2+}$ emerging from a tandem accelerator, have been demonstrated [3].

$[BO_2]^+$. High spatial resolution SIMS (secondary ion mass spectrometry) imaging has been demonstrated for the ion $[^{11}BO_2]^+$, prepared on a shiny $Fe_{75}B_{15}Si_{10}$ metallic glass surface structure after crystallization in a quartz tube at a residual Ar pressure of ca. 10^{-6} Torr [4]. A single previously unreported red band was observed at 6117.4 Å (16342 cm^{-1}) in a microwave discharge through BF_3 and excess oxygen. The only reasonable rotational analysis is compatible with a $^1\Sigma_u^+ \rightarrow {}^1\Sigma_g^+$ or a $^1\Sigma_g^- \rightarrow {}^1\Sigma_u^-$ transition. This band is due to symmetric O–B–O, and its singlet character favors its attribution to the $[BO_2]^+$ ion as carrier. The B value is

0.29482 and the B′ value is 0.22004 cm^{-1} for the lower and upper states, respectively. The corresponding **$[^{10}BO_2]^+$** band was also observed with B = 0.2930 and B′ = 0.2201 cm^{-1} [5].

The four lower-lying electronic states of $[BO_2]^+$ along with the two first doublet states of $[BO_2]^{\bullet}$ were studied by ab initio 6-31G*, 6-311G*, and 6-31+G* methods at different levels of accuracy. The ground state of $[BO_2]^+$ is of $^3\Sigma_g^-$ symmetry and lies at 0.7 eV below the singlet $^1\Sigma_g^+$ state. The dipole forbidden transition $^3\Sigma_u^+ \leftarrow {}^3\Sigma_g^-$ exhibits a calculated excitation energy of 1.95 ±0.2 eV. The $^1\Sigma_u^+ \leftarrow {}^1\Sigma_g^+$ excitation energy is predicted as 1.23±0.1 eV. The harmonic vibrational frequencies of $[BO_2]^+$ in both states $^3\Sigma_g^-$ and $^1\Sigma_g^+$, computed by HF/6-31G*, are ν_1 = 1042 and 1139 cm^{-1}, ν_2 = 179 and 348 cm^{-1}, and ν_3 = 938 and 1303 cm^{-1}, respectively. The $[BO_2]^+$ ion does not appear to be a candidate for the new absorption band centered at 2.03 eV (16324 cm^{-1}), that was recently reported by Bredohl and Dubois [5], as argued by Nguyen [6].

On the basis of Hartree-Fock-double-zeta-Roos-Siegbahn (HF/DZRS) calculations for the transition of the ground-state **$[BOH]^+$** to the excited-state **$[HBO]^+$**, an energy of isomerization of 16 kcal/mol and a barrier energy of isomerization of 50 kcal/mol are obtained [9].

There are **$[NaBO_2]^+$** and **$[Na_2(BO_2)_2]^+$** species present in the vapor above a molten B_2O_3–Na_2O–SiO_2–TeO_2 glass, based on mass spectrometric studies [10].

Mass spectral studies of the ion-molecule equilibrium in a saturated vapor of $CsBO_2$ were conducted between 828 and 943 K. The equilibrium constants and enthalpies were determined for the following reactions: $(CsBO_2)_c \rightleftharpoons Cs_g^+ + ([BO_2]^-)_g$, $Cs^+ + (CsBO_2)_c \rightleftharpoons$ **$[Cs_2BO_2]^+$**, and $[BO_2]^- + (CsBO_2)_c \rightleftharpoons [Cs(BO_2)_2]^-$. The heats of formation of the gaseous ions $[Cs_2BO_2]^+$ and $[Cs(BO_2)_2]^-$ are −419±19 and −1569±29 kJ/mol, respectively [11].

For the appearance potentials for $[B(OH)_2]^+$ and $[B(OH)_3]^+$, see Section 3.5.7.2 on p. 184.

Biuret ($H_2NC(O)NHC(O)NH_2$) reacts with H_3BO_3 to give tetra-coordinate solid compounds of the types **$[B(OH)_2(H_2NC(O)NHC(O)NH_2)]OH$** and (in the presence of $[B(C_6H_5)_4]^-$ ions) **$[B(OH)_2(H_2NC(O)NHC(O)NH_2)][B(C_6H_5)_4]$**, respectively. IR, NMR, and UV data suggest that biuret functions as a bidentate ligand. From small changes in the IR and UV spectra it is deduced that biuret coordinates through the oxygen atoms of the carbonyl groups [7].

Luminescence properties of bis(1,3-diketonato)boronium perchlorates $[B(O_2Z)_2][ClO_4]$ (cf. p. 239) with Z = CHR–CH_2–CHR′ (R, R′ = CH_3, C_6H_5) have been investigated; ^{11}B NMR data (in $CHCl_3$, versus $[BF_4]^-$): δ = 8.18 to 9.38 ppm; UV: λ = 296 to 382 nm. Luminescence maxima are at 77 K between 505 and 595 nm, and at 293 K between 525 and 540 nm [8].

References for 3.3:

[1] Karna, S. P.; Grein, F. (Mol. Phys. **61** [1987] 1055/62).
[2] Singh, M. (Astrophys. Space Sci. **140** [1988] 421/7).
[3] Galindo-Uribarri, A.; Lee, H. W.; Chang, K. H. (J. Chem. Phys. **83** [1985] 1210/4).
[4] Levi-Setti, R.; Crow, G.; Wang, Y. L. (Springer Ser. Chem. Phys. **44** [1986] 132/8).
[5] Bredohl, H.; Dubois, I. (J. Phys. B **18** [1985] 233/7).
[6] Nguyen, M. T. (Mol. Phys. **58** [1986] 655/8).
[7] Mukhopadhyay, D.; Sur, B.; Bhattacharyya, R. G. (Indian J. Chem. A **24** [1985] 425/6).
[8] Karasev, V. E.; Korotkikh, O. A. (Zh. Neorg. Khim. **30** [1985] 2269/72; Russ. J. Inorg. Chem. **30** [1985/86] 1290/2).
[9] Charkin, O. P.; Zyubina, T. S. (Koord. Khim. **12** [1986] 1011/37; Sov. J. Coord. Chem. **12** [1987] 583/97).
[10] Asano, M. (J. Nucl. Mater. **132** [1985] 288/90).

[11] Sidorova, I. V.; Gorkhov, L. N. (Teplofiz. Vys. Temp. **25** [1987] 1100/6 from C.A. **108** [1988] No. 121483).

3.4 Nonhydrated Anionic Structures and Species Derived Thereof

For previous data, see "Boron Compounds" 3rd Suppl. Vol. 2, 1987, pp. 41/70, and literature cited therein.

In the original literature, the alternative formulations MB_xO_y, $M(B_xO_y)$, and $M[B_xO_y]$ are often used for the same compound. Typically, compounds such as borates are neither fully homopolar nor fully heteropolar. In most cases the homopolar portion is not known, though some investigations have been done. We prefer $M[B_xO_y]$ instead of $M(B_xO_y)$ or MB_xO_y (also for systematic reasons), if there is no proof of a predominant homopolar structure. Thus, the writing $M[B_xO_y]$ does not permit a conclusion to the degree of a salt-like character of the compound.

A review containing 550 references has been prepared in conjunction with a survey of the structures of anhydrous borates (in addition to water-containing and other complex borates, e.g., polyborates and borotungstates) [1]. A review containing 265 references on the properties of small and macroanionic metal borates in crystalline, amorphous, and glassy states, in solution as well as molten and gaseous phases is also available. The variety of borates and many of their properties are discussed in terms of the configurations and conformations of the macroanions in relation to changes in external effects due to the flexibility of B–O–B bonds by virtue of the oxygen atoms with oscillating hybridization (involving boron atoms) [2]. A review of the chemistry of rare earth borates [3] and a book about alkaline earth metal borates [4] have been published.

Furthermore, a review (34 references) on the usefulness for ^{10}B and ^{11}B NMR spectroscopy for structural studies including borates has been compiled [5]. The occurrence and major uses of borates, e.g., in glassmaking, has been summarized (15 references) [6].

A review with 37 references gives a systematic classification of the borate series in terms of the structural types of the anionic groups. Calculations were made for the linear and nonlinear optical (NLO) properties for most of the important borate anionic groups such as in β-$Ba_3[B_3O_6]_2$ (BBO) or $K[B_5O_8]\cdot 4H_2O$, including second-order susceptibilities and absorption edges in the UV spectral region [7].

References for 3.4:

[1] Heller, G. (Top. Curr. Chem. **131** [1986] 39/86).
[2] Tarasevich, B. P.; Kuznetsov, E. V. (Usp. Khim. **56** [1987] 353/92 from C.A. **107** [1988] No. 16579).
[3] Leskela, M.; Niinisto, L. (Handb. Phys. Chem. Rare Earths **1986** Vol. 8, pp. 203/334 from C.A. **105** [1987] No. 16995).
[4] Gode, H. (Alkaline Earth Metal Borates Zinatne, Riga, USSR, 1986, pp. 1/167 from C.A. **106** [1987] No. 42979).
[5] Bray, P. J. (AIP Conf. Proc. No. 140 [1986] 142/67 from C.A. **105** [1987] No. 34119).
[6] Smith, R. A. (J. Non-Cryst. Solids **84** [1986] 421/32).
[7] Chen, Chuangtian; Wu, Yicheng; Li, Rukang (J. Cryst. Growth **99** [1990] 790/8).

3.4.1 Monoborate Ions and Related Molecules

3.4.1.1 The Ion $[BO]^-$ and MBO Molecules

The monoborate ion **$[BO]^-$** was calculated with a restricted Hartree-Fock (RHF) wave function in which the first spin contaminant has been annihilated self-consistently as the starting point (UHF) for subsequent extended Møller-Plesset (EMP) perturbation calculations. For a bond length of 122.7 pm for $[BO]^-$, the following energies were calculated with basis STO-3G: for EHF, −97.99911; EMP2, −98.10081; EMP3, −98.09861; and EMP4, −98.12154 hartree. With the basis 3-21G: for EHF, −98.98554; EMP2, −99.16411; EMP3, −99.15509; and EMP4, −99.17694 hartree. With the basis 6-31G*: for EHF, −99.53976; EMP2, −99.79467; EMP3, −99.79558; and EMP4, −99.81362 hartree. With the basis 6-31++G**: for EHF, −99.57653; EMP2, −99.84200; EMP3, −99.83950; and EMP4, −99.86144 hartree. With the basis 6-311++G**: for EHF, −99.60043; EMP2, −99.88630; EMP3, −99.88214; and EMP4, −99.90793 hartree. For the calculated values for the electron affinity of $[BO]^•$, see Section 3.2.1, pp. 8/9 [1].

The ground-state energies of the linear isomers **LiBO** and **LiOB** have been calculated using the self-consistent field (SCF), complete active space self consistent field (CASSCF) and configuration interaction (CI) approximations. If the 4σ-MO is kept doubly occupied in all configurations, the energy balance changes towards LiOB, but with proper correlation treatment LiBO is more stable than LiOB [2]. Potential surfaces for LiBO have been calculated using a model in the diatomic fragments-in-molecule (DIM) theory on the basis of generalized correlation relations [3]. Nonempirical calculations of the energy of the LiBO molecule associated with use of restricted AO basis sets show that the description of the potential surface of LiBO needs to take into consideration the electron correlation and the main contributions to the correlation corrections [4]. The quantum-chemical calculations of the total energy are compared with experimental mass spectrometric data; the heat of reaction has been determined by the second-order law procedure to be −276.2±4 kJ/mol [5], by calculations −258.6 kJ/mol for LiOB and −314.1 kJ/mol for LiBO, indicating the formation of a mixture of the isomers of comparable energies for the reaction $Li_g+(BO)_g \rightarrow (LiBO)_g$ [6]. The reaction $Li_g+(BO)_g \rightarrow (LiBO)_g$ was studied by Knudsen effusion mass spectrometry. The heats of reaction were determined for the linear structure Li–O–B to be −332.5±5.0 kJ/mol, for linear O–B–Li to be −317.4±5.0 kJ/mol, and for the bent structure (LiBO) −251.2±5.0 kJ/mol. The barrier height relative to B–O–Li is 81.4 kJ/mol [5].

An ab initio Hartree-Fock calculation was made for the ground state of BOK by using a potential function based on the isoelectronic molecule KCN and on the equilibrium internuclear distance of the BO radical $r_e(BO)=227.697$ Bohr. A total energy of the BO fragment of −99.552581 a.u. was obtained; the energy difference to the linear structure was determined to be ca. 2 kJ/mol. Therefore, a quasi-linear molecule structure is favored. The calculated geometrical parameters are $d_{KB}=5.788$ Bohr and the angle $\measuredangle(KBO)=50.1°$. The self-consistent field (SCF) dipole moment for the equilibrium structure is for the bent structure 11.67 D, for linear **KBO** 13.54 D and for linear **KOB** 13.59 D [7].

The potential energy surfaces for the isomerization **CuBO→CuOB** were calculated by ab initio SCF methods, as were the equilibrium geometrical parameters, the relative energies, and the electronic structure [8].

From a cold-pressed and sintered LaB_6 surface ionizer, the background is emitted predominantly by $[BO]^-$ and $[BO_2]^-$ [9]. The $[BO]^-$ ion was also produced by using a sputtered ion source and atomic and molecular ion transmission measured in a tandem accelerator [10].

References on p. 72

References for 3.4.1.1:

[1] Baker, J. (Chem. Phys. Lett. **152** [1988] 227/32).
[2] Nemukhin, A. V.; Stepanov, N. F. (Theor. Chim. Acta **67** [1985] 287/92).
[3] Nemukhin, A. V.; Lyudkovskii, S. V.; Stepanov, N. F. (Vestn. Mosk. Univ. Khim. **28** [1987] 332/6 from C.A. **108** [1988] No. 11366).
[4] Nemukhin, A. V.; Stepanov, N. F. (Teor. Eksp. Khim. **23** [1987] 728/33; Theor. Exp. Chem. [USSR] **23** [1987] 674/9 from C.A. **108** [1988] No. 227199).
[5] Neubert, A. (J. Chem. Phys. **82** [1985] 939/41).
[6] Ermilov, A. Yu.; Nemukin, A. V.; Stepanov, N. F. (Zh. Fiz. Khim. **62** [1988] 212/4; Russ. J. Phys. Chem. **62** [1988] 108/9).
[7] Rosmus, P. (Angew. Chem. **100** [1988] 1376/7).
[8] Musaev, D. G.; Yakobson, V. V.; Klimenko, N. M.; Charkin, O. P. (Koord. Khim. **13** [1987] 1188/97 from C.A. **108** [1988] No. 27199).
[9] Maartenson, B. M.; Wilhelmsson, S. O. (Int. J. Mass Spectrom. Ion Proc. **67** [1985] 179/89 from C.A. **104** [1986] No. 126139).
[10] Shima, K.; Takahashi, T.; Kakita, T.; Yamanouchi, M. (Natl. Lab. High Energy Phys. KEK-88-7 [1988] 46/52 from C.A. **110** [1989] No. 161805).

3.4.1.2 The Ion $[BO_2]^-$, MBO_2 Molecules, $M_2(BO_2)_2$ Molecules, $M_n[(BO_2)_n]$ Salts, and Related Compounds

3.4.1.2.1 Ions and Molecules

In the original literature, the alternative formulations MBO_2, $M(BO_2)$, and $M[BO_2]$ are used without regard to the predominant ionic or homopolar character of the compound (for the "percentage of ionic character", see p. 73). In this section, the writing mode is adapted to the real character of the compound as far as possible; otherwise, the formulation in the original literature is given.

Using the method of ion-molecular equilibria with the ion $[AlF_4]^-$, the standard enthalpy of formation of gaseous $\mathbf{[BO_2]^-}$ is determined to be -709 ± 16 kJ/mol [29].

By configuration interaction by perturbation selected iterations (CIPSI) calculations on the MP2 level, total energies (in hartree) for the $[BO_2]^-$ system at its MP2/6-31+G* optimized geometry in the fundamental states were calculated and are given in Table 3/7 on p. 15 [30].

The softness parameter $\sigma_x=0.94$ [1], derived from the equation $\sigma_x=[\sigma_B(BO_2^-)-\sigma_B(OH^-)/\sigma_A(H^+)]$, where σ_A and σ_B are other softness parameters, depends on the enthalpy of hydration and the ionization potential [2].

Gap modes were investigated for $[BO_2]^-$ centers in the lattice of KI single crystals by measuring the far-infrared absorption spectrum. According to group-theoretical considerations, two gap modes of symmetry A_{2u} at 81.5 cm^{-1} and symmetry E_u at 78.4 cm^{-1} could be ascertained experimentally [28]. Infrared spectra of MBO_2 molecules (M = Li, Na, K, Rb, and Cs) in a matrix of solid argon or krypton are reported [7].

Results of ab initio calculations at different levels of accuracy are reported for the $\mathbf{LiBO_2}$ molecule. The gas phase $LiBO_2$ has a linear $C_{\infty v}$ (in [3] erroneously $C_{\infty h}$) conformation, Li–O–B=O (∢LiOB = 180°). The cyclic C_{2v} structure, [–LiOBO–] (∢OBO = 149° to 153°), represents a saddle point for the lithium migration. No C_s structure, Li–O–B=O (∢LiOB about 90°), was found for a stationary point on the energy surface. Comparison between computed and observed vibrational frequencies suggests that a lowering of the molecular symmetry from $C_{\infty v}$

to C_s can occur under matrix conditions. Linear LiOBO is predicted to be a very stable species with regard to fragmentation into Li^+ and $[OBO]^-$ with a dissociation energy of 143.4 kcal/mol, computed on the basis of MP4SDQ/6-31+G* plus ZPE contributions, but the migration of lithium from one oxygen atom to another has an energy barrier of only 8.2 kcal/mol. Some molecular properties such as net charge, overlap population, and electric field gradient have been computed and discussed in conjunction with the dipole moment of 11.2 D and the eQq_{zz} value of 2.09 MHz (^{11}B nucleus), both at HF/6-31+G*; at 4-31G, the rotational constant B = 4966 MHz, the specific heat capacity $C_p^\circ = 57.6\ J \cdot mol^{-1} \cdot K^{-1}$, and the molecular entropy $S^\circ = 254.4\ J \cdot mol^{-1} \cdot K^{-1}$ were obtained [3].

The ionic character of the M–OBO bond in $LiBO_2$, **$NaBO_2$**, and **KBO_2** was estimated to be 43, 46, and 49%, respectively. The heat of dissociation of LiOBO and NaOBO was calculated by the extended Hückel molecular orbital (EHMO) method for 20% p character of the alkali metal atom to be 597.0 and 563.6 kJ/mol, respectively. The higher heat of dissociation for the lithium than for the sodium compound is related to the appreciable sp hybridization of the lithium atom. Linear ($C_{\infty v}$) and bent (C_s) conformations are discussed [4].

A mass-spectrometric study by the Knudsen effusion method showed that between 1070 and 1473 K the MBO_2 (M = Li, Na, or K) species are present in the vapor over a borosilicate glass such as Pyrex glass or IKhS-10 glass (ca. 69.5% SiO_2, 27.0% B_2O_3, 1.0% Al_2O_3, 0.4% CaO, 2.3% Na_2O, 0.5% Li_2O) [5]. Na_2O–B_2O_3 glasses evaporate congruently; the rate decreases with the time. After technical correction, the rate was seen to be independent of time. The transport of $NaBO_2$ into the vapor phase is the only rate-controlling step. A simple analysis of the data, however, does not directly give the vapor pressure of the volatile species [6].

In the vapor over B_2O_3–Na_2O–SiO_2 glasses the gaseous species $NaBO_2$ and **$Na_2(BO_2)_2$** were identified by a mass spectrometric Knudsen effusion method [10]. Between 806 and 1153 K, the gaseous species over a B_2O_3–Na_2O–Rb_2O–SiO_2 glass were found to be $NaBO_2$, $Na_2(BO_2)_2$, **$NaRb(BO_2)_2$**, **$Rb(BO_2)$**, and **$Rb_2(BO_2)_2$** [11]; between 845 and 1150 K, over a B_2O_3–Na_2O–Cs_2O–SiO_2 glass or over a Cs-containing sodium borosilicate glass, gaseous $NaBO_2$, $Na_2(BO_2)_2$, **$NaCs(BO_2)_2$**, **$Cs(BO_2)$**, and **$Cs_2(BO_2)_2$** have been identified [12, 13].

An effusion mass spectral electron-impact investigation of the thermodynamic properties of gaseous KBO_2 was performed between 1053 and 1240 K. The KBO_2 molecule fragments to more than 90% of its initial concentration at electron ionization energies 2 eV above its ionization potential of 8 ± 1 eV. The fragmentation processes were studied up to ionization energies as high as 70 eV, with the appearance of K^+, $[BO]^{2+}$, and $[BO]^+$ fragments. Results obtained for the sublimation process $(KBO_2)_s \rightarrow (KBO_2)_g$ and for the vaporization process $(KBO_2)_l \rightarrow (KBO_2)_g$ gave second- and third-law reaction heats. A third-law heat of −672.4 kJ/mol and a second-law value of −672.8 kJ/mol were obtained for gaseous KBO_2. The heat of melting at the melting point of 1220 K is 31.8 + 0.4 kJ/mol [14].

The enthalpy of solution of solid $CsBO_2$ in aqueous H_2SO_4 was measured by calorimetry and, in combination with the enthalpies of solution of H_3BO_3 and $Cs_2[SO_4]$ in the same solvent, the standard molar enthalpy of formation of $CsBO_2$ was derived as −976.8 ± 0.9 kJ/mol [15]. By adiabatic calorimetry between 9 and 346 K, the low-temperature heat capacity for defined mass fractions of $CsBO_2$ was measured to be 1.247 $J \cdot mol^{-1} \cdot K^{-1}$ at 8.76 K and calculated to be 69.49 $J \cdot mol^{-1} \cdot K^{-1}$ at 298.15 K. High-temperature enthalpy increments were measured by drop calorimetry; $H_T^\circ - H_{298.15}^\circ$ at 414.0 K is 8.554 kJ/mol, and at 671.2 K is 31.127 kJ/mol. From these results were derived (T = 298.15 K): $S_T - S_0^\circ = 105.38\ J \cdot mol^{-1} \cdot K^{-1}$, $H_T^\circ - H_0^\circ = 13.915$ kJ/mol, $-[G_T^\circ - H_0^\circ] = 58.71\ J \cdot mol^{-1} \cdot K^{-1}$; $\Delta H_f^\circ(T) = -976.800$ kJ/mol, and $\Delta G_f^\circ(T) = -919.917$ kJ/mol. The melting point in O_2 atmosphere is 989.7 ± 0.3 K, the enthalpy of fusion is 34.8 ± 1 kJ/mol [16]. Thermochemical research on reactor materials and fission products includes the formation properties of $CsBO_2$ [17].

References on pp. 75/6

Mass spectral studies of the ion-molecule equilibrium in a saturated vapor of **$CsBO_2$** (for its structure, cf. that of $BaBO_2$, below) were conducted between 828 and 943 K and gave the heat of formation of crystalline $CsBO_2$, $\Delta H_f^\circ = -966 \pm 17$ kJ/mol. The equilibrium constants and enthalpies were determined for the following reactions: $(CsBO_2)_c \rightleftharpoons Cs_g^+ + ([BO_2]^-)_g$, $Cs^+ + (CsBO_2)_c \rightleftharpoons [Cs_2BO_2]^+$, and $[BO_2]^- + (CsBO_2)_c \rightleftharpoons [Cs(BO_2)_2]^-$. The heats of formation of the gaseous ions $[Cs_2BO_2]^+$ and $[Cs(BO_2)_2]^-$ are -419 ± 19 and -1569 ± 29 kJ/mol, respectively [18].

By mass spectrometric Knudsen-effusion measurements in the B_2O_3–SrO system, the gaseous species BO, B_2O_2, B_2O_3, Sr, and **$SrBO_2$** radical were identified between 1281 and 1572 K [19], also over $Sr(BO_2)_2$ at temperatures between 1310 and 1442 K [20]. The structural model of gaseous $SrBO_2$ is assumed to be the same as that of $RbBO_2$, a bent Sr–O–B=O structure with a linear O–B=O group and an Sr–O–B angle of 90°. On the basis of their partial pressures, the enthalpy of formation and the dissociation energy for gaseous $SrBO_2$ were determined to be -636.5 ± 10.9 kJ/mol and 516.0 ± 9.7 kJ/mol, respectively. The molar enthalpies of formation for gaseous $SrBO_2$ were determined and compared with those for other gaseous MBO_2 and M_xO_y compounds (M = alkaline earth and alkali metal, respectively) [19].

The same method was also used for the study of the vaporization behavior of solid $Ba_3B_2O_6$ in the temperature range from 1434 to 1630 K. The gaseous species BO, B_2O_2, BO_2, B_2O_3, Ba, and **$BaBO_2$** radical (with a bent Ba–O–B=O structure and a linear –O–B=O group) were observed. From the partial pressures, the molar enthalpies of formation and dissociation for gaseous $BaBO_2$ and the molar enthalpies of the formation for solid $Ba_3B_2O_6$ from the elements and from the constituent oxides were determined [21].

3.4.1.2.2 Salts of the Type $M_n[(BO_2)_n]$ and Related Compounds

Because the polymeric state of the anions is not known in most cases, simplified formulations such as $Ca(BO_2)_2$ are used in this section.

IR and Raman spectra of the metaborate $LiBO_2$ with infinite chain structure (including ^6Li and ^{10}B labeled species) have been recorded and discussed in relation to the structure and possible coupling between the chains of the unit cell. As group theoretical methods and the isotopic substitutions show, there are only condensed BO_3 groups. There is no evidence for the existence of an out-of-plane O(2) vibration. On the low frequency side of the strong band at 349 cm^{-1}, there are three weak bands, so that all the seven predicted modes are observed [9]. The equilibrium diagram of the $LiBO_2$–ZnO quasi-binary section of the B_2O_3–Li_2O–ZnO ternary system has been constructed as result of an investigation by differential thermal analysis (DTA), X-ray diffraction and IR studies [23]. For tetragonal and monoclinic $LiBO_2$ at 77 and 300 K ^7Li and ^{11}B NMR data are reported; the ^{11}B chemical shift is 16 ± 2 ppm in the monoclinic phase and 0 ppm in the tetragonal phase [22].

$Ca(BO_2)_2$ is orthorhombic, space group Pbcn–D_{2h}^{14} (No. 60) with a = 1160.4, b = 428.5, and c = 621.4 pm [9]. The electron density distribution in $Ca(BO_2)_2$ has been also studied by X-ray diffraction experiments on two crystals, I with $R_w = 2.5\%$ for 603 and II with $R_w = 2.1\%$ for 1577 reflections, space group Pnca–D_{2h}^{14} (No. 60) with a = 620.46(3), b = 1158.65(7), and c = 427.47(3) pm; $D_c = 2.715$ g/cm³; the B–O distances are 132.00(2), 138.56(2), and 140.10(2) pm [8].

The IR and Raman vibrational spectra of the chain metaborates $Ca(BO_2)_2$ and the isotypic **$Sr(BO_2)_2$** were investigated by group theoretical methods and isotopic substitutions with ^{40}Ca,

^{44}Ca, ^{10}B, or ^{11}B. The spectra are discussed in relation to the structure and possible coupling between the chains in the unit cell [9]. The molar enthalpies of formation for solid $Sr(BO_2)_2$ were determined and compared with those for other alkaline earth metal metaborates [20].

The electrical properties of $\mathbf{AgBO_2}$ have been determined by a.c. impedance, d.c. polarization measurements with ionically blocking electrodes, and d.c. transference measurements with electronically blocking ionic conducting electrodes. $AgBO_2$ is identified as a predominantly ionic and conducting compound with high Ag^+ ion conductivity [25]. $AgBO_2$ reacts with compounds R_nSnCl_{4-n} (R = CH_3 or C_6H_5; n = 2, 3) to give $R_nSn(BO_2)_{4-n}$ [26].

In the system $\mathbf{LiBO_2}$–$\mathbf{LiBS_2}$, oxide-sulfide glasses were prepared over a wide range of compositions. The glass transition temperature, T_g, with a maximum near the composition containing 80 mol% $LiBO_2$ was observed. The electrical conductivity at 500 K ranged from 5×10^{-4} to 5×10^{-3} S/cm with the maximum near the composition containing 55 mol% $LiBO_2$. IR and Raman spectra showed that structural units with bridging oxygen and nonbridging sulfur atoms predominated [27].

$\mathbf{Bi_4ZnO_6(BO_2)_2}$ crystallizes in the monoclinic space group $P2_1/m-C^2_{2h}$ (No. 10) with a = 1643.0(8), b = 1027.0(6), and c = 560.7(3) pm; β = 95(1)°; Z = 4; D_m = 7.58 g/cm³ and D_c = 7.68 g/cm³. IR and crystal chemical results suggest that the compound contains $[B_4O_8]^{4-}$ units, consisting of two BO_4 tetrahedra and two BO_3 triangles connected by common oxygen vertices. Similar chain anions are encountered in the structures of the high-temperature modifications of $Ca(BO_2)_2$ and $Sr(BO_2)_2$, the IR spectra of which contain bands analogous to those for $Bi_4ZnO_6(BO_2)_2$ [24].

References for 3.4.1.2:

[1] Bard, A. J.; Parsons, R.; Jordan, J. (Standard Potentials in Aqueous Solutions, Dekker, New York 1985).
[2] Marcus, Y. (Thermochim. Acta **104** [1986] 389/94).
[3] Nguyen, M. T. (J. Mol. Struct. **136** [1986] 371/9 [THEOCHEM **29**]).
[4] Tsekhanskii, R. S.; Skvortsov, V. G.; Molodkin, A. K.; Vinogradov, L. I. (Zh. Neorg. Khim. **31** [1986] 1578/80; Russ. J. Inorg. Chem. **31** [1986] 903/4).
[5] Archakov, I. Yu.; Stolyarova, V. L. (Fiz. Khim. Stekla **14** [1988] 440/4 from C.A. **109** [1988] No. 175033).
[6] Cable, M.; Fernandes, M. H. V. (Glass Technol. **28** [1987] 135/40 from C.A. **107** [1987] No. 82452).
[7] Seshadri, K. S.; Nimon, L. A.; White, D. J. (J. Mol. Spectrosc. **30** [1969] 128/39).
[8] Kirfel, A. (Acta Crystallogr. B **43** [1987] 333/43).
[9] Rulmont, A.; Almou, M. (Spectrochim. Acta A **45** [1989] 603/10).
[10] Asano, M.; Yasue, Y. (J. Nucl. Mater. **138** [1986] 65/72).

[11] Asano, M.; Kou, T. (Phys. Chem. Glasses **30** [1989] 39/45).
[12] Asano, M.; Yasue, Y. (J. Nucl. Sci. Technol. **22** [1985] 1029/31 from C.A. **104** [1986] No. 77351).
[13] Asano, M.; Kou, T.; Yasue, Y. (J. Non-Cryst. Solids **92** [1987] 245/60).
[14] Farber, M.; Srivastava, R. D.; Moyer, J. W.; Leeper, J. D. (J. Chem. Soc. Faraday Trans. I **81** [1985] 913/8).
[15] Konings, R. J. M.; Cordfunke, E. H. P.; Ouweltjes, W. (J. Chem. Thermodyn. **19** [1987] 201/3).
[16] Cordfunke, E. H. P.; Konings, R. J. M.; Westrum, E. F. (Thermochim. Acta **128** [1988] 31/8).

[17] Cordfunke, E. H. P.; Konings, R. J. M.; Westrum, E. F. (J. Nucl. Mater. **167** [1988/89] 205/12 from C.A. **111** [1989] No. 203142).
[18] Sidorova, I. V.; Gorkhov, L. N. (Teplofiz. Vys. Temp. **25** [1987] 1100/6 from C.A. **108** [1988] No. 121483).
[19] Kou, T.; Asano, M. (High Temp. Sci. **24** [1987] 1/19; C.A. **109** [1988] No. 238161).
[20] Asano, M.; Kou, T. (J. Chem. Thermodyn. **20** [1988] 1149/56).

[21] Asano, M.; Kou, T. (J. Chem. Thermodyn. **21** [1989] 837/45).
[22] Vorotilova, L. S.; Dmitrieva, L. V.; Samoson, A. V. (Zh. Strukt. Khim. **30** [1989] 70/3 from C.A. **112** [1990] No. 110525).
[23] Buludov, N. T.; Kara'ev, Z. Sh.; Abdulla'ev, G. K. (Zh. Neorg. Khim. **31** [1986] 2108/11; Russ. J. Inorg. Chem. **31** [1986] 1215/7).
[24] Zargarova, M. I.; Kasumova, M. F.; Abdulla'ev, G. K. (Zh. Neorg. Khim. **32** [1987] 1211/4; Russ. J. Inorg. Chem. **32** [1987] 737/9).
[25] Koehler, B. U.; Jansen, M.; Weppner, W. (J. Solid State Chem. **57** [1985] 227/33).
[26] Ghose, B. N. (An. Quim. B **81** [1985] 5/6 from C.A. **103** [1985] No. 142109).
[27] Tatsumisago, M.; Yoneda, K.; Minami, T. (J. Am. Ceram. **71** [1988] 766/9).
[28] Kauschke, W.; Otto, J.; Fischer, F. (J. Phys. C **18** [1985] 1263/7).
[29] Semenikhin, V. I.; Mina'eva, I. I.; Sorokin, I. D.; Nikitin, M. I.; Rudnyi, E. B.; Sidorov, L. N. (Teplofiz. Vys. Temp. **25** [1987] 660/70; High Temp. [USSR] **25** [1987] 497/501).
[30] Mota, F.; Novoa, J. J.; Ramirez, A. C. (J. Mol. Struct. **166** [1988] 153/8 [THEOCHEM **43**]).

3.4.1.3 The Ion $[BO_3]^{3-}$ and Related Structures

In the original literature, the alternative formulations $M[BO_3]$, $M(BO_3)$, and MBO_3 are often used for the same borate. We prefer $M[BO_3]$ (also for systematic reasons), cf. Chapter 3.4, p. 70.

3.4.1.3.1 The Ion $[BO_3]^{3-}$

The nature of the unoccupied molecular orbitals of the boron-oxygen bonds in $\mathbf{[BO_3]^{3-}}$ has been studied by molecular orbital (MO) calculations, X-ray absorption near edge (XANES), electron transmission (ETS), and NMR spectroscopy. The lowest energy unoccupied MO (LUMO) for three-coordinate boron is essentially a boron 2pπ nonbonding or weakly antibonding orbital of A_2 symmetry within the D_{3h} point group. This orbital generates an absorption about 4 eV below the boron 1s ionization potential in XANES and a scattering resonance about 4 eV above the threshold in ETS. It is the final state for the lowest energy UV absorptions of the $[BO_3]^{3-}$ group, and excitations to it from the E′ B–O bonding MO dominate the paramagnetic contribution to the ^{11}B NMR chemical shift. The only other unoccupied orbital observed for three-coordinate boron is the E′ B–O antibonding orbital that appears above the threshold in both XANES and ETS; the energy of this E′ antibonding orbital is highly distance dependent and, thus, may provide useful information on B–O distances in amorphous materials [1].

Ab initio Hartree-Fock-Roothaan calculations with large polarized basis sets gave ^{17}O nuclear quadrupole coupling constants, e^2qQ/h, and values of the NMR shielding constant σ and the electric-field gradient q in planar $[BO_3]^{3-}$ ions which were compared with experimental values. The results indicate that accurate calculations on simple molecular models can reproduce the trends in σ and q observed in borate glasses [2].

Hartree-Fock calculations for $[BO_3]^{3-}$ (D_{3h} symmetry), referred to H_3BO_3, with HF/STO-3G gave E=−244.4448 a.u. and r=141.81 pm; with HF/6-31G*, E=−248.6183 a.u. and r=141.87 pm, as well as the vibrational frequencies $\nu_1(A_1')$ at 848, $\nu_2(A_2'')$ at 876, $\nu_3(E')$ at 1215, and $\nu_4(E')$ at 581 cm^{-1} [3]. By extrapolation of the existing experimental and (by STO-3G) calculated values for $[BO_3]^{3-}$ (and additional isoelectronic molecular ions), the heat of formation of the O_4 molecule with D_{3h} symmetry was obtained [4].

3.4.1.3.2 $M^I_3(BO_3)$, $M^IM^{II}(BO_3)$, and $M^{III}(BO_3)$ Compounds

3.4.1.3.2.1 Fe[BO$_3$] and ^{57}Fe[BO$_3$]

In the H_3BO_3–Fe_3O_4 system at 700 °C, Fe[BO$_3$] is formed with planar $[BO_3]^{3-}$ groups, as was characterized by IR spectra [36]. Highly perfect single crystals of ^{57}Fe[BO$_3$] (enriched to 95%) with diameters of the platelets up to 18 mm were grown from an Fe_2O_3–B_2O_3–PbO/PbF_2 flux; the B_2O_3 concentration should be higher than 70 mol%, the Fe_2O_3 concentration so small that the solid phase appears at a temperature lower than 1123 K. The optimum range of slow cooling is from 1108 to 1073 K [37]. The morphological characteristics of this growth of Fe[BO$_3$] crystals have been discussed. The conditions for the formation of peculiarities in the shape of Fe[BO$_3$] single crystals were investigated. The solubility of Fe[BO$_3$] in the solvent 53.3 B_2O_3-32.7 PbO-14.0 PbF_2 is 7.2 wt% at 854 °C. When supercooling at a rate of 3 to 7 °C/h, the single crystal faces increase linearly along the (111) direction [38]. Reactive radiofrequency sputtering was used in order to deposit ferromagnetic-ferroelectric Fe[BO$_3$] films by using α-Fe_2O_3 and B_2O_3 as targets in an atmosphere containing a mixture of Ar, O_2, and H_2. By controlling the H_2 content during sputtering, reproducible ferroelectric characteristics were achieved [39].

The spectrum of inhomogeneous magnetic oscillations and the dispersion dependence of the magnetostatic modes (MSM) in a canted antiferromagnetic slab of rhombohedral Fe[BO$_3$] were investigated at 77 K and 30 to 45 GHz. Experimental dispersion and angle dependence of volume MSM agree with calculated ones. The coupled oscillations of the magnetic and the electrodynamic subsystems were observed in a plate of Fe[BO$_3$] [40, 41].

The excitation of magnetostatic oscillations and waves in rhombohedral weak ferromagnetic Fe[BO$_3$] has been studied between 77 and 300 K for fields of 20 to 35 Oe by antiferromagnetic ESR spectroscopy. The excited oscillations and waves are not homogeneous. A theoretical calculation of the vibrational linewidth variation with temperature agrees with the experimental observations [42]. The effect of an electric field on the antiferromagnetic ESR spectra of Fe[BO$_3$] crystals was studied. Shifts of the resonance peaks are observed which depend on the angle between the trigonal axis and the magnetic field. The results are discussed in terms of the effect of the electric field on the magnetic anisotropy [43].

The acoustic NMR of Fe[BO$_3$] easy-plane crystals has been studied. Oscillations were observed due to magnetoelastic interactions; this phenomenon is labeled magnetoelastic NMR. Possible mechanisms are described [47]. Magnetoelastic NMR spectroscopy of ^{57}Fe in Fe[BO$_3$] has been studied for combined acoustic- and nuclear-resonance frequencies. The radiofrequency field excited two types of nuclear precession: uniform (ordinary NMR) and non-uniform (magnetoelastic NMR) [44]. A decay or decrease in the magnetoelastic NMR strength is observed in the acoustic resonance region, and the minimum coincides with the acoustic resonance [45]. Effects connected with the excitation of the magnetoelastic oscillations in the Fe[BO$_3$] crystals by pulses of the magnetic fields with radiofrequency were studied. The characteristic oscillations of samples were investigated and found to depend on the acoustic stress. In the absence of the acoustic stress, the observed signal was significantly wider. The

References on pp. 87/91

signal of the magnetoacoustic parametric echo was recorded in addition to the magnetoelastic oscillations [46]. The magnetoelastic oscillations in excited thin plates of antiferromagnetic $Fe[BO_3]$ were studied for fields between 40 and 300 Oe in pulses of 1 to 10 μs with frequencies between 60 and 80 MHz. The results can be explained in terms of paramagnetic echo effects [48]. The effect of the ^{57}Fe nuclear spin system on the attenuation of magnetoelastic oscillations in $Fe[BO_3]$ at 77 K was studied by NMR spectroscopy. Coupling of the nuclear spins with acoustic oscillations observed at NMR frequencies increases the spin-spin relaxation time and shifts the magnetoacoustic echo [49].

Magnetoelastic resonance peaks were also observed in the ^{57}Fe NMR spectrum of $Fe[BO_3]$ during application of an alternating magnetic or ultrasonic field between 4.2 and 77 K. A temperature-induced shift in the peak position is observed. The resonance amplitude decreases as the temperature increases [50]. The ^{57}Fe NMR signals of $Fe[BO_3]$ at 77 K are detected from both within the domains and at the domain boundary walls due to differences in the shift dynamics [51]. The free induction signal in this system with an anomalously narrow NMR line is explained by assuming an anisotropy of the hyperfine interaction related to the sublattice magnetization emerging from the basis plane and assuming that the NMR signal is due to the domain walls [52].

The motion of single domain walls in $Fe[BO_3]$ depends on the amplitude of the controlling magnetic field at room temperature. The path of the domain wall depends also on the interval of time between the front of a pulse of magnetic field and a light pulse. At room temperature, the amplitude-frequency dependence of oscillations of single domain walls in $Fe[BO_3]$ was measured. The results support a theory by which a high limiting rate of motion of domain walls (nonlinear dynamics) and strong appearance of magnetoelastic interaction are expected in easy-plane ferromagnets. Thus, an effect on the rate of motion of domain walls is exerted as a function of the magnitude of the external magnetic field [53]. The domain-wall dynamics of $Fe[BO_3]$ was studied in fields between 0 and 78 Oe and between 100 and 293 K. The velocity of the domain-wall motion increases as field strength increases [54].

The magnetoelastic resonances of Bloch domain walls were calculated for easy-plane weak ferromagnets under pressure by using the Landau-Lifshitz equation. A critical pressure is predicted at which the resonance vanishes. Calculations are given for $Fe[BO_3]$ [55].

The domain structure of nonbasal planes of isometric crystals of $Fe[BO_3]$ studied by measuring the powder shapes is similar to that of thin-layer garnet materials. The sample thickness influences the structure only insignificantly. The labyrinth structure was observed for thicker crystals [56]. Surface magnetism (a macroscopic transition magnetic layer due to surface magnetic anisotropy) is observed and investigated in the natural nonbasal faces of $Fe[BO_3]$ crystals by the magnetooptical technique. The anisotropy is the result of a symmetry change of the surface magnetic-ion surroundings. Erasure of surface magnetism at a face of the $(10\bar{1}4)$ type occurs in a field of ca. 1.6 kOe. The surface anisotropy energy for the $(10\bar{1}4)$ and $(11\bar{2}0)$ faces was calculated. The theory describes the symmetry of the magnetic anisotropy, and the calculated values of the field, to an order of magnitude, are the same as those measured [57]. Magnetooptical methods were also used to study the surface magnetic anisotropies of $Fe[BO_3]$ crystals of $(10\bar{1}4)$, $(11\bar{2}0)$, and $(11\bar{2}3)$ orientations between 0 and 350 K. The anisotropies are discussed in terms of magnetic dipole interactions with the magnetic ions [58].

An energy-level model is given to describe the induction of magnetic anisotropy in $Fe[BO_3]$ on exposure to light. The results agree with experimental determinations [59].

Studies of the nonlinear Faraday effect of $Fe[BO_3]$ in superstrong magnetic fields of up to 800 T showed the absence of saturation of the polarization of rotation plane on fields. This effect may be related to the height of exchange interaction in magnetic fields and also to the effects of fields on electron transitions in ions of Fe [60].

Spectra were studied of optical absorption, magnetic linear dichroism (MLD), and circular dichroism (CD) in the weak ferromagnetic $Fe[BO_3]$ for the interconfiguration transitions $^6A_{1g} \rightarrow {}^4T_{1g}$, $A_{1g} \rightarrow {}^4T_{2g}$ in the temperature interval from 30 to 300 K. The temperature dependence was obtained for the magnetooptical activity of narrow lines and broad bands. A relation was experimentally demonstrated between magnetic linear dichroism and exciton-magnon processes [61].

The hardness phenomenon involved in the parametric excitation of magnons in $Fe[BO_3]$ was investigated between 1.2 and 7 K at pumping frequencies of 35.6 and 26.3 GHz. The effect of nonequilibrium magnons of other frequencies and of the proper magnetoelastic oscillations excited in the sample on the hardness of the magnon parametric excitation process was studied. The hardness phenomenon is due to improper relaxation of the parametrically excited magnons [62].

Low-temperature relaxation of magnons in $Fe[BO_3]$ was investigated for pumping frequencies $\omega_p/2\pi = 26.2$ and 35.4 GHz between 1.2 and 20 K. The temperature and field dependences of the magnon relaxation parameter were found to be the main contribution to this parameter between 1.2 and 4.2 K (from 3-particle magnon-phonon interactions). The characteristic magnon-phonon interaction energy is ca. 10^{-14} erg. The probability for scattering of magnons by phonons at low temperatures should increase in the presence of magnetic impurity ions [63]. The excitation of transverse phonons with frequencies $\omega_p/2 = 2\pi$ (300 to 600) MHz was carried out by applying a microwave magnetic field $h \cdot \cos \omega_p t$ in a condition of parallel pumping with a field strength between 50 and 700 Oe and temperatures between 4.2 and 77 K. The lowering of the threshold of parametric phonon excitation is observed when an integral number of half-waves (n = 140 to 350) fits into the thickness of a plate of single crystal $Fe[BO_3]$. The experimental findings make it possible to prove the validity of theoretical expressions that link the threshold field amplitude with the excited phonon relaxation rate and the experimental parameters [64]. During neutron scattering from $Fe[BO_3]$ in a magnetic field, peaks are observed which are due to magnetoelastic vibrations. The peaks decrease in intensity on heating above the Néel point of 348 K. The results are interpreted in terms of magnetoacoustic resonances [65]. The intensity of magnetic, nuclear, and mixed scattering of neutrons by a perfect weakly ferromagnetic crystal $Fe[BO_3]$ oscillates with the variation of the optical thickness of the specimen. The positions of the oscillations correspond to the predictions of the dynamic theory. The magnitude of the external magnetic field affects the oscillation contrast, whereas the direction of the field affects their position in the case of magnetic or mixed scattering. The accuracy of the determination of the structural factors on the basis of the pendulum bands is ca. 1% [66]. The effect of high-frequency magnetization on the X-ray scattering in $Fe[BO_3]$ crystals was also studied. Resonance peaks were observed in the X-ray scattering which depend on the frequency. These peaks are not observed above the Néel point; they, therefore, appear to be due to magnetoacoustic effects [67].

The effect of fast magnetization reversal induced by external radiofrequency fields was studied in $Fe[BO_3]$ using the Mößbauer technique. The frequency collapse and sideband effects were investigated as a function of intensity at 62 and 36 MHz. Because of the relatively short switching times, the magnetization reversal must be of rotational character [68].

The anomalous transmission of resonant γ-radiation through a perfect single crystal of $^{57}Fe[BO_3]$ was studied in the nearly pure nuclear Bragg reflection (330) at room temperature. The comparison of the obtained Mößbauer reflection spectra with computer calculations yield the result that this transmission does not follow the temperature law that is valid for the anomalous transmission of X-rays. Evidence was obtained that the lattice vibrations at room temperature do not affect the suppression effect or do this much less than they affect the related Borrmann effect [69]. Mößbauer spectra of purely nuclear diffraction of resonance γ-radiation of ^{57}Fe in $Fe[BO_3]$ crystals in the vicinity of phase transition from the antiferromag-

References on pp. 87/91

netic to the paramagnetic state show sensitivity to the onset of antiferromagnetic ordering [70]. The angular dependence of the anomalous transmission through a $^{57}Fe[BO_3]$ crystal for a pure nuclear Laue diffraction shows a peak of ca. 8.5″ half width [71]; the cases of intermediate and weak resonance absorption were realized at energy distances $E = \pm 2\gamma$ and $\pm 6\gamma$ off-resonance, where γ is the natural linewidth [72].

Pure nuclear resonance diffraction by single crystals of $^{57}Fe[BO_3]$ has been studied with conventional ^{60}Co sources [73]. Experimental studies were also conducted on the time path (speed up) of the response of collective nuclear states in a crystal during their selective excitation by a collimated beam of γ-quanta in the vicinity of the Bragg angle. The measurements were carried out with a nanosecond magnetic shutter. A curtailment of about sixfold was observed in the characteristic response time of nuclear excitation at the exact position of the Bragg angle in comparison with the duration of the response of an isolated nucleus. The optical scheme is shown for an experiment with 14.4 keV Mößbauer radiation of the ^{57}Fe nucleus in a 2-crystal geometry involving Si(111) and $^{57}Fe[BO_3](222)$ [74]. The pulsed magnetization and remagnetization were studied of $Fe[BO_3]$ crystals, and nearly harmonic oscillations, to which magnetoelastic and magnetoacoustic ones contribute, are observed [75].

The angular dependence of the anomalous transmission of polarized Mößbauer radiation through a $^{57}Fe[BO_3]$ crystal shows rocking curves of the Laue-reflected and of the Laue-transmitted beams for the pure nuclear reflection (001). The asymmetry between the intensities of the transmitted and reflected beams revealed a strongly asymmetric composition of the wave-field experiencing anomalous low resonance absorption [76]. The spectra of the nearly backward nuclear reflections (3 3 11) and (5 5 12) show the characteristic features of dynamic diffraction like those previously observed in the comparable forward reflections (111) and (222) of $^{57}Fe[BO_3]$. The antiferromagnetic pure nuclear backward reflection (3 3 11) is a promising candidate for the filtering of Mößbauer lines from white synchrotron radiation [77].

A strongly speeded-up coherent decay of a collective nuclear (Bragg scattering) excitation was observed in pure nuclear resonance diffraction of synchrotron-radiation pulses by a perfect single crystal of antiferromagnetic $^{57}Fe[BO_3]$. There appeared a superimposed quantum-beat spectrum exhibiting the interference of four hyperfine transitions. The angular dependence of the time spectrum was verified with measurements above, below, and at the exact Bragg position. The off-Bragg measurements sensitively reveal the weak quadrupole interaction in $^{57}Fe[BO_3]$ [78].

The delayed-time spectra were observed of the coherently scattered radiation following nuclear Bragg diffraction of incident synchrotron radiation pulses on $^{57}Fe[BO_3]$ crystals. Although the intention of these measurements was only to show effects such as speedup and quantum beats, the spectra also provide the possibility of a sensitive direct measure of the hyperfine interaction parameters. It is demonstrated that even with a counting rate of only 1 Hz it was possible to obtain spectra of sufficient statistics to derive precise hyperfine parameters within 1 to 2 h [79]. Speedup and quantum beats were discussed in a simple dynamic approach introducing the following assumptions: (a) no polarization mixing for the scattering, (b) pure nuclear reflection, no electronic scattering contribution, and (c) isotropic nuclear response: the amplitude for scattering $k_0 \rightarrow k_1$ is equal to $k_0 \rightarrow k_0$ or $k_1 \rightarrow k_1$. Even with these assumptions the characteristic behavior of the diffraction can be studied [80].

During pulse photo-excitation of $Fe[BO_3]$, there is observed an increase in the ferromagnetic vector and change by several degrees of its equilibrium orientation in a magnetic field. Possible mechanisms are discussed [81].

A phenomenological expression has been obtained for the anisotropy of the part of the energy of a crystal depending on the direction of the spin system at the moment of illumination.

With the help of antiferromagnetic resonance, the sign and magnitude of the field of uniaxial photo-induced magnetic anisotropy was established experimentally in Ni-doped $Fe[BO_3]$ [82]. Dynamic magnetic superstructure photo-induced by light in Ni-doped $Fe[BO_3]$ was studied magnetooptically. The direction of propagation of magnetic wave excitations was determined by the crystal structure only. An averaged direction of the ferromagnetic vector is deviated from the wave vector of oscillations. Experimental data show interactions between photo-induced Fe^{3+}, the impurity system, and the magnetic system [83].

The magnitude of the photo-induced linear optical anisotropy, arising in the basal plane of a magnetically ordered $Fe[BO_3]$ crystal, is determined by the presence of Ni impurities in the low-spin state. The results are discussed in terms of the ordering model for Jahn-Teller distortions at impurity ions due to the interaction with spin-orbit distortions at the photo-excited Fe^{3+} ions. The calculations are compared with experimental angular dependences of the photo-induced linear optical anisotropy, and the model chosen yields agreement [84]. The photo-induced phase transition in $Fe[BO_3]$ with 0.1% Ni to the modulated superstructure state owing to indirect coupling of the Ni impurities through magnetoelastic waves is also considered. An external magnetic field causes a decrease of the temperature of the phase transition and decreases the space period of the modulated state [85].

The transition temperature to the modulated state in an illuminated magnetically ordered $Fe[BO_3]$ crystal with Ni impurities decreases with increasing magnetic field strength and is dependent on its direction in the basal plane. The dependence of the magnitude of the wave vector of the modulated phase on the external magnetic field strength was studied. A mechanism was suggested for the transition to the modulated phase due to photo-induced indirect interaction of the impurity O–Ni complexes via the magnetoacoustic waves [86].

Main properties of the photo-induced dynamic structure in $Fe[BO_3]$:Ni were studied magnetooptically. The oscillations of the ferromagnetic moment follow a quasi-harmonic law. The deviation of the equilibrium direction of the ferromagnetic moment m in the structure from the direction of the applied magnetic field and the amplitude of its oscillations decrease and the cyclic frequency of the oscillations increase nonlinearly, when the magnetic field applied along the wave vector of the structure during illumination is growing. The phase velocity of the photo-induced structure increases linearly, when the intensity of the excited illumination grows. In $Fe[BO_3]$:Ni, nickel is considered to be present as Ni^{3+} ions in the crystal, which interact with photo-excited ions of the Fe^{3+} matrix [87]. The photo-induced magnetic structures of $Fe[BO_3]$:Ni are heterogeneous; the Cotton-Mouton effect is described [88]. The susceptibility and loss of $Fe[BO_3]$:Ni between 80 and 350 K show a minimum at 180 K. Photomagnetization effects and the possibility of a magnetic transition are discussed [89].

The effect of laser irradiation on the magnetic properties of $Fe[BO_3]$ doped with 0.5 wt% NiO was studied by measuring the magnetic resonance under magnetic-saturation conditions. The temperature dependence of the resonance field in the dark is the same as that of pure $Fe[BO_3]$. In the presence of laser irradiation with $\lambda = 1.06\ \mu m$, an erosion of the maximum in the temperature dependence of the resonance field takes place. The width of the resonance curve in the dark and under irradiation was ca. 140 Oe and did not change appreciable between 4 and 70 K. The results are attributed to reconstruction of light-sensitive structural complexes in these crystals [90]. Photo-induced large changes of the magnetic resonance in pure and Ho-doped $Fe[BO_3]$ single crystals under laser irradiation with $\lambda = 1.064\ \mu m$ are observed at 4 K. For both crystals the temperature dependences of the resonance field before and after irradiation coincide at ca. 170 K. The results can be interpreted by a photo-induced change of the magnetic crystallographic anisotropy in the basal plane, because of the redistribution of the photoactive centers on positions which are nonequivalent with respect to an internal magnetic field direction [91].

References on pp. 87/91

3.4.1.3.2.2 Additional Compounds

The electric properties of **$Ag_3[BO_3]$** were determined by a.c. impedance, d.c. polarization measurements with ionically blocking electrodes, and d.c. transference measurements with electronically blocking ionic conducting electrodes. $Ag_3[BO_3]$ shows predominantly electronic (semiconducting) transport with high partial conductivity of Ag^+ ions as the minority charge carriers [35].

$NaLi_2[BO_3]$ is prepared from Na_2O, Li_2O, and B_2O_3 and consists of monoclinic, colorless-transparent, and prismatic single crystals, space group $P2_1/c-C_{2h}^5$ (No. 14) with a=950.7(4), b=1203.7(4), c=493.0(3) pm; β=104.0(10)°; Z=8; D_m=2.28 g/cm³ and D_c=2.32 g/cm³; R=7.52% and R_w=4.66% for 1294 reflections. The two crystallographic independent $(BO_3)^{3-}$ groups are nonplanar with deviation angles of 1.3(9)° and 1.8(9)°, respectively; the B–O distances are 137, 138, and 140 pm. The Madelung part of the lattice energy and effective coordination numbers are calculated [17].

$KLi_2[BO_3]$ is prepared from the oxides as orthorhombic, colorless-transparent, and columnar single crystals, space group $Pnma-D_{2h}^{16}$ (No. 62) with a=797.1(4), b=643.2(3), c=645.7(3) pm; Z=4; D_m=2.20 g/cm³ and D_c=2.25 g/cm³. The crystal structure has been solved by 4-cycle diffractometry from 1129 measured and 522 symmetry independent reflections to R=5.78% and R_w=3.83%. The $(BO_3)^{3-}$ group is nearly planar with r(BO)=137.6, 137.6, and 137.9 pm. The angles ∢(OBO) are 118.7°, 118.7°, and 122.6°. It hydrolyzes in the presence of atmospheric moisture. Effective coordination numbers were calculated using mean effective ionic radii and are discussed [18].

$K_2Li[BO_3]$ is prepared as monoclinic, colorless-transparent, round single crystals, space group $C2-C_2^3$ (No. 5) with a=876.1(3), b=608.1(2), c=735.4(3) pm; β=102.57(5)°; Z=4; D_m=2.42 g/cm³ and D_c=2.50 g/cm³. The crystal structure has been solved from 1099 symmetry independent reflections to R=9.38% and R_w=5.35%. Effective coordination numbers and the Madelung part of the lattice energy are calculated via mean fictive ionic radii and discussed; r(BO)=140, 140, and 138 pm [19].

$LiMg[BO_3]$ crystallizes in the monoclinic space group $C2/c-C_{2h}^6$ (No. 15) with a=516.1(1), b=888.0(2), and c=991.1(2) pm; β=91.29(2)°; Z=8; refined to R=4.2% using 368 reflections; the lithium positions are disordered [14].

α-$LiCd[BO_3]$ was prepared between 600 and 620°C from a 1:1 mixture of $LiBO_2$ and $Cd[CO_3]$ in the solid state. At 620±10°C, the α-compound is transformed to **β-$LiCd[BO_3]$** witch melts incongruently at 867±5°C. The preparation conditions and the second harmonic generation (SHG) effect in α-$LiCd[BO_3]$ powder were examined which has a approximately threefold larger SHG effect than $[NH_4][H_2PO_4]$, but β-$LiCd[BO_3]$ has no SHG effect [15]. The α- and β-$LiCd[BO_3]$ (melting point 835°C) were found by differential thermal analysis (DTA), infrared, and X-ray phase analysis data in addition to $Li_3CdB_3O_7$ (see p. 122), obtained in the $LiBO_2$–CdO system (0 to 85 mol%). The α-compound is triclinic, space group $P\bar{1}-C_i^1$ (No. 2) with a=611.8(4), b=848.6(3), and c=525.7(2) pm; α=91.46(3)°, β=89.64(4)°, and γ=104.85(4)°; Z=4; D_m=4.50 g/cm³. The β-compound is hexagonal, space group $P6/m-C_{6h}^1$ (No. 175) with a=832.4 and c=326.4 pm; Z=3; D_m=4.46 g/cm³ [16].

$Na(UO_2)[BO_3]$ is prepared from U_3O_8, $Na_2[CO_3]$, and B_2O_3 (mole ratio 1U:2Na:11B) at 1373 K in 15 h. It is orthorhombic, space group $Pcam-D_{2h}^{11}$ (No. 57) with a=1071.2(3), b=578.0(1), and c=686.2(2) pm; Z=4; D_c=5.5 g/cm³; refined for 1478 reflections to R=7.5% and R_w=9.0%. The $(BO_3)^{3-}$ triangles join polyhedral uranyl chains through an edge and a corner; the B–O distances are 134.7(20), 138.7(20), and 138.7(20) pm; the O–B–O angles are 124(2)°, 124(2)°, and 112(1)° [22].

Sc[BO$_3$] is trigonal, space group R$\bar{3}$c–D_{3d}^6 (No. 167) with a=474.8(1) and c=1526.2(2) pm; Z=6; D_c=3.47 g/cm³ with a final R of 1.7% and R_w=2.7% for 444 reflections; the distance r(BO)=137.52(5) pm, the bond angle ∢(BOB) is 120° and the bond angle ∢(OScO) is 92.28(1)° in the calcite structure [6]. Raman spectra of an Sc[BO$_3$] single crystal were obtained, and five lines were observed. One of these lines is assigned to the A_{1g} and four to the E_g modes based on their selection rules [7].

The vaterite-type structure of **Lu[BO$_3$]** crystallizes in the rhombohedral space group P6$_3$/mmc–D_{6h}^4 (No. 194) with Z=2, the calcite-type structure of the same compound in the trigonal space group R3c–C_{3v}^6 (No. 161) with Z=6 [8].

Surface structural investigations were carried out on the (100) and (110) faces of **La[BO$_3$]** crystals grown from the PbO–B_2O_3 flux system [9].

Solid solutions **Rh$_{1-x}$M$_x$[BO$_3$]** (M=Fe, Cr, Ga, Sc, In; x=0 to 0.4) were prepared from mixtures of Rh, Rh_2O_3, M_2O_3, and $Na_3[B_3O_6]$ or $Na_2[B_4O_7]$ or H_3BO_3; they are hexagonal, of calcite-structure type, and belong to the space group R$\bar{3}$c–D_{3d}^6 (No. 167) with a between 458 and 469 pm and c between 1424 and 1485 pm. In pure **Rh[BO$_3$]**, a=461.5(3) and c=1434.9(5) pm. The solid solutions have high hardness and are semiconductors [10]. In [5] the same values for a and c were found, but the calculated density is given as 2.80 g/cm³; decomposition occurs at nearly 1000°C, and the electrical resistance at 298 K is 5×10⁴ Ω·cm.

The electronic structure of Ce^{3+} and Eu^{3+} in crystalline **YBO$_3$** was theoretically simulated. The characteristics of the single-electron states were studied in the cluster approximation. By using the spin-polarization X_α method of discrete variations, the calculations were made for the clusters $(B_3O_9)^{9-}$, $(Ce(B_3O_9)_2)^{15-}$, and $(Eu(B_3O_9)_2)^{15-}$. The transition between the valence and vacancy bands of the single-electron states determines the optical and luminescent characteristics of these compounds. The high efficiency of the given phosphors in the near-UV region is caused by the high-energy B 2s and B 2p states, vacant in the orthoborates [11]. The luminescence and energy transfer was also studied for rare earth-doped **(Y,Gd)BO$_3$** [12]. The Pr^{3+} ion is used in order to sensitize the Gd^{3+} sublattice in **GdBO$_3$** [13].

The magnetic phase diagram was calculated for **Fe$_{1-x}$Cr$_x$[BO$_3$]** based on a model of competing antisymmetric exchange and weak ferrimagnetism for different values of x [92].

A Mößbauer study was carried out on the effect of substitution of Al^{3+}, Ga^{3+}, and Cr^{3+} in ferromagnetic Fe[BO$_3$] close to the Curie temperature. The Curie temperatures, T_c, of Fe[BO$_3$], **Fe$_{0.9}$Al$_{0.1}$[BO$_3$]**, **Fe$_{0.9}$Ga$_{0.1}$[BO$_3$]**, and **Fe$_{0.9}$Cr$_{0.1}$[BO$_3$]** are 352, 317, 319, and 335 K, respectively. Anomalous spectra appear 1 K below the Curie temperature for Fe[BO$_3$], but 15 K below the Curie temperature for substituted borates. The inadequacy of static models in explaining these spectra is pointed out. These spectra are simulated by using relaxation models and estimations of activation energy obtained for spin-flip processes. The results are attributed to dominant superparamagnetic relaxation effects due to the presence of clusters. Mößbauer spectra of Fe[BO$_3$], substituted by a magnetic ion like Cr^{3+}, differ significantly from those of the other samples [93].

3.4.1.3.3 $M^I_6(BO_3)_2$, $M^{II}_3(BO_3)_2$, $M^{II}M^{IV}(BO_3)_2$, and $M^{VI}(BO_3)_2$ Compounds

CsLi$_5$(BO$_3$)$_2$ is prepared from $CsO_{0.48}$, Li_2O, and B_2O_3 (mole ratio 1.2 : 2.2 : 1) as monoclinic, colorless-transparent, and columnar single crystals, space group C2/c–C_{2h}^6 (No. 15) with a=1179.5(2), b=943.3(2), and c=809.6(3) pm; β=132.76(8)°; Z=4; D_m=2.80 g/cm³ and D_c=

References on pp. 87/91

2.86 g/cm³; refined from 1317 independent reflections to R=5.66% and R_w=3.82%. The $(BO_3)^{3-}$ group is nearly planar with a deviation of 0.7° and with B–O distances of 137, 138, and 140 pm. The compound is sensitive to atmospheric moisture [21].

The structure of $\mathbf{Ca_3[BO_3]_2}$ and of the isostructural $\mathbf{Sr_3[BO_3]_2}$ is difficult to describe using the traditional cation-centered polyhedra model. However, the structure can be described in a simple and more elegant way as anticorundum Ca_3X_2 (X = BO_3), by considering the alternative model [24] of an anion-stuffed cation array [25].

$\mathbf{Zn_3(BO_3)_2}$ is included in the major crystallization phases of B_2O_3–ZnO–SiO_2 passivation glass for semiconductor devices [26].

The low-temperature α-form and the high-temperature β-form of $\mathbf{Sr_2Cu(BO_3)_2}$ were prepared and their structures established. The α-phase crystallizes in the monoclinic space group $P2_1/c-C_{2h}^5$ (No. 14) with a=570.7(1), b=879.6(2), and c=602.7(1) pm; β=116.98(1)°; Z=2; from 1039 reflections refined to R=3.7% and R_w=5.1%. The structure is composed of sheets of isolated CuO_4 square planes that are rotated out of the b-c plane and connected by BO_3 and SrO_7 units; it is isomorphous with $Na_2Cu[CO_3]_2$. The β-phase crystallizes in the orthorhombic space group Pnma–D_{2h}^{16} (No. 62) with a=761.2(3), b=1085.4(7), and c=1350.3(4) pm; Z=8; refined from 1235 reflections to R=3.0% and R_w=3.9%. This structure is composed of isolated $Cu_2(BO_3)_4$ units which are built from CuO_4 distorted square planes and triangular $(BO_3)^{3-}$ groups. These units are bridged by three strontium ions [27].

$\mathbf{BaSn(BO_3)_2}$ was prepared from $Ba[CO_3]$, SnO_2, and B_2O_3 between 500 and 1150°C and sintering between 1000 and 1150°C; the maximal electric resistivity of the sintered product is 7.3×10^{14} Ω·cm [28].

$\mathbf{U(BO_3)_2}$ is prepared from U_3O_8, $Sr[CO_3]$, and B_2O_3 (mole ratio 1U:1Sr:12B) at 1423 K in 15 h. It is monoclinic, space group $C2/c-C_{2h}^6$ (No. 15) with a=1250.4(3), b=418.3(1), and c= 1045.3(3) pm; β=122.18(3)°; Z=4; D_c=5.10 g/cm³; refined to R=6.6% and R_w=8.2% from 1707 contributing reflections. The $(BO_3)^{3-}$ triangles lie in chains between the UO_6 octahedra arrays; the B–O distances are 132.2(14), 137.7(10), and 139.3(8) pm; the O–B–O angles are 128.5(6)°, 117.1(7)°, and 114.1(7)° [23].

3.4.1.3.4 $\mathbf{M_9^I[BO_3]_3}$ and $\mathbf{M_3^{II}M^{III}[BO_3]_3}$ Compounds

$\mathbf{Na_4Li_5[BO_3]_3}$ is prepared from Na_2O, Li_2O, and B_2O_3 (mole ratio 4.4 : 5.5 : 3) as monoclinic, colorless-transparent, and prismatic single crystals, space group $C2/c-C_{2h}^6$ (No. 15) with a= 1238.8(5), b=729.6(3), and c=973.8(3) pm; β=107.29(4)°; Z=4; D_m=2.37 g/cm³ and D_c= 2.397 g/cm³; refined from 1221 independent reflections to R=9.02% and R_w=5.29%. The borate anions correspond to two crystallographic independent, planar $(BO_3)^{3-}$ groups with B–O distances of 134.6, 138.7, and 139.9, or 136.7, 136.8, and 136.6 pm, respectively. The compound is sensitive to atmospheric moisture; hydrolysis of powdered samples takes one hour [20].

$\mathbf{Sr_3Sc[BO_3]_3}$ was prepared and structurally characterized. It crystallizes in the rhombohedral space group $R\bar{3}-C_{3i}^2$ (No. 148) with a=1213.5(1) and c=918.4(1) pm. Two crystallographically independent Sc^{3+} ions occupy trigonally elongated octahedral sites. When doped with Cr^{3+}, it exhibits a luminescence band centered at 750 nm [29].

3.4.1.3.5 $M^{II}_3M^{III}_2[BO_3]_4$ and $M^{III}_4[BO_3]_4$ Compounds

$Ba_3Nd_2[BO_3]_4$ was prepared from Nd_2O_3, BaO, B_2O_3, and BaF_2 at 1230°C (9 h), and slowly cooling to 850°C. It is orthorhombic, space group Pnma–D^{16}_{2h} (No. 62) with a=771.43(11), b=1677.90 (34), and c=894.78(14) pm; Z=4; refined from 2935 reflections to R=7.15% and R_w=7.93%. The basic structure is composed of three sets of planar $[BO_3]^{3-}$ triangles which are not linked to one another but bridge two types of MO_8 dodecahedra (randomly distributed; M=Ba, Nd). There is also one isolated NdO_8 polyhedron (see **Fig. 3-8**). The B–O distances are 135, 137, 137; 136, 136, 137; 135, 135, and 126 pm. Crystals of laser quality have been grown [32].

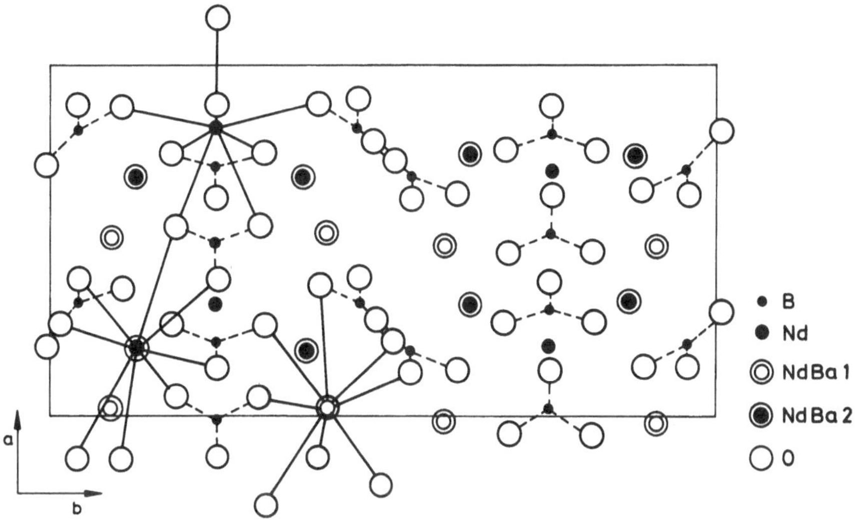

Fig. 3-8. a-b Planar projection of crystal structure of $Ba_3Nd_2[BO_3]_4$ [32].

Crystals of laser quality have been grown from **$Er_xY_{1-x}Al_3[BO_3]_4$** [102, 103]. For x=0.6 the material crystallizes in huntite-type space group R32–D^7_3 (No. 155) with a=926 and c=718 pm, and decomposes at 1279°C [102].

The Raman spectrum of monoclinic **$NdAl_3[BO_3]_4$** (NAB) crystals was recorded; a factor group analysis of the space group showed that there are 25 A_g and 26 B_g Raman-active modes, and 34 A_u and 35 B_u infrared-active modes [109]. Optical phonons of a crystal of the compound were studied by Raman spectroscopy and the fluorescence quenching mechanism is discussed. The compound has less quenching than any other host [110]. Crystals of laser quality have been grown [94, 96 to 101]. The pulsed-laser performance of the $NdAl_3[BO_3]_4$ crystal, with a maximal energy output of 422 mJ when it was pumped with a single xenon flashlamp, is reported [101]. The laser threshold was 67 mJ, the laser pulse width was 8 ns, and the output laser beam was linearly polarized with a beam divergence angle of ca. 2 mrad and a wavelength of 1063 nm [111]. The maximal laser output of a rod of 3 mm×18 mm size was determined to be 1194 mJ per pulse, when it was pumped with a xenon flashlamp of 3 mm×25 mm. The overall and slope efficiencies are 1.51% and 1.6%, respectively [112].

Growth of **$Nd_xY_{1-x}Al_3[BO_3]_4$** (NYAB) crystals of laser quality is reported in [105 to 108]. Performance of emissions at 1063 and 532 nm is presented for an $Nd_xY_{1-x}Al_3[BO_3]_4$ crystal for flash-lamp pumped (pulsed) as well as dye-laser and diode-laser pumped continuous-wave

References on pp. 87/91

emissions [107]. The oscillation threshold was measured to be 78 mJ, the pulse duration 100 ns, the laser beam polarization 90%, and the laser beam divergence 2 mrad [108]. Using dye-laser as pumping light source, excited emission and the laser self-frequency-doubling effect of $Nd_xY_{1-x}Al_3[BO_3]_4$ from 1060 to 530 nm was obtained [106]; the threshold energy is 4 mJ [106] or below 2 mJ [113], the conversion efficiency is 6.2% [106] or above 10% [113], and the self-frequency-doubling green light output energy is 2.5 mJ [106] or 5 mJ [113]. The title compound is a new excited emission nonlinear multifunctional rare earth borate crystal. The effective nonlinear coefficient of type-I phase matching is fourfold that of the $K[H_2PO_4]$ crystal, proving that phase-matching conditions are satisfied. The measured value obtained by using the second harmonic generation is in good agreement with the value cited above. Using a tunable dye laser, the threshold energy of the pump is 2 mJ, the maximal conversion efficiency is 14.3%, and the output energy of green light at 530 nm is 5 mJ [114].

$\mathbf{Eu_xY_{1-x}Al_3[BO_3]_4}$ crystals alter their crystal parameters from x = 0.1 with a = 928.33(7) and c = 723.31(6) pm (refined to R = 2.2% from 687 reflections) to x = 0.5 with a = 929.45(7) and c = 724.91(7) pm (refined to R = 3.2% from 677 reflections) [33]. For x = 0.1 the polarized absorption spectrum of the huntite-type crystals with the space group R32−D_3^7 (No. 155) was analyzed between 16000 and 24300 cm^{-1}; it shows three σ $^5D_1 \leftarrow {}^7F_1$ transitions [115]. The $^5D_1 \leftarrow {}^7F_0$ transition appears as a positive A_1 signal, whereas the $^5D_2 \leftarrow {}^7F_0$ transition corresponds to two A_1 terms with opposite sign [116]. A crystal field analysis of the D_3 site symmetry groups provided information about the energy levels and corresponding wave functions. Absorption, fluorescence, and magnetic circular dichroism (CD) spectra are reported, and special emphasis is placed on the proof of this technique for assessing the intensity mechanisms responsible for the transitions between energy levels [33]. The polarized absorption ($^7F_6 \rightarrow {}^7F_n$) and fluorescence spectra ($^5D_4 \rightarrow {}^7F_n$) of $\mathbf{TbAl_3[BO_3]_4}$ are also given; a crystal field fitting has been performed, the overall symmetry being lowered from D_{3h} to D_3 by a distortion angle of 8.14° [117].

$\mathbf{EuAl_3[BO_3]_4}$ (EuAB) shows thermally activated energy migration among the $^5D_0$ excited level down to the lowest temperatures, whereas in $EuMg[B_5O_{10}]$ one-dimensional migration was observed. For short Eu-Eu distances and low temperatures, the $Eu^{3+} \rightarrow Eu^{3+}$ transfer occurs by exchange [118]. The energy migration has been observed also for $EuAl_3[BO_3]_4$, activated with Tb^{3+}, Gd^{3+}, and Y^{3+} [119], and $\mathbf{Gd_xY_{1-x}Al_3[BO_3]_4}$, activated with Eu^{3+}, Dy^{3+}, and Cr^{3+} [120].

Additional rare-earth compounds of the type $\mathbf{REAl_3[BO_3]_4}$ have been used in the growth of crystals of laser quality [104].

$(BO_3)^{3-}$ structures are also found in Nd-Al-Ba borate (NABB) samples; crystals for lasers have been grown [95].

For compounds of the types $\mathbf{MSc_3[BO_3]_4}$ with M = Ce, Pr, Nd, Sm, diffractograms and IR spectra of the powders (melted at 1100°C) are given, also some IR data at 300 K. The materials seem to crystallize in the trigonal space group R32−D_3^7 (No. 155) [30].

In the system $Sc[BO_3]$-$Eu[BO_3]$, the compound $\mathbf{EuSc_3[BO_3]_4}$ of the huntite structure type, space group R32−D_3^7 (No. 155), and solid solution regions based on $Sc[BO_3]$ ($Eu[BO_3]$ up to 15 mol%) and on $Eu[BO_3]$ ($Sc[BO_3]$ up to 30 mol%) were found [31].

3.4.1.3.6 $Al_6[BO_3]_5F_3$

$Al_6[BO_3]_5F_3$ was prepared from $2Al_2O_3 \cdot B_2O_3 \cdot 5H_2O$ and B_2O_3 (mole ratio 1:3) in the presence of a fluorine-containing mineralizer between 555 and 665 °C and 800 to 1000 bar. It is hexagonal, space group $P6_3/m-C_{6h}^2$ (No. 176) with a = 855.2(4), c = 818.7(3) pm; Z = 12; D_c = 3.28 g/cm³; R = 3.88%. The fluorine atoms are located at the vertices of the aluminium octahedra and are not bound to the $[BO_3]^{3-}$ triangles [34].

References for 3.4.1.3:

[1] Tossell, J. A. (Am. Mineral. **71** [1986] 1170/7).
[2] Tossell, J. A.; Lazzeretti, P. (J. Non-Cryst. Solids **99** [1988] 267/75).
[3] Hotokka, M.; Pyykkö, P. (Chem. Phys. Lett. **157** [1989] 415/8).
[4] Jubert, A. H.; Varetti, E. L. (An. Quim. A **82** [1986] 227/30 from C.A. **106** [1987] No. 108266).
[5] Shaplygin, I. S.; Prosychev, I. I.; Lazarev, V. B. (Zh. Neorg. Khim. **31** [1986] 2870/5; Russ. J. Inorg. Chem. **31** [1986] 1649/52).
[6] Keszler, D. A.; Sun, Hongxing (Acta Crystallogr. C **44** [1988] 1505/7).
[7] Bogachev, G.; Iliev, M.; Petrov, V. (Phys. Status Solidi B **152** [1989] K29/K31).
[8] Hölsä, J. (Inorg. Chim. Acta **139** [1987] 257/9).
[9] Kotru, P. N.; Jain, A.; Razdan, A. K.; Wanklyn, B. M. (J. Mater. Sci. **24** [1989] 1413/20 from C.A. **111** [1989] No. 15538).
[10] Shaplygin, I. S.; Lazarev, V. B.; Varlamov, N. V. (Zh. Neorg. Khim. **33** [1988] 2217/20; Russ. J. Inorg. Chem. **33** [1988] 1266/8).

[11] Ryzhkov, M. V.; Dotsenko, V. P.; Efryushina, N. P.; Gubanov, V. A. (Opt. Spektrosk. **67** [1989] 988/90 from C.A. **112** [1990] No. 44740).
[12] Sobieraj, M. T. (Diss. Polytech. Univ., Brooklyn, New York 1989, pp. 1/172 from C.A. **112** [1990] No. 44724).
[13] Srivastava, A. M.; Sobieraj, M. T.; Ruan, S. K.; Banks, E. (Mater. Res. Bull. **21** [1986] 1455/63 from C.A. **106** [1987] No. 92855).
[14] Norrestam, R. (Z. Kristallogr. **187** [1989] 103/10).
[15] Yin, X.-D.; Huang, Q.-Z.; Ye, Sh.-S.; Lei, Sh.-R.; Chen, Ch.-T. (Huaxue Xuebao **43** [1985] 822/6 from C.A. **104** [1986] No. 60945).
[16] Buludov, N. T.; Kara'ev, Z. Sh.; Abdulla'ev, G. K. (Zh. Neorg. Khim. **30** [1985] 1523/6; Russ. J. Inorg. Chem. **30** [1985] 868/70); Sokolova, E. V.; Boronikhin, V. A.; Smirnov, G. V.; Belov, N. V. (Dokl. Akad. Nauk SSSR **246** [1979] 1126/9).
[17] Miessen, M.; Hoppe, R. (Z. Anorg. Allg. Chem. **545** [1987] 157/68).
[18] Miessen, M.; Hoppe, R. (Z. Anorg. Allg. Chem. **521** [1985] 7/14).
[19] Miessen, M.; Hoppe, R. (Rev. Chim. Miner. **22** [1985] 331/43).
[20] Miessen, M.; Hoppe, R. (Z. Anorg. Allg. Chem. **536** [1986] 101/13).

[21] Miessen, M.; Hoppe, R. (Z. Anorg. Allg. Chem. **536** [1986] 92/100).
[22] Gasperin, M. (Acta Crystallogr. C **44** [1988] 415/6).
[23] Gasperin, M. (Acta Crystallogr. C **43** [1987] 2031/3).
[24] White, T. J.; Hyde, B. G. (Phys. Chem. Miner. **8** [1982] 55/63; Acta Crystallogr. B **39** [1983] 10/7).
[25] Vegas, A. (Acta Crystallogr. C **41** [1985] 1689/90).
[26] Zhou, S.; Zhou, P.; Zhang, L. (Huadong Huagong Xueyuan Xuebao **1984** No. 4, pp. 455/62 from C.A. **103** [1985] No. 41199).
[27] Smith, R. W.; Keszler, D. A. (J. Solid State Chem. **81** [1989] 305/13).

[28] Iizumi, K.; Mochizuki, M.; Kudaka, K. (Nippon Kagaku Kaishi **1988** 2048/50 from C.A. **110** [1989] No. 43541).
[29] Thompson, P. D.; Keszler, D. A. (Chem. Mater. **1** [1989] 292/4 from C.A. **110** [1989] No. 224358).
[30] Magunov, I. R.; Voevudskaya, S. V.; Zhirnova, A. P.; Zhikhareva, E. A.; Efryushina, N. P. (Izv. Akad. Nauk SSSR Neorg. Mater. **21** [1985] 1532/4; Inorg. Mater. [USSR] **21** [1985] 1532/4).

[31] Efryushina, N. P.; Zhikhareva, E. A.; Magunov, I. R.; Zakolodyazhnaya, O. V.; Magunov, R. L. (Dopov. Akad. Nauk Ukr. RSR B **1986** No. 5, pp. 36/8 from C.A. **105** [1986] No. 85882).
[32] Yan, J. F.; Hong, H. Y.-P. (Mater. Res. Bull. **22** [1987] 1347/53).
[33] Görller-Walrand, C.; Vandevelde, P.; Hendrickx, I.; Porcher, P.; Krupa, J. C.; King, G. S. D. (Inorg. Chim. Acta **143** [1988] 259/70).
[34] Sokolova, E. V.; Egorov-Tismenko, Yu. K.; Kargal'tsev, S. V.; Klyakhin, V. A.; Urnsov, V. S. (Vestn. Mosk. Univ. Geol. **1987** No. 3, pp. 82/4 from C.A. **108** [1988] No. 67644).
[35] Koehler, B. U.; Jansen, M.; Weppner, W. (J. Solid State Chem. **57** [1985] 227/33).
[36] Kalinskaya, T. V.; Lobanova, L. B.; Ka'ekina, M. L. (Zh. Neorg. Khim. **30** [1985] 1513/8; Russ. J. Inorg. Chem. **30** [1985] 863/6).
[37] Kotrbova, M.; Kadeckova, S.; Novak, J.; Bradler, J.; Smirnov, G. V.; Shvyd'ko, Yu. V. (J. Cryst. Growth **71** [1985] 607/14).
[38] Petrakovskii, G. A.; Rudenko, V. V.; Sosnin, V. M.; Stepanov, G. N. (Fiz. Svoistva Magnitodielektrikov, Krasnoyarsk **1987** 126/32 from C.A. **109** [1988] No. 15040; Acta Phys. Hung. **61** [1987] 243/6 from C.A. **107** [1987] No. 209127).
[39] Hiroshima, K.; Kajima, A.; Fujii, T. (Nippon Oyo Jiki Gakkaishi **11** [1987] 267/70 from C.A. **107** [1987] No. 146094).
[40] Pankrats, A. I.; Petrakovskii, G. A.; Smyk, A. F. (Solid State Commun. **59** [1986] 657/60; Dokl. Akad. Nauk SSSR **294** [1987] 1097/101).

[41] Pankrats, A. I.; Smyk, A. F. (Fiz. Svoistva Magnitodielektrikov, Krasnoyarsk **1987** 114/25 from C.A. **109** [1988] No. 31047).
[42] Pankrats, A. I.; Yus'kiv, I. S.; Vasil'ev, A. A. (Fiz. Tverd. Tela [Leningrad] **27** [1985] 2784/6 from C.A. **103** [1985] No. 188233).
[43] Bichurin, M. I.; Petrov, V. M. (Fiz. Tverd. Tela [Leningrad] **29** [1987] 2509/10 from C.A. **107** [1987] No. 227704).
[44] Petrov, M. P.; Paugurt, A. P.; Pleshakov, I. V.; Ivanov, A. V. (Pis'ma Zh. Tekh. Fiz. **13** [1987] 193/6 from C.A. **106** [1987] No. 206514).
[45] Ivanov, A. V.; Korneev, V. R.; Paugurt, A. P.; Pleshakov, I. V. (Pis'ma Zh. Tekh. Fiz. **14** [1988] 2049/52 from C.A. **110** [1989] No. 106833).
[46] Paugurt, A. P.; Pleshakov, I. V.; Ivanov, A. V. (Fiz. Tverd. Tela [Leningrad] **29** [1987] 2959/65 from C.A. **108** [1988] No. 160062).
[47] Petrov, M. P.; Ivanov, A. V.; Paugurt, A. P.; Pleshakov, I. V. (Fiz. Tverd. Tela [Leningrad] **29** [1987] 1819/25 from C.A. **107** [1987] No. 125678).
[48] Petrov, M. P.; Paugurt, A. P.; Pleshakov, I. V.; Ivanov, A. V. (Pis'ma Zh. Tekh. Fiz. **11** [1985] 1204/7 from C.A. **104** [1986] No. 27663).
[49] Paugurt, A. P.; Pleshakov, I. V.; Khomchenkov, I. M.; Ivanov, A. V. (Pis'ma Zh. Tekh. Fiz. **13** [1987] 587/90 from C.A. **107** [1987] No. 189288).
[50] Bogdanova, Kh. G.; Bagautdinov, R. A.; Golenishchev-Kutuzov, V. A.; Enikeeva, G. R.; Medvedev, L. I. (Pis'ma Zh. Eksp. Teor. Fiz. **44** [1986] 219/21 from C.A. **105** [1986] No. 201826).

[51] Bagautdinov, R. A.; Bogdanova, Kh. G.; Golenishchev-Kutuzov, V. A.; Enikeeva, G. R.; Medvedev, L. I. (Fiz. Tverd. Tela [Leningrad] **28** [1986] 924/6 from C.A. **104** [1986] No. 198622).
[52] Bogdanova, Kh. G.; Golenishchev-Kutuzov, V. A.; Medvedev, L. I.; Kurkin, M. I.; Turov, E. A. (Zh. Eksp. Teor. Fiz. **95** [1989] 613/20 from C.A. **110** [1989] No. 204419).
[53] Chetkin, M. V.; Shcherbakov, Yu. I.; Gadetskii, S. N.; Tereshchenko, V. D. (Zh. Tekh. Fiz. **55** [1985] 207/9 from C.A. **102** [1985] No. 141999).
[54] Chetkin, M. V.; Tereshchenko, V. D. (Kristallografiya **33** [1988] 1311/3; Sov. Phys. Crystallogr. **33** [1988] 781/3 from C.A. **109** [1988] No. 221158).
[55] Shamsutdinov, M. A.; Farztdinov, M. M.; Baitimerov, I. R. (Fiz. Tverd. Tela [Leningrad] **27** [1985] 3450/2 from C.A. **104** [1986] No. 44796).
[56] Prokopov, A. P.; Seleznev, V. N.; Strugatskii, M. B.; Yagupov, S. V. (Zh. Tekh. Fiz. **57** [1987] 2051/3 from C.A. **108** [1988] No. 141842).
[57] Zubov, V. E.; Krinchik, G. S.; Seleznev, V. N.; Strugatskii, M. B. (Zh. Eksp. Teor. Fiz. **94** [1988] 290/300 from C.A. **110** [1989] No. 49900).
[58] Zubov, V. E.; Krinchik, G. S.; Seleznev, V. N.; Strugatskii, M. B. (Fiz. Tverd. Tela [Leningrad] **31** [1989] 273/5 from C.A. **111** [1989] No. 145474).
[59] Zabluda, V. N.; Malakhovskii, A. V.; Edel'man, I. S. (Fiz. Tverd. Tela [Leningrad] **27** [1985] 133/9 from C.A. **102** [1985] No. 175228).
[60] Druzhinin, V. V.; Pavlovskii, A. I.; Pisarev, R. V.; Tatsenko, O. M.; Platonov, V. V. (Pis'ma Zh. Eksp. Teor. Fiz. **43** [1986] 282/4 from C.A. **104** [1986] No. 195746).

[61] Zvezdin, A. K.; Mukhin, A. A. (Kratk. Soobshch. Fiz. **1988** No. 5, pp. 20/2 from C.A. **109** [1988] No. 221159).
[62] Kotyuzhanskii, B. Ya.; Prozorova, L. A.; Svistov, L. E. (Zh. Eksp. Teor. Fiz. **93** [1987] 1140/50 from C.A. **107** [1987] No. 248682).
[63] Kotyuzhanskii, B. Ya.; Prozorova, L. A.; Svistov, L. E. (Zh. Eksp. Teor. Fiz. **92** [1987] 238/47 from C.A. **106** [1987] No. 148085).
[64] Andrienko, A. V.; Podd'yakov, L. V. (Zh. Eksp. Teor. Fiz. **95** [1989] 2117/24 from C.A. **111** [1989] No. 145544).
[65] Kvardakov, V. V.; Somenkov, V. A.; Tyugin, A. B. (Pis'ma Zh. Eksp. Teor. Fiz. **48** [1988] 396/8 from C.A. **109** [1988] No. 242728).
[66] Zelepukhin, M. V.; Kvardakov, V. V.; Somenkov, V. A.; Shil'shtein, S. Sh. (Zh. Eksp. Teor. Fiz. **95** [1989] 1530/6 from C.A. **111** [1989] No. 107625).
[67] Kvardakov, V. V.; Somenkov, V. A. (Fiz. Tverd. Tela [Leningrad] **31** [1989] 235/7 from C.A. **111** [1989] No. 16495).
[68] Kopcewicz, M.; Engelmann, H.; Stenger, S.; Smirnov, G. V.; Gonser, U.; Wagner, H. G. (Appl. Phys. A **44** [1987] 131/4).
[69] Smirnov, G. V.; Shvyd'ko, Yu. V.; Van Buerck, U.; Mößbauer, R. L. (Phys. Status Solidi B **134** [1986] 465/75).
[70] Smirnov, G. V.; Zelepukhin, M. V.; Van Buerck, U. (Pis'ma Zh. Eksp. Teor. Fiz. **43** [1986] 274/7 from C.A. **105** [1986] No. 32347).

[71] Van Buerck, U.; Smirnov, G. V.; Maurus, H. J.; Mößbauer, R. L. (J. Phys. C **19** [1986] 2557/66).
[72] Van Buerck, U.; Smirnov, G. V.; Maurus, H. J.; Mößbauer, R. L. (J. Phys. C **19** [1986] 2567/73).
[73] Smirnov, G. V.; Van Buerck, U. (Hyperfine Interact. **27** [1986] 203/19).
[74] Smirnov, G. V.; Shvyd'ko, Yu. V. (Pis'ma Zh. Eksp. Teor. Fiz. **44** [1986] 431/5; JETP Lett. **44** [1986] 556/61).

[75] Kolotov, O. S.; Pogozhev, V. A.; Smirnov, G. V.; Shvyd'ko, Yu. V. (Fiz. Tverd. Tela [Leningrad] **29** [1987] 2548/9 from C.A. **107** [1987] No. 227632).
[76] Smirnov, G. V.; Van Buerck, U.; Mößbauer, R. L. (J. Phys. C **21** [1988] 5835/42).
[77] Van Buerck, U.; Smirnov, G. V.; Mößbauer, R. L. (J. Phys. C **21** [1988] 5843/51).
[78] Van Buerck, U.; Mößbauer, R. L.; Gerdau, E.; Rueffer, R.; Hollatz, R.; Smirnov, G. V.; Hannon, J. P. (Phys. Rev. Lett. **59** [1987] 355/8).
[79] Hollatz, R.; Rueffer, R.; Gerdau, E. (Hyperfine Interact. **42** [1988] 1141/4).
[80] Rueffer, R.; Hollatz, R.; Gerdau, E.; Van Buerck, U.; Hannon, J. P. (Hyperfine Interact. **42** [1987] 1161/4).

[81] Borowiec, M.; Garmonov, A. A.; Rudov, S. G.; Fedorov, Yu. M. (Pis'ma Zh. Eksp. Teor. Fiz. **50** [1989] 431/3 from C.A. **112** [1990] No. 47327).
[82] Fedorov, Yu. M.; Pankrats, A. I.; Leksikov, A. A.; Seleznev, V. N.; Prokopov, A. R. (Fiz. Tverd. Tela [Leningrad] **27** [1985] 289/91 from C.A. **102** [1985] No. 104750).
[83] Fedorov, Yu. M.; Leksikov, A. A.; Vorotynova, O. V. (Solid State Commun. **55** [1985] 987/9; Phys. Lett. A **123** [1987] 145/7).
[84] Fedorov, Yu. M.; Leksikov, A. A.; Aksenov, A. E. (Zh. Eksp. Teor. Fiz. **89** [1985] 2099/112 from C.A. **104** [1986] No. 77945).
[85] Sadreev, A. F.; Fedorov, Yu. M. (Phys. Lett. A **123** [1987] 148/50).
[86] Fedorov, Yu. M.; Sadreev, A. F.; Leksikov, A. A. (Zh. Eksp. Teor. Fiz. **93** [1987] 2247/56 from C.A. **108** [1988] No. 139865).
[87] Fedorov, Yu. M.; Leksikov, A. A.; Vorotynova, O. V. (J. Magn. Magn. Mater. **68** [1987] 383/90 from C.A. **108** [1988] No. 67418).
[88] Fedorov, Yu. M.; Vorotynova, O. V.; Leksikov, A. A. (Fiz. Tverd. Tela [Leningrad] **31** No. 5 [1989] 192/7 from C.A. **111** [1989] No. 122983).
[89] Chzhan, A. V.; Fedorov, Yu. M.; Isaeva, T. I. (Phys. Status Solidi A **115** [1989] K101/K103 from C.A. **112** [1990] No. 15361).
[90] Petrakovskii, G. A.; Patrin, G. S.; Volkov, N. V. (Phys. Status Solidi A **87** [1985] K153/K156).

[91] Patrin, G. S.; Petrakovskii, G. A.; Rudenko, V. V. (Phys. Status Solidi A **99** [1987] 619/23).
[92] Moskvin, A. S.; Vigura, M. A. (Fiz. Tverd. Tela [Leningrad] **28** [1986] 2259/61 from C.A. **105** [1986] No. 125611).
[93] Vithal, M.; Jagannathan, R. (J. Solid State Chem. **63** [1986] 16/22).
[94] Thirumavalavan, M.; Kumar, J.; Gnanam, F. D.; Ramasamy, P. (J. Mater. Sci. Lett. **6** [1987] 1241/2).
[95] Kumar, J.; Thirumavalavan, M.; Gnanam, F. D.; Ramasamy, P. (Cryst. Res. Technol. **23** [1988] 1337/41 from C.A. **110** [1989] No. 163749).
[96] Chen, Changkang; Zhang, G.-F. (Guisuanyan Xuebao **14** [1986] 219/25 from C.A. **105** [1986] No. 200682).
[97] Chen, Changkang (Huaxue Xuebao **44** [1986] 775/80 from C.A. **105** [1986] No. 162425).
[98] Chen, Changkang (J. Cryst. Growth **89** [1988] 295/33).
[99] Oishi, S.; Mori, M.; Nishikawa, C.; Tate, I. (Chem. Express **2** [1987] 5/8 from C.A. **106** [1987] No. 93809).
[100] Oishi, S.; Nishikawa, C.; Tate, I. (Chem. Express **2** [1987] 647/50 from C.A. **107** [1987] No. 246958).

[101] Luo, Zundu; Jiang, Aidong; Huang, Yichuan; Qiu, Minwang (Chin. Phys. Lett. **3** [1986] 541/4 from C.A. **106** [1987] No. 165576).

[102] Yu, Y.-Q.; Zhang, X.-M.; Lei, M.; Liu, Sh.-Z. (Zhongguo Jiguang **14** [1987] 176/8 from C.A. **107** [1987] No. 165842).
[103] Liu, Mingguo; Lu, Baosheng; Pan, Hengfu; Song, Qiang (Guangxue Xuebao **8** [1988] 1079/84 from C.A. **111** [1989] No. 68138).
[104] Belokeneva, E. L.; Leonyuk, N. I.; Pashkova, A. V.; Timchenko, T. I. (Kristallografiya **33** [1988] 1287/8; Sov. Phys. Crystallogr. **33** [1988] 765/7).
[105] Gulyaev, Yu. V.; Ivanov, S. N.; Kotelyanskii, I. M.; Leonyuk, N. I.; Makletsov, A. N.; Medved, V. V.; Potkin, L. I.; Timchenko, T. I. (Pis'ma Zh. Tekhn. Fiz. **12** [1986] 18/21 from C.A. **104** [1986] No. 119184).
[106] Lu, Baosheng; Wang, Jun; Pan, Hengfu; Jiang, Minhua; Liu, E.-Q.; Hou, X.-Y. (Chin. Phys. Lett. **3** [1986] 413/6 from C.A. **107** [1987] No. 48993).
[107] Luo, Zundu; Lin, J.-T.; Jiang, Aidong; Juang, Yichuan; Qiu, Minwang (Proc. SPIE-Int. Soc. Opt. Eng. No. 1104 [1989] 132/41 from C.A. **112** [1990] No. 27686).
[108] Luo, Zundu; Jiang, Aidong; Huang, Yichuan; Qiu, Minwang (Chin. Phys. Lett. **6** [1989] 440/3 from C.A. **112** [1990] No. 45097).
[109] Ha, Kim; Shi, Shan (Chin. Phys. Lett. **5** [1988] 1/4 from C.A. **108** [1988] No. 176144).
[110] Ha, Kim; Shi, Shan (J. Lumin. **40/41** [1988] 698/9 from C.A. **108** [1988] No. 176426).

[111] Huang, Yichuan; Qiu, Minwang; Chen, Guang; Chen, Jiming; Luo, Zundu (Zhongguo Jiguang **14** [1987] 524/8 from C.A. **108** [1988] No. 84844).
[112] Qiu, Minwang; Huang, Yichuan; Chen, Guang; Chen, Jinhua; Chen, Jiming; Luo, Zundu (Rengong Jingti **16** [1987] 201/5 from C.A. **111** [1989] No. 204875).
[113] Lu, Baosheng; Wang, Jun; Pan, Hengfu; Jiang, Minhua (Rengong Jingti **16** [1987] 195/200 from C.A. **112** [1990] No. 87549).
[114] Liu, Enquan; Hou, Xueyuan; Lu, Baosheng; Wang, Jun; Pan, Hengfu; Jiang, Minhua (Guangxue Xuebao **7** [1987] 139/43 from C.A. **108** [1988] No. 29023).
[115] Görller-Walrand, C.; Vandevelde, P. (Chem. Phys. Lett. **122** [1985] 276/9).
[116] Görller-Walrand, C.; Vandevelde, P. (Bull. Soc. Chim. Belg. **94** [1985] 873/82 from C.A. **104** [1986] No. 158364).
[117] Görller-Walrand, C.; Vandevelde, P.; Hendrickx, I.; Porcher, P.; Krupa, J. C. (Inorg. Chim. Acta **139** [1987] 277/9).
[118] Berdowski, P. A. M.; Buijs, M.; Blasse, G. (J. Phys. Colloq. [Paris] **46** [1985] C7-31/ C7-34 from C.A. **104** [1986] No. 98615).
[119] Blasse, G. (Recl. J. R. Neth. Chem. Soc. **105** [1986] 143/9 from C.A. **105** [1986] No. 32257).
[120] De Vries, A. J.; Minks, B. P.; Blasse, G. (J. Lumin. **39** [1988] 153/60 from C.A. **108** [1988] No. 195081).

3.4.1.4 The Ion $[BO_4]^{5-}$ and Structures Containing $(BO_4)^{5-}$ Units

The IR spectra of ^{24}MgAl[^{10}BO$_4$], ^{24}MgAl[^{11}BO$_4$], ^{26}MgAl[^{10}BO$_4$], and ^{26}MgAl[^{11}BO$_4$] show stretching vibrations of the BO_4 group for ν_1 at 800 (to 780) and for ν_3 at 1100 (to 930) cm^{-1}, but no significant ^{10}B-^{11}B isotopic shift is observed at lower wavenumbers. Thus, the identification of the isolated BO_4 bending vibrations is impossible; these vibrations are probably mixed with lattice vibrations. Several translational modes of the Mg^{2+} ions were identified in the 300 to 500 cm^{-1} region with the help of the ^{24}Mg-^{26}Mg isotopic shift. The large splitting of the ν_3 vibration is discussed in connection with the vibrational behavior and the structure of olivine-type compounds [1].

References on pp. 92/3

$MgFe[BO_4]$ and **$MgFe_{0.9}Ga_{0.1}[BO_4]$** single crystals were grown by using the flux method [2]. The sign of the quadrupole splitting Δ is positive, a probability distribution $P(\Delta)$ was determined (mean value $\bar{\Delta}=0.83$ mm/s at 30 K) and the axis of the mean electric field gradient (asymmetry $\bar{\eta}=0.58\pm0.05$) lies preferentially in the crystallographic a-b plane. The quasi-one-dimensional insulator compound $MgFe[BO_4]$ has been studied by Mößbauer powder and single-crystal spectra. The compound is orthorhombic, space group Pnam$-V_h^{16}$ (?) with a=925.8, b=942.7, and c=310.4 pm and, according to the degree of crystallographic site inversion, should be described as $(Fe_{1-x}Mg_x)(Fe_xMg_{1-x})[BO_4]$ with $x\approx0.15$ [3].

The magnetic structure, the spin correlation functions, and the spin dynamics were studied in $MgFe[BO_4]$ spin glass by elastic neutron scattering. Long-range order is observed, and the Néel point is given along with the spin-glass transition temperatures [4]. The spin relaxation process in the frustrated quasi-1D spin-glass $MgFe[BO_4]$ in a large time domain (between 6×10^{-13} and 10^{-9} s) is interpreted in terms of an Edwards-Anderson-type relaxation order parameter q. The freezing temperature does not correspond to the appearance of q, but to the freezing of the low-frequency spin fluctuations [5].

$Nb[BO_4]$ has been obtained by reaction of H_3BO_3 and Nb_2O_5 (2:1) in a boron nitride crucible at pressures above 25 kbar and temperatures above 800°C. It is tetragonal, space group $I4_1/amd-D_{4h}^{19}$ (No. 141) with a=621.41(6), and c=547.6(1) pm; Z=4; $D_c=5.27$ g/cm³. It has a zircon ($ZrSiO_4$) structure with the following distances (in pm): r(BO)=146.9(2), r(OO)=222.2(3), 235.6(3), 248.4(3), 266.3(3), and 293.8(3); see **Fig. 3-9** [6].

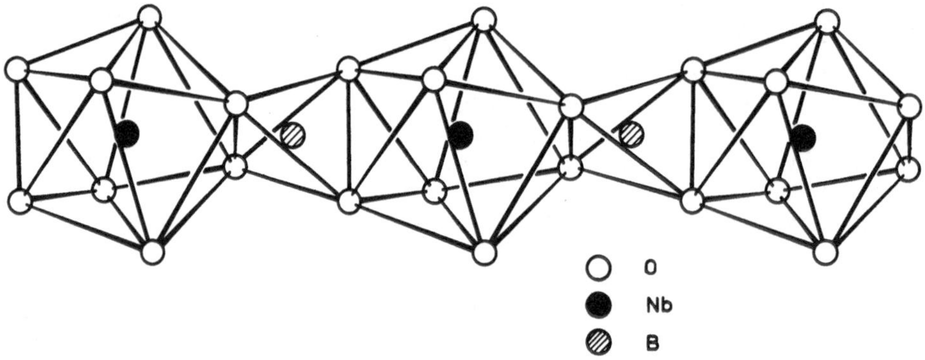

Fig. 3-9. Edge-linked NbO_8 dodecahedra and BO_4 tetrahedra in the crystal structure of $Nb[BO_4]$ (O–O distances in the linking edges are 222.2 pm) [6].

Hydrothermal syntheses gave access to **$LiSi[BO_4]$**, space group $I\bar{4}-S_4^2$ (No. 82) with a=439.9 and c=683.7 pm; Z=2 (refined to R=10.5% from 189 reflections), and to the isostructural **$LiGe[BO_4]$** with a=450.1 and c=689.8 pm; Z=2 (refined to R=3.3% from 170 reflections); r(BO)=147 pm and r(OO)=229 pm [7].

$Bi_{12}Mn_{0.5}B_{0.5}O_{20}$, a new sillenite-type compound, prepared between 700 and 750°C in a corundum crucible, forms a structure containing $(MnO_4)^{3-}$ and $(BO_4)^{5-}$ units [8].

References for 3.4.1.4:

[1] Tarte, P.; Cahay, R.; Rulmont, A.; Werding, G. (Spectrochim. Acta A **41** [1985] 1215/9).
[2] Wanklyn, B. M.; Watts, B. E.; Wondre, F. R.; Davison, W. (J. Mater. Sci. Lett. **5** [1986] 499/502).

[3] Pankhurst, Q. A.; Thomas, M. F.; Wanklyn, B. M. (J. Phys. C **18** [1985] 1255/61).
[4] Wiedenmann, A.; Gunsser, W. (Spez. Ber. Kernforschungsanlage Juelich Juel-Spez-31b [1985] 232/3 from C.A. **104** [1986] No. 44735).
[5] Wiedenmann, A.; Mezei, F. (J. Magn. Magn. Mater. **54/57** [1986] 103/4 from C.A. **104** [1986] No. 121428).
[6] Range, K. J.; Wildenauer, M.; Heyne, A. M. (Angew. Chem. **100** [1988] 973/5).
[7] Klaska, R.; Selker, P.; Jendryan, R. (Z. Kristallogr. **174** [1986] 112/4).
[8] Speer, D.; Jansen, M. (Z. Anorg. Allg. Chem. **542** [1986] 153/6).

3.4.1.5 $[O|(BO_3)]^{5-}$, $[O_2|(BO_3)]^{7-}$, and Related Structures

Vonsenite, $\mathbf{Fe^{II}_2Fe^{III}[O_2|(BO_3)]}$ has been synthesized under dry conditions from a charge consisting of $4Fe+7Fe_2O_3+3B_2O_3$ at temperatures ranging from 300 to 550 °C and pressures from 10^{-2} to 1000 bar with magnetite as an intermediate product [1].

The structural hierarchy of the borate minerals kotoite, fluoborite $\mathbf{Mg_3(F,OH)_3[BO_3]}$, painite, jeremejevite, warwickite $\mathbf{(Mg,Ti)_2[O|(BO_3)]}$, ludwigite $\mathbf{Mg_2Fe^{III}[O_2|(BO_3)]}$, azoprototite $\mathbf{Mg_2(Mg,Fe^{III},Ti)[O_2|(BO_3)]}$, bonaccordite $\mathbf{Ni_2Fe^{III}[O_2|(BO_3)]}$, vonsenite $\mathbf{Fe^{II}_2Fe^{III}[O_2|(BO_3)]}$, pinakiolite $\mathbf{Mg_2Mn^{III}[O_2|(BO_3)]}$, orthopinakiolite $\mathbf{Mg_2Mn^{III}[O_2|(BO_3)]}$, takéuchite $\mathbf{(Mg_2Mn^{III})[O_2|(BO_3)]}$, hulsite $\mathbf{(Mg,Fe^{II})_2(Fe^{III},Sn)[O_2|(BO_3)]}$, and wightmanite $\mathbf{Mg_5(OH)_5[O|(BO_3)]\cdot 2H_2O}$ has been classified, see **Fig. 3-10**, p. 94. In all of these types, the framework stoichiometry can be modified by ordered vacancy substitutions along the length of the octahedral edge-sharing chains [2].

IR spectra of synthetic samples belonging to the ludwigite-vonsenite series, containing different amounts of Mg^{2+} and Fe^{2+}, are recorded. The series is completely isomorphous. Factor group and site analyses were made based on structural data. From these analyses, the selection rule, symmetry, and the number of various vibration modes were obtained. The correlation analysis of internal modes shows a positive correlation between the spectra and the symmetry of $[BO_3]^{3-}$ ions, and their distribution in normal frequencies [3].

Single-crystal X-ray diffraction data have been used to classify the regular intergrowths of Sn-rich borates from the endogenic boron ore deposits into three groups:

1) the twinned structure of magnesiohulsite,
2) the regular intergrowth of magnesiohulsite and Sn-rich ludwigite, and
3) the regular intergrowth of magnesiohulsite, Sn-rich ludwigite, and an unidentified mineral [4].

The crystal structure of a takéuchite mineral $\mathbf{Mg_{1.71}Mn^{II}_{0.37}Mn^{III}_{0.92}[O_2|(BO_3)]}$ was solved by a combined single-crystal X- ray and high-resolution transmission electron microscope (HRTEM) study. The space group is Pnnm – D^{12}_{2h} (No. 58) with a = 2758.4(4), b = 1256.1(3), and c = 602.7(3) pm; Z = 24; refined to R = 6.3% and R_w = 6.7% from 1492 reflections. The crystal structure can be interpreted as a chemical twinning of the basis structure type pinakiolite. As for other members of the ludwigite-orthopinakiolite structural family, some of the cation positions (Mn^{3+}) are disordered [5].

The crystal structure of a magnesium-aluminium ludwigite, $\mathbf{Mg_{2.11}Al_{0.31}Fe_{0.53}Ti_{0.05}Sb_{0.01}[O_2|(BO_3)]}$, has also been determined by a combined single-crystal X-ray and high-resolution transmission electron microscopy (HRTEM) study. The space group is Pbam – D^{9}_{2h} (No. 55) with a = 923.24(8), b = 1222.47(13), and c = 299.72(3) pm; Z = 4; D_c = 3.58 g/cm³; refined to R = 1.8%

 References on p. 95

and $R_w = 2.6\%$ from 1015 reflections. The structure is well-ordered and has a metal ion charge distribution in agreement with that found in other members of the pinakiolite family; the B–O distances are 138.0, 138.1, and 138.8 pm [6].

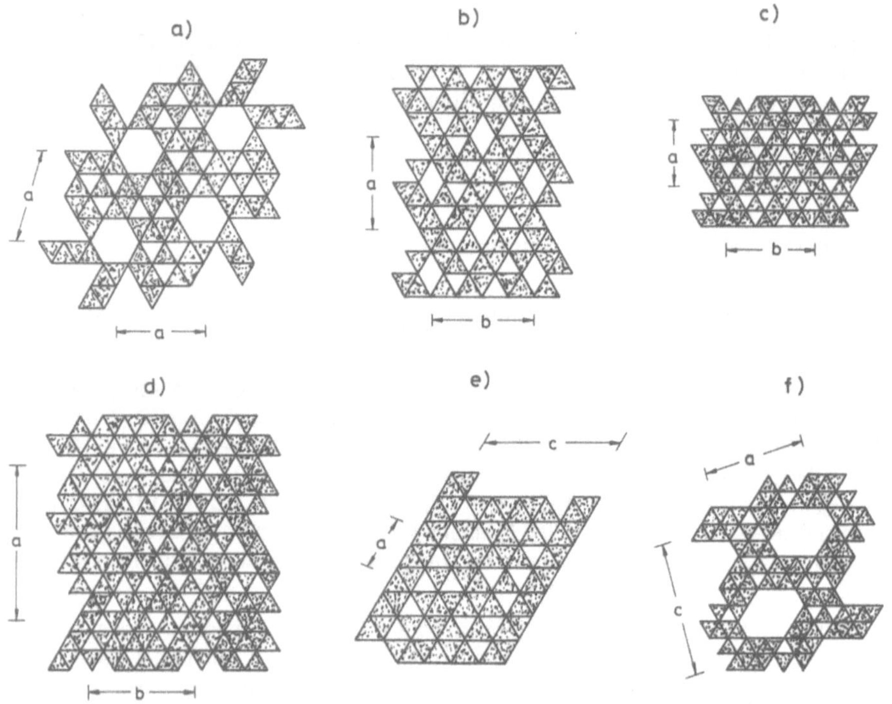

Fig. 3-10. The "3 Å wallpaper structures" of $[M_x(TL_y)L_z]$ minerals [2]. M = six-coordinate cation, T = three-coordinate cation, L = unspecified ligand.

a) Fluoborite type (orthogonal to the plane of the net), $[M_3(TL_3)L_3]$;

b) warwickite type, $[M_2(TL_3)L]$;

c) ludwigite type, $[M_3(TL_3)L_2]$;

d) orthopinakiolite type, $[M_3(TL_3)L_2]$;

e) hulsite type, $[M_3(TL_3)L_2]$;

f) wightmanite type, $[M_5(TL_3)L_6]$.

The structures of two other synthetic ludwigites were investigated by single-crystal X-ray diffraction analysis. $\mathbf{Mg_{1.93}Mn^{II}_{0.07}Mn^{III}[O_2|(BO_3)]}$ crystallizes in the space group Pbam–D^9_{2h} (No. 55) with a = 920.2(2), b = 1253.2(2), and c = 299.3(1) pm; Z = 4; $D_c = 3.78$ g/cm³; refined to R = 3.5% and $R_w = 3.8\%$ from 645 significant reflections. The scheme of the cation ordering could be obtained; the B–O distances are 138.1(5), 138.4(5), and 139.3(5) pm. $\mathbf{Co^{II}_2Co^{III}[O_2|(BO_3)]}$ also crystallizes in the space group Pbam–D^9_{2h} (No. 55) with a = 927.5(1), b = 1214.6(1), and c = 302.65(3) pm; Z = 4; $D_c = 5.214$ g/cm³; refined to R = 3.8% and $R_w = 4.1\%$ from 670 significant reflections. As indicated by estimated bond valence distributions, the $Co^{II}:Co^{III}$ ratio is 2.1; the B–O distances are 137.5(7), 138.4(7), and 139.1(7) pm [7].

Crystals with the composition $\mathbf{Mg_{1.5}Mn_{1.5}[O_2|(BO_3)]}$, prepared from stoichiometric amounts of MgO, Mn_3O_4, and hydrated B_2O_3 at 1300°C in sealed Pt tubes, possess the ludwigite structure with very few inherent planar defects, as was shown by high-resolution transmission electron microscopy (HRTEM). After heating in an electron beam, the electron diffraction patterns and electron micrographs showed that the ludwigite structure rearranged to the orthopinakiolite structure with the same nominal stoichiometry. Loss of oxygen to the surrounding vacuum of the electron microscope column would necessarily change the $Mn^{II}:Mn^{III}$ ratio of the crystals; this may stabilize the orthopinakiolite structure [8].

Two new oxyborate compounds were prepared by thermal reactions of MgO, Mn_3O_4, and B_2O_3 (1:0.75:0.33) during a study of the phase relationships between the pinakiolite-ludwigite series and the compounds. The structural topologies of these previously unreported materials were determined by comparing calculated with observed electron microscope images. The structures are very similar and also closely related to pinakiolite, which consists of flat walls of edge-sharing octahedra connected to zigzag chains of octahedra via triangular BO_3 groups. The two new structures contain similar infinite walls which are separated by slabs of octahedra that are wider than the zigzag chain found in pinakiolite. A new series $M_{3+x}[BO_{5+x}]$ (M = Mg or Mn; x = 1, 2, 3, etc.) of structurally related compounds can be envisaged [9].

The phase conversions upon heating to 1500°C were studied in the **$(Mg_{1-x}Fe^{II}_x)_2Fe^{III}$-$[O_2|(BO_3)]$** group with $0<x<0.25$ (magnesioludwigite), $0.25<x<0.50$ (ferroludwigite), and $0.50<x<1$ (vonsenite) by thermal and X-ray phase analyses, Mößbauer and IR spectroscopy. At ca. 650°C, the borates of each series retain their structure. The final thermolysis products at 1500°C are $MgFe[O|(BO_3)]$, MgO, and unreacted substance for magnesioludwigite, $Mg[Fe_2O_4]$ and $Mg_2[B_2O_5]$ for ferroludwigite, and $MgFe[O|(BO_3)]$ and Fe_2O_3 for vonsenite [10]. For the stability of mangan-axinite in H_3BO_3 containing systems, see [11].

Structural investigations of two new synthetic warwickites show B–O distances between 135.3 and 141.0 pm. The undistorted orthorhombic **$MgSc[O(BO_3)]$** crystallizes in the space group Pnam–D^{16}_{2h} (No. 62) with a = 949.0(1), b = 942.2(1), and c = 321.89(4) pm; Z = 4; D_c = 3.325(1) g/cm³; refined to R = 4.3% and R_w = 5.9% from 258 reflections. The ratios of $Mg^{2+}:Sc^{3+}$ at the two different metal sites are about 1:3 and 3:1, respectively. The monoclinic distorted **$Mg_{0.76}Mn^{II}_{0.24}Mn^{III}[O(BO_3)]$** crystallizes in the space group $P2_1/a$–C^5_{2h} (No. 14) with a = 937.0(2), b = 927.9(2), and c = 318.1(1) pm; β = 85.65(2)°; Z = 4; D_c = 3.887(1) g/cm³; refined to R = 4.8% and R_w = 5.7% from 385 reflections. The lowering of the symmetry is apparently due to structural effects caused by the Mn^{3+} ions [12].

References for 3.4.1.5:

[1] Barrese, E.; Burragato, F.; Flamini, A. (Neues Jahrb. Mineral. Monatsh. **1984** 483/9).
[2] Hawthorne, F. C. (Can. Mineralog. **24** [1986] 625/42).
[3] Peng, W.-Sh.; Xie, H.-S.; Zhang, Y.-M.; Xu, H.-G. (Kuangwu Xuebao **5** [1985] 43/7 from C.A. **107** [1987] No. 10577).
[4] Yang, Guangming; Peng, Zhizhong (Kuangwu Xuebao **6** [1986] 289/97 from C.A. **107** [1987] No. 239902).
[5] Norrestam, R.; Bovin, J.-O. (Z. Kristallogr. **181** [1987] 135/49).
[6] Norrestam, R.; Dahl, S.; Bovin, J.-O. (Z. Kristallogr. **187** [1989] 201/11).
[7] Norrestam, R.; Nielsen, K.; Soetofte, I.; Thorup, N. (Z. Kristallogr. **189** [1989] 33/41).
[8] Cooper, J. J.; Tilley, R. J. D. (J. Solid State Chem. **58** [1985] 375/82).
[9] Cooper, J. J.; Tilley, R. D. J. (J. Solid State Chem. **63** [1986] 129/38).
[10] Kononova, G. N.; Gonchar, S. V.; Dara, O. M.; Kolotyrkin, P. Ya. (Zh. Neorg. Khim. **32** [1987] 1986/90; Russ. J. Inorg. Chem. **32** [1987/88] 1170/2).

[11] Kurshakova, L. D. (Ocherki Fiz. Khim. Petrol. No. 12 [1984] 170/9 from C.A. **104** [1986] No. 37053).
[12] Norrestam, R. (Z. Kristallogr. **189** [1989] 1/11).

3.4.1.6 Fe_3BO_6

In the H_3BO_3–Fe_3O_4 system at 600°C, $\mathbf{Fe_3BO_6}$ was formed with a tetrahedral structure of the boron polyhedra (see also "Boron Compounds" 1st Suppl. Vol. 1, 1980, p.161); at 700°C, Fe_3BO_6 was converted to $Fe[BO_3]$ [1].

The diffraction of Mößbauer γ–radiation by the weakly ferromagnetic crystal Fe_3BO_6 was investigated. The shape of the spectra of pure nuclear Bragg reflection of resonance γ-radiation of the ^{57}Fe isotope with an energy of 14.4 keV depends on the nature of scattering interference on the ^{57}Fe nuclei in the 4c and 8d positions. Possible applications of interference of resonance nuclear levels in structure investigations were discussed [2].

The energetics of the spectra of Laue diffraction of Mößbauer γ-rays in crystals of Fe_3BO_6 in the vicinity of the spin-reorientation phase-transition temperature have been studied. The resonance peaks observed in the reflection spectra were explained by a combination of hyperfine interactions. The effects were considered as spin-reorientation and interferences of resonance lines on the shape of the diffracted radiation [3]. Pure nuclear Laue diffraction of Mößbauer radiation on an Fe_3BO_6 crystal above and in the vicinity of the Néel phase transition was investigated [4]. The study in both Bragg and Laue geometries showed that the Néel phase transition occurs differently in different layers of the crystal. Near the surface, spin relaxation processes play an important role, whereas they do not appear in the bulk [5].

An analysis of the Mößbauer γ-radiation in the Laue geometry (14.4 keV) is presented for the effect of the spin flip phase transition in the $\mathbf{^{57}Fe_3BO_6}$ single-crystal on the interference phenomena [6]. Effects were also studied of the nuclear and Rayleigh interference on the narrowing of the Mößbauer diffraction resonance line in single-crystals of $^{57}Fe_3BO_6$ [7].

Nuclear quadrupole maxima in the diffraction of Mößbauer γ-radiation in an Fe_3BO_6 crystal at 520 K under conditions of hyperfine quadrupole splitting during purely nuclear scattering were observed. The analysis of the reflection spectra gave the relationship between the signs of structure amplitudes for the iron nuclei in 4c and 8d positions [8]. A flow-type proportional counter was used for conversion-electron Mößbauer spectroscopy and records the characteristic X-rays at temperatures between 100 and 700 K. The observed spectra showed that the proportional counter can be used for studying the properties of surface layers of a material [9, 10].

Antiferromagnetic resonance in Fe_3BO_6 was investigated by using a quasi-optical method in the submillimeter-wave band. Depending on the conditions of excitation, resonance modes are observed. One mode which is quasi-ferromagnetic, was investigated both in phases Γ_1 and Γ_3. The results favor the 2-sublattice model [11].

References for 3.4.1.6:

[1] Kalinskaya, T. V.; Lobonova, L. B.; Kajekina, M. L. (Zh. Neorg. Khim. **30** [1985] 1513/8; Russ. J. Inorg. Chem. **30** [1985] 863/6).

[2] Kovalenko, P. P.; Labushkin, V. G.; Ovsepyan, A. K.; Sarkisov, E. R.; Smirnov, E. V.; Tolpekin, I. G. (Zh. Eksp. Teor. Fiz. **88** [1985] 1336/47; Sov. Phys.-JETP **61** [1985] 793/800 from C.A. **102** [1985] No. 228932).

[3] Tolpekin, I. G.; Kovalenko, P. P.; Labushkin, V. G.; Ovchinnikova, E. N.; Sarkisov, E. R.; Smirnov, E. V. (Pis'ma Zh. Eksp. Teor. Fiz. **43** [1986] 474/6 from C.A. **105** [1986] No. 69501).

[4] Tolpekin, I. G.; Kovalenko, P. P.; Labushkin, V. G.; Sarkisov, E. R. (Metody Apparatura dlya Toch. Izmerenii Parametrov Ionizir. Izluch. M. **1987** 81/6 from C.A. **110** [1989] No. 32709 and 87/91 from C.A. **109** [1988] No. 203426).

[5] Kovalenko, P. P.; Labushkin, V. G.; Sarkisov, E. R.; Tolpekin, I. G. (Fiz. Tverd. Tela [Leningrad] **29** [1987] 593/6 from C.A. **107** [1987] No. 16418).
[6] Tolpekin, I. G.; Kovalenko, P. P.; Labushkin, V. G.; Ovchinnikova, E. N.; Sarkisov, E. R.; Smirnov, E. V. (Zh. Eksp. Teor. Fiz. **94** [1988] 329/43; Sov. Phys.-JETP **67** [1988] 404/12 from C.A. **108** [1988] No. 140174).
[7] Kovalenko, P. P.; Labushkin, V. G.; Ovsepyan, A. K.; Sarkisov, E. R.; Smirnov, E. V.; Tolpekin, I. G. (Fiz. Tverd. Tela [Leningrad] **29** [1987] 2569/72 from C.A. **108** [1988] No. 13412).
[8] Tolpekin, I. G.; Labushkin, V. G.; Ovchinnikova, E. N.; Smirnov, E. V. (Pis'ma Zh. Tekh. Fiz. **14** [1988] 2024/7 from C.A. **110** [1989] No. 144159).
[9] Kamzin, A. S.; Rusakov, V. P. (Prib. Tekh. Eksp. **1988** No. 5, pp. 56/8 from C.A. **110** [1989] No. 31129).
[10] Kamzin, A. S.; Rusakov, V. P.; Grigor'ev, L. A. (Mater. 2 Vses. Soveshch. po Yader.-Spektroskop. Issled Sverkhtonk Vzaimodeistvii, Groznyi, 1987, [1988] 88/91 from C.A. **111** [1989] No. 204632).

[11] Harutunyan, V. E.; Kocharyan, K. N.; Martirosyan, R. M.; Voronkov, V. D.; Bezmaternich, L. N. (Int. J. Infrared Millimeter Waves **10** [1989] 841/5 from C.A. **111** [1989] No. 207919).

3.4.2 Diborates

For $H_4B_2O_5$ (C_2 symmetry) and $H_8B_2O_7$ (C_v symmetry), see Section 3.5.11, p. 256.

3.4.2.1 The Ion $[B_2O_5]^{4-}$

$\mathbf{Mg_2[B_2O_5]}$ single crystals were grown using the flux method [1]. Suanite, $\mathbf{(Mg_{1.99}Fe_{0.00}Mn_{0.01})_{2.00}[B_2O_5]}$, crystallizes in the monoclinic space group $P2_1/a-C_{2h}^5$ (No. 14) with a = 1231.9(2), b = 312.1(1), and c = 921.0(2) pm; β = 104.45(1)°; Z = 4 [2]. Problems of asbestos, glass fibers, aluminosilicate fibers, and pitch-based carbon fibers were discussed along with the properties and merits of a new whisker-like fiber of $Mg_2[B_2O_5]$ [3]. The effects of NaCl on the growth of the fibrous form of $Mg_2[B_2O_5]$ from a flux with an Mg/B ratio between 0.5 and 1.0 was studied [4]. The effect of a KCl flux was investigated by thermogravimetry (TG) and differential thermal analysis (DTA), X-ray powder diffraction analysis, and scanning electron microscopy (SEM). The very low solubility of $Mg_2[B_2O_5]$ in the KCl flux suggested that the needle-like $Mg_2[B_2O_5]$ was formed by the ripening mechanism in the liquid phase of KCl [5]. Acicular $Mg_2[B_2O_5]$ whiskers are formed by the reaction of a fine $MgO/H_3BO_3/KCl$ mixture at 850 °C by spray drying. Large crystals (20 to 40 μm × 1 to 3 μm) were obtained by seeding the raw-material mixture with small $Mg_2[B_2O_5]$ crystals [6].

Differences were examined in the IR and Raman spectra of the following alkaline earth element pyroborates: **L-$\mathbf{Mg_2[B_2O_5]}$**, monoclinic space group $P2_1/a-C_{2h}^5$ (No. 14) with a = 1210, b = 312, and c = 936 pm; β = 104.20°; Z = 4; **H-$\mathbf{Mg_2[B_2O_5]}$**, triclinic space group $P\bar{1}-C_i^1$ (No. 2) with a = 618, b = 912, and c = 311 pm; α = 90.4°, β = 92.13°, and γ = 104.32°; Z = 2; $\mathbf{Ca_2[B_2O_5]}$, monoclinic space group $P2_1/a-C_{2h}^5$ (No. 14) with a = 1150, b = 516, and c = 720 pm; β = 92.92°; Z = 4; $\mathbf{Sr_2[B_2O_5]}$, monoclinic space group $P2_1/a-C_{2h}^5$ (No. 14) with a = 1185, b = 535, and c = 771 pm; β = 92.6°; Z = 4; $\mathbf{Cd_2[B_2O_5]}$, triclinic space group $P\bar{1}-C_i^1$ (No. 2) with a = 636, b = 995, and c = 345 pm; α = 90.83°, β = 91.97°, and γ = 105.47°; Z = 2 [7].

$Mg(UO_2)[B_2O_5]$, or MgB_2UO_7, has been prepared from U_3O_8, $Mg[CO_3]$, and B_2O_3 (1:14:18) at 1473 K; it is orthorhombic, space group Pcam$-D_{2h}^{11}$ (No. 57) with a=974.7(3), b=731.5(2), and c=791.1(2) pm; Z=4; D_c=4.7 g/cm³. The structure was refined to R=3.4% and R_w=4.3% from 1997 contributing reflections. All the atoms, except O(3), are located in sheets perpendicular to [00z]; U^{6+} and Mg^{2+} are surrounded by octahedra; the boron atoms are in triangular B_2O_5 groups with the following distances (in pm): 135.0(8), 135.3(9), and 140.2(8); 133.4(8), 134.1(8), and 143.8(8) [8].

$CaO(UO_2)_2[B_2O_5]$, or $CaB_2U_2O_{10}$, was prepared from a $U_3O_8/Ca[CO_3]/B_2O_3$ mixture at 1423 K in air for 15 h; it is monoclinic, space group C2$-C_2^3$ (No. 5) with a=1651.2(3), b=816.9(2), and c=658.2(1) pm; β=96.97° (3); Z=4; D_c=5.26 g/cm³. The structure was refined to R=2.8% from 3719 reflections and to R_w=3.3% from 3516 reflections. The Ca^{2+} ions are surrounded by seven oxygen atoms, the U^{6+} ions by 4+2, 2+4, or 2+5 oxygen atoms. The boron atoms are in B_2O_5 triangular groups with the following distances (in pm): 135.1(8), 135.7(8), and 141.1(8); 133.8(8), 133.8(9), and 138.5(8) [9].

3.4.2.2 The Ions $[B_2O_8]^{10-}$ and $[B_2O_{10}]^{14-}$

$Mg_3Zr[B_2O_8]$ (warwickite structure), **$Mg_5Ti[B_2O_{10}]$** (ludwigite structure), and **$Mg_5Sn[B_2O_{10}]$** (orthopinakiolite structure) have been prepared from the corresponding metal oxides and H_3BO_3. The luminescence of **$Mg_3(Ti,Zr)[B_2O_8]$** and **$Mg_5(Ti,Sn)[B_2O_{10}]$** has been measured. The results are discussed in terms of the delocalization in the clusters of the titanate octahedra. A general scheme for the luminescence of Ti^{4+}-activated compounds is given. The compound $Mg_5Sn[B_2O_{10}]$:Ti is a very luminescent material for a titanium concentration of ca. 20 mol% [10].

$Ni_5Ti[B_2O_{10}]$ is orthorhombic, space group Pbam$-D_{2h}^9$ (No. 55) with a=1222.1(2), b=919.9(2), and c=299.6(1) pm; Z=2; D_m=5.14 g/cm³ and D_c=5.16 g/cm³. The final R value was 2.1% for 1042 reflections. The Ti^{4+} ion partially occupies the four smallest octahedral sites. The (Ni,Ti) octahedra share edges and corners in order to form a three-dimensional framework, in which boron occupies triangular interstices. The oxygen atoms are approximately cubic close packed as in the NiO structure. The B–O distances are 137.5(5), 137.9(3), and 139.0(3) pm [11].

Blue-black **$Ni_5V^{IV}[B_2O_{10}]$** and dark green **$Ni_5Mn^{IV}[B_2O_{10}]$** are prepared by using a flux growth of V_2O_5/NiO (1:8) or MnO_2/Ni (1:5), respectively, at 1300°C in a great excess of B_2O_3. The obtained crystals are isotypic to $Ni_5Ti[B_2O_{10}]$ (see above) with Z=2 and a=981.3(5), b=1218.9(8), and c=298.9(5) pm for the V^{IV} compound refined to R=4.1% from 747 reflections; as well as a=913(1), b=1216(2), and c=297(2) pm for the Mn^{IV} compound to R=4.8% from 239 reflections. One of the metal point positions is occupied statistically by Ni^{2+} and V^{4+} or Mn^{4+}, respectively. The partial disorder in respect to the metal distribution is confirmed by calculations of the Coulomb term of the lattice energy. The B–O distances are 136.6(8), 139.3(8), 139.5(8) pm for the V^{IV} compound as well as 132(3), 141(3), and 143(3) pm for the Mn^{IV} compound [12]. **$Ni_5Ge[B_2O_{10}]$** and **$Ni_5Zr[B_2O_{10}]$** have the same structural type as $Ni_5Ti[B_2O_{10}]$ [13].

$Ni_5Sn^{IV}[B_2O_{10}]$ with a=930.2(1), b=608.9(2), and c=1228.0(2) pm crystallizes in the orthorhombic space group Pnma$-D_{2h}^{16}$ (No. 62) with Z=4. The compound has a hitherto unobserved, strictly ordered metal distribution. It is not isotypic but related to $Ni_5Ti[B_2O_{10}]$ and was prepared by the melting of NiO with SnO_2 in a ratio of 5:1 in a B_2O_3 flux at 1450°C, cooling within 4 h to 800 °C, crystallization and washing with hot water. The structure has been refined from 1071 reflections to R = 5.8%. The B(1)–O distances are 138, 140, 141 pm, the B(2)–O distances are 132, 139, and 143 pm [14]. **$Ni_5Hf[B_2O_{10}]$** has the same structure with a = 932.8, b = 612.0, and c = 1233.4 pm. In contrast to the $Ni_5Ti[B_2O_{10}]$ type, all metal point positions are strongly ordered. The partly statistical metal distribution within $Ni_5M^{IV}[B_2O_{10}]$ (M^{IV} = Ti, Zr, Ge) was not exactly ascertained [15].

References for 3.4.2:

[1] Wanklyn, B. M.; Watts, B. E.; Wondre, F. R.; Davison, W. (J. Mater. Sci. Lett. **5** [1986] 499/502).
[2] Lisitsyn, A. E.; Rudnev, V. V.; Yurkina, K. V. (Mineral. Zh. **7** No. 5 [1985] 32/40 from C.A. **104** [1986] No. 92226).
[3] Kitamura, T. (Kinoshi Kenkyu Kaishi No. 24 **1985** [1986] 33/41 from C.A. **106** [1987] No. 86400).
[4] Kitamura, T.; Sakane, K.; Wada, H. (J. Mater. Sci. Lett. **7** [1988] 467/9 from C.A. **109** [1988] No. 139401).
[5] Sakane, K.; Kitamura, T.; Ogawa, J. (Gypsum Lime No. 216 [1988] 281/7 from C.A. **110** [1989] No. 100680).
[6] Kitamura, T.; Sakane, K.; Wada, H.; Suzue, M. (5th Int. Symp. Agglom., Brighton, U.K., 1989, pp. 187/95 from C.A. **112** [1990] No. 82657).
[7] Il'in, Yu. N.; Kravchenko, V. V.; Petrov, K. I.; Golovin, Yu. M. (Zh. Neorg. Khim. **33** [1988] 2145/8; Russ. J. Inorg. Chem. **33** [1988/89] 1224/6).
[8] Gasperin, M. (Acta Crystallogr. C **43** [1987] 2264/6).
[9] Gasperin, M. (Acta Crystallogr. C **43** [1987] 1247/50).
[10] Konijnendijk, W. L.; Blasse, G. (Mater. Chem. Phys. **12** [1985] 591/9 from C.A. **103** [1985] No. 114938).

[11] Armbruster, T.; Lager, G. A. (Acta Crystallogr. C **41** [1985] 1400/2).
[12] Bluhm, K.; Mueller-Buschbaum, H. (Z. Anorg. Allg. Chem. **579** [1989] 111/5).
[13] Bluhm, K.; Mueller-Buschbaum, H. (J. Less-Common Met. **147** [1989] 133/6).
[14] Bluhm, K.; Mueller-Buschbaum, H. (Monatsh. Chem. **120** [1989] 85/9).
[15] Bluhm, K.; Mueller-Buschbaum, H. (Z. Anorg. Allg. Chem. **575** [1989] 26/30).

3.4.3 Triborates

For $H_3B_3O_6$ (C_{3h} symmetry), see Section 3.5.11, p. 256.

3.4.3.1 The Ion $[B_3O_5]^-$

Differential thermal analysis (DTA), IR spectroscopy, and X-ray diffraction phase analysis studies in the system B_2O_3–Na_2O–V_2O_5 show formation of **$Na_3[(VO_3)_2(B_3O_5)]$**, or $3Na_2O \cdot 2V_2O_5 \cdot 3B_2O_3$, melting point 840 °C, in which VO_4 tetrahedra are linked in metavanadate chains [1].

$Li[B_3O_5]$ (LBO) is recognized as a new nonlinear optical (NLO) crystal. Single crystals were grown in a B_2O_3 flux at an $Li_4[B_2O_5]$: B_2O_3 mole ratio of 2 : 1. The growth rate was limited by the high viscosity of the flux. Transparent crystals of 60×10×10 mm or 35×30×15 mm size are obtained by a modified flux and flux-pulling method [2]. Single crystals of good quality and large size were also grown by a modified flux, cooling from 834 °C with a rate of 0.2 to 2 °C/day or by a flux-pulling method. LBO crystals are orthorhombic, space group $Pna2_1-C^9_{2v}$ (No. 33) with a=844.7(1), b=737.89(8), and c=514.08(6) pm; D_c=2.47 g/cm³. The structure was refined to R=1.88% and R_w=1.90% for 1130 reflections. There are a large number of (B_3O_7) B–O bonds with triangular and tetragonal coordination of B; along the c axis, they are linked with each other into a spiral structure. In the B–O bridge bonds, the spirals are interconnected with each other and form the crystal structure. The deformation electron density map shows vacant 2p orbitals populated by electrons belonging to lone pairs of oxygen atoms. A one-dimensional highly anisotropic Li^+ conductivity is observed between 300 and 750 K and discussed in terms of the crystal structure. The bond structures in $Li[B_3O_5]$ and β-$Ba_3[B_3O_6]_2$ are very similar. Preliminary measurements indicate that the $Li[B_3O_5]$ crystals have a high damage threshold, a wide transparency range (165 to 3200 nm), a large acceptance angle for phase-matching and a high frequency-doubling efficiency. Type-I and type-II phase-matching with $Li[B_3O_5]$ crystals can be performed at room temperature. The crystals also possess good mechanical properties, chemical stability, and moisture resistance [3 to 7]. For $Li[B_3O_5]$ crystals used as a frequency doubler, see also [8].

An optically perfect $Li[B_3O_5]$ single crystal, grown by the high-temperature flux method, is transparent from 160 nm to 2.6 μm. The nonlinearity follows theoretical calculations of the second harmonic generation (SHG) coefficients using the anionic group theory and the complete neglect of differential overlap (CNDO) approximation in order to obtain the localized wave functions of component groups. The optically perfect crystal has an SHG coefficient comparable with that of β-$Ba_3[B_3O_6]_2$ as well as two other outstanding advantages: a high damage threshold of 25 GW/cm² (at 1.064 μm, 0.1 ns) and a wide acceptance angle of 25 mrad for $\theta \neq 90°$ and 95 mrad for $\theta = 90°$ with a crystal of 6 mm length [9].

$Li[B_3O_5]$ single crystal fibers were also grown by the laser-heated pedestal growth method. Some $Li[B_3O_5]$ single crystal fibers were obtained by using the seeds of different orientations, different source rods, and different fluxes. The $Li[B_3O_5]$ single crystal fibers obtained by using the seeds with phase-matching direction (10 mm in length, 200 μm in diameter) give an output of 532 nm green light from one polished end-face when an Nd:YAG laser beam of 1.064 μm is incident on the other polished end-face, and are used as the second harmonic generation device of an Nd:YAG laser. The double frequency effect was preliminarily tested [10].

The $Li[B_3O_5]$ crystal was also grown by the flux seeding method; the measured density was 2.47 g/cm³ and found to be an excellent UV nonlinear optical (NLO) material [11]. The relationship between the anionic groups of β-$Ba_3[B_3O_6]_2$ and $Li[B_3O_5]$ NLO crystals and their bonding, electronic structure, and transmission cutoffs has also been studied. Only β-Ba_3-$[B_3O_6]_2$ consists of $(B_3O_6)^{3-}$ groups with trigonally coordinated boron; these groups are isolated in the crystal structure. $Li[B_3O_5]$ (LBO), however, is based on $(B_3O_7)^{5-}$ groups with boron either trigonally or tetrahedrally coordinated by oxygen; these groups are linked throughout the crystal. These structural differences between β-$Ba_3[B_3O_6]_2$ and $Li[B_3O_5]$ lead to a larger bandgap energy in $Li[B_3O_5]$. Bandgap, absorption edge, and volume bands were studied experimentally, using vacuum UV spectroscopy and valence band X-ray photoemission spectroscopy [12].

IR and polarized single crystal Raman spectra of $Li[B_3O_5]$ are analyzed; all four (A_1, A_2, B_1, and B_2) symmetry species of the C_{2v} point group isomorphic to the C^9_{2v} space group are Raman active in distinct crystal polarization experiments. A complete set of symmetry assignments based on a factor group analysis is presented. Experimentally, more peaks than theoretically predicted have been observed. An assignment of the peaks has been made based on the following: there are four triborate rings in the unit cell. Each ring has 18 internal normal modes of vibration. Each of these 18 vibrations of species A in the site group is split into a quartet with one component each of species A_1, A_2, B_1, and B_2 in the factor group. A clear line at approximately 500 cm^{-1} separates the behavior of near degenerate internal modes with multiple species representation from cases at lower frequency where only a single entry appears. Frequencies at 1554 (A_1), 1061 (B_1), 1023 (B_2), and 805 (A_2) cm^{-1} were assigned as combinations (or overtone) bands; three bands at 1395, 1383, and 1346 cm^{-1} are assigned as being due to the minor ^{11}B isotopic species. The motions corresponding to relative movement of the individual rings and lattice modes should occur as $8A_1+9A_2+8B_1+8B_2$ for a total of 33 vibrational bands. A comparison of the internal frequencies of various borate species suggests a correlation between the B–O stretching frequency and the nonlinear optical efficiency of $Li[B_3O_5]$ [13].

The band structure and the interband optical conductivity of an $Li[B_3O_5]$ crystal with excellent nonlinear optical properties were studied by means of a first-principles method. The electronic structure is characterized by highly localized bands with a direct gap of 7.37 eV and a very low static dielectric constant of 2.7. The optical absorption curve shows several prominent structures, but with very small directional anisotropy. The calculated frequency-dependent refractive indices are in good agreement with experimental data [14].

The nonlinear optical coefficients of $Li[B_3O_5]$ at 1.079 μm relative to d_{36} of $K[H_2PO_4]$ were determined: $d_{31}=\mp2.51(1\pm0.09)$, $d_{32}=\pm2.69(1\pm0.12)$, $d_{33}=\pm0.15(1\pm0.10)$, $d_{15}=\mp2.30(1\pm0.22)$, and $d_{24}=\pm2.52(1\pm0.20)$. The phase-matching conditions, effective second harmonic generation coefficients, walk-off angles, and acceptance angles for SHG at 1.079 μm were investigated. The figures of merit of $Li[B_3O_5]$ in different cases were compared with those of $K[H_2PO_4]$, β-$Ba_3[B_3O_6]_2$, and $K[TiO][PO_4]$ crystals. All these figures combined with the damage thresholds of the crystals show that an $Li[B_3O_5]$ crystal may operate better than $K[H_2PO_4]$ and β-$Ba_3[B_3O_6]_2$ crystals for frequency conversions of a tightly focused Nd:$YAlO_3$ or Nd:YAG laser and also a dye laser [15].

The anisotropic thermal expansion of $Li[B_3O_5]$ has been studied along the principal crystallographical direction over the temperature range from 16 to 790°C by means of high-temperature X-ray powder diffraction analysis. This orthorhombic crystal exhibits strongly anisotropic expansion with coefficients of 108.2×10^{-6} K^{-1} for the (100), 33.6×10^{-6} K^{-1} for the (010), and -88.0×10^{-6} K^{-1} for the (001) direction at 50°C. The coefficients tend towards zero when the temperature approaches the crystal's peritectic reaction point [16].

A type-I $Li[B_3O_5]$ optical parametric amplifier pumped by 15 ps, 355 nm laser pulses is tested. With proper design, $Li[B_3O_5]$ can be as efficient as β-$Ba_3[B_3O_6]_2$ for such an application [17].

The potential was evaluated of temperature-tuned $Li[B_3O_5]$ to achieve type-I noncritical phase-matching for ps high-power second harmonic generation and tunable optical parametric amplification (OPA). Pumped at 35 ps, 1.064 μm laser pulses, a conversion efficiency of 65% was obtained from SHG. The output of OPA, pumped by 532 nm, was tunable from 0.75 to 1.8 μm with an efficiency better than 20% [18].

A critical review (43 references) on compact blue laser devices based on nonlinear frequency upconversion includes lithium triborate [19]. Over 40 references were reviewed on the basic concepts for the nonlinear optical effect calculations of the SHG coefficients in borate crystals including β-$Ba_3[B_3O_6]_2$ and $Li[B_3O_5]$ [20].

A visible optical parametric oscillator (OPO) using an $Li[B_3O_5]$ crystal as the nonlinear element has an energy efficiency as high as 22% for a signal wave, can generate an output energy of 2.7 mJ, and can be continuously tuned from 435 to 1922 nm [21]. Tunable coherent

References on pp. 102/4

vacuum UV radiation down to 187.7 nm can be generated by means of type-I sum-frequency mixing with the outputs of the fifth harmonic of an Nd:YAG laser and an OPO pumped by the second harmonic of the same Nd:YAG laser in an $Li[B_3O_5]$ crystal [22].

An efficient OPO of tunable high-power picosecond pulses with phase-matching lengths between 187.7 and 2650 nm, energy conversion efficiencies from 22 to 70%, inherent linewidths of 0.3 nm, output power of 8 mW at 500 Hz, oscillation threshold of 2.5, and internal pump depletions of 20 to 64% was investigated since 1991 by several groups [23 to 38].

The optical characteristics of $Li[B_3O_5]$ were also evaluated for second harmonic generation (SHG) in the femtosecond region. The use of $Li[B_3O_5]$ for autocorrelation measurements of ultrashort optical pulses is shown. Pulse widths down to 40 fs can be measured accurately with crystal thicknesses varying from 0.195 to 4.165 mm. The effect of the crystal length on the autocorrelation shape was also investigated [39].

Using an $Li[B_3O_5]$ crystal, temperature-tuned noncritical phase-matched SHG from 1.025 to 1.253 μm is achieved in a temperature range from +190 to −3°C. The noncritical phase-matching temperature for 1.064 μm radiation is 148.0 ± 0.5°C with a temperature acceptance bandwidth of 3.9°C·cm [40].

The optical properties of $Li[B_3O_5]$ (LBO), β-$Ba_3[B_3O_6]_2$ (BBO), and $K[D_2PO_4]$ (KD*P) were compared for second and third harmonic generation (THG) of Nd:YAG laser radiation. Experimentally, the conversion efficiency was measured as a function of the energy density of 8 ns long laser pulses, generated by a commercial Nd:YAG oscillator-amplifier system. In $Li[B_3O_5]$ and β-$Ba_3[B_3O_6]_2$, the second harmonic generation saturates at an energy density of approximately 1.5 J/cm^2 at efficiencies between 55 and 60%, while in $K[D_2PO_4]$ densities of 2.0 to 2.6 J/cm^2 are required for efficiencies between 40 and 55%. Similar results are obtained for frequency tripling. In $Li[B_3O_5]$ and β-$Ba_3[B_3O_6]_2$, saturated efficiencies between 20 and 25% are measured at an energy density of approximately 1.5 J/cm^2, while in $K[D_2PO_4]$ efficiencies of 20% are obtained at energy densities above 2 J/cm^2. The phase-matching is calculated for doubling and tripling of Nd:YAG laser radiation and for frequency conversion of tunable laser light. In $Li[B_3O_5]$ and β-$Ba_3[B_3O_6]_2$, phase-matched sum-frequency mixing of UV and IR laser light generates tunable radiation at 160 and 190 nm, respectively [41].

Highly efficient generation of UV radiation at 355 nm was achieved in an $Li[B_3O_5]$ crystal through the frequency mixing of the fundamental and second-harmonic radiation of an Nd:YAG laser. An energy conversion efficiency of 60% was obtained under special experimental conditions [42]. Sum-frequency mixing of UV laser light from 266 and 231 nm and tunable coherent IR light from 1.2 to 2.6 μm generates in an $Li[B_3O_5]$ crystal a UV radiation between 188 and 242 nm. The UV radiation at 266 and 231 nm is produced by the 4th and 5th harmonic of a pulsed Nd:YAG laser, respectively. The IR light is generated with an optical parametric oscillator (OPO) of β-$Ba_3[B_3O_6]_2$. The phase-matching angle is measured as a function of the radiation between 188 and 242 nm and compared with calculated values. For UV radiation between 173 and 213 nm, the calculations predict an extension of the tuning range of the sum-frequency generated between 188 and 242 nm to wavelengths as short as the $Li[B_3O_5]$ transmission cutoff at 160 nm [43].

References for 3.4.3.1:

[1] Buludov, N. T.; Mamedaliev, F. D.; Kara'ev, Z. Sh.; Abdulla'ev, G. K. (Zh. Neorg. Khim. **30** [1985] 1816/8; Russ. J. Inorg. Chem. **30** [1985/86] 1032/4).

[2] Jiang, Aidong; Chen, Tianbin; Zheng, Yong; Chen, Zusheng; Wang, Shengsheng (Guisuanyan Xuebao **17** [1989] 189/90 from C.A. **111** [1989] No. 184431).

[3] Zhao, Shuqing; Huang, Chaoen; Zhang, Hongwu (J. Cryst. Growth **99** [1990] 805/10).

[4] Khodyakov, A. A.; Dzhafarov, M. Kh.; Kurilenko, L. N.; Saunin, E. I.; D'yakov, V. A. (Zh. Fiz. Khim. **65** [1991] 2561/3; Russ. J. Phys. Chem. **65** [1991/92] 1356/7).
[5] Radaev, S. F.; Genkina, E. A.; Lomonov, V. A.; Maksimov, B. A.; Pisarevskii, Yu. V.; Chelnikov, M. N.; Simonov, V. I. (Kristallografiya **36** [1991] 1419/26).
[6] Radaev, S. F.; Maksimov, B. A.; Simonov, V. I.; Andreev, B. V.; D'yakov, V. A. (Acta Crystallogr. B **48** [1992] 154/60).
[7] Radaev, S. F.; Sorokin, N. I.; Simonov, V. I. (Fiz. Tverd. Tela [St. Petersburg] **33** [1991] 3597/600).
[8] Zhao, Shuqing; Zhang, Hongwu; Huang, Chaoen; Liu, Youchen; Liu, Wei; Hao, Zhiwu; Su, Zhenzhen (Rengong Jingti **18** [1989] 9/17 from C.A. **111** [1989] No. 222329).
[9] Chen, Chuangtian; Wu, Yicheng; Jiang, Aidong; Wu, Bochang; You, Guiming; Li, Rukang; Lin, Shujie (J. Opt. Soc. Am. B **6** [1989] 616/21 from C.A. **110** [1989] No. 239715).
[10] Ji, Yangyang; Zaho, Shuqing; Huo, Yujing; Zhang, Hongwu; Li, Ming; Huang, Chaoen (J. Cryst. Growth **112** [1991] 283/6).

[11] Wu, Yicheng; Chen, Chuangtian (Reza Kenkyu **19** [1991] 941/9 from C.A. **116** [1992] No. 161275); Wu, Yicheng; Jiang, Aidong; Lu, Shaofang; Chen, Chuangtian; Shen, Yusheng (Rengong Jingti Xuebao **19** [1990] 33/8 from C.A. **116** [1992] No. 204933).
[12] French, R. H.; Ling, J. W.; Ohuchi, F. S.; Chen, Chuangtian (Phys. Rev. [3] B **44** [1991] 8496/502).
[13] Shang, Qanhuan; Hudson, B. S.; Huang, Chaoen (Spectrochim. Acta A **47** [1991] 291/8).
[14] Xu, Yongnian; Ching, W. Y. (Phys. Rev. [3] B **41** [1990] 5471/4).
[15] Lin, Shujie; Sun, Zhaoyang; Wu, Baichang; Chen, Chuangtian (J. Appl. Phys. **67** [1990] 634/8).
[16] Lin, Wei; Dai, Guiqing; Huang, Qingzhen; Zhen, An; Liang, Jingkui (J. Phys. D **23** [1990] 1073/5).
[17] Zhang, J. Y.; Huang, J. Y.; Shen, Y. R.; Chen, Chuangtian; Wu, Baichang (Appl. Phys. Lett. **58** [1991] 213/5).
[18] Huang, J. Y.; Shen, Y. R.; Chen, Chuangtian; Wu, Baichang (Appl. Phys. Lett. **58** [1991] 1579/81).
[19] Risk, W. P.; Lenth, W. (Proc. SPIE-Int. Soc. Opt. Eng. **1104** [1989] 23/32 from C.A. **112** [1990] No. 65500).
[20] Chen, Chuangtian (ACS Symp. Ser. **455** [1991] 360/79 from C.A. **115** [1991] No. 59616).

[21] Wang, Yunping; Xu, Zuyan; Deng, Daoqun; Zheng, Wanhua; Wu, Baichang; Chen, Chuangtian (Appl. Phys. Lett. **59** [1991] 531/3).
[22] Wu, Baichang; Xie, Fali; Chen, Chuangtian; Deng, Daoqun; Xu, Zuyan (Opt. Commun. **88** [1992] 451/4).
[23] Xie, Fali; Wu, Baichang; You, Guiming; Chen, Chuangtian (Opt. Lett. **16** [1991] 1237/9).
[24] Lin, Shujie; Wu, Baichang; Xie, Falie; Chen, Chuangtian (Appl. Phys. Lett. **59** [1991] 1541/3).
[25] Lin, Shujie; Huang, J. Y.; Ling, Jiwu; Chen, Chuangtian; Shen, Y. R. (Appl. Phys. Lett. **59** [1991] 2805/7).
[26] Shen, Jinhui; He, Huijuan; Liu, Yupu; Zhang, Yinghua; Wang, Zhijiang; Wu, Bochang (Guangxue Xuebao **11** [1991] 1068/73 from C.A. **116** [1992] No. 244717).
[27] Hanson, F.; Dick, D. (Opt. Lett. **16** [1991] 205/7).
[28] Robertson, G.; Henderson, A.; Dunn, M. H. (Opt. Lett. **16** [1991] 1584/6).
[29] Robertson, G.; Henderson, A.; Dunn, M. H. (Appl. Phys. Lett. **60** [1992] 271/3).
[30] Tang, Y.; Cui, Y.; Dunn, M. H. (Opt. Lett. **17** [1992] 192/4).

[31] Ebrahimzadeh, M.; Robertson, G.; Dunn, M. H. (Opt. Lett. **16** [1991] 767/9).
[32] Ebrahimzadeh, M.; Hall, G. J.; Ferguson, A. I. (Appl. Phys. Lett. **60** [1992] 1421/3).
[33] Ebrahimzadeh, M.; Hall, G. J.; Ferguson, A. I. (Opt. Lett. **17** [1992] 652/4).
[34] Malcolm, G. P. A.; Ebrahimzadeh, M.; Ferguson, A. I. (IEE J. Quantum Electron. **28** [1992] 1172/8 from C.A. **116** [1992] No. 265024).
[35] Malcolm, G. P. A.; Persaud, M. A.; Ferguson, A. I. (Opt. Lett. **16** [1991] 983/5).
[36] Yang, S. T.; Pohalski, C. C.; Gustafson, E. K.; Byer, R. L.; Feigelson, R. S.; Raymakers, R. J.; Route, R. K. (Opt. Lett. **16** [1991] 1493/5).
[37] Skripko, G. A.; Bartoshevich, S. G.; Mikhnyuk, I. V.; Tarasevich, I. G. (Opt. Lett. **16** [1991] 1726/8).
[38] Krause, H. J.; Daum, W. (Appl. Phys. Lett. **60** [1992] 2180/2).
[39] Pelouch, W. S.; Ukachi, T.; Wachman, E. S.; Tang, C. L. (Appl. Phys. Lett. **57** [1990] 111/3).
[40] Ukachi, T.; Lane, R. J.; Bosenberg, W. R.; Tang, C. L. (Appl. Phys. Lett. **57** [1990] 980/2).

[41] Borsutzky, A.; Bünger, R.; Huang, Chaoen; Wallenstein, R. (Appl. Phys. B **52** [1991] 55/62).
[42] Wu, Baichang; Chen, Nong; Chen, Chuangtian; Deng, Daoqun; Xu, Zuyan (Opt. Lett. **14** [1989] 1080/1 from C.A. **111** [1989] No. 243513).
[43] Borsutzky, A.; Bünger, R.; Wallenstein, R. (Appl. Phys. B **52** [1991] 380/4).

3.4.3.2 $[B_3O_6]^{3-}$ Structures

3.4.3.2.1 $Ba_3[B_3O_6]_2$

In technical literature, the formulas β-BaB_2O_4, β-$Ba(B_2O_4)$, or β-$Ba[B_2O_4]$ may be found instead of β-$Ba_3[B_3O_6]_2$. In this section, irrespective of the original writing, β-$Ba_3[B_3O_6]_2$ is used.

$Ba_3[B_3O_6]_2$ exists in two crystalline phases, a high-temperature α-form and a low-temperature β-form (BBO), with a first-order transition temperature of 920±10°C [1] or near 925°C [2, 3]. $Ba_3[B_3O_6]_2$ melts congruently at 1095°C [3].

As was discovered in 1984 [4] and reviewed [5] (see also "Boron Compounds" 3rd Suppl. Vol. 2, 1987, p. 58/9), the β-$Ba_3[B_3O_6]_2$ crystals show second harmonic generation (SHG) after treatment with a copper vapor laser. These crystals are an important new nonlinear optical (NLO) material of good pyroelectric properties (cf. also under $Li[B_3O_5]$, pp. 99/102).

α-$Ba_3[B_3O_6]_2$ crystallizes in the centrosymmetric space group $R\bar{3}c-D_{3d}^6$ (No. 167) [7]. Surprisingly, [8, 9] quote the space group incorrectly as $R3c-C_{3v}^6$ (No. 161) as in β-$Ba_3[B_3O_6]_2$ and state that only the coordination of the Ba^{2+} ions is different. Furthermore, they state that "powder second harmonic generation tests on the quenched high-temperature phase gave signals about as large as the low-temperature phase" [9]. Therefore, [10] repeated an SHG test with an X-ray controlled powder sample of the pure high-temperature form of $Ba_3[B_3O_6]_2$. In contrast to the observation in [8, 9], no SHG was detected for α-$Ba_3[B_3O_6]_2$. The α-form can be stabilized by partial substitution of barium by strontium [11]. A single crystal of the isostructural solid solution $Ba_{3-x}Sr_x[B_3O_6]_2$ (x = 0.93) did not show any longitudinal piezoelectric effect along the c axis. The SHG test on a powdered fragment of that crystal gave no SHG signal at all. All these negative tests are only consistent with the centrosymmetric space group $R\bar{3}c-D_{3d}^6$ (No. 167) for α-$Ba_3[B_3O_6]_2$ [10].

β-$Ba_3[B_3O_6]_2$ crystallizes in the trigonal space group $R3c-C_{3v}^6$ (No. 161) [9, 12, 13, 107] with a = 1253.16(3) and c = 1272.85(9) pm; Z = 18; D_c = 3.850 g/cm³; R = 1.29 and R_w = 1.30%.

The structure is composed of Ba^{2+} cations and planar $B_3O_6^{3-}$ anions. The valence-electron distribution of barium is elongated along the c axis; the charge of barium is +0.74(19). Significant differences in the charges are observed between two independent boron atoms and among four independent oxygen atoms. Of the two crystallographically independent $B_3O_6^{3-}$ anions, the one situated closer to barium has a smaller charge than the other [107]. The space group R3−C_3^4 (No. 146) given elsewhere [14] is used as a basis for some theoretical calculations. But investigations of the electro-optic and piezoelectric effects of β-$Ba_3[B_3O_6]_2$ [15] confirm the point symmetry 3m, consistent with the space group R3c, because the piezoelectric component is $d_{111}=0$ and the electro-optic component is $r_{111}=0$ within the limits of error. Furthermore, no optical activity has been detected, not even with plates of 12 mm thickness. Optical activity should generally be present if the symmetry is 3, but is forbidden if it is 3m. All these arguments clearly require the space group R3c for β-$Ba_3[B_3O_6]_2$ [10].

β-$Ba_3[B_3O_6]_2$ has the hexagonal dimensions a=1254.7(6) and c=1273.6(9) pm with Z=6; R was refined to 3.5% for 2969 data points [9]. β-$Ba_3[B_3O_6]_2$ consists of nearly planar $[B_3O_6]^{3-}$ rings (see "Boron Compounds" 3rd Suppl. Vol. 2, 1987, p. 60, Fig. 3-12), perpendicular to the polar trigonal axis, bonded together through Ba^{2+} ions. The structure has much in common with the graphite structure [7]. β-$Ba_3[B_3O_6]_2$ has a layered structure; the average distance between the $[B_3O_6]^{3-}$ rings is 318 pm. This structure can easily account for the strong growth rate anisotropy where $[B_3O_6]^{3-}$ is a growth unit. The dimensional increase along the a axis or b axis will be four or five times that along the c axis, when one $[B_3O_6]^{3-}$ ring has been added. Therefore, the c axis was expected to be the slow growth direction, and this was, in fact, found to be the case [14].

The electronic structure and the electric dipole matrix elements of a $[B_3O_6]^{3-}$ cluster were calculated by the multiple scattering X_α method. The UV absorption of a β-$Ba_3[B_3O_6]_2$ crystal was determined by this calculation; the absorption edge of 200 nm is in good agreement with experimental results [15].

High-purity β-$Ba_3[B_3O_6]_2$ single crystals were grown at 1048°C by the direct Czochralski method with a temperature gradient of 125°C/mm near the crystal-growing point [108, 109], by the traveling solvent-zone melting method [110], by the top-seeded pulling technique [111 to 113], and by the temperature solution stable growth (TSSG) technique [114, 115] using Na_2O, BaO, and B_2O_3 in the flux. The growth direction was [001] [108]. The development of the β-$Ba_3[B_3O_6]_2$ optical parametric oscillator (OPO) is described in a dissertation. Recent results are reported of growth experiments in the Na_2O flux that increase the size and quality of β-$Ba_3[B_3O_6]_2$ crystals suitable for optical applications. 15 to 23 mm long crystals were produced using large melt volumes and oriented seeds. Decreasing the cooling rate of the 150 mL solution from 3.7 to 1.5°C per day reduces dramatically the density of inclusions in the grown material [116].

Bulk crystals of β-$Ba_3[B_3O_6]_2$ were successfully grown by the top seeded growth method with [2] or without [16] pulling. The flux seeding methods are preferable to the flux pulling methods. Efforts made to find the best flux for the growth led to the liquid temperature curves and to the spontaneous crystallization temperature curves of the $Ba_3[B_3O_6]_2$–BaF_2, Ba_3-$[B_3O_6]_2$–$BaCl_2$, and $Ba_3[B_3O_6]_2$–$BaO\cdot B_2O_3\cdot Na_2O$ systems. Large transparent single-crystals with a size up to 76 mm in diameter and 15 mm thickness were grown by the flux seeding method combined with slow cooling. The crystals were cut into several pieces in order to make second harmonic generation devices [17].

High optical quality β-$Ba_3[B_3O_6]_2$ single-crystal fibers of 100 to 1000 μm diameter and up to 1 cm in length, with orientation in the c axis, were grown by the CO_2 laser-heated pedestal growth (LHPG) method from a solution containing 3 to 25 mol% Na_2O with a growth rate in the

References on pp. 117/22

0.1 to 0.15 mm/min range or from a solution containing 44 wt% B_2O_3 with a growth rate between 0.25 and 0.3 mm/min. Inclusions were observed in some β-$Ba_3[B_3O_6]_2$ fibers and were thought to be produced by constitutional supercooling. In order to grow inclusion-free fibers, stable laser power, constant melting composition, and slow growth rates must be used, along with a smooth growth interface, convex toward the liquid [3]. In another publication, β-$Ba_3[B_3O_6]_2$ crystals were grown by using the $Ba_3[B_3O_6]_2$–$BaO \cdot B_2O_3 \cdot Na_2O$ system. Experimental crystal yields and published phase diagrams quantitatively agree [18].

Microscopic characterization of the inclusions found in the crystal boule is presented, and the applicability of the Na_2O flux in the growth of these crystals is discussed [18]. A dissertation on the growth and applications of these crystals has also been published [19]. β-$Ba_3[B_3O_6]_2$ single-crystals were also obtained by the Czochralski and Stepanov methods from the reaction of H_3BO_3 and $Ba[CO_3]$ by spontaneous crystallization. The increased viscosity and possibility of attaining a high degree of supercooling of the melt allowed easy control of the shape of the sample being crystallized [20, 21]; its luminescence has been studied [22].

The crystal growth of β-$Ba_3[B_3O_6]_2$ was described, and the phase diagrams of the related secondary and ternary section systems $Ba_3[B_3O_6]_2$–Na_2O or $Ba_3[B_3O_6]_2$–$Na_3[B_3O_6]$ [23], $Ba_3[B_3O_6]_2$–Li_2O or $Ba_3[B_3O_6]_2$–$LiBO_2$ [24], $Ba_3[B_3O_6]_2$–K_2O or $Ba_3[B_3O_6]_2$–$K_3[B_3O_6]$ [25], $Ba_3[B_3O_6]_2$–SrO or $Ba_3[B_3O_6]_2$–$Sr(BO_2)_2$ [26], and $Ba_3[B_3O_6]_2$–LiF or $Ba_3[B_3O_6]_2$–$LiBO_2$–LiF [27] were studied.

The enthalpy of fusion of $Ba_3[B_3O_6]_2$ was estimated from the limiting liquidus steps of the $Ba_3[B_3O_6]_2$–$Na_3[B_3O_6]$ and $Ba_3[B_3O_6]_2$–$K_3[B_3O_6]$ systems. The optimized thermodynamic data for these two systems were derived from the experimental phase diagrams. In optimization, the ionic fractions and equivalent fractions were introduced to represent the ideal mixing entropy and excess Gibbs free energy. With these additions, systems such as those without Ba^{2+} ions were compared [28].

The phase relations in the $Ba_3[B_3O_6]_2$–NaCl, $Ba_3[B_3O_6]_2$–CaF_2, and $Ba_3[B_3O_6]_2$–$Na_2[SO_4]$ systems were investigated by X-ray and thermal analysis. The systems of $Ba_3[B_3O_6]_2$ were reviewed in conjunction with other alkali and alkaline earth metal inorganic salts. Rules governing the selection of fluxes for single-crystal growth of β-$Ba_3[B_3O_6]_2$ have been summarized [30].

A review (containing 33 references) of experiments on the bulk growth of β-$Ba_3[B_3O_6]_2$ (BBO) using the top-seeded solution growth method with and without pulling, using Na_2O as a solvent, has been given. The methods yielded relatively large boules; solvent inclusions, however, limit the yield of high-optical-quality crystals from these boules. Oriented, single-crystal fibers have also been grown, from both Na_2O and B_2O_3 solutions, using the laser-heated pedestal growth technique. B_2O_3 solutions are stable under laser heating and yield high-optical-quality single-crystal fibers with minimal inclusions [31]. Such crystals were grown under high and low thermal gradient conditions. High thermal gradients allowed faster growth rates and reduced the density of inclusions which are a persistent problem with solution-grown β-$Ba_3[B_3O_6]_2$. Crystals grown along the y axis fracture along the basal cleavage phases during cooling, but otherwise yield useful crystal dimensions and good optical quality material. Inclusions usually contain the solvent phase but voids were also observed. Interface breakdown resulting from constitutional supercooling is thought to be related to the restricted mass transport in these highly viscous solutions [32].

β-$Ba_3[B_3O_6]_2$ was also grown in a new flux composition consisting of NaCl and Na_2O. Crystal growth rate was very fast in a pure NaCl flux with well-developed facets. Na_2O was added as a retardant to slow down both the growth rate and spurious nucleation. The crystal habit also changes from long-prismatic shape to a more semispherical one. As the Na_2O concentration was increased, the crystal clarity was reduced because of more severe flux inclusion. It is believed that this is due to the contamination by carbonate in the flux since $Na_2[CO_3]$ was used as a precursor of Na_2O [33, 34]. Crystal growth by a micro-Czochralski technique of ca. 7 mm length and ca. 30 μm thickness single crystals of β-$Ba_3[B_3O_6]_2$ is discussed [35]. Experimental evidence is obtained that both α-and β-$Ba_3[B_3O_6]_2$ can be grown directly from pure melts. Growth of α-phase material onto a Pt rod is initiated by α-phase structural centers present in the melts which are stable at temperatures near or just above the melting point. Overheated melts in which solid nuclei are destroyed have modified freezing characteristics and form β-phase material which can be produced onto a Pt rod at temperatures well above the reported phase transition temperature [36].

Starting from a general quantum-mechanical perturbation theory on the nonlinear optical (NLO) effects in crystals, the "anionic group theory for the NLO effect of crystals" was developed, and approximations were discussed. For the second harmonic generation coefficients, calculations have been performed and the theoretical values have been compared with experimental values obtained on both powdered and single-crystal materials; they suffice to show the feasibility of the theoretical treatment and calculation methods. On this basis, borate ions of various structure types are classified, and systematic calculations are carried out for the NLO susceptibilities of some typical borate crystals with good prospects of application in optoelectronics; the best NLO effects are shown for β-$Ba_3[B_3O_6]_2$ crystals [37].

A new technique of double resonance with spin mixing by continuous coupling was used for the observation of β-$Ba_3[B_3O_6]_2$ NMR-NQR spectra. The quadrupole coupling data were discussed in the terms of the crystal structures [117]. Nonlinear optical properties of β-$Ba_3[B_3O_6]_2$ such as laser harmonic generations are given. Its laser damage thresholds, energy conversion efficiencies, peak power conversion efficiencies, output powers, tuning wavelengths from 415 to 2411 nm and repetition rates as well as all typical applications in nonlinear optics and quantum electronics are described in numerous recent publications [118 to 142].

At 85 K, the photoluminescence (PL), thermally stimulated luminescence (TSL) and their excitation spectra, optical absorption spectra, and X-ray luminescence spectra in the UV and visible regions were studied of undoped crystals of α-$Ba_3[B_3O_6]_2$. The characteristic absorption edge was established at 195 nm and the beginning of the effective excitation of the band-band transitions by light with a wavelength of 165 nm. It is proposed that the absorption between 175 and 180 nm is related to excitons. The PL spectra contain several overlapping bands in the range between 290 and 500 nm. The fundamental peaks of TSL are observed at 280 and 340 K [22].

The IR reflection spectra (100 to 400 cm^{-1}) of two different polarized components of β-$Ba_3[B_3O_6]_2$ single-crystals at room temperature show transverse and longitudinal frequencies of the fundamental vibration modes at 200 cm^{-1}. The spectra were analyzed in terms of the layer-molecular structure model with reference to the Raman spectra. The assignment of the inner fundamental vibration modes of the crystals was achieved [38].

Raman spectra of β-$Ba_3[B_3O_6]_2$ at 84 K show the directional dispersion of extraordinary phonons and the vibration-intensity relations between the $[B_3O_6]^{3-}$ ring and the crystal [143]. A high-pressure Raman study of β-$Ba_3[B_3O_6]_2$ reveals four phase transitions [144]. Room temperature Raman spectra of α- and β-$Ba_3[B_3O_6]_2$ [145] as well as polarized IR reflection spectra are also studied [146].

Lattice vibrations of a β-$Ba_3[B_3O_6]_2$ crystal were analyzed by using the group theory. It was (wrongly) assumed that a β-$Ba_3[B_3O_6]_2$ crystal belongs to the space group R3$-C_3^4$ (No. 146). Its vibration modes were classified by the site-symmetry method. All modes of A and E symmetry are both Raman and IR active. Using a model called the layer-molecular structure, the

References on pp. 117/22

lines of the Raman spectra observed in different geometric configurations can be explained satisfactorily with a preliminary assignment of the internal and external modes of the β-Ba_3-$[B_3O_6]_2$ crystal [39].

Raman and infrared spectra of crystalline β-$Ba_3[B_3O_6]_2$ were completely assigned with all external and internal vibration modes of the metaborate ring of symmetry A: at 1540, 1525 cm^{-1}, and 771 to 788 cm^{-1} (Raman), at 733 and 702 cm^{-1} (IR), and at 638, 620, and 599 cm^{-1} (Raman); and of symmetry E: at 1508, 1480, 1444, 1405, 1280, 1263, 1240, 1191, 964, and 951 cm^{-1} (IR), as well as at 493, 482, 393, and 382 cm^{-1} (Raman). The site group effect, the isotopic effect, and the dynamic coupling between the external and internal motion of the $[B_3O_6]^{3-}$ ring and the Ba^{2+} ion were discussed. Force constants were also calculated with the prediction that there is one internal vibration mode of the $[B_3O_6]^{3-}$ ring at a frequency of the lattice modes. Multipolar coupling between the $[B_3O_6]^{3-}$ rings in this crystal was also analyzed with the generalization of the dipolar coupling theory [40]. The dipolar coupling method for crystals was used to calculate the induced dipole moment of the $[B_3O_6]^{3-}$ ring of the β-Ba_3-$[B_3O_6]_2$ crystal. The electric moment of the B_3O_6 unit is multipolar and cannot be represented by a point dipole moment at the center of the cyclic structure. When the electric moment of the B_3O_6 unit is approximated by the three dipole moments at the positions of the three boron atoms, the calculated electronic polarizabilities along the optic axis and the optic plane are the same. As the wavelength of light is shorter, its electronic polarizability becomes larger [41].

α- and β-$Ba_3[B_3O_6]_2$ crystals were also studied by Raman spectroscopy between 25 and 400 °C. As the temperature rises, the frequencies of the internal modes are unaffected, but those of the lattice modes decrease by ca. 3 cm^{-1}. The phase transition occurs over a wide temperature range with no significant spectroscopic changes as the temperature rises except enhancement of the translational mode of the Ba^{2+} ions at 86 cm^{-1}. Most of the enhanced intensity is of E symmetry. The enhancement is most drastic around 217 and 317 °C. The phase transition probably is mainly due to the rearrangement of Ba^{2+} in the crystal. As the temperature rises, more Ba^{2+} ions are shifted to the sites of C_3 and D_3 symmetries. The role of Ba^{2+} is discussed along with its relationship with the second harmonic generation property of the β-$Ba_3[B_3O_6]_2$ crystals [42].

In studying the intermolecular forces from the pressure dependency of Raman scattering, the Raman spectra of single crystals of β-$Ba_3[B_3O_6]_2$ were investigated as a function of the hydrostatic pressure from zero to 65 kbar. Around 50 kbar, a pressure-induced structural phase transition was observed. External modes at ambient pressure have the following frequencies (in cm^{-1}): 60, 73, 87, 96, 114, 120, 144, 161, 180, 200, and 245 (E modes); 58, 73, 99, 124, 172, and 197 (A modes). These external modes are more sensitive to pressure than the internal E mode lines at 382, 390, 480, and 490 cm^{-1} (all E'' out-of-plane modes); 644 and 695 cm^{-1} (both E' in-plane ring-angle bending modes); 969, 1303, and 1403 cm^{-1} (all E' in-plane ring-stretching modes), and 1229 cm^{-1} (scattering band of two phonons) or the internal A mode lines at 598 and 620 cm^{-1} (both A_2' in-plane ring-angle bending modes); 638, 770, and 788 cm^{-1} (all A_1' in-plane stretching modes); 1499 cm^{-1} (A_1' in-plane ring-angle bending mode); 1526 and 1541 cm^{-1} (both A_1' in-plane ring-stretching modes); 1331, 1429, and 1698 cm^{-1} (all scattering bands of two phonons). Near 50 kbar, the frequencies of the most internal mode peaks have sudden changes, which enables one to identify the external and internal mode lines and relates the internal mode lines with the vibration modes of the $[B_3O_6]^{3-}$ ring. The softening of the mode at 664 cm^{-1} with the pressure is attributed to the transfer of the bond charge. Also, the scaling behavior of the crystal was analyzed [43].

Dielectric and pyroelectric properties of single crystals of β-$Ba_3[B_3O_6]_2$ are measured between +50 and −190 °C, while thermal expansion and piezoelectric properties are measured at room temperature. Piezoelectric and electromechanic coupling coefficients in these crystals are low whereas the pyroelectric figure of merit p/K is quite attractive for some device applications [44].

The relationship between pyroelectric properties and structures of polar crystals containing six-membered borate rings such as β-$Ba_3[B_3O_6]_2$ was studied. These compounds have low dielectric constants, a stable polarity, and small piezoelectric effects [45]. The electrical properties of β-$Ba_3[B_3O_6]_2$ were studied by experimental determinations of the pyroelectric coefficients, the dielectric constants, the loss tangent, the d.c. resistivity, the heat capacity, and the optical transmittance. The possible use of this material as a high-efficiency infrared pyroelectric detector was discussed [46]. When an infrared pulse laser beam is incident upon the center of a thin slab of a β-$Ba_3[B_3O_6]_2$ crystal cut along the c axis, the variation of the temperature in the slab with time and space was studied. On this basis, and on the basis of measurements of various parameters of the crystal, the prospect of the application of a β-Ba_3-$[B_3O_6]_2$ crystal to a side-electrode high power infrared pulse laser energy meter is discussed [47].

Data from dielectric measurements, differential thermal analysis (DTA), and the temperature-dependence of the second harmonic generation (SHG) indicated the presence of phase transitions at 700 and 1153 K. In β-$Ba_3[B_3O_6]_2$ crystals, the SHG was obtained by a Cu laser with the wavelengths $\lambda = 257.2$ and 514.5 nm; the intensity of the second harmonic output was negligible. Damage to the crystal by the laser beam was not observed [20].

Experimental results are reported on upper harmonic conversion with picosecond pulses using an $Li[NbO_3]$ and a β-$Ba_3[B_3O_6]_2$ crystal to quadrupole the free-electron laser light up into the visible and near-infrared region. The effects of finite linewidth, birefrigent walk-off, and group velocity walk-off on conversion efficiency are discussed with reference to the experimental results [48].

A review with 14 references describing the properties, advantages, and limitations of β-Ba_3-$[B_3O_6]_2$ (BBO) in laser technology was given [49]. Notes on the laser techniques for obtaining resonance Raman spectra in the vacuum far-UV region are discussed with emphasis on the use of β-$Ba_3[B_3O_6]_2$ for harmonic generations [50]. Another review shows that β-$Ba_3[B_3O_6]_2$, a major contributor in the development of the field of nonlinear optical materials (NLO), will find its niche in UV generation on the basis of its phase-matching properties [51]. A critical review is given (43 references) presenting features of nonlinear crystals of β-$Ba_3[B_3O_6]_2$, $Li[B_3O_5]$, $K[TiO][PO_4]$, $K[NbO_3]$, $MgO:Li[NbO_3]$, and YAG (yttrium aluminium garnet) and comparing them for various applications such as compact blue laser devices [52].

β-$Ba_3[B_3O_6]_2$ crystals are apparently about an order of magnitude more resistant to thermal fracture than the common nonlinear and electro-optic materials $K[D_2PO_4]$, $Li[IO_3]$, and $Li[NbO_3]$ with fracture temperatures between 3 and 20°C [8]. For β-$Ba_3[B_3O_6]_2$, some constants have been measured: the optical absorption, the refractive indices from the UV to the near-infrared, the thermo-optic coefficients, the nonlinear optical (NLO) coefficients (for example, $d_{11} = 1.6 \pm 0.4$ pm/V, relative to the value d_{36} of $K[D_2PO_4]$ for the second harmonic generation of 1064 nm), the resistance to laser damage, the elastic constants (for instance, $K_{11} = 6.7$ and $r_{11} = 2.7$ pm/V), and the fracture toughness 150 $kPa \cdot m^{0.5}$. These data are used in order to evaluate β-$Ba_3[B_3O_6]_2$ for a variety of applications, especially in the 250 nm UV which permits deep UV generation [9]. The optical transmission characteristics of β-$Ba_3[B_3O_6]_2$ were studied between 185 and 220 nm; the results were compared with those of $[NH_4][H_2PO_4]$ (ADP) and urea crystals in the same wavelength region. The optical transmission was found to depend on surface polishing [53].

A number of experimentally determined dispersions of ordinary and extraordinary refractive indices in β-$Ba_3[B_3O_6]_2$ crystals gave the dispersion curves for the phase synchronism of the

References on pp. 117/22

first and second type. Both curves are nonmonotonous, and in both instances the angle of phase synchronism sharply increases in vicinity of the crystal intrinsic absorption edge. The results can be useful for selecting optimal conditions for the harmonic generation in β-Ba_3-$[B_3O_6]_2$ crystals [147].

All components of the unclamped linear electro-optic tensor $[r^{\sigma}_{ijk}]$ of low-temperature β-Ba_3-$[B_3O_6]_2$ and the linear electrostrictive tensor $[d_{ijk}]$ were determined at 293 K and $\lambda = 633$ nm: $r_{113} = 0.27$, $r_{222} = -2.41$, $r_{333} = 0.29$, $r_{131} = 1.7$ pm/V; $d_{311} = -1.17$, $d_{222} = 2.30$, $d_{333} = 3.4$, $d_{113} = -9.6$ pm/V. In its electro-optic and electrostrictive behavior, low-temperature β-$Ba_3[B_3O_6]_2$ resembles the tetraborates $Li_2[B_4O_7]$ and $Pb[B_4O_7]$. The electro-optic effect is only about 1/4th of that found in $K[H_2PO_4]$. This reduces the importance of β-$Ba_3[B_3O_6]_2$ for electro-optic applications. However, its electrostrictive effect is four times that of α-quartz [148].

β-$Ba_3[B_3O_6]_2$ might also be an excellent candidate for use in a high average power Pockels cell. The linear electro-optic coefficients had been measured in order to evaluate this possibility. The clamped (S) and unclamped (T) values are: $|r^S_{yyy}| = (0.24 \pm 0.03) \cdot r^S_{xyz}$ of $K[D_2PO_4]$ and $|r^S_c| = (0.013 \pm 0.002) \cdot r^S_{xyz}$ of $K[D_2PO_4]$ as well as $|r^T_{yyy}| = 2.5 \pm 0.1$ pm/V and $|r^T_c| = 0.17 \pm 0.02$ pm/V, where $r_c = r_{zzz} - (n_o/n_e)3r_{xxz}$. The magnitude of the electro-optic coefficients is due mainly to the electronic nonlinearity [54]. The expected half-wave voltages and thermomechanical parameters of β-$Ba_3[B_3O_6]_2$ plates in various configurations are calculated and compared with those of other materials of interest. Its resistance to thermal fracture, high damage threshold, and its wide range of transparency, make β-$Ba_3[B_3O_6]_2$ an excellent candidate for use in a high average power Pockels cell, and as an intracavity laser Q-switch [55].

The optically uniaxial crystals of β-$Ba_3[B_3O_6]_2$ have a number of important properties. They exhibit large effective second harmonic generation coefficients (with d_{eff} being six times the d_{eff} of $K[H_2PO_4]$ at 1.06 μm), exhibit wide transparent waveband (transparent range extends from 190 to 3500 nm), show large birefringence and low dispersion. These properties allow phase-matching for harmonic generation from 200 to 1500 nm. Furthermore, it possesses a high damage threshold (13.5 ± 2.0 GW/cm² for 1 ns pulses at 1.06 μm, and 7.0 ± 1.0 GW/cm² for 250 ps pulses at 0.532 μm), high optical homogeneity ($\Delta n \approx 10^{-6}/cm^{-1}$), nondeliquescence, and good mechanical properties [3].

β-$Ba_3[B_3O_6]_2$ is a crystal with great promise for device applications. Particular attractive features are its UV transparency, its large birefringence, large nonlinear optical (NLO) coefficients, and good mechanical and thermal properties. Recent experiments demonstrate its effectiveness as a frequency converter into UV. It is also a leading candidate for use in high-power electro-optic and frequency conversion devices, where low linear absorption and high fracture temperatures are important [18]. The largest linear electro-optical coefficient is $r_{22} = 2.5$ pm/V, approximately one fourth the size of the value r_{63} of $K[H_2PO_4]$. The dispersion of the electro-optical coefficients was measured and was in agreement with predicted values using the anharmonic oscillator model. Implications of the result on the point group symmetry of β-$Ba_3[B_3O_6]_2$ are discussed [56].

β-$Ba_3[B_3O_6]_2$ was evaluated for its use in the measurement of ultrashort optical pulses. Pulses as short as 50 fs can be measured accurately with a crystal of 1.8 mm thickness. The advantages of using β-$Ba_3[B_3O_6]_2$ rather than $K[H_2PO_4]$ (KDP) or $[NH_4][H_2PO_4]$ (ADP) for autocorrelation measurements are discussed [57]. It has a relatively large group velocity dispersion which tends to broaden the generated femtosecond pulses and make the pulse width measurements difficult unless extremely thin crystals are used. Such crystals, in turn, are difficult to fabricate and polish. The result of an intracavity doubling experiment leads to nearly complete recovery of the normal fundamental output power in the form of UV pulses at the second harmonic generation. Pulses as short as 43 fs at a 108 Hz repetition rate and outputs for somewhat longer pulses as high as 20 mW per arm of the femtosecond laser on a continuous-wave basis can be achieved. The UV pulse widths were determined through detailed cross-correlation measurements based on sum-frequency mixing to 210 nm in ultrathin β-Ba_3-$[B_3O_6]_2$ crystals [58].

β-$Ba_3[B_3O_6]_2$, an optical parametric oscillator (OPO) tunable in the visible and in the near infrared, was pumped at 354.7 nm and was tunable from 0.45 to 1.68 μm. A maximal total energy conversion efficiency of 9.4% was measured from the OPO containing a 10.5×10.5×11.5 mm size, 30° cut β-$Ba_3[B_3O_6]_2$ crystal [59]; this was grown as described elsewhere [18]. Optical parametric luminescence was used in order to measure the 354.7 and 266 nm pumped OPO tuning curves (type-I and type-II) in β-$Ba_3[B_3O_6]_2$ from ca. 0.3 to 3.3 μm. The implication of these results on the validity of the published Sellmeier equations was discussed [59]; the best was reported in [9].

The operation is reported of a β-$Ba_3[B_3O_6]_2$ singly resonant optical parametric oscillator (OPO) broadly tunable in the UV, visible and IR regions of the spectrum. The OPO was pumped at 266 nm and was continuously tunable throughout 0.33 to 1.37 μm. A new cavity design was implemented which has applicability to other parametric oscillators. The oscillation threshold of this device was measured and compared to the calculated results [60]. A new, 2-crystal, walk-off-compensated OPO design using β-$Ba_3[B_3O_6]_2$ is described, which significantly improves the performance of the device. The oscillator is pumped at 354.7 nm and is tunable throughout 0.42 to 2.3 μm with overall conversion efficiencies as high as 32%. A demonstration is given of linewidth narrowing in the same β-$Ba_3[B_3O_6]_2$ oscillator, obtaining linewidths as narrow as 0.3 Å [61].

Laser-induced damage in β-$Ba_3[B_3O_6]_2$ was investigated at 1064, 532, and 354.7 nm. The single-shot bulk damage threshold is 50 GW/cm² at 1064 nm, 48 GW/cm² at 532 nm, and 25 GW/cm² at 354.7 nm. Damage by multiple pulse irradiation was also studied and the probabilistic nature of the damage is discussed [62].

Three new absorption steps are reported appearing at the long wavelength side of the UV intrinsic absorption edge of the β-$Ba_3[B_3O_6]_2$ crystal at low temperatures. These new absorption steps should be attributed to the transitions of the indirect excitons bound in the centers of certain neutral impurities contained in the crystal specimens [63]. The characteristics of the intrinsic absorption edge spectra (from 50 to ca. 350 nm) and their temperature dependence can be well explained with the theory of the indirect interband transition. The indirect absorption edge of this crystal is $E_0=6.24$ eV at room temperature, and the average temperature coefficient of the indirect band gap is -6.5×10^{-4} eV/K (from 288 to ca. 363 K). The types and Debye temperatures of the phonons participating in the indirect transition were also determined [64].

Both calculated and experimental results show that second harmonics of all spectral lines of Ar^+ laser light can be produced simply by angular tuning of β-$Ba_3[B_3O_6]_2$ crystals at room temperature. This is used in the autocorrelation measurement of modelocked Ar^+ laser pulses, and the Ar^+ laser pulses, with four stronger spectral lines, were measured to be 180 to ca. 230 ps at room temperature, overcoming the shortcoming of the previous second harmonic generation (SHG) method at low temperatures and cross-correlation method [65].

SHG coefficients and the UV absorption edge of β-$Ba_3[B_3O_6]_2$ were calculated using the extended Hückel molecular orbital (EHMO) theory. The calculated and experimental values are in good agreement. The UV absorption edge was also calculated by using the more accurate self-consistent field X_α scattered wave (SCF-X_α-SW) method, and the calculated results support the EHMO calculation. The factors which determined the positions of the UV absorption edge and the values of the SHG coefficients are discussed [66].

References on pp. 117/22

SHG coefficients for a β-$Ba_3[B_3O_6]_2$ crystal were also calculated by using complete neglect of differential overlap (CNDO/S) approximations. The calculated values agree satisfactorily with the experimental data for both the absolute values and the relative sign of the above coefficients. The result shows that the conjugate π orbitals of $[B_3O_6]^{3-}$ groups are major contributors to the SHG coefficients. The CNDO/S approximation is a suitable method for calculating molecular orbitals of covalent-bonded $[B_3O_6]^{3-}$ groups in crystals [67]. The frequency dependent second-order hyperpolarizability tensor of β-$Ba_3[B_3O_6]_2$ is computed for the $[B_3O_6]^{3-}$ ion using the conventional semiempirical perturbation theory with variably-sized basis sets generated by CNDO/S, CNDO-CI, as well as by intermediate neglect of differential overlap approximations (INDO/2-CI). The principal source of the second-order nonlinear optical (NLO) properties response arises from *O* to *B* charge transfer states. The second-order susceptibility of the crystal is deduced using the oriented gas mode for the ions in the crystal structure [68].

A general theory for three-wave-mixing, valid for both type-I and type-II $Ba_3[B_3O_6]_2$ crystals is presented. Analytical expressions for the second harmonic generation (SHG) efficiencies were compared with the numerically exact solution. The effects of beam walk-off and spatial profile overlap for various polarizations are included. The integrated efficiency for a divergent pumped beam was calculated for various ratios of beam divergence and crystal angular acceptance width. Experimental data for SHG of 1064 nm (YAG laser) are presented and compared for type-I and type-II $Ba_3[B_3O_6]_2$ crystals. From the measured data, it is concluded that the simple model based upon the ratio of laser power and beam quality or inverse-square of the angular sensitivity overestimates the effects of beam quality on the efficiency. A rigorous model including the complicated coupling between laser power, beam quality, and the efficiency was presented which correctly explains the data [69].

Phase matching angle curves of SHG, third harmonic generation (THG), sum-frequency mixing and optical parametric oscillator (OPO) for various crystals are generated based on the available refractive indices. The new crystal β-$Ba_3[B_3O_6]_2$ is emphasized whose basic properties, applications, and some of the recent experimental data are shown [70].

According to the equations of three interacting plane waves and the small-signal approximation, spectral and angular half-widths are calculated with β-$Ba_3[B_3O_6]_2$ for frequency doubling, tripling, and quadrupling (λ_1 = 1.064 μm). The calculated enhancement factor $2N^3 + N/3$ of the second harmonic generation (SHG) shows that the processes of side-band sum-frequency are important for modelocked laser pulse frequency conversion [71].

The nonlinear optical properties of β-$Ba_3[B_3O_6]_2$ are demonstrated in the generation of 2nd through 5th harmonics at 1.06 μm Nd laser radiation and in an optical parametric oscillator (OPO) pumped by 532 nm radiation. The species is particularly useful for high-average-power applications and nonlinear frequency generation in the UV to wavelengths as short as 200 nm. An internal energy conversion efficiency of 84% for 1064 to 532 nm SHG, and cascade harmonic conversion to the 213 nm fifth harmonic with 11% overall conversion were obtained. The observed performance agrees well with that predicted by modeling [72]. Also, a visible $Ba_3[B_3O_6]_2$ OPO pumped by a (single-axial-mode) 355 nm source was demonstrated; an average output power of 140 mW with a signal-wave-conversion efficiency of 13% and an idler-conversion efficiency of 11% for a total conversion efficiency of 24% was achieved. The oscillator has continuously tuned 412 to 2550 nm, limited by the infrared transmission range of the crystal. Through injection seeding, single-axial-mode OPO operation with a corresponding OPO linewidth of 3 GHz was obtained [73].

$Ba_3[B_3O_6]_2$ was also used to generate the 5th harmonic at 212.8 nm with an output pulse energy of 20 mJ and a pulse width of 5 ns [74]. A $Ba_3[B_3O_6]_2$ crystal was optical parametric oscillator pumped at 532 nm; the pump source was the second harmonic output of a Q-switched TEM_{00} mode Nd:YAG laser. The OPO had an output energy of 1 mJ with a peak power of 80 kW tuning between 0.94 and 1.22 μm [75]. High-power tunable UV laser radiation between 275 and 320 nm was obtained by second harmonic generation in a β-$Ba_3[B_3O_6]_2$ crystal with a maximal frequency doubling efficiency of 41% and a UV output power of 3.7 MW [76]. High-power tunable UV coherent radiation was also achieved by sum-frequency generation (SFG) in β-$Ba_3[B_3O_6]_2$ crystals. The maximal output power obtained is ca. 3 MW (near 368 nm) and 0.75 MW (near 222.5 nm) [77].

Efficient and tunable coherent UV generation between 360 and 390 nm in β-$Ba_3[B_3O_6]_2$ crystals is presented using type-I phase matching at room temperature. The phase matching angle was characterized by a $Be(AlO_2)_2$ laser with a wavelength tuning range between 725 and 785 nm. The crystal angular band width of 0.9 mrad·cm and spectral band width of 1100 nm ·cm were measured. A UV output pulse energy of 105 mJ at 378 nm with 31% energy conversion efficiency was achieved [78].

The type-II phase-matched third harmonic generation (THG) in a β-$Ba_3[B_3O_6]_2$ crystal was studied experimentally. A passively modelocked Nd:phosphate glass laser was used as a pump source. At a pump pulse peak intensity of the input pump peak intensity $I_{10}=5\times10^{10}$ W/cm^2, a third harmonic conversion efficiency of 1.0% was obtained. A theoretical discussion of phase-matched THG in crystals of the symmetry group of β-$Ba_3[B_3O_6]_2$ (trigonal class 3) is given. The effective nonlinear susceptibility for type-II phase THG is $\chi_{eff}=(6.4\pm2.8)\times10^{-23}$ m^2/V^2 [79]. The damage threshold is ca. 10^{12} W/cm^2 for single pulses of 5 ps duration. At pump pulse intensities slightly below the damage threshold, very high conversion efficiencies are expected [80].

Single-pass spontaneous parametric emission and amplification was demonstrated in relatively short (6.5 mm) β-$Ba_3[B_3O_6]_2$ crystals. The observed angle tuning curves, in close agreement with the calculated ones, revealed a wide (20000 cm^{-1}) tuning range. For subsequent amplification, the spectral width of the parametric emission was also studied. Again, a relatively good agreement between theory and experiment was obtained. From these data, an optical parametric source (OPS) can be constructed which fulfills the particular spectral needs of a given application. With the output of the OPS frequency doubled in a subsequent β-$Ba_3[B_3O_6]_2$ crystal, the entire wavelength region between 205 nm and 2.6 μm can be covered using the third harmonic of an Nd laser system as the pump source [81]. The usefulness of β-$Ba_3[B_3O_6]_2$ for the generation of high pulse energies at ca. 200 nm by THG of a narrow-band pulsed dye laser was demonstrated. The crystal showed efficiencies higher than $K[B_5O_8]\cdot4H_2O$ and allowed phase-matched THG to a short wavelength limit of 197 nm [82].

Coupled wave equations were solved for large signals by means of numerical integration. A numerical method for calculating second harmonic generation (SHG) and third harmonic generation (THG) efficiencies of beams with Gaussian temporal profiles is given. A high-efficiency frequency tripling scheme of 1.06 μm laser radiation is presented for low and moderate power levels. With type-I β-$Ba_3[B_3O_6]_2$ crystals used in doubling and type-II $K[D_2PO_4]$ crystals used in tripling, energy conversion efficiencies of THG of ≤41.8% were achieved at 365 mJ energy [83]. The frequency of 492 to 550 nm Coumarin 500 dye laser light was multipled by using a β-$Ba_3[B_3O_6]_2$ crystal, then the SHG was mixed with the 1064 nm light generated by an Nd:YAG laser; 200 to 218 nm tunable UV radiation was obtained [84]. Thus, β-$Ba_3[B_3O_6]_2$ is phase-matchable for SHG down to 204.8 nm at room temperature. Sellmeier's equations are highly accurate from 1.064 μm to 205 nm [85]. The 5th harmonic generation by mixing the Nd:YAG laser fundamental and its 4th harmonic radiation using a β-$Ba_3[B_3O_6]_2$ crystal at room temperature is reported. The energy conversion efficiencies of the 4th harmonic to the 5th harmonic of ca. 48% were achieved. A numerical method was used to

References on pp. 117/22

analyze the process of mixing 1064 and 266 nm, and its results agree with the experimental results [86].

A β-$Ba_3[B_3O_6]_2$ crystal was also used to obtain wavelengths between 188.9 and 197 nm by type-I sum-frequency generation (SFG). The fundamental beams were supplied by a pulsed dye laser tuned between 780 and 950 nm and a second laser operating at 248.5 nm. Four different sources for the 248.5 nm wavelength were used. The output at the ArF excimer wavelength at 193 nm was demonstrated [87]. SFG in a cooled β-$Ba_3[B_3O_6]_2$ crystal was studied by mixing the unpolarized output of the excimer laser-pumped tunable dye laser (Rhodamine 6G or Rhodamine B, both in methanol) with its second harmonic. The shortest wavelength obtained at 95 K was 195.3 nm. A lower limit of 194.4 ± 0.2 nm is expected at the temperature of liquid helium [88]. Wavelengths between 188.9 and 197 nm were also obtained by type-I SFG in β-$Ba_3[B_3O_6]_2$. The fundamental beams were supplied by pulsed dye lasers, one of which was tuned between 780 and 950 nm and the other frequency-doubled at 497 nm. The possibility of shifting the excimer wavelength 248.5 nm to the excimer wavelength 193 nm was achieved by replacing the frequency-doubled dye laser by KrF excimer lasers of different beam properties. By means of a picosecond excimer laser, a second harmonic generation efficiency of 7% has been observed [89]. By frequency-doubling of a pulsed dye laser in β-$Ba_3[B_3O_6]_2$, UV radiation of ca. 205 nm can be generated with high efficiency. A simple method of SFG extends the wavelength range close to 197 nm, in a cooled crystal even to 195 nm. The most interesting excimer laser wavelength 193 nm can be obtained by SFG using an IR dye laser and an excimer laser operating at 248 nm. The lowest wavelength obtained in this way is 189 nm [90].

The optical characteristics of β-$Ba_3[B_3O_6]_2$, including the group-velocity dispersion and the nonlinear coefficients in the femtosecond region, are investigated through second harmonic generation (SHG). The performances of this crystal relative to $K[H_2PO_4]$ were evaluated from both calculations and measurements. From the measurements using crystals of different thicknesses, the influence of the correlation width is revealed, and is smaller than that of $K[H_2PO_4]$. Overall performance of the SHG conversion efficiency of β-$Ba_3[B_3O_6]_2$ including all affecting factors is evaluated to show higher values by a factor of 10 to 10^2 than $K[H_2PO_4]$ depending on operating conditions [91]. Noncollinearly phase-matched SHG is reported in a β-$Ba_3[B_3O_6]_2$ crystal and is compared with existing $K[H_2PO_4]$ (KDP) and $K[D_2PO_4]$ (KD*P) crystals using Nd:YAG radiation [92].

Second harmonic generation from orthogonally polarized Nd:YAG laser radiation was studied in a β-$Ba_3[B_3O_6]_2$ crystal as a function of the degree of noncollinearity between the fundamental beams. The experimental phase-matching angles agree well with predicted values [93]. Generation of tunable IR radiation between 3 and 3.4 μm uses difference-frequency mixing of the SHG of Nd:YAG laser radiation with Rhodamine B dye laser radiation in a β-$Ba_3[B_3O_6]_2$ crystal. The interaction is phase-matchable, and calculations showed that the overall efficiency of conversion in the IR can be 10% compared to less than 1% in the $AgGaAs_2$ system. The Sellmeier coefficients for β-$Ba_3[B_3O_6]_2$ in the formula for the refractive-index dispersion $n^2 = A + B(1 - C\lambda^{-2})^{-1} + D(1 - E\lambda^{-2})^{-1}$ are given in Table 3/13 [94].

Table 3/13
Sellmeier Coefficients for β-$Ba_3[B_3O_6]_2$ [94].

polarization	A	B	C	D	E
O	1.46357	1.26172	0.01628	0.00166	30
E	1.40567	0.95869	0.01431	0.01644	30

The third harmonic generation (THG) of an Nd:YAG laser was generated in a β-$Ba_3[B_3O_6]_2$ (BBO) crystal by noncollinear sum-frequency mixing of the second harmonic generation (SHG) with the fundamental beam from an Nd:YAG laser. Tunable UV radiation is also generated by the second harmonic of different dye lasers in this crystal [95].

THG was achieved by sum frequency mixing the SHG of the 23 W YAG laser at 532 nm with the fundamental at 1064 nm in a type-I angle-tuned (critically) phase-matched β-$Ba_3[B_3O_6]_2$ crystal with a power density of 250 MW/cm². A THG (355 nm) power of nearly 2 W was demonstrated from a continuous wave mode-locked YAG laser (76 MHz), and the successful utilization is described of this new capability as a pump source for a dispersion compensated two jet Stilbene 3 dye laser. In the mixing process, the third harmonic power P_{355} is proportional to the product of the power in the fundamental P_{1064} and the second harmonic power P_{532} [97].

The first cascading THG is reported for an alexandrite laser in β-$Ba_3[B_3O_6]_2$. UV radiation was generated by mixing the fundamental laser beam with the second harmonic. The calculated effective nonlinear constant for type-I mixing is nearly three times larger than that for type-II. Using type-I mixing, UV generation between 244 and 259 nm was achieved with 24% mixing efficiency at maximal wavelength [98].

Efficient and tunable deep-UV (205 to 310 nm) generation by frequency coupling in β-Ba_3-$[B_3O_6]_2$ crystals was studied using type-I phase matching at room temperature. A second harmonic generation (SHG) efficiency of 10% was obtained at 206 nm wavelength. The maximal SHG efficiency was 36% for a fundamental dye laser power of 150 kW; the SHG efficiency at $\lambda=310$ nm was 4 to 6 times larger than in $[NH_4][H_2PO_4]$ [96].

In order to obtain an efficient and high-repetition-rate UV light, the 255.3 nm second harmonic of a copper vapor laser was generated in $Ba_3[B_3O_6]_2$ (nonlinear in the UV region). A high quality laser beam from an unstable resonator was focused in the crystal to achieve high power density. The conversion efficiency of 9% was obtained at an average input power of 2.5 W. The conversion efficiency depends on several parameters such as the peak power, the beam size and divergence, the focal length of the lens, and the crystal length. A parametric study was made to find the optimal operation conditions. A new theory was introduced which is valid for the harmonic generation of partially coherent beams. The temporal behavior of the process was studied [99].

A versatile tunable, continuous-wave UV source for any spectroscopy is described. An angle-tuned β-$Ba_3[B_3O_6]_2$ frequency doubler was placed inside a doubly resonant buildup cavity. By compensating for double refraction, this scheme provided efficient SHG in a pure TEM_{00} mode [100].

An optical parametric oscillator (OPO) in β-$Ba_3[B_3O_6]_2$ was demonstrated using synchronous pumping by the frequency doubled output train from an actively modelocked Nd:YAG laser. The parametric oscillator converts up to ca. 30% of the pump train to produce broadly tunable pulses (0.68 to 2.4 μm) of ca. 75 ps duration, with peak idler powers of up to ca. 1.6 MW. The tuning range was extended up to 0.53 μm by frequency doubling the idler output with up to ca. 3% efficiency [101].

A tunable OPO in a β-$Ba_3[B_3O_6]_2$ crystal was achieved with a 308 nm XeCl pumped laser source. Signal wavelengths between 422 and 477 nm were generated by angle tuning at energy-conversion efficiencies of as much as ca. 10% [102].

Other applications for $Ba_3[B_3O_6]_2$ are performance as a corrosion inhibitor [103] in alkyl resin paints, its insecticidal activity against cockroaches [104], and the densification of $Ba[TiO_3]$ in the presence of its eutectic liquor for describing the electrical properties of ferroelectric ceramics [105].

References on pp. 117/22

$BaO \cdot 4B_2O_3$ glasses crystallize on heat treatment; the two immiscible phases were identified as B_2O_3 and as $Ba_3[B_3O_6]_2$ by X-ray and electron diffraction analyses [106].

The structure of the B_2O_3–BaO–Fe_2O_3 system has been studied by ^{57}Fe Mößbauer spectroscopy; the only crystalline borate found was $Ba_3[B_3O_6]_2$ [6]. Vitreous and crystalline phases in the systems $Ba_3[B_3O_6]_2$–$Sr(BO_2)_2$ and $Ba_3[B_3O_6]_2$–$Pb(BO_2)_2$ were studied by differential thermal analysis (DTA) and X-ray powder diffraction analysis [149].

3.4.3.2.2 Additional Compounds

$\mathbf{Ba_{3-x}Sr_x[B_3O_6]_2}$ (x = 1.16) has a structure like that of α-$Ba_3[B_3O_6]_2$ [150]. A single crystal (with x = 0.93) did not show any longitudinal piezoelectric effect along the c axis. The second harmonic generation (SHG) test on a powdered fragment of that crystal gave no SHG signal at all. These are consistent with the centrosymmetric space group $R\bar{3}c-D_{3d}^6$ (No. 167) of α-Ba_3-$[B_3O_6]_2$ [10].

The enthalpy interaction parameters for $\mathbf{Na_3[B_3O_6]}$ in SiO_2 at 1394 K change from −30 kJ/mol in pure $Na_3[B_3O_6]$ to −20 kJ/mol at 67% SiO_2 [151]. For the enthalpies of mixing of the liquid mixtures of $Na_3[B_3O_6]$ and B_2O_3 or $Na_4[P_2O_7]$, measured by liquid-liquid calorimetry, and tentative structural models of the melts, see Section 3.2.5.3, p. 42.

The $\mathbf{Na_3[B_3O_6]}$–$\mathbf{K_3[B_3O_6]}$ binary system was determined by means of differential thermal analysis (DTA) and X-ray diffraction methods to be a complete solid solution system with a minimal melting point of 836 °C and a minimal composition of 0.375 mol $K_3[B_3O_6]$. The solidus was calculated from the measured liquidus by the generalized Kohler and Pelton method. The miscibility gap of the solid solution was predicted thermodynamically and confirmed by high-temperature X-ray diffraction experiments [29].

The liquidus surface of the $\mathbf{Na_3[B_3O_6]}$–**NaCl**–$\mathbf{Na_2[MoO_4]}$ system has been given. Triangulation along quasibinary sections of each of the ternary systems results in a division into two subsystems, the phase diagrams of which are formed by the intersection of the three second-order surfaces [152].

The luminescence of several rare earth ions, an s^2 ion, and a transition metal ion in crystals of the composition $\mathbf{Al[B_3O_6]}$ was compared with the luminescence of these ions in lanthanum borate glasses [153].

$\mathbf{Gd[B_3O_6]}$ is a very promising luminescent material for lamp phosphors [154]. It is an excellent host lattice to obtain efficient photoluminescent materials. For this purpose, a sensitizer (S) of the Gd^{3+} sublattice and an activator (A) are needed. Energy transport from S to A occurs via the Gd^{3+} sublattice. The S and A ions were investigated, separated in $\mathbf{La[B_3O_6]}$ and $Gd[B_3O_6]$ for determination of the luminescence phenomena in single-doped $Gd[B_3O_6]$ with S = Ce^{3+}, Sb^{3+}, Bi^{3+} and A = Eu^{3+} or Tb^{3+}. Thereafter, double-doped $Gd[B_3O_6]$ was studied. Results confirm the transport process S → Gd → $(Gd)_n$ → A [155]. The mechanism of the total transport from S to A in $Gd[B_3O_6]$: Bi,Tb has been elucidated; Gd^{3+} trap emission was observed at low temperatures. Not only shallow (ca. 50 cm^{-1}), but also deep (ca. 500 cm^{-1}) traps were found. The decay of the intrinsic Gd^{3+} ions showed that the migration among the Gd^{3+} lattice was of the fast-diffusion type [156]. The energy migration in Eu^{3+} compounds depends on the dimensionality and on the Eu^{3+}-Eu^{3+} distance [157].

For $Bi[B_3O_6]$, no luminescence was observed down to liquid helium temperatures; this was attributed to quenching by defect centers of an unknown nature. The reasons for electron delocalization being of no importance was discussed [158].

$[Ag(NH_3)_2]_3[B_3O_6]$ is an ingredient of an antibacterial toothpaste for children [159].

References for 3.4.3.2:

[1] Liang, Jingkui; Zhang, Y.-L.; Huang, Qingzhen (Huaxue Xuebao **40** [1982] 994/1000).
[2] Tang, D.-Y.; Lin, S.-T.; Dai, G.-Q.; Zheng, W.-R.; Huang, B.-Q. (Rengong Jingti **14** [1985] 149/50).
[3] Tang, D.-Y.; Route, R. K.; Feigelson, R. S. (J. Cryst. Growth **91** [1988] 81/9).
[4] Zhang, Gunniyin; Jin, Ch.-Y.; Lin, F.-Ch.; Chen, Chuangtian; Wu, Baichang (Guangxue Xuebao **4** [1984] 513/6).
[5] Chen, Chuangtian (Laser Focus World **25** No. 11 [1989] 129/30, 132, 134/7 from C.A. **112** [1990] No. 27282).
[6] Melzer, K.; Knauf, O. (Int. Wiss. Kolloq. Tech. Hochsch. Ilmenau **31** [1986] 123/6 from C.A. **106** [1987] No. 73282).
[7] Mighell, A. D.; Perloff, A.; Block, S. (Acta Crystallogr. **20** [1966] 819/23).
[8] Eimerl, D.; Marion, J.; Graham, E. K. (Proc. SPIE-Int. Soc. Opt. Eng. **736** [1987] 54/9 from C.A. **107** [1987] No. 105943).
[9] Eimerl, D.; Davis, L.; Velsko, S.; Graham, E. K.; Zalkin, A. (J. Appl. Phys. **62** [1987] 1968/83).
[10] Liebertz, J. (Z. Kristallogr. **182** [1988] 307/8).

[11] Liebertz, J.; Froehlich, R. (Z. Kristallogr. **168** [1984] 293/7).
[12] Froehlich, R. (Z. Kristallogr. **168** [1984] 109/12).
[13] Liebertz, J.; Staehr, S. (Z. Kristallogr. **165** [1983] 91/3).
[14] Lu, Shaofang; Ho, Meiyun; Huang, Jinling (Wuli Xuebao **31** [1982] 948/55).
[15] Huang, Jinling; Wang, Dingsheng (Chinese Phys. Lett. **3** [1986] 157/60 from C.A. **106** [1987] No. 185314).
[16] Jiang, Aidong; Cheng, Fen; Lin, Qi; Cheng, Zusheng; Zheng, Yong (Rengong Jingti **14** [1985] p. 148).
[17] Jiang, Aidong; Cheng, Fen; Lin, Qi; Cheng, Zusheng; Zheng, Yong (J. Cryst. Growth **79** [1986] 963/9).
[18] Cheng, K. L.; Bosenberg, W.; Tang, C. L. (J. Cryst. Growth **89** [1988] 553/9).
[19] Cheng, Lap Kin Kevin (Diss. Cornell Univ., Ithaca, NY, 1988, pp. 1/129 from C.A. **110** [1989] No. 125814).
[20] Ivleva, L. I.; Kiselev, D. T.; Kuz'minov, Yu. S.; Polozkov, N. M. (Izv. Akad. Nauk SSSR Neorg. Mater. **24** [1988] 1153/7; Inorg. Mater. [USSR] **24** [1988] 982/6).

[21] Ivleva, L. I.; Gordadze, I. G.; Kuz'minov, Yu. S.; Voronov, V. V.; Ivanovskaya, V. M. (Izv. Akad. Nauk SSSR Neorg. Mater. **25** [1989] 804/7; Inorg. Mater. [USSR] **25** [1989] 682/5).
[22] Valbis, J.; Ivleva, L. I.; Kuz'minov, Yu. S.; Puyats, V. (Opt. Spektrosk. **66** [1989] 308/11; Opt. Spectrosc. [USSR] **66** [1989] 177/8 from C.A. **110** [1989] No. 221824).
[23] Huang, Qingzhen; Liang, Jingkui (Wuli Xuebao **30** [1981] 559/64).
[24] Huang, Qingzhen; Wang, Guofu; Liang, Jingkui (Wuli Xuebao **33** [1984] 76/85).
[25] Wang, Guofu; Huang, Qingzhen (Wuli Xuebao **34** [1985] 562/6 from C.A. **103** [1985] No. 28068).
[26] Wang, Guofu; Huang, Qingzhen; Liang, Jingkui (Huaxue Xuebao **42** [1984] 503/8).

[27] Huang, Qingzhen; Liang, Jingkui; Gan, Wuer (Kexue Tongbao [Foreign Lang. Ed.] **31** [1986] 610/4 from C.A. **105** [1986] No. 67451).
[28] Rao, G. H.; Liang, Jingkui; Qiao, Z. Y.; Huang, Qingzhen (CALPHAD **13** [1989] 169/75 from C.A. **111** [1989] No. 84872).
[29] Rao, G. H.; Liang, Jingkui; Qiao, Z. Y. (CALPHAD **13** [1989] 177/82 from C.A. **111** [1989] No. 84873).
[30] Huang, Qingzhen; Liang, Jingkui (J. Cryst. Growth **97** [1989] 720/4).

[31] Feigelson, R. S.; Route, R. K. (Proc. SPIE-Int. Soc. Opt. Eng. No. 1104 [1989] 142/55 from C.A. **112** [1990] No. 27277).
[32] Feigelson, R. S.; Raymakers, R. J.; Route, R. K. (J. Cryst. Growth **97** [1989] 352/66).
[33] Chai, B. H. T.; Gualtieri, D. M.; Randles, M. H. (Proc. SPIE-Int. Soc. Opt. Eng. No. 968 [1988] 69/72 from C.A. **110** [1989] No. 183193).
[34] Gualtieri, D. M.; Chai, B. H. T.; Randles, M. H. (J. Cryst. Growth **97** [1989] 613/6).
[35] Ohnishi, Norio (Seramikkusu **24** [1989] 319/24 from C.A. **111** [1989] No. 105987).
[36] Ovanesyan, K. L.; Petrosyan, A. G.; Shirinyan, G. O. (Cryst. Res. Technol. **24** [1989] 859/63 from C.A. **111** [1989] No. 222359).
[37] Chen, Chuangtian; Wu, Yicheng; Li, Rukang (Int. Rev. Phys. Chem. **8** No. 1 [1989] 65/91 from C.A. **112** [1990] No. 86 911; J. Cryst. Growth **99** [1990] 790/8).
[38] Yang, Yanyong; Zhang, Gunnyin; Wu, Baichang (Wuli Xuebao **36** [1987] 395/400 from C.A. **106** [1987] No. 185249).
[39] Zhang, Gunnyin; Yang, Yanyong; Wu, Baichang (Guangxue Xuebao **5** [1985] 548/56 from C.A. **104** [1986] No. 118914).
[40] Tian, Bogang; Wu, Guozhen; Xu, Ruiya (Spectrochim. Acta A **43** [1987] 65/71).

[41] Wu, Guozhen (J. Mol. Sci. [Int. Ed.] **4** [1986] 49/53 from C.A. **105** [1986] No. 125366).
[42] Lai, Xaojun; Wu, Guozhen (Spectrochim. Acta A **43** [1987] 1423/6).
[43] Lu, Junqing; Lan, Guoxiang; Li, Bing; Yang, Yanyong; Wang, Huafu; Wu, Baichang (J. Phys. Chem. Solids **49** [1988] 519/27).
[44] Guo, R.; Bhalla, A. S. (J. Appl. Phys. **66** [1989] 6186/8).
[45] Shi, Zikong (Rengong Jingti **16** [1987] 336/9 from C.A. **109** [1988] No. 84451).
[46] Shi, Zikong; Jiang, Aidong; Ye, Haitao; Chen, Chuangtian (Wuli Xuebao **34** [1985] 140/4).
[47] Shi, Z.-K. (Hongwai Yanjiu **4** [1985] 198/202 from C.A. **103** [1985] No. 112916).
[48] Hooper, B. A.; Benson, S. V.; Cutolo, A.; Madey, J. M. J. (Nucl. Instrum. Methods Phys. Res. A **272** [1988] 96/8 from C.A. **109** [1988] No. 200795).
[49] Hudson, B. S. (Spectroscopy [Eugene, Oreg.] **2** No. 4 [1987] 33, 36/7 from C.A. **106** [1987] No. 185009).
[50] Hudson, B. S. (J. Lumin. **40/41** [1988] 827/8 from C.A. **108** [1988] No. 158 122).

[51] Adhav, R. S.; Adhav, A. C.; Pelaprat, J. M. (Laser Focus [Littleton, Mass.] **23** No. 9 [1987] 88, 90, 92, 94, 96, 98 from C.A. **107** [1987] No. 186052).
[52] Risk, W. P.; Lenth, W. (Proc. SPIE-Int. Soc. Opt. Eng. No. 1104 [1989] 23/32 from C.A. **112** [1990] No. 65500).
[53] Huang, Deru; Yang, Shubing; Chen, Bo (Wuli **14** [1985] 307/8 from C.A. **103** [1985] No. 186152).
[54] Ebbers, C. A. (Appl. Phys. Lett. **52** [1988] 1948/9).
[55] Ebbers, C. A. (Proc. SPIE-Int. Soc. Opt. Eng. No. 968 [1988] 66/8 from C.A. **110** [1989] No. 144333).
[56] Nakatani, H.; Bosenberg, W. R.; Cheng, K. L.; Tang, C. L. (Appl. Phys. Lett. **52** [1988] 1288/90).

[57] Cheng, K. L.; Bosenberg, W. R.; Wise, F. W.; Walmsley, I. A.; Tang, C. L. (Appl. Phys. Lett. **52** [1988] 519/21).
[58] Edelstein, D. C.; Wachman, E. S.; Cheng, L. K.; Bosenberg, W. R.; Tang, C. L. (Appl. Phys. Lett. **52** [1988] 2211/3).
[59] Cheng, K. L.; Bosenberg, W. R.; Tang, C. L. (Appl. Phys. Lett. **53** [1988] 175/7).
[60] Bosenberg, W. R.; Cheng, L. K.; Tang, C. L. (Appl. Phys. Lett. **54** [1989] 13/5).

[61] Bosenberg, W. R.; Pelouch, W. S.; Tang, C. L. (Appl. Phys. Lett. **55** [1989] 1952/4).
[62] Nakatani, H.; Bosenberg, W. R.; Cheng, L. K.; Tang, C. L. (Appl. Phys. Lett. **53** [1988] 2587/9).
[63] Zhang, Gunnying; Yang, Yanyong; Zhang, Chunping (Appl. Phys. Lett. **53** [1988] 1019/21).
[64] Yang, Yanyong; Zhang, Gunnying; Zhang, Chunping; Chen, Chuangtian; Wu, Baichang (Guangxue Xuebao **6** [1986] 1105/10 from C.A. **106** [1987] No. 92804).
[65] Guan, Xinan; Yuan, Shuzhong; Wu, Baichang (Zhongguo Jiguang **13** [1986] 771/2 from C.A. **106** [1987] No. 223869).
[66] Zhu, Jikong; Zhang, Bing; Liu, Songhao (Guangxue Xuebao **5** [1985] 217/24 from C.A. **103** [1985] No. 131930).
[67] Li, Rukong; Chen, Chuangtian (Wuli Xuebao **34** [1985] 823/7 from C.A. **103** [1985] No. 79044).
[68] Willand, C. S.; Albrecht, A. C. (Opt. Commun. **57** [1986] 146/52 from C.A. **104** [1986] No. 138956).
[69] Lin, J. T. (Proc. SPIE-Int. Soc. Opt. Eng. No. 895 [1988] 162/9 from C.A. **109** [1988] No. 119152).
[70] Lin, J. T. (Proc. Int. Conf. Lasers **1986/87** 262/4 from C.A. **108** [1988] No. 213415).

[71] Qiu, Z.-R.; Cai, X.-J.; Wang, Z.-J. (Guangxue Xuebao **7** [1987] 1063/8 from C.A. **108** [1988] No. 213450).
[72] Chen, Chuangtian; Fan, Yuang Xuan; Eckardt, R. C.; Byer, R. L. (Proc. SPIE-Int. Soc. Opt. Eng. No. 681 [1987] 12/9 from C.A. **109** [1988] No. 223841).
[73] Fan, Yuan Xuan; Eckardt, R. C.; Byer, R. L.; Nolting, J.; Wallenstein, R. (Appl. Phys. Lett. **53** [1988] 2041/6).
[74] Lago, A.; Wallenstein, R.; Chen, Chuangtian; Fan, Yuan Xuan; Byer, R. L. (Opt. Lett. **13** [1988] 221/3 from C.A. **108** [1988] No. 195396).
[75] Fan, Yuan Xuan; Eckardt, R. C.; Byer, R. L.; Chen, Chuangtian; Jiang, A. D. (IEEE J. Quantum Electron. **25** [1989] 1196/9 from C.A. **111** [1989] No. 86856).
[76] Xu, Zuyan; Deng, Daoqun; Zhao, Tienan; Shou, Hansen (Chin. Phys. Lett. **5** [1988] 389/92).
[77] Xu, Zuyan; Deng, Daoqun; Zhao, Tienan; Wu, Biachang; You, Guiming (Chin. Phys. Lett. **6** [1989] 68/71).
[78] Chen, Da Wun; Yeh, Jen Jye (Opt. Lett. **13** [1988] 808/10).
[79] Qiu, P.; Penzkofer, A. (Appl. Phys. B **45** [1988] 225/36).
[80] Penzkofer, A.; Qiu, P.; Ossig, F. (Springer Proc. Phys. **36** [1988/89] 312/20 from C.A. **111** [1989] No. 183676).

[81] Vanherzeele, H.; Chen, Chuangtian (Appl. Opt. **27** [1988] 2634/6).
[82] Glab, W. L.; Hessler, J. P. (Appl. Opt. **26** [1987] 3181/2).
[83] You, Chenhua; Lu, Zukang; Fan, Qikang (Guangxue Xuebao **6** [1986] 413/9 from C.A. **105** [1986] No. 161807).
[84] You, Chenhua; Lu, Zukang; Fan, Qikang; You, Guiming (Zhongguo Jiguang **16** [1989] 327/30 from C.A. **111** [1989] No. 243484).

[85] Kato, K. (IEEE J. Quantum Electron. **22** [1986] 1013/4 from C.A. **105** [1986] No. 51789).
[86] You, Chenhua; Fan, Qikang; Lu, Zukang; You, Guming (Guangxue Xuebao **9** [1989] 401/6 from C.A. **111** [1989] No. 204882).
[87] Mueckenheim, W.; Lokai, P.; Burghardt, B.; Basting, D.; Essary, M. F. (AIP Conf. Proc. No. 172 [1987/88] 221/3 from C.A. **109** [1988] No. 240141).
[88] Lokai, P.; Burghardt, B.; Mueckenheim, W. (Appl. Phys. B **45** [1988] 245/7).
[89] Mueckenheim, W.; Lokai, P.; Burghardt, B.; Basting, D. (Appl. Phys. B **45** [1988] 259/61).
[90] Lokai, P.; Burghardt, B.; Basting, D.; Mueckenheim, W. (Proc. SPIE-Int. Soc. Opt. Eng. No. 1017 [1989] 150/4 from C.A. **110** [1989] No. 222120).

[91] Ishada, Y.; Yajima, K. (Opt. Commun. **62** [1987] 197/200 from C.A. **107** [1987] No. 49023).
[92] Bhar, G. C.; Das, S.; Chatterjee, U. (Appl. Phys. Lett. **54** [1989] 1383/4).
[93] Bhar, G. C.; Das, S.; Chatterjee, U. (J. Phys. D **22** [1989] 562/3).
[94] Bhar, G. C.; Das, S.; Chatterjee, U. (Appl. Opt. **28** [1989] 202/4).
[95] Bhar, G. C.; Chatterjee, U.; Das, S. (J. Appl. Phys. **66** [1989] 5111/3).
[96] Miyazaki, K.; Sakai, H.; Sato, T. (Opt. Lett. **11** [1986] 797/9).
[97] Negus, D. K.; Couillaud, B. C.; Brady, R. (Springer Ser. Chem. Phys. **48** [1988] 97/100).
[98] Imai, Sh.; Yamada, T.; Fujimori, Y.; Ishikawa, K. (Appl. Phys. Lett. **54** [1989] 1206/8).
[99] Kuroda, K.; Omatsu, T.; Shimura, T.; Chihara, M.; Ogura, I. (Proc. SPIE-Int. Soc. Opt. Eng. No. 1041 [1989] 60/6 from C.A. **111** [1989] No. 86889).
[100] Zimmermann, C.; Kallenbach, R.; Haensch, T. W.; Sandberg, J. (Opt. Commun. **71** [1989] 229/34 from C.A. **111** [1989] No. 30974).

[101] Bromley, L. J.; Guy, A.; Hanna, D. C. (Opt. Commun. **67** [1988] 316/20 from C.A. **109** [1988] No. 119186).
[102] Komine, H. (Opt. Lett. **13** [1988] 643/5).
[103] Dalton, D. L.; Stevens, L. C. (Mod. Paint Coat. **75** No. 11 [1985] 52/4 from C.A. **104** [1986] No. 70328).
[104] Gomaa, A. Z.; Metwally, M. M.; Moustafa, M.; Abd El-Ghaffar, M. A. (Pigm. Resin Technol. **18** No. 4 [1989] 4/8 from C.A. **111** [1989] No. 116862).
[105] Kolar, D.; Trontelj, M.; Marsch, L. (J. Phys. Colloq. [Paris] **47** [1986] C1-447/C1-450 from C.A. **104** [1986] No. 191508).
[106] Fahmy, M. F.; Subramanian, K. N. (Phys. Chem. Glasses **28** [1987] 1/3).
[107] Itoh, K.; Marumo, F.; Ohgaki, M.; Tanaka, K. (Rep. Res. Lab. Eng. Mater. Tokyo Inst. Technol. No. 15 [1990] 1/12 from C.A. **114** [1991] No. 33384).
[108] Itoh, K.; Marumo, F.; Kuwano, Y. (J. Cryst. Growth **106** [1990] 728/31).
[109] Kozuki, Y.; Itoh, M. (J. Cryst. Growth **114** [1991] 683/6; Denzaiken Giho **9** [1991] 26/33 from C.A. **116** [1992] No. 224949).
[110] Hengel, R. O.; Fischer, F. (J. Cryst. Growth **114** [1991] 650/60).

[111] Tang, Dingyuan; Lin, Sitai; Dai, Guigin; Lin, Qi; Zeng, Wenrong; Zhao, Qinglan; Huang, Yisen (Rengong Jingti Xuebao **19** [1990] 21/7 from C.A. **116** [1992] No. 204731).
[112] Ferrand, B.; Chartier, I.; Coatantiec, C.; Couchaud, M.; Rolland, G.; Wyon, C.; Aubert, J. J.; Doizi, D. (J. Phys. IV **1** [1991] C7-753/C7-756 from C.A. **116** [1992] No. 224250).
[113] Polgar, K.; Peter, A.; Schmidt, F. (Cryst. Prop. Prep. **36/38** [1991] 209/15 from C.A. **116** [1992] No. 224965).
[114] Nikolov, V.; Peshev, P. (J. Solid State Chem. **96** [1992] 48/52).
[115] Nikolov, V.; Peshev, P.; Khubanov, Kh. (J. Solid State Chem. **97** [1992] 36/40).

[116] Bosenberg, W. R. (Diss. Cornell Univ., Ithaca, NY, USA, 1990, 185 pp. from C.A. **114** [1991] No. 111380); Bosenberg, W. R.; Lane, R. J.; Tang, C. L. (J. Cryst. Growth **108** [1991] 394/8).
[117] Molchanov, S. V.; Anferov, V. P.; Svirkst, J.; Gode, H. (Latv. PSR Zinat. Akad. Vestis Kim. Ser. **1990** 192/5).
[118] Nikogosyan, D. N. (Appl. Phys. A **52** [1991] 359/68; Electron. Tekhn. Ser. 11 **1990** No. 2, pp. 3/13 from C.A. **116** [1992] No. 30695).
[119] Hou, Jianguo; Tan, Qiguang; Zhao, Qinnan; Wu, Baichang; Chen, Chuangtian (Appl. Phys. Lett. **58** [1991] 1149/51).
[120] Wang, Yunping; Xu, Zuyan; Deng, Daoqun; Zheng, Wanhua; Liu, Xiang; Wu, Baichang; Chen, Chuangtian (Appl. Phys. Lett. **58** [1991] 1461/3).

[121] Huang, J. Y.; Zhang, J. Y.; Shen, Y. R.; Chen, Chuangtian; Wu, Baichang (Appl. Phys. Lett. **57** [1990] 1961/3).
[122] Haub, J. G.; Johnson, M. J.; Orr, B. J.; Wallenstein, R. (Appl. Phys. Lett. **58** [1991] 1718/20).
[123] Borsutzky, A.; Bünger, R.; Huang, Chaoen; Wallenstein, R. (Appl. Phys. B **52** [1991] (1) 55/62).
[124] Bhar, G. C.; Das, S.; Datta, P. K. (Phys. Status Solidi A **119** [1990] K173/K176).
[125] Bhar, G. C.; Chatterjee, U.; Datta, P. K. (Appl. Phys. B **51** [1990] 317/9).
[126] Bhar, G. C.; Chatterjee, U.; Das, S. (Appl. Phys. Lett. **58** [1991] 231/3).
[127] Müschenborn, H. J.; Theiss, W.; Demtröder, W. (Appl. Phys. B **50** [1990] 365/9).
[128] Sukowski, U.; Seilmeier, A. (Appl. Phys. B **50** [1990] 541/5).
[129] Stankov, K. (Appl. Phys. Lett. **58** [1991] 2203/4).
[130] Stankov, K.; Tsolov, V.; Mirkov, M. (Opt. Lett. **16** [1991] 639/41 and 1119/21).

[131] Bosenberg, W. R.; Tang, C. L. (Appl. Phys. Lett. **56** [1990] 1819/21).
[132] Kölsch, H. J.; Rairoux, P.; Wolf, J. P.; Wöste, L. (Appl. Opt. **28** [1989] 2052/6).
[133] Banfi, G. P.; Ghigliazza, M.; Di Trapani, P. (Opt. Commun. **89** [1992] 63/7).
[134] Li, Gang; Hao, Hailin (Beijing Gongye Daxue Xuebao **18** [1992] 17/22 from C.A. **117** [1992] No. 16741).
[135] Watanabe, M.; Hayasaka, K.; Imajo, H.; Urabe, S. (Opt. Lett. **17** [1992] 46/8).
[136] Ellingson, R. J.; Tang, C. L. (Opt. Lett. **17** [1992] 343/5).
[137] Komine, H.; Long, W. H.; Stappaerts, E. A. (Proc. Int. Conf. Lasers **1990** [1991] 612/8 from C.A. **16** [1992] No. 161958).
[138] Danielius, R.; Piskarskas, A.; Podenas, D.; Di Trapani, P.; Varanavicius, A.; Banfi, G. P. (Opt. Commun. **87** [1992] 23/7).
[139] Joosen, W.; Agostini, P.; Petite, G.; Chambaret, J. P.; Antonetti, A. (Opt. Lett. **17** [1992] 133/5).
[140] Joosen, W.; Bakker, H. J.; Noordam, L. D.; Muller, H. G.; Van Linden van den Heuvell, H. B. (J. Opt. Soc. Am. B **8** [1991] 2537/43 from C.A. **116** [1992] No. 161846).

[141] Pini, R.; Salimbeni, R.; Toci, G.; Vannini, M. (Appl. Opt. **31** [1992] 2747/51).
[142] Taira, Yoichi (Jpn. J. Appl. Phys. **31** Pt. 2 [1992] L682/L684 from C.A. **117** [1992] No. 58261).
[143] Wang, Yufang; Lan, Guoxiang; Wang, Huafu (Spectrochim. Acta A **48** [1992] 181/91).
[144] Lin, Yuankun; Cai, Qingrui; Lan, Guoxing; Wang, Huafu (Spectrochim. Acta A **48** [1992] 653/7).
[145] Roussigne, Y.; Farhi, R.; Dugautier, C.; Godard, J. (Solid State Commun. **82** [1992] 287/93).

[146] Adamiv, V. T.; Berko, T. I.; Kityk, I. V.; Burak, Ya. V.; Dzhala, V. I.; Dovgii, Ya. O.; Moroz, I. E. (Ukr. Fiz. Zh. [Russ. Ed.] **37** [1992] 368/73 from C.A. **116** [1992] No. 225182).
[147] Adamiv, V. T.; Burak, Ya. V.; Dovgii, Ya. O.; Kityk, I. V. (Kristallografiya **36** [1991] 229/31).
[148] Bohaty, L.; Liebertz, J. (Z. Kristallogr. **192** [1990] 91/5).
[149] Almou, M.; Rulmont, A.; Tarte, P. (Eur. J. Solid State Inorg. Chem. **25** [1988] 387/97 from C.A. **111** [1989] No. 16628).
[150] Huang, Qingzhen; Huang, Liangren; Dai, Guiqin; Liang, Jingkui (Acta Crystallogr. C **48** [1992] 539/41).

[151] Julsrud, S.; Kleppa, O. J. (Geochim. Cosmochim. Acta **50** [1986] 1201/4).
[152] Mokhosoev, M. V.; Shurdumov, G. K.: Khakulov, Z. L.; Koshkarov, Zh. A. (Zh. Neorg. Khim. **33** [1988] 1615/6; Russ. J. Inorg. Chem. **33** [1988/89] 917/8).
[153] Verwey, J. W. M.; Blasse, G. (Proc. SPIE-Int. Soc. Opt. Eng. No. 1128 [1989] 205/8 from C.A. **112** [1990] No. 107775).
[154] Blasse, G. (J. Less-Common. Met. **112** [1985] 79/82).
[155] Hao, Z.-R.; Blasse, G. (Mater. Chem. Phys. **12** [1985] 257/74).
[156] De Vries, A. J.; Blasse, G. (J. Phys. Colloq. [Paris] **46** [1985] C7-109/C7-112).
[157] Berdowski, P. A. M.; Buijs, M.; Blasse, G. (J. Phys. Colloq. [Paris] **46** [1985] C7-31/C7-34).
[158] Blasse, G., Oomen, E. W. J. L.; Liebertz, J. (Phys. Status Solidi B **137** [1986] K77/K81).
[159] Lezovic, J.; Filak, M.; Koprivova, L.; Koudelova, L.; Kasa, J.; Huda, A. (Czech. 221049/50 [1981/86] 1/3; Czech. 221189 [1981/86] 1/3 from C.A. **104** [1986] Nos. 192952 to 192954).

3.4.3.3 $K_3[B_3O_3F_6]$ and $Li_3CdB_3O_7$

Formation of **$K_3[B_3O_3F_6]$**, melting point 560°C, was observed in the B_2O_3–KF–$K[BF_4]$ system; it crystallizes orthorhombically with a = 1012.6(4), b = 1488.0(12), and c = 966.2(3) pm; Z = 8; D_c = 2.861 g/cm³ [1].

Differential thermal analysis (DTA), IR spectroscopy, and X-ray phase analysis data in the system B_2O_3–Li_2O–CdO were used to construct the phase diagram. Besides α- and β-LiCd[BO_3] (see p. 82), **$Li_3CdB_3O_7$**, melting point 780°C, has been characterized. It crystallizes in the monoclinic space group $P2_1/c-C_{2h}^5$ (No. 14) with a = 1023(5), b = 1074(5), and c = 528(3) pm; β = 118.0(5)°; Z = 4; D_c = 3.74 g/cm³ [2].

References for 3.4.3.3:

[1] Andriiko, A. A.; Parkhomenko, N. I.; Antishko, A. N. (Zh. Neorg. Khim. **33** [1988] 729/33; Russ. J. Inorg. Chem. **33** [1988] 410/3).
[2] Buludov, N. T.; Kara'ev, Z. Sh.; Abdulla'ev, G. K. (Zh. Neorg. Khim. **30** [1985] 1523/6; Russ. J. Inorg. Chem. **30** [1985/86] 868/70).

3.4.4 Tetraborates

For tetraborates, see also p. 203, for $H_2B_4O_7$, see Section 3.5.11, p. 256.

3.4.4.1 The Ion $[B_4O_7]^{2-}$

The mineral diomignite, $\mathbf{Li_2[B_4O_7]}$, occurring in Manitoba, crystallizes in the tetragonal space group $I4_1cd-C_{4v}^{12}$ (No. 110) with a = 947.0(4) and c = 1027.9(5) pm; Z = 8; D_m = 2.44 g/cm³ and D_c = 2.437 g/cm³ [1]. Synthetic $Li_2[B_4O_7]$ has a = 947.9(3) and c = 1029.0(4) pm; the structure was determined by direct methods and refined by least-squares to R = 1.33%; the crystal packing and electron distribution are described [2].

The structure and superstructure of $Li_2[B_4O_7]$ were studied between 80 and 400 K by X-ray diffraction analysis. The a lattice parameter increases and the c parameter decreases (except at 350 to 370 K) as the temperature increases; several phase transitions were observed. An incommensurate phase is observed at room temperature; a second-order transition occurs at 95 K [3]. The effect of thermal cycling between 80 and 400 K on the structure and superstructure of $Li_2[B_4O_7]$ was also studied. Phase transitions are observed at 95, 118.5, and 147 K. On repeated thermal cycling, other transition points are observed, which are associated with incommensurate effects [4]. The phase transition into an incommensurate state can be also observed as a result of changes of pressure and temperature field or during an action of periodic changes of the temperature field [5].

A good quality $Li_2[B_4O_7]$ single-crystal of 23 × 80 mm was grown at a crystal pulling rate of 1.3 mm/h and the seed rotation rate of six rotations per minute by an automatic diameter control of the Czochralski method. Cores generated in growing crystals largely depend on the crystal pulling rate, rapid temperature change during the crystal growth, and purity of starting materials [6, 7]. Two- and three-inch diameter single-crystals were also grown by this method; the localized dislocation densities are less than 3000/cm² [8]. The optimum conditions for the Czochralski crystal growth in this case are discussed; the major type of defect is a gas bubble [9]. The effect of B–O network structure on the growth of these crystals and the formation of a core in the crystal during growth was discussed [10]. Piezoelectric thin films of $Li_2[B_4O_7]$ were fabricated by radiofrequency sputter deposition or by vapor deposition under a low gas pressure. The last method is better for c axis oriented films with good piezoelectric characteristics [11].

A 45°-rotated x axis, which is (110), z-cut $Li_2[B_4O_7]$ single-crystal up to three inches in diameter was grown by the Czochralski method. Wafers for surface acoustic wave (SAW) device substrates were fabricated. Aluminium strips on the SAW substrate can reflect SAW more efficiently than most other substrates. A low-loss wide-band-gap 280 MHz SAW resonator was developed by using this substrate with a new configuration [12]. Configuration suppressing surface-to-bulk wave conversion was discussed. Results prove that the aluminium strips on an $Li_2[B_4O_7]$ substrate reflect efficiently, compared with those on the other substrates, and that a newly developed configuration improves the resonator Q by at least a factor of two. Furthermore, a low insertion loss of 2.5 decibels is achieved for a paging receiver filter [13].

Acoustic modes propagating in x-cut, z-propagating $Li_2[B_4O_7]$ substrates were also analyzed theoretically and experimentally. The theoretical predictions were made using a computer program written in order to search for all possible acoustic modes propagating in arbitrary cut of piezoelectric materials. A new photolithographic process was developed in order to fabricate interdigital transducers (IDT) on $Li_2[B_4O_7]$ to avoid using acids which attack and damage the substrates. Using this process, split-electrode surface acoustic wave IDT were fabricated with a period of 16 μm. The frequencies of the observed modes agreed to better than 0.5% with the theoretical predictions. The insertion loss of the surface acoustic wave delay time was less than 10 dB at 219 MHz and the temperature coefficient of delay was measured to be about 6 ppm/°C. These data suggest that $Li_2[B_4O_7]$ is a candidate to replace quartz as a substrate for narrow-band surface acoustic wave filter applications [14].

References on pp. 130/3

Lateral-field coupling factor calculations have been performed on $Li_2[B_4O_7]$ crystals of the symmetry 4mm [15]. All dielectric, elastic, and piezoelectric constants and their first- and second-order temperature coefficients were determined by measuring resonant and antiresonant frequencies of bars and plates with various orientations. Based upon these data, Rayleigh surface acoustic wave (SAW), leaky SAW, and bulk wave properties were studied from both theoretical and experimental points of view. The results show the usefulness of $Li_2[B_4O_7]$ for bulk wave or SAW applications [16].

Both theoretical and experimental results showed that Rayleigh SAW possesses a particularly useful cut with the zero temperature coefficient of delay (TCD) at 0, 78, and 90 K. A new temperature-compensated (TC) leaky SAW orientation (v=4120 m/s) has a coupling coefficient k^2=1.6% [17]. The existence of an orientation with zero TCD has been predicted by the computer analyses and confirmed experimentally for the leaky SAW. A new temperature-compensated leaky SAW orientation has been located around (0°, 75°, 75°) with v=4235 m/s and a coupling coefficient of k^2=1.8% [18].

The material constants of $Li_2[B_4O_7]$ crystals, related to the bulk and surface acoustic wave propagation show temperature-compensated cuts for rotated y-cut bulk wave devices around 32 K for the quasilongitudinal mode and around 53 K for the quasishear bulk mode [7, 17].

Pyroelectric, piezoelectric, and dielectric properties of single crystal $Li_2[B_4O_7]$ were measured from 50°C down through −150°C; at room temperature, it has a dielectric constant K_3 of ca. 10 (at 100 kHz) and a dielectric $\tan\delta$ of ca. 1 (at 10 kHz). At 78 K, the value of $\tan\delta$ decreases to ca. 10^{-3} (at 100 kHz). The pyroelectric coefficient at room temperature is ca. 30, which increases to −120 $\mu C \cdot m^{-2} \cdot K^{-1}$ at −150°C. Primary and secondary components of pyroelectricity are separated, and the results suggest that the major contribution comes from the primary effect. Piezoelectric measurements show high values of $g_{33} \approx 0.27\ V \cdot m \cdot N^{-1}$ and $g_h \approx 0.1\ V \cdot m \cdot N^{-1}$ indicating a potential application of $Li_2[B_4O_7]$ in the area of pressure sensors [19].

Anomalies in the elastic stiffness in $Li_2[B_4O_7]$ and the corresponding peaks in the attenuation of ultrasound were observed between 75 and 215 K; changes in the interlayer force constants are considered [20]. A set of first- and second-order temperature coefficients of the elastic stiffness and compliance was calculated for the constant normal electric field displacement conditions on the face of the plate of crystals [21]. The effect of thermal cycling on the elasticity of $Li_2[B_4O_7]$ was studied. The crystals were cooled and heated at average rates of 0.3 K/min. The elastic properties were determined using ultrasonic probes. The hysteresis in the elastic properties shifts to higher temperatures as the number of cycles increases [22].

Band structure calculations (modified LCAD) are presented for $Li_2[B_4O_7]$, separating the fragments BO_3, BO_4, LiO_4, and LiO_6, which have a substantial contribution to the electronic structure. The effect of the surroundings is taken into account as a perturbation. Hartree and Slater exchange-correlation correlations were introduced. The band parameters are adjusted using dielectrical permittivity spectra [23].

The absorption and luminescence characteristics of $Li_2[B_4O_7]$ (crystals grown as described [3]) were studied between 77 and 450 K. They are transparent between 165 and 600 nm with the most longwave peak of fundamental absorption distributed at ca. 133 nm. The X-ray luminescence spectra consist of bands at ca. 365 nm which are ascribed to the luminescence of autolocalized excitons. According to the curvature of the thermoluminescence, the depth of levels of capture of current carriers were determined for maxima at 110 and 273 K (0.11 and 0.05 eV, respectively) [24].

The localized dislocation densities in ca. 2-inch diameter $Li_2[B_4O_7]$ single crystals are less than 3000/cm². The dissolution rate can be suppressed below 10 Å/min in UV lithography; the $Li_2[B_4O_7]$ crystals do not dissolve in deep UV lithography. The resonant frequency deviation of substrates from different ingots is ±310 ppm. Design parameters for an $Li_2[B_4O_7]$ resonator, i.e., the attenuation constants, the normalized susceptance, and the surface acoustic wave (SAW) reflectivity in the equivalent network model are obtained by matching the measured value for the 90 and 850 MHz resonator with the calculated value. It is estimated that the chip area of a 90 MHz SAW resonator can be reduced by 60% in comparison to a chip on $Li[TaO_3]$ [8].

Color centers are formed in $Li_2[B_4O_7]$ by 0.4 and 1.3 electrons. Annealing of these centers is observed at temperatures higher than 370 K with complete erasure at ca. 470 K. The centers appear to contain oxygen vacancies [25]. The defects formed in $Li_2[B_4O_7]$ by irradiation with neutrons of more than 0.1 MeV and electrons of 1.3 MeV were studied by optical spectroscopy. The formation of color centers and regions of disorder is observed. An atomic displacement mechanism is proposed [26].

The electrical conductivity was studied for $Li_2[B_4O_7]$ single crystals as a function of the temperature between 295 and 673 K and the crystal orientations [100] or [001]. At 295 K, the electrical conductivity in the [100] direction is $7\times10^{-5}\ \Omega^{-1}\cdot cm^{-1}$; electrons are responsible for the conductivity. In the [001] direction, the Li^+ ions are responsible for the conductivity, and the value is $4.7\times10^{-10}\ \Omega^{-1}\cdot cm^{-1}$. The activation energies of conductivity are estimated, and a model for the conductivity in the [001] direction is suggested [27].

A mass spectrometric method based on the thermal ionization of $Li_2[B_4O_7]$ has been developed for the isotopic analysis of lithium [28]. For the analytical use of $Li_2[B_4O_7]$, see Section 3.2.5.5, p. 65.

Photons in the energy range 15 keV to 3 MeV with an error of less than ±40% were measured with thermoluminescent $Li_2[B_4O_7]$ dosimeters [29]; for the thermoluminescence of $CuCl_2$ (also Pd^{2+} and Tb^{3+}) doped $Li_2[B_4O_7]$, see [30]. Thermally stimulated exoelectron emission (TSEE) of $Li_2[B_4O_7]$ glass (doped with Cu^{2+}, Mn^{2+}, Ag^+, or Tb^{3+}) after X-ray irradiation [31, 32] and of ceramic and sintered $Li_2[B_4O_7]$/LiCl mixtures was determined [33].

New values for the following properties of tetragonal $Li_2[B_4O_7]$ are reported: dielectric, electrostrictive, electro-optic, elastic, thermoelastic, and piezoelastic constants, thermal expansion coefficients, and refractive indices. The data are considered to be of superior precision as compared to earlier published results. Most elastic constants possess positive temperature deviations; this anomalous behavior is not reflected in the pressure derivations. The electro-optic effects are not large enough to compete with other materials used for technical applications [34].

The following dosimetric properties of $Li_2[B_4O_7]$: Mn were studied: the structure of the glow curve, lower dose detection threshold, sensitivity for high-energy photon radiation, dose characteristics, and fading properties [35].

$Li_2[B_4O_7]$: Cu was prepared by the sintering technique. The following items were investigated: glow curves, thermoluminescence (TL) response versus dose, fading, fatigue of the material, TL response versus γ and electron energy, emission spectrum, unstable low temperature peaks, TL at different heating rates, temperature of the main glow peak, fractional glowing, anomalous fading, and fading under daylight exposure [36].

Samples of $Li_2[B_4O_7]$ activated either by transition elements or rare earths were also prepared and tested. Cu is the best activator for use in radioprotection and radiotherapy [37]. $Li_2[B_4O_7]$: Cu was synthesized; it is promising as a tissue equivalence phosphor, giving good energy response. $Li_2[B_4O_7]$ was doped with $CuCl_2$ in ethanol, fused at 950°C for 1h, quenched, ground, and annealed at 500°C for 2 h and at 880 to 890°C for 1.5 h. The $CuCl_2$ concentration dependence and superlinearity of the luminescence intensity dose dependence

References on pp. 130/3

were studied. The glow curve consisted of two peaks at 220 and 270°C with activation energies of 1.14 and 1.56 eV, respectively. The Cu site in the crystal was analyzed using ESR data [38]. The TL was also investigated for its trap distribution characteristics. The analysis, based on a fractional glow technique, shows that TL processes involve a number of trap groups with activation energies between 0.9 and 1.73 eV distributed conspicuously in the 80 to 335°C glow temperature region, unlike that of other common TL phosphors. The mean glow peak region between 170 and 260°C with a peak at 220°C was linked to the presence of predominantly deep trapping levels having an energy of 1.65±0.08 eV [39].

Thermoluminescence (TL) characteristics of $Li_2[B_4O_7]$ glass and glass-ceramics doped with Cu as activator are markedly affected by the change in the crystallizing temperature which results in the change of the Cu^+ content in the crystal. TL intensities did not simply increase with increasing Cu content, but showed maximal values at ca. 0.05 mol% Cu. The shape of the 70 and 150°C glow peaks were explained by the first-order kinetics model, and the activation energies were estimated to be 0.8 and 1.1 eV, respectively. The emission mechanism of TL is proposed as follows: electron holes created by irradiation are trapped by Cu^+ centers and after liberation from the trap centers by thermal energy, the holes combine with electrons at electron trap centers, giving rise to the emission of TL [40].

The thermal neutron response of the $Li_2[B_4O_7]$:Cu thermoluminescence dosimeter was measured. The results (obtained by using the Panasonic UD-806 dosimeter and UD-854A holder) yield a free-in-air response to 3.3±0.1R ^{60}Co equivalent per mSv of thermal neutrons [41]. Also, other basic dosimetric characteristics of $Li_2[B_4O_7]$:Cu in the Panasonic UD-802 dosimeter were measured. The TL dose response linearity was determined over the useful range from 0.005 to 10 mGy, and the minimal measurable doses were calculated. The phosphor is UV sensitive and, therefore, reassessment is not applicable [42]. A dosimeter, consisting of two $Li_2[B_4O_7]$:Cu elements and one $Ca[SO_4]$:Tm element, is proposed for detecting noble gas releases (principially ^{133}Xe) from nuclear power plants [43].

$Li_2[B_4O_7]$:Tm shows a TL effect. Optical absorption measurements between 2850 and 300 nm showed bands due to the Tm^{3+} characteristic electronic transitions. The glassy $Li_2[B_4O_7]$:Tm material is more resistant to radiation damage, but its TL sensitivity is lower by a factor of 4.5 than that of the polycrystalline form. The TL emission spectrum has a dominant peak at 455 nm which falls into the useful photomultiplier detection range of the common TL readers. The TL glow curves consist of three peaks for polycrystalline and glassy samples. The polycrystalline material has a dosimetric peak at 288°, a linear TL output in the 0.1 mGy to 103 Gy dose range, and post-irradiation fading at ambient temperature of less than 10% in three months [44].

The thermally stimulated exoelectron emission (TSEE) glow patterns of sintered $Li_2[B_4O_7]$ depend on the kind of radiation used and, therefore, indicate a possibility of dose determination for each kind of radiation [45]. TSEE was applied to measurements of the depth-dose curve for γ-ray fields; with 0.4 mm thin LiF single-crystals and 0.18 mm thin $Li_2[B_4O_7]$ glass ceramics as dosimeter elements, the absorbed dose was determined as a function of depth to be between 0.0264 and 2.64 g/cm² [46]. Simultaneous TL and TSEE dosimeter measurements were carried out with $Li_2[B_4O_7]$:Cu glass ceramics. The TSEE responses are stable and reliable, but the TL response shows a very unstable behavior, probably since the implanted Cu^+ cannot find stable sites in crystals under the heat treatment condition adopted in this experiment; this problem was solved by addition of SiO_2 and Pt [47]. The TSEE glow patterns of a magnesium borate glass-ceramic and an $Li_2[B_4O_7]$ glass-ceramic depend on the radiation used and the heat resistance of the material. The TSEE glow patterns of the magnesium compound indicate a possibility for use in dose measurement for each kind of radiation in a mixed radiation field [48].

The thermally stimulated exoelectron emission (TSEE) of ion-implanted samples like LiF : Mg^+, MgO : H^+, and $Li_2[B_4O_7]$: Cu^+ glass-ceramics show a different glow curve from that of a nonimplanted one, and the total TSEE sensitivity was a function of the implanted element, implantation energy, and implanted dose [49]. The TSEE intensity of glass-ceramics increased when they were prepared in a reducing atmosphere or annealed in vacuum. This result suggests that oxygen vacancies have a role as an origin of exoelectron emission [50]. The TSEE glow curves for nonimplanted $Li_2[B_4O_7]$ glass-ceramics, irradiated by X-rays, were compared with Cu^+-implanted $Li_2[B_4O_7]$ glass-ceramics (400 keV, 1013 ions/cm^2); the latter show a greater response to X-ray irradiation [51]. TSEE and thermoluminescence (TL) of an $Li_2[B_4O_7]$ single crystal previously quenched to 77 K were measured simultaneously; the quenched crystal shows appreciable TSEE and a slight TL even without previous X-ray irradiation; no clear correspondence was observed with peak temperatures between TSEE and thermoluminescence (TL) in general [52].

A survey on the properties of a new surface acoustic wave (SAW) device material, an $Li_2[B_4O_7]$ single-crystal, is given [53].

Another noteworthy property is the catalytic behavior for the continuous oxidative dehydrogenation of methane between 600 and 720°C with 14% $Li_2[B_4O_7]$ on an Al_2O_3 support [54].

Some γ-irradiated sodium and lithium diborate compounds have both paramagnetic and thermoluminescent centers. ESR-TL correlation measurements indicate that the paramagnetic centers can only be related to the thermoluminescent processes taking place at ca. 150°C, when the heating rate is ca. 2°C/min [55].

The sensitivity of thermoluminescence dosimeters containing **$Mg[B_4O_7]$** to fast neutrons inside a water container is much higher than in free air [56]. Graphite-mixed $Mg[B_4O_7]$ pellet dosimeters are good for TL [57]. A β-dosimeter prepared from graphite mixed with sintered $Mg[B_4O_7]$: Dy has been presented [58]. Thin TL detectors like graphite-mixed sintered $Mg[B_4O_7]$: Dy pellets satisfy the requirements for measuring β-rays [59].

The practical applicability of graphite-mixed and boron-diffused thermoluminescence dosimeters was also studied for skin dose assessment. Experimental data are presented on effective dosimeter thickness, β-ray energy response, and lowest detectable dose. Promising results were obtained from graphite-mixed $Mg[B_4O_7]$: Dy dosimeters, showing a high sensitivity combined with a nearly tissue-equivalent, photon-energy response. Furthermore, data are presented on directional dependence and energy response of a modified Risoe personnel TL dosimeter badge with improved capability for β-dose measurements [60]. The dosimetric performances of improved magnesium borate TL dosimeters were evaluated under operational conditions. The wide range of application of this type of dosimeter for different radiation protection purposes is presented also in [61]. The TL efficiency of $Mg[B_4O_7]$: Dy was investigated under laser heating; a new annealing procedure is suggested which makes this material a viable candidate for applications in TL dosimetry [62].

The photon energy dependence of different thermoluminescence phosphors like $Mg[B_4O_7]$: Dy was measured by using X-ray radiation (48 to 205 keV) and a ^{60}Co source. The relative sensitivity, linearity, fading, and the UV light-induced TL signal were investigated in the dose range from several mGy to several Gy [63]. $Mg[B_4O_7]$: Dy shows a higher intrinsic sensitivity to UV radiation when it is annealed at 300°C between exposures. The response is very linear and covers the UV radiant exposures that might be of interest in biological research. Its use will, however, require an appropriate fading, correction, and matching of UV response characteristic of individual dosimeter samples. Better stability might be obtained by first annealing out the low-temperature peaks before evaluation or by taking only the height of the 200°C peak. There is a good correlation between phototransferred TL and prior γ-ray

References on pp. 130/3

absorbed dose. This will allow γ-ray dose re-estimation, but only for doses much higher than those routinely encountered in personnel dosimetry [64]. The advantage of $Mg[B_4O_7]$:Dy include good tissue equivalence, since the material has an effective atomic number of 7.8, increased sensitivity, excellent linearity of response in the low-dose range of interest in personnel and environmental monitoring, and minimal annealing requirements. Readout alone apparently clears the phosphor completely. There is also a potential of its development for UV dosimetry. The secondary activator quantities strongly affect the final dosimetric characteristics of this phosphor [65].

The effect of composition and temperature was studied on the photoluminescence brightness, emission, and excitation spectra of Ce-Tb coactivated $Mg[B_4O_7]$. Ce^{3+} is a good sensitizer for Tb^{3+} emission under excitation of 254 nm UV light. The optimal sensitization effect is observed when the Ce : B_2O_3 mole ratio is 0.11; the concentration quenching occurs when the Tb : B_2O_3 mole ratio is 0.08. The mechanism of energy transfer from Ce^{3+} to Tb^{3+} is a resonance transfer; the Ce^{3+} and Tb^{3+} emission decreases with increasing temperature between 20 and 200 °C [66].

Thermoluminescence (TL) glow curves were also measured for Dy- and Tm-activated $Mg[B_4O_7]$ polycrystalline sintered disks. The band assignment to dopant transitions is discussed. Both materials are of interest to TL X-ray dosimetry [67]. The thermoluminescence and the thermally stimulated exoelectron emission (TSEE) of the same materials are compared with those of **$Ca[B_4O_7]$**; the TL is related to the F^+ centers, TSEE is a surface effect [68].

A comparative study of TL of X-ray irradiated samples is reported for sintered and glassy $Ca[B_4O_7]$ with and without Cu dopant. The TL intensity of Cu emission centers depends not only on the concentration of Cu, but also on the method of material processing. The TL response of the 180 °C glow peak of $Ca[B_4O_7]$ with 0.1 mol% $CuCl_2$ sintered twice increases linearly up to 2×10^3 R and becomes saturated with higher doses. This peak seems to be suitable for radiation dosimetry. Additionally, TL glow peaks were found at 70 and 120 °C [69]. The X-ray diffraction of samples, made by sintering of $Ca[B_4O_7]$ glass compacts, shows that the samples always contain crystals in addition to the glass. The samples thus made showed TL glow peaks at ca. 75, 115, 150, and 180 °C. The 180 °C glow peak which can be used for TL dosimetry is characteristic for $Ca[B_4O_7]$ crystals [70]. Thermoluminescence glow curves in sintered and X-ray irradiated $Ca[B_4O_7]$: $CuCl_2$: $EuCl_3$ consist of two peaks at 90 to 100 °C and 180 to 190 °C, and the spectral composition of both peaks was characterized by maxima at 280 and 445 nm. The former band was assigned to an unidentified impurity, and the emission at 445 nm was attributed to Eu^{2+} ions. The mechanism of TL of this material was proposed assuming that the valence of Eu^{3+} is reduced in the sintering process, and Eu^{2+} changes Cu^{2+} into Cu^+ by the effect of irradiation [71].

Thermoluminescence (TL) and thermally stimulated exoelectron emission (TSEE) peaks for sintered $Ca[B_4O_7]$ doped with $EuCl_3$ or $DyCl_3$ were observed at ca. 103 °C (like the Cu^{2+} compound), and, doped with PbO, at ca. 120 °C, and for all species at ca. 325 °C, indicating that these TSEE peaks are characteristic of the sintered $Ca[B_4O_7]$ matrix. All the TSEE peaks except for the one at 325 °C for PbO-doped ones are accompanied by the TL peaks, although the TSEE peaks often shift to higher temperatures than the corresponding TL peaks. Not all the TL peaks are accompanied by TSEE peaks; these are considered to be caused by thermal release of a trapped hole [72]. Sintered $Ca[B_4O_7]$, doped with activators such as Eu, Dy, Cu, or Pb, show TL glow peaks (for a heating rate of 20 °C/min) in three temperature regions: 75 to 110 °C, 90 to 170 °C, and 175 to 255 °C. Their emission spectra depend on the activator and its valence. The activators relate to the luminescence in the following way: Eu^{3+} to 440 nm, Dy^{3+} to 480 and 580 nm, Cu^+ to 540 nm, and Pb^{2+} to 275 nm, respectively. Thermally stimulated current (TSC) was observed, corresponding to all the temperatures at which TL glow peaks are observed. This behavior indicated that the thermally released charge carriers recombine with the luminescent centers through conduction and/or valence bands, and not by a tunneling process. None of the TL glow peaks show supralinearity before they saturated at an exposure of ca. 104 R [73].

Single crystals of **$Sr[B_4O_7]$** (40 × 40 × 10 mm) were grown by the Czochralski method; due to the high viscosity of the melt, the pulling rate was only ca. 0.2 mm/h. The linear electro-optical tensors are: $r_{113}=0.58(3)$, $r_{223}=0.78(3)$, $r_{333}=0.39(3)$, $r_{131}=0.22(4)$, $r_{232}=0.33(4)$; the tensors of the linear electrostriction are: $d_{311}=1.16(7)$, $d_{322}=1.61(6)$, $d_{333}=-0.39(7)$, $d_{113}=0.26(8)$, $d_{223}=0.97(10)$; the unit is 10^{-12} m/V (T = 295 K, λ = 633 nm) [74].

The accuracy of pressure measurements in a diamond anvil cell was improved by using the shift of the $Sr[B_4O_7]:Sm^{2+}$ fluorescence (6.854 Å at 1 atm) rather than that of the R_1 line of ruby's doublet at 200 kbar. Due to the small temperature effects on position and linewidth, the accidental temperature change was ignored. The evolution of the optical sensors excitation at higher pressure was investigated [75]. The 0.255 Å/kbar linear pressure shift of the intense, narrow, and isolated emission line of $Sr[B_4O_7]:Sm^{2+}$ is suggested as a new optical pressure gauge for the diamond anvil cell; the other characteristics of the emission spectrum are promising for simultaneous measurements of pressure and temperature (at least up to 400 °C) and for further research on other rare earth-doped compounds. The possibility of pressure coefficients larger than that of the R_1 ruby line is shown by the 0.45 Å/kbar value measured for another emission line of $Sr[B_4O_7]:Sm^{2+}$ [76]. An energy-level diagram for $Sr[B_4O_7]:Eu^{2+}$ was obtained showing the relative ordering of the energy levels of defects and impurities with respect to the top of the valence band of the pure material. The optical and thermoluminescence (TL) data available for this system are explained with the use of this theoretical energy-level diagram. Factors affecting the valence state of activators were explored [77].

The luminescence and energy migration in **$Sr_{1-x}Eu_x[B_4O_7]$** was also studied. The energy migration in $Eu[B_4O_7]$ occurs over the $^6P_{7/2}$ level of Eu^{2+} below 20 K. The decay curves of the Eu^{2+} emission were analyzed using a random walk model for two-dimensional energy migration. From the analysis, a value of 107 s^{-1} for the transfer probability of excitation energy between two nearest Eu^{2+} neighbors has been calculated [78, 79]. The temperature dependence of the energy migration process was studied, and explained by phonon assistance for below 20 K and by the change from two-dimensional energy migration over the $^6P_{7/2}$ level of Eu^{2+} to three-dimensional energy migration over the $4f^65d$ level of Eu^{2+} below 20 K. The two-dimensional migration is exchange-mediated, the three-dimensional migration is due to dipole-dipole interaction [80].

Solid solutions **$Zn_xCd_{1-x}[B_4O_7]$** were prepared from the appropriate stoichiometric amounts of ZnO, $Cd(OH)_2$, and H_3BO_3 between 523 and 623 K and 10^{-4} to 10^{-5} Torr. The X-ray powder diffraction analysis shows the orthorhombic space group Pbca–D_{2h}^{15} (No. 61). The lattice parameters (in Å) are given for x = 1 with a = 13.714, b = 8.091, and c = 8.631; for x = 0.75 with a = 13.891, b = 8.102, and c = 8.637; for x = 0.5 with a = 13.983, b = 8.127, and c = 8.646; for x = 0.25 with a = 14.109, b = 8.171, and c = 8.688; for x = 0 with a = 14.176, b = 8.229, and c = 8.704. IR spectra and thermal data are discussed on the basis of the stereochemistry of boron atoms in the structure [81].

Analogously, **$Cd_xHg_{1-x}[B_4O_7]$** was prepared from HgO, $Cd(OH)_2$, and H_3BO_3 between 523 and 623 K and 10^{-4} to 10^{-5} Torr. This method of synthesis can pretentiously be applied to the isolation of related metal borates at temperatures several hundred degrees below those cited in the literature. $Cd_xHg_{1-x}[B_4O_7]$ was characterized by X-ray powder diffraction analysis, IR, thermogravimetry (TG), and differential thermogravimetry (DTG). The lattice parameters (in Å)

References on pp. 130/3

are given for x=1 with a=8.704, b=14.176, and c=8.229; for x=0.75, a=8.729, b=14.168, and c=8.242; for x=0.5, a=8.758, b=14.162, and c=8.303; for x=0.25, a=8.769, b=14.147, and c=8.348; for x=0, a=8.792, b=14.107, and c=8.382. All the samples crystallize in the orthorhombic space group Pbca−D_{2h}^{15} (No. 61) [82].

3.4.4.2 The Ions $[B_4O_9]^{6-}$ and $[B_4O_{10}]^{8-}$

Eu^{2+}-activated **$Sr_3[B_4O_9]$** is contained in the fluorescent layer of a UV lamp [83]. Computer simulation based on the minimization of thermodynamic functions was used in order to determine structural units in the B_2O_3–Al_2O_3 system, for instance the ion $[B_4O_9]^{6-}$ [84].

$LiSr_2Sc[B_4O_{10}]$ was synthesized and characterized by X-ray diffraction methods. The compound is monoclinic, space group $P2_1/m-C_{2h}^2$ (No. 11) with a=1254.3(1), b=522.01(6), c=1363.5(2) pm; β=116.69(1)°; Z=4; R=4.1% and R_w=4.7%. The structure is a new type having layers of pyroborate groups with the associated cations residing in interlayer interstices. The strontium atoms occupy sites in sevenfold coordination and the Sc atom occupies a distorted octahedron. Five-coordinate lithium atoms reside in oblong tunnels extending along the direction [010] [85].

References for 3.4.4:

[1] London, D.; Zolensky, M. E.; Roedder, E. (Can. Mineral. **25** [1987] 173/80).
[2] Radaev, S. F.; Muradyan, L. A.; Malakhova, L. F.; Burak, Ya. V.; Simonov, V. I. (Kristallografiya **34** [1989] 1400/7).
[3] Zaretskii, V. V.; Burak, Ya. V. (Fiz. Tverd. Tela [Leningrad] **31** No. 6 [1989] 80/4 from C.A. **111** [1989] No. 106168).
[4] Zaretskii, V. V.; Burak, Ya. V. (Pis'ma Zh. Eksp. Teor. Fiz. **49** [1989] 198/201 from C.A. **110** [1989] No. 240515).
[5] Zhigadlo, N. D.; Zaretskii, V. V. (Pis'ma Zh. Eksp. Teor. Fiz. **49** [1989] 498/500 from C.A. **111** [1989] No. 88252).
[6] Adachi, M.; Shiosaki, T.; Kawabata, A. (Jpn. J. Appl. Phys. **24** Suppl. **3** [1985] 72/5).
[7] Shiosaki, T.; Adachi, M.; Kawabata, A. (Proc. 6th IEEE Int. Symp. Appl. Ferroelectr., Lehigh Univ., Bethlehem, Pa., 1986, pp. 455/64 from C.A. **108** [1988] No. 22984).
[8] Abe, H.; Saitou, H.; Ohmura, M.; Yamada, T.; Miwa, K. (Ultrason. Symp. Proc. **1987** 91/4 from C.A. **109** [1988] No. 240923).
[9] Balakireva, T. P.; Lebol'd, V. V.; Nefedov, V. A.; Provotorov, M. V.; Maier, A. A. (Izv. Akad. Nauk SSSR Neorg. Mater. **25** [1989] 524/6; Inorg. Mater. [USSR] **25** [1989] 462/4).
[10] Liu, Xingzhong; Zhang, Shunxing (Rengong Jingti **18** [1989] 30/4 from C.A. **111** [1989] No. 244541).

[11] Uno, Takehiko (Jpn. J. Appl. Phys. **27** Suppl. 1 [1987/88] 120/2 from C.A. **109** [1988] No. 220716).
[12] Matsumara, S.; Omi, T.; Yamaji, N.; Ebata, Y. (Ultrason. Symp. Proc. **1987** 247/52 from C.A. **109** [1988] No. 240924).
[13] Ebata, Y.; Koshino, M. (Jpn. J. Appl. Phys. **26** Suppl. 1 [1986/87] 123/5 from C.A. **108** [1988] No. 104782).
[14] Ishak, W. S.; Flory, C. A.; Auld, B. A. (Ultrason. Symp. Proc. **1987** 241/5 from C.A. **109** [1988] No. 161775).

[15] Ballato, A.; Hatch, E. R.; Lukaszek, T.; Mizan, M. (Ultrason. Symp. Proc. **1986** 339/42 from C.A. **107** [1987] No. 210294).
[16] Shiosaki, T.; Adachi, M.; Kobayashi, H.; Araki, K.; Kawabata, A. (Jpn. J. Appl. Phys. **24** Suppl. 1 [1984/85] 25/7).
[17] Adachi, M.; Shiosaki, T.; Kobayashi, H.; Ohnishi, O.; Kawabata, A. (Ultrason. Symp. Proc. **1985** 228/32 from C.A. **106** [1987] No. 147775).
[18] Adachi, M.; Yamamichi, S.; Ohira, M.; Shiosaki, T.; Kawabata, A. (Jpn. J. Appl. Phys. **28** Suppl. 2 [1989] 111/3 from C.A. **112** [1990] No. 170226).
[19] Bhalla, A. S.; Cross, L. E.; Whatmore, R. W. (Jpn. J. Appl. Phys. **24** Suppl. 2 [1985] 727/9).
[20] Sehery, A. A.; Somerford, D. J. (J. Phys. Condens. Matter **1** [1989] 2279/81 from C.A. **110** [1989] No. 240474).

[21] Zelenka, J. (Ferroelectrics **81** [1987/88] 335/8 from C.A. **109** [1988] No. 201882).
[22] Sil'vestrova, I. M.; Senyushchenkov, P. A.; Lomonov, V. A.; Pisarevskii, Yu. V. (Fiz. Tverd. Tela [Leningrad] **31** [1989] 311/3 from C.A. **112** [1990] No. 67036).
[23] Burak, Ya. V.; Dovgii, Ya. O.; Kityk, I. V. (Fiz. Tverd. Tela [Leningrad] **31** [1989] 275/8 from C.A. **112** [1990] No. 12105).
[24] Antonyak, O. T.; Burak, Ya. V.; Lyseiko, I. T.; Pidzyrailo, N. S.; Khapko, Z. A. (Opt. Spektrosk. **61** [1986] 550/3 from C.A. **105** [1986] No. 199580).
[25] Burak, Ya. V.; Kopko, B. N.; Lyseiko, I. T.; Matkovskii, A. O.; Slipetskii, R. R.; Ulmanis, U. A. (Izv. Akad. Nauk SSSR Neorg. Mater. **25** [1989] 1226/8; Inorg. Mater. [USSR] **25** [1989] 1038/9 from C.A. **111** [1989] No. 124147).
[26] Matkovskii, A. O.; Sugak, D. Y.; Burak, Ya. V.; Lyseiko, I. T.; Mironova, N.; Skvortsova, V.; Slipetskii, R.; Ulmanis, U. A. (Latv. PSR Zinat. Akad. Vestis Fiz. Teh. Zinat. Ser. **1989** No. 6, pp. 20/4 from C.A. **112** [1990] No. 108937).
[27] Burak, Ya. V.; Lyseiko, I. T.; Garapin, I. V. (Ukr. Fiz. Zh. [Russ. Ed.] **34** [1989] 226/8 from C.A. **110** [1989] No. 183803).
[28] Chan, L. H. (Anal. Chem. **59** [1987] 2662/5).
[29] Jossen, H. (PTB-Dos-14 [1986] 142/9 from C.A. **107** [1987] No. 163664).
[30] Kutomi, Y.; Takeuchi, N. (Zairyo **34** No. 380 [1985] 586/90 from C.A. **103** [1985] No. 44408; Radiat. Eff. Lett. Sect. **85** No. 4 [1985] 163/8 from C.A. **103** [1985] No. 13247; J. Mater. Sci. Lett. **5** [1986] 51/3 from C.A. **104** [1986] No. 58806).

[31] Kutomi, Y.; Tomita, A.; Takeuchi, N. (Phys. Status Solidi A **97** [1986] K169/K172; Radiat. Prot. Dosim. **17** [1986] 499/502 from C.A. **106** [1987] No. 185367; J. Mater. Sci. Lett. **6** [1987] 301/2 from C.A. **106** [1987] No. 223573; Radiat. Prot. Dosim. **17** [1986] 519/22 from C.A. **107** [1987] No. 14336).
[32] Kutomi, Y.; Ariyasu, T.; Takeuchi, N. (Yogyo Kyokaishi **95** [1987] 686/92 from C.A. **107** [1987] No. 82480).
[33] Tomita, A.; Takeuchi, N. (J. Mater. Sci. Lett. **7** [1988] 1157/9 from C.A. **110** [1989] No. 241073).
[34] Bohaty, L.; Haussühl, S.; Liebertz, J. (Cryst. Res. Technol. **24** [1989] 1159/63 from C.A. **112** [1990] No. 127931).
[35] Ramain, A.; Surjanto, U. (Maj. BANTAN **18** No. 4 [1985] 9/25 from C.A. **107** [1987] No. 14361).
[36] Visocekas, R.; Lorrain, S.; Marinello, G. (Ho Tzuk'o Hsueh **22** [1985] 61/6 from C.A. **103** [1985] No. 130860).
[37] Lorrain, S.; David, J. P.; Visocekas, R.; Marinello, G. (Radiat. Prot. Dosim. **17** [1986] 385/92 from C.A. **106** [1987] No. 222471).
[38] Harada, H.; Ogata, Y.; Hama, Y. (Rikogaku Kenkyusho Hokoku Waseda Daigaku No. 121 [1988] 49/55 from C.A. **110** [1989] No. 66079).

[39] Srivastava, J. K.; Supe, S. J. (J. Phys. D **22** [1989] 1537/43).
[40] Tsurumi, T.; Saigoh, H.; Hoshino, Y. (Nippon Seramikkusu Kyokai Gakujutsu Ronbunshi **97** No. 5 [1989] 525/32 from C.A. **111** [1989] No. 27457).

[41] Gauld, I. C.; Harvey, J. W.; Kennett, T. J.; Prestwich, W. V. (Nucl. Instr. Methods Phys. Res. A **251** [1986] 380/5 from C.A. **105** [1986] No. 180125).
[42] Ben-Shachar, B.; Catchen, G. L.; Hoffman, J. M. (Radiat. Prot. Dosim. **27** [1989] 121/4 from C.A. **111** [1989] No. 66302).
[43] Marcinowski, F.; Plato, P. (Radiat. Prot. Manage. **5** [1988] (5), 63/9 from C.A. **109** [1988] No. 199544).
[44] Rzyski, B. M.; Morato, S. P. (IPEN-Pub-111 [1987] 1/9, IPEN-Pub-114 [1987] 1/9 from C.A. **107** [1987] Nos. 143345 and 143346).
[45] Kawamoto, T.; Kikuchi, R.; Lee, C. H.; Yanagisawa, H.; Kawanishi, M. (Osaka Kogyo Gijutsu Shikensho Kiho **36** [1985] 229/31 from C.A. **105** [1986] No. 10618).
[46] Oda, K.; Kagiage, T.; Yamamoto, T.; Kawanishi, M. (Jpn. J. Appl. Phys. **24** Suppl. 4 [1985/86] 229/32 from C.A. **105** [1986] No. 31627).
[47] Kawamoto, T.; Tomita, A.; Kawanishi, M. (Jpn. J. Appl. Phys. **24** Suppl. 4 [1985/86] 242/5 from C.A. **105** [1986] No. 31629).
[48] Kawamoto, T.; Yanagisawa, H.; Nakamichi, H.; Kikuchi, R.; Kawanishi, M. (Osaka Kogyo Gijutsu Shikensho Kiho **37** [1986] 122/4 from C.A. **105** [1986] No. 213055).
[49] Kawanishi, M.; Yamamoto, T.; Oda, K.; Kakiage, T.; Nakasaku, S.; Kawamoto, T.; Chubaci, J. F. D. (Acta Univ. Wratislav. Mat. Fiz. Astron. No. 51 [1987] 9/18 from C.A. **107** [1987] No. 227383).
[50] Yanagisawa, H.; Kawamoto, T. (Chem. Express **4** [1989] 229/32 from C.A. **110** [1989] No. 217738).

[51] Yanagisawa, H.; Kawamoto, T. (Osaka Kogyo Gijutsu Shikensho Kiho **39** [1988] 15/7 from C.A. **109** [1988] No. 78320).
[52] Kutomi, Y.; Tomita, A.; Takeuchi, N. (Radiat. Eff. Express **1** No. 2 [1987] 63/8 from C.A. **107** [1987] No. 248275).
[53] Meng, Xianlin (Rengong Jingti **17** No. 2 [1988] 136/44 from C.A. **112** [1990] No. 14643).
[54] Yates, D. J. C.; Zlotin, N. E.; McHenry, J. A. (J. Catal. **117** [1989] 290/4 from C.A. **111** [1989] No. 9220).
[55] Sanchez, G.; Spano, F.; Caselli, E.; Meurisse, E.; Cuttela, M.; Mansant, M. (J. Phys. Condens. Matter **1** [1989] 2235/40).
[56] Scarpa, G. (Ho Tzu K'o Hsueh **22** [1985] 114/7 from C.A. **104** [1986] No. 11869).
[57] Uchrin, G. (Radiat. Prot. Dosim. **17** [1986] 99/104 from C.A. **106** [1987] No. 222432).
[58] Prokic, M. S. (Phys. Med. Biol. **30** [1985] 323/9 from C.A. **102** [1985] No. 193571).
[59] Christensen, P.; Prokic, M. S. (Radiat. Prot. Dosim. **17** [1986] 341/50 from C.A. **106** [1987] No. 222428).
[60] Christensen, P. (NUREG CP-0050 [1984] 15/24 from C.A. **103** [1985] No. 77992).

[61] Prokic, M. S. (Radiat. Prot. Dosim. **17** [1986] 393/6 from C.A. **106** [1987] No. 222472).
[62] Abtahi, A.; Haugan, T.; Kelly, P. (Radiat. Prot. Dosim. **21** [1987] 211/7 from C.A. **108** [1988] No. 194432).
[63] Ranogajec-Komor, M.; Osvay, M. (Radiat. Prot. Dosim. **17** [1986] 379/84 from C.A. **106** [1987] No. 222470).
[64] Richmond, R. G.; Ogunleye, O. T.; Cash, B. L.; Jones, K. L. (Appl. Radiat. Isot. **38** [1987] 313/4 from C.A. **107** [1987] No. 29919).
[65] Ogunleye, O. T.; Richmond, R. G.; Cash, B. L. (Health Phys. **49** [1985] 527/32 from C.A. **103** [1985] No. 149241).

[66] Guo, F.-Y.; Lin, H. (Zhongguo Xitu Xuebao **5** No. 2 [1987] 15/20 from C.A. **107** [1987] No. 105459).
[67] Fukuda, Y.; Takeuchi, N. (J. Mater. Sci. Lett. **8** [1989] 1001/2 from C.A. **111** [1989] No. 204533).
[68] Fukuda, Y.; Takeuchi, N. (Phys. Status Solidi A **114** [1989] K245/K247).
[69] Fukuda, Y.; Takeuchi, N. (J. Mater. Sci. Lett. **4** [1985] 94/6 from C.A. **102** [1985] No. 102808).
[70] Fukuda, Y.; Takeuchi, N. (Zairyo **34** No. 380 [1985] 533/5 from C.A. **103** [1985] No. 44407).

[71] Fukuda, Y.; Takeuchi, N. (J. Mater. Sci. Lett. **5** [1986] 379/80 from C.A. **104** [1986] No. 196188).
[72] Fukuda, Y.; Tomita, A.; Takeuchi, N. (Phys. Status Solidi A **99** [1987] (2) K135/K138).
[73] Fukuda, Y.; Mizuguchi, K.; Takeuchi, N. (Radiat. Prot. Dosim. **17** [1986] 397/401 from C.A. **106** [1987] No. 222473).
[74] Bohaty, L.; Liebertz, J.; Staehr, S. (Z. Kristallogr. **172** [1985] 135/8).
[75] Lacam, A.; Genotelle, M.; Chateau, C. (Compt. Rend. Acad. Sci. [2] **303** [1986] 547/52 from C.A. **105** [1986] No. 199600).
[76] Lacam, A.; Chateau, C. (J. Appl. Phys. **66** [1989] 366/72).
[77] Mishra, K. C.; Berkowitz, J. K.; DeBoer, B. G.; Dale, E. A.; Johnson, K. H. (Phys. Rev. [3] B **37** [1988] 7230/7).
[78] Meijerink, A.; Nuyten, J.; Blasse, G. (Proc. Electrochem. Soc. **88**-24 [1988] 227/34 from C.A. **110** [1989] No. 104236).
[79] Meijerink, A.; Nuyten, J.; Blasse, G. (Excited States Transition Elem. Proc. 1st Int. Sch., Ksiaz Castle, Pol., 1988 [1989], pp. 360/7 from C.A. **112** [1990] No. 13643).
[80] Meijerink, A.; Nuyten, J.; Blasse, G. (J. Lumin. **44** [1989] 19/31 from C.A. **112** [1990] No. 27490).

[81] Laureiro, Y.; Camps, M. D.; Veiga, M. L.; Jerez, A.; Pico, C. (Eur. J. Solid State Inorg. Chem. **25** [1988] 381/6 from C.A. **110** [1989] No. 241536).
[82] Laureiro, Y.; Veiga, M. L.; Jerez, A.; Pico, C. (Polyhedron **8** [1989] 1567/70).
[83] Wolff, F. (Eur. Appl. 190 536 [1985/86] 1/8 from C.A. **106** [1987] No. 110987).
[84] Bukhtoyarov, O. I.; Lepinskikh, B. M.; Kurlov, S. P. (Fiz. Khim. Issled. Metall. Protsessov No. 15 [1987] 53/6 from C.A. **111** [1989] No. 101802).
[85] Thompson, P. D.; Keszler, D. A. (Solid State Ionics **32/33** [1988/89] 521/7 from C.A. **112** [1990] No. 29720).

3.4.5 Pentaborates

3.4.5.1 The Ion $[B_5O_9]^{3-}$

$NaCa[B_5O_9]$ is monoclinic, space group $P2_1/c-C^5_{2h}$ (No. 14) with a = 646.3(5), b = 1393.2(11), and c = 785.8(6) pm; β = 109.55(5)°; Z = 4; D_c = 2.601 g/cm³. The final R = 6.4% was refined for 594 reflections. The structure consists of complex metaborate sheets with a B_5O_9 building block which contains BO_4 and BO_3 groups in the ratio 2:3 in two rings. Sodium and calcium are partially ordered in channels between the metaborate sheets [1].

References on pp. 135/6

3.4.5.2 $[B_5O_9Cl]^{4-}$ and $[B_5O_9(OH)]^{4-}$ Structures

$[B_5O_9Cl]^{4-}$ and $[B_5O_9(OH)]^{4-}$ are not discrete ions, but have to be considered as structural units only.

A new nomenclature for the borate mineral group hilgardite-tyretskite has been given. The crystal structures of the various monoclinic and triclinic mineral phases belonging to this group are based on the linkage in chains or layers of the pentaborate ion $[B_5O_{12}]^{9-}$, which can exist in two stereoisomeric configurations. The three known mineral species with the general formula **$Ca_2[B_5O_9Cl]\cdot H_2O$** are designated as hilgardite-1Tc (triclinic with Z=1 and V=205.8 Å^3, previously reported in 1977 [2] as an unnamed phase), hilgardite-4M (monoclinic with Z=4 and V=817.8 Å^3, previously named hilgardite), and hilgardite-3Tc (triclinic with Z=3 and V=614.8 Å^3, previously named parahilgardite). The single known mineral species with the formula **$Ca_2[B_5O_9(OH)]\cdot H_2O$** is designed tyretskite-1Tc (triclinic with Z=1 and V=203.3 Å^3, previously named tyretskite). Accommodation of the mineral phases strontiohilgardite-1Tc [3] and Cl-tyretskite [4] as varieties of hilgardite-1Tc is discussed [5].

Hilgardite-4M, $Ca_2[B_5O_9Cl]\cdot H_2O$, "$H_2[Ca_2(BO_2)_5Cl]$", occurs as euhedral crystals in evaporites of New Brunswick. It is monoclinic, space group Aa$-C_s^4$ (No. 9) with a=1147.0(3), b=1132.1(4), and c=632.1(5) pm; β=90.02(7)°; Z=1; D_m=2.67(3) g/cm^3 and D_c=2.676(2) g/cm^3 [6].

In the triclinic mineral kurgantaite, **$CaSr[B_5O_9Cl]\cdot H_2O$**, "$CaSr[\mathit{B}_3B_2O_9]Cl\cdot H_2O$" (*B* is four-coordinate), a supergroup symmetry has been found [7].

3.4.5.3 The Ion $[B_5O_{10}]^{5-}$

$Na_3Ca[B_5O_{10}]$ is triclinic, space group $P\bar{1}-C_i^1$ (No. 2) with a=746.0(3), b=744.5(4), and c=1112.4(6) pm; α=120.58(4)°, β=61.94(3)°, and γ=120.17(3)°; Z=2; D_c=2.437 g/cm^3. The final R value is 6.9% for 2487 reflections. The structure contains two triangular BO_3 units; the rings are connected by a shared tetrahedral BO_4 unit [8].

$MgNd[B_5O_{10}]$ crystals are grown from high-temperature solutions of $K_2[B_4O_7]$ flux. According to the measured solubility, ca. 183 g of $MgNd[B_5O_{10}]$ were dissolved in 100 g of $K_2[B_4O_7]$ at 1273 K, heating the mixture for 5 h, followed by cooling to ca. 973 K at a rate of 2 K/h. Crystalline plates of up to 3.7×2.5×0.5 mm in size were grown with well-developed (102) faces in shape, reddish purple in color and transparent. The melting point is 1346±5 K, D_m=4.08(3) g/cm^3 [9].

$CoYb[B_5O_{10}]$ is monoclinic, space group $P2_1/m-C_{2h}^2$ (No. 11) with a=852.9(3), b=760.7(4), and c=941.6(3) pm; β=93.90(2)°; Z=4; D_c=4.88 g/cm^3. The structure was refined to R=4.3% and consists of infinite chains of $[B_5O_{10}]^{5-}$, composed of three BO_4 tetrahedra and two BO_3 triangles joined by common vertices; the Co^{2+} and Yb^{3+} ions are in distorted octahedral and decahedral coordination, respectively, with oxygen atoms [10].

$LiBa_2[B_5O_{10}]$ is monoclinic, space group $P2_1/m-C_{2h}^2$ (No. 11) with a=441.4(1), b=1457.6(2), c=669.7(2) pm; β=104.26(2)°; Z=2. The final R value was 2.1% and R_w was 4.2% for 2373 reflections. The structure exhibits a unique one-dimensional polyborate ion, built from two crystallographic independent BO_3 groups and a distinct BO_4 group that share vertices. The polyanions are bridged by an Li^+ ion occupying a distorted tetrahedral site and a Ba^{2+} ion occupying an irregular eightfold coordination site [11].

$MgGd[B_5O_{10}]$ is a very promising luminescent material for lamp phosphors [12]. Trace impurities in $MgGd[B_5O_{10}]$ can act as luminescent centers, e.g., Ce^{3+} strongly absorbs short wave UV [13]. In sensitized phosphors like $MgGd[B_5O_{10}]$, energy transfer processes occur and extremely efficient luminescence is shown [14]. Pb^{2+} ions may sensitize the Gd^{3+} sublattice in $MgGd[B_5O_{10}]$, and energy transfer from Pb^{2+} to Mn^{2+} or Tb^{3+} via the Gd^{3+} sublattice may occur [15].

The crystal structures of **$MgGd_{1-x}Eu_x[B_5O_{10}]$** compounds with $0<x\leq1$ are determined [16]; the decay curves and the concentration quenching of the Eu^{3+} emission in **$MgEu[B_5O_{10}]$** show that excitation migration in this compound is one-dimensional [17]. The energy migration in doped $MgEu[B_5O_{10}]$ depends on the dimensionality and the Eu^{3+}-Eu^{3+} distance [18]. The energy transfer processes in $MgEu[B_5O_{10}]$ doped with Nd^{3+} or Ni^{2+} occur down to low temperatures, since the host lattice contains linear Eu^{3+} chains, and at low temperatures, mainly Eu^{3+} traps act as acceptors, whereas at higher temperatures the intentionally added impurities (and defects of an unknown nature) dominate. The transfer probabilities in and between the linear Eu^{3+} chains were estimated and differed by orders of magnitude. The critical interaction distances between regular Eu^{3+} ions as donors and Eu^{3+} traps, Nd^{3+} ions, and Ni^{2+} ions as acceptors were determined. The temperature dependence of the transfer processes is explained by phonon assistance [19].

Energy transfer processes in undoped **$MgTb[B_5O_{10}]$** and in $MgTb[B_5O_{10}]$ doped with Eu^{3+} or Mn^{2+} were evaluated. This host lattice contains linear Tb^{3+} chains. Energy migration occurs to defects of an unknown nature and to the intentionally added impurities at all temperatures from 1.3 to 300 K and is quasi-one-dimensional. At 100 K, the energy migration is enhanced by phonon assistance, whereas it is slowed down at higher temperatures due to a decrease of the spectral overlap between neighboring Tb^{3+} ions [20].

The fluorescence spectrum of Eu^{3+}-doped **$MgLa[B_5O_{10}]$** was studied at 77 and at 300 K under UV and visible dye laser excitation. A phenomenological crystal field analysis was conducted on the basis of the observed 34 $^7F_{0-5}$ level energies. A descending symmetry method from D_{2d} over C_{2v} to C_2 symmetries was used in the simulation. D_{2d} symmetry was inadequate to describe the experimental energy level scheme, whereas for C_{2v} symmetry a satisfactory deviation of 7.1 cm^{-1} was obtained. The decrease in symmetry to C_2 did not improve the simulation. The weak $^5D_0\rightarrow{}^7F_0$ transition intensity correlated with the mixing of the $^7F_{2,4}$ wave functions with the $^7F_{0,0}$ one [21].

Pure and mixed rare earth and alkaline earth metal metaborates activated by Ce and Tb have been synthesized, and their luminescent properties were studied. The La-Ga-Mg metaborates activated by Ce and Tb are effective green components of phosphor mixtures for lamps with large light parameters [22].

3.4.5.4 The Ion $[B_5O_{12}]^{9-}$

Luminescence of **$Bi_3[B_5O_{12}]$** crystals with the maximum at 460 nm was observed at temperatures of ca. 150 K; the maximum of the corresponding excitation band was at 265 nm. The Stokes shift of the emission at 4.2 K amounted to 17000 cm^{-1} [23].

References for 3.4.5:

[1] Fayos, J.; Howie, R. A.; Glasser, F. P. (Acta Crystallogr. C **41** [1985] 1394/6).

[2] Rumanova, I. M.; Yorish, Z. I.; Belov, N. V. (Dokl. Akad. Nauk SSSR **236** [1977] 91/4).

[3] Braitsch, O. (Beitr. Mineral. Petrogr. **6** [1959] 233/47).
[4] Hodenberg, R. V.; Kühn, R. (Kali Steinsalz **7** [1988] 165/70).
[5] Ghose, S. (Am. Mineral. **70** [1985] 636/7).
[6] Rachlin, A. L.; Mandarino, J. A.; Murowchik, B. L.; Ramik, R. A.; Dunn, P. J.; Back, M. E. (Can. Mineral. **24** [1986] 689/93).
[7] Rumanova, I. M.; Razmanova, Z. P. (Kristallokhim. Rentgenogr. Mineralov **1987** 106/12 from C.A. **108** [1988] No. 41194).
[8] Fayos, J.; Howie, R. A.; Glasser, F. P. (Acta Crystallogr. C **41** [1985] 1396/8).
[9] Oishi, Sh.; Miyashita, K.; Tate, I. (Nippon Kagaku Kaishi **1986** No. 1, pp. 39/42 from C.A. **104** [1986] No. 99704).
[10] Dzhafarov, G. G.; Abdulla'ev, G. K.; Mamedov, Kh. S. (Azerb. Khim. Zh. **1985** No. 1, pp. 110/3 from C.A. **103** [1985] No. 62921).

[11] Smith, R. W.; Keszler, D. A. (Mater. Res. Bull. **24** [1989] 725/31 from C.A. **111** [1989] No. 124263).
[12] Blasse, G. (J. Less-Common Met. **112** [1985] 79/82).
[13] Blasse, G. (Chem. Mag. [Rijswijk, Neth.] **1986** 717/9 from C.A. **106** [1987] No. 164999).
[14] Blasse, G.; Kiliaan, H. S.; De Vries, A. J. (J. Less-Common Met. **126** [1986] 139/46).
[15] Shen, Chonghui; Hao, Zhiren (J. Lumin. **40/41** [1988] 663/4 from C.A. **108** [1988] No. 176419).
[16] Vlasse, M.; Fouassier, C. (J. Solid State Chem. **34** [1980] 271/2).
[17] Buijs, M.; Blasse, G. (Chem. Phys. Lett. **113** [1985] 384/6).
[18] Berdowski, P. A. M.; Buijs, M.; Blasse, G. (J. Phys. Colloq. [Paris] **46** [1985] C7-31/C7-34 from C.A. **104** [1986] No. 98615).
[19] Buijs, M.; Blasse, G. (J. Lumin. **34** [1986] 263/78 from C.A. **104** [1986] No. 98687).
[20] Buijs, M.; Van Vliet, J. P. M.; Blasse, G. (J. Lumin. **35** [1986] 213/22 from C.A. **105** [1986] No. 161325).

[21] Holsa, J.; Leskelä, M. (Mol. Phys. **54** [1985] 657/67).
[22] Ratner, I. M.; Martsokha, V. I.; Mirokhina, L. L. (Sb. Nauchn. Tr. Vses. Nauchno-Issled. Inst. Lyuminoforov Osobo Chist. Veshchestv No. 30 [1986] 51/4 from C.A. **108** [1988] No. 46195).
[23] Blasse, G.; Oomen, E. W. J. L.; Liebertz, J. (Phys. Status Solidi B **137** [1986] K77/K81).

3.4.6 Hexaborates

3.4.6.1 The Ion $[B_6O_{10}]^{2-}$

The preparation and luminescence properties of manganese-activated $\mathbf{Sr[B_6O_{10}]}$ were studied. Mn^{2+} in the $Sr[B_6O_{10}]$ matrix emits a green light at 512 nm; this emission is weak when it was excited with UV radiation, but can be improved with sensitizers like Pb^{2+} or Sn^{2+}, having emission in the UV region at 312 nm and in the blue region at 430 nm. Pb^{2+} is a better sensitizer than Sn^{2+}, and the greatest emission brightness is obtained with a Mn^{2+} concentration of 4% and a Pb^{2+} concentration of ca. 15%. Complete energy transfer from the sensitizer to the activator was not obtained with either Pb^{2+} or Sn^{2+}. The emission wavelength of $\mathbf{Sr_{0.8}Mn_{0.04}Pb_{0.16}\text{-}[B_6O_{10}]}$ at 512 nm is somewhat too short for a low-pressure fluorescence lamp containing three phosphors. The wavelength can be slightly lenthened by the addition of Ba, Ca, or Cd, but then there is a dramatic decrease in the emission intensity of Mn [1].

The luminescence properties of Eu^{3+}, Sn^{2+}, and Pb^{2+} in $Sr[B_6O_{10}]$ and $\mathbf{(Sr_{1-x}Mn^{II}_x)[B_6O_{10}]}$ were studied both at 293 and 4.2 K, and the decay times of Sn^{2+} and Pb^{2+} in this matrix were measured and analyzed. There seem to be three different cation sites in $Sr[B_6O_{10}]$. Eu, Sn, and Pb were also used as sensitizers for Mn^{2+} and the energy transfer within different Eu^{2+} centers. The sensitization action of Sn^{2+} and Pb^{2+} on Mn^{2+} was different because Pb-Pb energy transfer occurs, even at 4.2 K, but Sn-Sn transfer can be neglected. A fast diffusion model for the Pb^{2+} system is suggested [2].

The luminescent properties of Ce^{3+} and Tb^{3+} in $Sr[B_6O_{10}]$ were studied including Ce^{3+}- and Tb^{3+}-codoped samples. The excitation spectrum of Ce^{3+} contains several bands between 260 and 320 nm; the emission spectrum of Ce^{3+} also has several maxima between 300 and 360 nm indicating that the Ce^{3+} ion may occupy several sites. The 4f-5d excitation band of Tb^{3+} lies below 210 nm; the emission of Tb^{3+} shows typical peaks from the $^5D_4$ level, having the most intense line at 542 nm. Sensitization of Tb^{3+} emission with Ce^{3+} is possible in these matrices, and a bright green emission was observed. Addition of alkali metal ions for charge compensation significantly affects the emission intensities [3].

When excited by Hg-low-pressure radiation, $\mathbf{(Sr,Ca)[B_6O_{10}]:Pb^{2+},Gd^{3+}}$ emits in the UV region with maxima of the Pb emission at 303 nm ($Sr[B_6O_{10}]$) and 284 nm ($Ca[B_6O_{10}]$), respectively, and of Gd emission at 311 to 313 nm. The ratio of intensities of the Gd^{3+} to the Pb^{2+} emission depends on the ratio of the concentrations of the activators and on the phosphor matrix and is greatest at 243 nm in the case of $Ca[B_6O_{10}]$: 0.015 mol Pb^{2+} or $Ca[B_6O_{10}]$: 0.25 mol Gd^{3+}. The Pb^{2+}-Gd^{3+} energy transfer is assumed to be an exchange mechanism [4].

3.4.6.2 $[B_6O_{12}]^{6-}$ Structures

The paramagnetic centers formed by UV-irradiation of $\mathbf{Zn_4O[B_6O_{12}]}$ (= $Zn_4O(BO_2)_6$), were studied by ESR. The crystals build dodecahedra with T_d symmetry and crystallize in the space group $I\bar{4}3m-T_d^3$ (No. 217) with a = 746.59(3) pm. Mn^{2+} impurity centers are observed along with oxygen vacancies; the thermal stabilities of these centers are described [5].

References for 3.4.6:

[1] Koskentalo, T.; Leskelä, M.; Niinistö, L. (Mater. Res. Bull. **20** [1985] 265/74).

[2] Leskelä, M.; Koskentalo, T.; Blasse, G. (J. Solid State Chem. **59** [1985] 272/9).

[3] Koskentalo, T.; Niinistö, L.; Leskelä, M. (J. Less-Common Met. **112** [1985] 67/70; C.A. **104** [1986] No. 12544).

[4] Tews, W.; Becker, P.; Herzog, G.; Kuenzler, G. (Z. Phys. Chem. [Leipzig] **268** [1987] 985/92).

[5] Otero de la Gandara, M. J.; Conesa, J. C.; Soria, J. (Phys. Status Solidi A **87** [1987] K81-K84).

3.4.7 Heptaborates and Boracites

3.4.7.1 $[B_7O_{13}]^{5-}$ and Related Structures

$\mathbf{CaLa[B_7O_{13}]}$ and $\mathbf{CaCe[B_7O_{13}]}$ were prepared, based on the study of the B_2O_3-rich portion of the ternary diagram. Indexing of the X-ray diffraction pattern of the compounds shows that these two borates are isostructural and monoclinic. Under UV excitation, Eu^{3+}, Tb^{3+}, and Ce^{3+}

in the borate exhibit red, green, and violet fluorescence, respectively. The green emission of Tb^{3+} is considerably sensitized by Ce^{3+}, and the energy transfer from Ce^{3+} to Tb^{3+} in $CaLa[B_7O_{13}]$ is efficient [1].

$Pb_4Bi_3O_6[B_7O_{13}]$, or $Pb_4Bi_3[B_7O_{19}]$, was prepared from an H_3BO_3–Bi_2O_3–PbO melt at 1073 K in an electrical Pt furnace and cooled in air, melting point 893±10 K (congruently). It is orthorhombic, space group $P222_1-D_2^2$ (No. 17) with a=1576(7), b=743(3), and c=1061(5) pm; Z=4; D_m=9.64 g/cm³ and D_c=9.73 g/cm³. It is characterized also by the IR spectrum between 400 and 2000 cm⁻¹, which confirms the boracite-like ion $[B_7O_{13}]^{5-}$ with four BO_3 and three BO_4 groups [2].

3.4.7.2 Boracites with $[B_7O_{13}X]^{6-}$ Structures (X=Cl, Br, or I)

The improper ferroelectric phase transition of **$Mg_3[B_7O_{13}Cl]$** (Mg-Cl boracite) has been reinvestigated [3]. The Raman scattering strength of this phase transition has been measured, and the Landau theory was extended to predict polarization-optic coefficients [4]. A single ferroelectric domain of the orthorhombic phase of the mixed **$Mg_3[B_7O_{13}Cl]:Mn^{2+}$** boracite crystal allowed one to correlate the three inequivalent EPR sites with the three crystallographic ones by Buerger precession studies [5].

The **$Cr_3[B_7O_{13}I]$** (Cr-I boracite) is cubic, space group $F\bar{4}3c-T_d^5$ (No. 219) with a=1221.4(1) pm; Z=8; D_c=4.13 g/cm³; final R=2.3% refined from 109 reflections. The shortest B–O distance is 143.9(3) pm [6].

The cubic→orthorhombic phase transition temperature was measured by polarized light microscopy in the pyramidal growth sector for **$Mn_3[B_7O_{13}I]$**, **$Ni_3[B_7O_{13}Br]$**, **$Cu_3[B_7O_{13}Br]$**, and **$Zn_3[B_7O_{13}I]$** (Mn-I, Ni-Br, Cu-Br, and Zn-I boracites) [7].

The trigonal room-temperature modification of the **$Fe_3[B_7O_{13}Cl]$** (Fe-Cl boracite) crystallizes in the rhombohedral space group $R3c-C_{3v}^6$ (No. 161) with a_{rh}=860.35(7) pm, α_{rh}=60.15(1)°; a_{hex}=862.31(5) pm and c_{hex}=2105.03(5) pm; Z=6; D_c=3.576 g/cm³; the final R=3.8% and R_w=5.4% for 1544 reflections. Both bond lengths and angles do not differ significantly from those reported [8] for the isostructural compound **$Fe_{2.4}Mg_{0.6}[B_7O_{13}Cl]$**, but are more accurate [9]. Glass-bonded crystalline boracite composites give higher pyroelectric p/K values than single crystals of **$Fe_3[B_7O_{13}I]$** (Fe-I boracite) [10].

Optical, electric, and dielectric studies were done on **$Co_3[B_7O_{13}Cl]$** (Co-Cl boracite) [11]. Linear and quadratic magnetoelectric effects α_{23}, α_{32}, and β_{311} were measured versus temperature for **$Co_3[B_7O_{13}Br]$** (Co-Br boracite); they are consistent with the symmetry m′m2′ down to 4 K [12]. The spontaneous birefrigence of **$Co_3[B_7O_{13}I]$** (Co-I boracite) was also measured between 10 and 200 K [14]. $Co_3[B_7O_{13}Br]$, $Co_3[B_7O_{13}I]$, and $Ni_3[B_7O_{13}Br]$ (Co-Br, Co-I, and Ni-Br boracites) present a succession of phases: $F\bar{4}3c1'\rightarrow Pca2_11'\rightarrow m'm2'$; the principal linear spontaneous magnetic birefrigence of the three compounds was measured in the ferroelectric ($Pca2_11'$) and weakly ferromagnetic (m′m2′) phases; the experimental data were compared with expressions of birefrigence deduced from group theory analysis [13].

Various anomalies of Faraday rotation, magnetoelectric, and magneto-optical coefficients were measured on $Ni_3[B_7O_{13}Br]$ (Ni-Br boracite) [15]. The infrared (20 to 4000 cm⁻¹) reflectivity of $Ni_3[B_7O_{13}Br]$ and **$Cu_3[B_7O_{13}Cl]$** (Ni-Br and Cu-Cl boracites) was measured in the ferroelectric and paraelectric phases [16]. A theory of the magnetic and magnetoelectric properties of boracites like latent antiferromagnetism, weak ferromagnetism, and improper magnetostructural couplings is developed, using the Landau-Dzialoshinskii approach. Monoclinic **Ni_3-[$B_7O_{13}I$]** (Ni-I boracite) is a magnetic material which can be recognized macroscopically by an $M \approx (T_c - T)^{3/2}$ variation law [17]. A theoretical model of the 61.5 K transition in Ni-I boracite is based on the symmetry of the transition order parameter [18].

Dielectric constants, spontaneous polarization, and birefrigence were measured between 10 and 290 K on single domain samples of $Cu_3[B_7O_{13}Br]$ (Cu-Br boracite) [19]. Yellowish brown **$Cu_3[B_7O_{13}I]$** (Cu-I boracite) is cubic, space group $F\bar{4}3c-T_d^5$ (No. 219) with a=1202.03(7) pm; Z=8; D_c=4.608 g/cm³; for 110 unique reflections R=1.4% and R_w=1.6%. It contains tetrahedral boron. This new boracite was prepared by high-pressure synthesis. The structure is comparable with that of other cubic boracites. No structural phase transition could be observed between 15 and 1265 K [20].

Defect $[BO_3]^{3-}$ centers induced by X-ray irradiation in Zn boracites, **$Zn_3[B_7O_{13}Cl]$**, **Zn_3-[$B_7O_{13}Br$]**, and $Zn_3[B_7O_{13}I]$, were observed by ESR spectroscopy [21].

References for 3.4.7:

[1] Cheng, Risheng; Huang, Jinggen; Lu, Ling; Xu, Yan (Mater. Res. Bull. **23** [1988] 1699/704).
[2] Shuster, N. S.; Zargarova, M. I.; Abdulla'ev, G. K. (Zh. Neorg. Khim. **30** [1985] 1890/2; Russ. J. Inorg. Chem. **30** [1985/86] 1073/4).
[3] Arakelian, H. E. (Diss. Stevens Inst. Techn., Hoboken, N.J., U.S.A., 1986, pp. 1/150 from C.A. **106** [1987] No. 76972).
[4] Arakelian, H. E.; Hart, T. R. (Ferroelectrics **74** [1987] 13/21 from C.A. **107** [1987] No. 166863).
[5] Rivera, J. P.; Bill, H.; Lacroix, R. (Ferroelectrics **80** [1987/88] 31/4 from C.A. **109** [1988] No. 162114).
[6] Monnier, A.; Berset, G.; Schmid, H.; Yvon, K. (Acta Crystallogr. C **43** [1987] 1243/5).
[7] Rossignol, J. F.; Rivera, J. P.; Schmid, H. (Jpn. J. Appl. Phys. **24** Suppl. 2 [1985] 574/6 from C.A. **105** [1986] No. 33790).
[8] Dowty, E.; Clark, J. R. (Z. Kristallogr. **138** [1973] 64/99).
[9] Mendoza-Alvarez, M. E.; Yvon, K.; Depmeier, W.; Schmid, H. (Acta Crystallogr. C **41** [1985] 1551/2).
[10] Bhalla, A. S.; Cross, L. E.; Newnham, R. E. (Jpn. J. Appl. Phys. **24** Suppl. 2 [1985] 454/6 from C.A. **105** [1986] No. 16161).

[11] Mendoza-Alvarez, M. E.; Rivera, J. P.; Schmid, H. (Jpn. J. Appl. Phys. **24** Suppl. 2 [1985] 1057/9 from C.A. **105** [1986] No. 16241).
[12] Clin, M.; Rivera, J. P.; Schmid, H. (Ferroelectrics **79** [1987/88] 173/6 from C.A. **109** [1988] No. 202947).
[13] Clin, M.; Rivera, J. P.; Schmid, H. (Helv. Phys. Acta **60** [1987] 287/93).
[14] Clin, M.; Rivera, J. P.; Schmid, H. (Jpn. J. Appl. Phys. **24** Suppl. 2 [1985] 1054/6 from C.A. **105** [1986] No. 16788).
[15] Rivera, J. P.; Schaefer, F. J.; Kleemann, W.; Schmid, H. (Jpn. J. Appl. Phys. **24** Suppl. 2 [1985] 1060/2 from C.A. **105** [1986] No. 16789).
[16] Moopenn, A.; Coleman, L. B. (Jpn. J. Appl. Phys. **24** Suppl. 2 [1985] 344/6 from C.A. **105** [1986] No. 69182).
[17] Tolédano, P.; Schmid, H.; Clin, M.; Rivera, J. P. (Phys. Rev. B **32** [1985] 6006/36 from C.A. **103** [1985] 225900).
[18] Tolédano, P.; Schmid, H.; Clin, M.; Rivera, J. P. (Jpn. J. Appl. Phys. **24** Suppl. 2 [1985] 179/81 from C.A. **105** [1986] 71004).

[19] Berset, G.; Clin, M.; Rivera, J. P.; Schmid, H. (Ferroelectrics **79** [1987/88] 177/8 from C.A. **109** [1988] No. 181700).
[20] Berset, G.; Depmeier, W.; Boutellier, R.; Schmid, H. (Acta Crystallogr. C **41** [1985] 1694/6).

[21] Haddad, M.; Vignaud, G.; Berger, R.; Levasseur, A. (J. Phys. Chem. Solids **46** [1985] 997/1005).

3.4.8 Higher Borates

3.4.8.1 $[B_8O_{13}]^{2-}$ Structures

Stability fields in the phase diagram **$Na_2[B_8O_{13}]$**–SiO_2 are plotted. No significant stable liquation was detected, even though metastable liquation was observed [1].

A discrete model for energy transfer, describing luminescent efficiencies of sensitizers and activators, is presented and applied to octaborates. The luminescence of Ce^{3+} and Tb^{3+} in **$Ba[B_8O_{13}]$** was studied including Ce^{3+}- and Tb^{3+}-codoped samples. The excitation spectrum of Ce^{3+} contains several bands between 260 and 320 nm; the emission spectrum of Ce^{3+} also has several maxima between 300 and 360 nm and, in addition, a broad band at 515 nm. The excitation band of Tb^{3+} lies below 210 nm; the emission of Tb^{3+} shows typical peaks, having the most intense line at 542 nm. Sensitization of Tb^{3+} emission with Ce^{3+} is possible, and a bright green emission was observed [2]. The Ce^{3+}-Tb^{3+} energy transfer in $Ba[B_8O_{13}]$ can also be described by a discrete model considering S-A and S-S transfers. Under UV excitation, the red and the green luminescence bands were observed in Ce^{3+}- and Mn^{2+}-codoped $Ba[B_8O_{13}]$. The former is attributed to the Mn^{2+} centers which are located at T_d sites, and the latter is assigned to Mn^{2+} ions which are situated in O_h positions. The Ce^{3+}-Mn^{2+} energy transfer is discussed [3].

3.4.8.2 $[B_9O_{16}]^{5-}$ Structures

The preparation and luminescence of **$BaRE[B_9O_{16}]$** (RE = rare earth element) is reported. Under UV excitation, **$BaLa[B_9O_{16}]:Ce^{3+}$** shows an emission band peaking at 350 nm. Its quantum efficiency is almost independent of the Ce^{3+} concentration [4].

For the compound **$BaCe_{0.7}Tb_{0.3}[B_9O_{16}]$** and for analogous compounds with higher Tb contents, energy transfer from Ce^{3+} to Tb^{3+} is efficient due to the overlap of the Ce^{3+} emission band with the most intense f-f Tb^{3+} absorption lines. The Tb^{3+} emission sensitized by Ce^{3+} shows a very intense green color; its stability in an Hg vapor lamp was checked [4].

The luminescence of Eu^{3+} in **$BaLa[B_9O_{16}]$**, **$BaGd[B_9O_{16}]$**, and **$BaY[B_9O_{16}]$** was also investigated. Under UV excitation, the Eu^{3+}-activated La and Gd compounds show a bright red luminescence at room temperature, whereas the Eu^{3+}-activated Y borate is characterized by a red-orange luminescence which indicates a change in the symmetry of the rare earth sites. As a consequence of the high energy of the charge transfer band, the Eu^{3+} emission has a high efficiency under 253.7 nm excitation, a characteristic favorable for lighting applications. The sensitization of the Eu^{3+} emission by Bi^{3+} was examined. Energy transfer from Bi^{3+} to Eu^{3+} is observed, but in the presence of Bi^{3+} the efficiency of excitation through the Eu^{3+} charge transfer band is reduced. The luminescence of Eu^{2+} was also studied. **$BaGd[B_9O_{16}]:Eu^{2+}$** shows an intensive blue emission band under UV excitation [5].

The phosphor host **$BaLa[B_9O_{16}]$** was investigated by analytical electron microscopy, electron diffraction patterns, high-resolution electron micrographs, and X-ray powder diffractometry. The crystal is monoclinic with a *c* center lattice and the following unit cell constants: a = 1369.7, b = 790.7, and c = 1631.5 pm; β = 106.251°. Microtwins were also observed in the specimen [6].

3.4.8.3 $BaNd_2Al_2(B_{12}O_{25})$

$BaNd_2Al_2(B_{12}O_{25})$ is hexagonal, space group $P\bar{6}2m-D^3_{3h}$ (No. 189) with a = 456.6(2) and c = 2492(1) pm; Z = 1; D_m = 3.778 g/cm³ and D_c = 3.72 g/cm³. The crystal structure has been solved by direct methods and refined to R = 8.80%. In the structure, 2/3 of the boron atoms are in tetrahedral coordination which gives the formula of the ion $[B_2O_5]^{4-}$; the other 1/3 of the boron atoms forms neutral groups $(B_2O_3)^0$ which link with other cations by van der Waals bonds. The obvious cleavage of $BaO \cdot Nd_2O_3 \cdot Al_2O_3 \cdot 6B_2O_3$ crystals perpendicular to the c axis can be explained by this structure characteristic [7].

References for 3.4.8:

[1] Epel'baum, M. B. (Geokhimiya **1988** 1513/7 from C.A. **109** [1988] No. 234430).
[2] Huang, Jinggen (J. Lumin. **40/41** [1988] 669/70 from C.A. **109** [1988] No. 13949).
[3] Chen, Risheng; Huang, Jinggen; Xu, Yan (Fudan Xuebao Ziran Kexueban **28** [1989] 304/9 from C.A. **112** [1990] No. 168255).
[4] Fu, W. T.; Fouassier, C.; Hagenmuller, P. (Mater. Res. Bull. **22** [1987] 389/97).
[5] Fu, W. T.; Fouassier, C.; Hagenmuller, P. (Mater. Res. Bull. **22** [1987] 899/909).
[6] Shen, C. F.; Ximen, L. L.; Zong, X. F. (Mater. Res. Bull. **24** [1989] 1223/30; C.A. **112** [1990] No. 14553).
[7] Li, Deyu; Tu, Hengyong; Xu, Yueying; Kong, Huashuang; He, Chongfan (Wuji Huaxue **4** [1988] 23/9 from C.A. **110** [1989] No. 48919).

3.4.9 Additional Compounds

The partial enthalpy of **$(NaBO_2)_2$** in liquid $(NaBO_2)_2/B_2O_3$ mixtures at 975 ± 2°C is −168 kJ/mol; at the stoichiometry of the diborate, the enthalpy of mixing is consistent with the formation of triborate groups, $(B_3O_5)^-$, containing tri- and four-coordinate boron. The enthalpy of solution of $(NaBO_2)_2$ in $Na_4[P_2O_7]$ is approximately −9 kJ/mol [1].

CdB_2O_4 crystallizes tetragonally with a = 620.6 and c = 628.8 pm. It properly contains BO_3 and BO_4 units [2].

CuB_2O_4 crystallizes tetragonally in the space group $I\bar{4}2d-D^{12}_{2d}$ (No. 122) with a = 1150.6 and c = 564.4 pm; the IR spectrum is given [3].

From the partial pressures the molar enthalpies of the formation for solid **$Ba_3B_2O_6$** from the elements and from the constituent oxides were determined [4].

References on p. 142

References for 3.4.9:

[1] Julsrud, S.; Kleppa, O. J. (Z. Naturforsch. **42a** [1987] 463/70).
[2] Laureiro, Y.; Pico, C.; Jerez, A. (Syn. React. Inorg. Met.-Org. Chem. **18** [1988] 119/26 from C.A. **108** [1988] No. 215157).
[3] Fomina, A. Z.; Sedmale, G. P.; Sedmalis, Yu. Ya. (Neorg. Stekla Pokrytiya Mater. No. 8 [1987] 67/74).
[4] Asano, M.; Kou, T. (J. Chem. Thermodyn. **21** [1989] 837/45).

3.5 Boron-Oxygen Compounds with Hydrogen and/or Organyl Moieties

For a previous review, see "Boron Compounds" 3rd Suppl. Vol. 2, 1987, pp. 71ff. For hydrated borates, which for practical reasons are represented separately, see Chapter 3.6, p. 258.

3.5.1 Oxygen-Donor Adducts of BH_3 and BR_3

H_3B–OH_2 is one of the assumed reaction products in the gas-phase ion molecular reactions of organic anions with B_2H_6, based on investigations using ion-cyclotron resonance spectroscopy. Ab initio calculations (4-31+G basis set) have been performed in order to obtain the relative stability of the product. The calculated values depend on using additionally flat s and p functions [1]. H_3B–OH_2 was characterized for the first time by means of the UV-He(I) spectrum, and the ionization energies were assigned to the various orbitals based on calculations. The He(I) spectrum shows two bands at 9.7 and 10.6 eV, which are much lower than the lowest ionization energies of water (12.6 eV) and B_2H_6 (11.9 eV). There are also bands at 11.8, 13.2, and 14.4 eV. The 3-21G calculations result in a dissociation energy of 94 kJ/mol for the staggered configuration. The optimized geometry gave r(OH) = 96 pm, r(BH) = 120 pm, and r(BO) = 171 pm; the angles are ∢(HBH) = 115.28°, ∢(HOB) = 114.98°, and ∢(HOH) = 113.22°, with a point group of C_s and a tetrahedral arrangement for the OBH_3 moiety [2].

The triplet oxygen system **H_3B–O^3** has been studied by ab initio MO-LCAO-SCF calculations with the 3-21G and the 6-31G** basis sets. The former one was used in the preliminary study of equilibrium geometries and stability, mainly to distinguish (in H_3B–A systems generally) the stable systems from the unstable ones. For the configuration C_{3v} of H_3B–O^3, r(BH) = 119.9 pm, r(BO) = 168.1 pm, and α = 100.40°; the interaction energy is -13.948×10^{-3} a.u.. For the planar configuration with bidentate oxygen, belonging to the C_{2v} symmetry group, r(B–H(1)) = 118.7 pm, r(B–H(2)) = 118.7 pm, r(BO) = 345.5 pm, and α = 119.94°; the interaction energy is -0.695×10^{-3} a.u.. A counterpoise correction for the basis set superposition errors changed the energies from -13.948×10^{-3} to -2.005×10^{-3} a.u. and for the C_{2v} symmetry group from -0.695×10^{-3} to $+2.353\times10^{-3}$ a.u.. From 6-31G** calculations with C_{3v}, r(BH) = 121.2 pm, r(BO) = 145.9 pm, α = 103.47°, the interaction energy is -9.192×10^{-3} a.u. (the value for correction is -1.668×10^{-3} a.u.), and the charge on oxygen is 0.102. Results of 6-31G**/MP2 calculations at 6-31G**/HF geometries gave -47.236×10^{-3} a.u. for the interaction energy. The values suggest the existence of associative interactions [3].

$(CH_3)_3B$–O_2 is a labile intermediate complex. Its formation has been studied by LCAO, MO-SCF, INDO, and MINDO/3 calculations. The bond lengths are (for C_{3v} geometry; in pm):

r(OO)=114.0, r(BO)=165.5, r(BC)=155.7; and the angle ∢(CBO) is 100.6°. The effective charges are: O(1)=+0.149, O(2)=+0.050, B=+0.022, and CH_3=−0.074; the symmetry is nearly C_{3v} and the enthalpy of formation is $\Delta H_f^\circ = -37$ kJ/mol. It is readily rearranged into the peroxide CH_3–OO–$B(CH_3)_2$, with an activation barrier which does not exceed 25 kJ/mol with respect to the intermediate [4, 24].

H_3B–OCH_2. The formation of the formaldehyde complex of BH_3 has been calculated by 3-21G and 6-31G* basis sets and corrections for electron correlation with the use of the Møller-Plesset (MP) perturbation method. Bent C_s symmetry (sp^2) has been calculated, with the linear structure 6 to 10 kcal/mol higher in energy, and an out-of-plane π structure still higher in energy. The coordination energies are with 3-21G, 17.1 kcal/mol; with 6-31G*, 7.33 kcal/mol; with MP2(FC)/6-31G*, 15.8 kcal/mol; with MP2(FC)/6-31G**, 16.0 kcal/mol; with MP3(FC)/6-31G*, 16.4 kcal/mol; and with MP2(full)/6-31G*, 14.4 kcal/mol. For the MP2(FC)/6-31G* basis, the rotational barriers are ΔH_{inv}=11.4 kcal/mol and ΔH_{rot}=1.37 kcal/mol; the energy of the complex is −140.65463 hartree. The energy as a function of the C–O–B angle is given in **Fig. 3-11.** For the MP2/6-31G* basis, the bond lengths are (in pm): r(CO)=123.3, r(BO)=168.6, r(BH)=120.1 and 120.9; the bond angles are: ∢(COB)=120.6°, ∢(OCH)=120.8°, ∢(OBH)=101.2° and 102.7° [5].

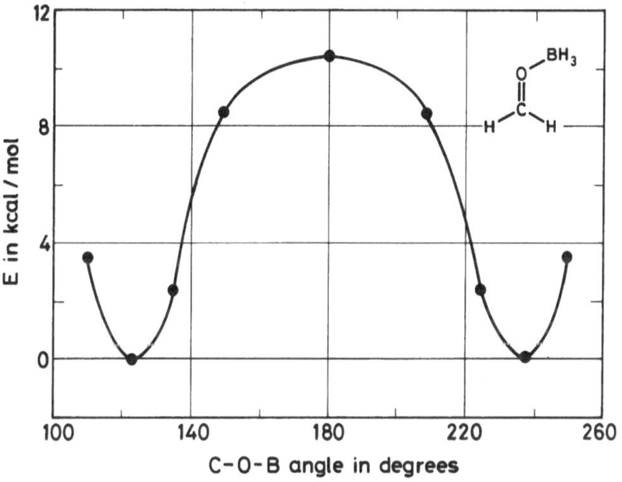

Fig. 3-11. Energy of the H_3B–OCH_2 complex as a function of C–O–B angle (MP3/6-31G*//6-31G*) [5].

H_3B–$O(CH_2)_4$, the tetrahydrofuran complex of BH_3, was used in the hydroboration of nonatriene [6], in order to prepare chiral Lewis acid complexes of juglone [7], and in the reduction of unsaturated nitro compounds (catalyzed by $Na[BH_4]$) [8]; it reacts with transition metal-substituted phosphanes in order to give the thermally stable *P*-borane adducts [9]. H_3B–$O(CH_2)_4$ was used in the hydrogenation of (E)-HOOC–CH=CH–$COOC_2H_5$ in order to give (E)-$HOCH_2$–CH=CH–$COOC_2H_5$ [10]; it reacts with CH_3NH_2 to give a mixture of the eee and eea isomers of 1,3,5-trimethylcycloborazane, $[CH_3N(H)BH_2]_3$ [11]. H_3B–$O(CH_2)_4$ reacts with aminoarsines $(CH_3)_2AsNR_2$ [12] and with $(\eta^5\text{-}C_5H_5)[P(C_6H_5)_3]Co(H_5C_2\text{–}C{\equiv}C\text{–}C_2H_5)$ as a dual reagent [13, 14]. It has been used for preparing phenylenediamine-borane adducts [15], five-membered N→B coordinated heterocycles [16], and it reacts with representative dithianes, diazolidines, thiazolidines, benzothiazolidines, and benzothiazoles [17]. H_3B–$O(CH_2)_4$ reduces phenylmalonic acid exceptionally slowly to form intermediates [18] and reacts with $U(CH_3)$-

References on pp. 144/5

(η^5-C_5H_5)$_3$ in toluene to give U(BH_4)(η^5-C_5H_5)$_3$ [19]. H_3B–O(CH_2)$_4$ reacts with $WH_3[P(CH_3)_3]_3$-B_3H_8 and was used in order to homologate *arachno*-2-tungstaboranes [20].

Na[H_3B–$OCOCH_3$], "sodium acetoxyborohydride", was prepared from Na[BH_4] and CH_3COOH in tetrahydrofuran at 0 °C; it is a simple selective hydroborating agent for olefins [21].

(CH_3)$_3$B–OC(CH_3)$_2$ formation has been investigated by modified neglect of diatomic overlap (MNDO) calculations: for a linear C–O–B bond, $\Delta H_f = -20.03$ kcal/mol; for the bent bond, $\Delta H_f = -28.85$ kcal/mol for the angle 131.6° [22].

A complex between the anion $[CH_2{=}CHO]^-$ and BH_3 was studied by MNDO-SCF calculations. The *O*-bonded complex, **$[CH_2{=}CH{-}O{-}BH_3]^-$**, is less stable than the *C*-bonded complex, **$[H_3B{-}CH_2CH{=}O]^-$** [23].

References for 3.5.1:

[1] Eisenstein, O.; Kayser, M.; Roy, M.; McMahon, T. B. (Can. J. Chem. **63** [1985] 281/7).
[2] Pradeep, T.; Sreekanth, C. S.; Hegde, M. S.; Rao, C. N. R. (J. Mol. Struct. **194** [1989] 163/70).
[3] Forcada, M. L.; Moscardó, F.; San-Fabián, E. (J. Mol. Struct. **166** [1988] 293/9 [THEOCHEM **43**]).
[4] Aleksandrov, Yu. A.; Vyshinskii, N. N.; Kokorev, V. N.; Al'ferov, V. A.; Chikinova, N. V.; Makin, G. I. (J. Organomet. Chem. **332** [1987] 259/69).
[5] LePage, T. J.; Wiberg, K. B. (J. Am. Chem. Soc. **110** [1988] 6642/50).
[6] Brown, H. C.; Negishi, E.; Dickason, W. C. (J. Org. Chem. **50** [1985] 520/7).
[7] Kelly, T. R.; Whiting, A.; Chandra Kumar, N. S. (J. Am. Chem. Soc. **108** [1986] 3510/2).
[8] Mourad, M. S.; Varma, R. S.; Kabalka, G. W. (J. Org. Chem. **50** [1985] 133/5).
[9] Maisch, R.; Ott, E.; Buchner, W.; Malisch, W.; Colquhoun, I. J.; McFarlane, W. (J. Organomet. Chem. **286** [1985] C31/C35).
[10] Kende, A. S.; Fludzinski, P. (Org. Synth. **64** [1986] 104/7).

[11] Narula, C. K.; Janik, J. F.; Duesler, E. N.; Paine, R. T.; Schaeffer, R. (Inorg. Chem. **25** [1986] 3346/9).
[12] Kanjolia, R. K.; Krannich, L. K.; Watkins, C. L. (J. Chem. Soc. Dalton Trans. **1986** 2345/8).
[13] Jiang, Feilong; Fehlner, T. P.; Rheingold, A. L. (J. Am. Chem. Soc. **109** [1987] 1860/1; J. Chem. Soc. Chem. Commun. **1987** 1385/6).
[14] Housecroft, C. E.; Buhl, M. L.; Long, G. J.; Fehlner, T. P. (J. Am. Chem. Soc. **109** [1987] 3323/9).
[15] Camacho, C.; Paz-Sandoval, M. A.; Contreras, R. (Polyhedron **5** [1986] 1723/32).
[16] Farfán, N.; Contreras, R. (J. Chem. Soc. Perkin Trans. II **1988** 1787/91).
[17] Contreras, R.; Morales, H. R.; De Lourdes Mendoza, M.; Dominguez, C. (Spectrochim. Acta A **43** [1987] 43/9).
[18] Choi, Y. M.; Emblidge, R. W.; Kucharczyk, N.; Sofia, R. D. (J. Org. Chem. **52** [1987] 3925/7).
[19] Porchia, M.; Brianese, N.; Ossola, F.; Rosetto, G.; Zanella, P. (J. Chem. Soc. Dalton Trans. **1987** 691/4).
[20] Grebenik, P. D.; Leach, J. B.; Green, M. L. H.; Walker, N. M. (J. Organomet. Chem. **345** [1988] C31/C34).

[21] Narayana, C.; Periasamy, M. (Tetrahedron Lett. **26** [1985] 1757/60).
[22] Nelson, D. J. (J. Org. Chem. **51** [1986] 3185/6).
[23] Jiang, D.-L.; Yu, J.-G.; Ju, C.-X.; Liu, R.-H. (Huaxue Xuebao **45** [1987] 1061/6 from C.A. **109** [1988] No. 93095).

[24] Kokorev, V. N.; Aleksandrov, Yu. A.; Vyshinskii, N. N.; Maslennikov, V. P.; Abronin, I. A. (Izv. Akad. Nauk SSSR Ser. Khim. **1985** 1788/94; Bull. Acad. Sci. USSR Div. Chem. Sci. **34** [1985] 1637/43).

3.5.2 Oxoborane and Its Organic Derivatives

3.5.2.1 Oxoboranes, HBO, BOH, and (–BHO–)

Rotational and vibrational constants in the ground state of **HBO** were calculated using the Møller-Plesset perturbation theory at the MP4SDQ level with two large Gaussian basis sets: $V_e = -100.52699$ a.u. and r(BO) = 120.75 pm [1]. Experimentally was found r(BH) = 116.81 pm and r(BO) = 120.04 pm [3]. The values were used to correct ab initio values for $[BO]^-$ [1].

The IR spectrum of the transient species HBO was observed in the gas phase by using diode laser spectroscopy. The use of the discharge modulation method was found to be essential in suppressing the strong absorption lines of chemically stable molecules, especially those of the precursor B_2H_6. The HBO molecule is formed in a d.c. discharge plasma of a B_2H_6/O_2 or a B_2H_6/NO mixture in He; its lifetime is less than 100 ms. The ν_3 band origin and the rotational constants (B_n) of **$H^{11}BO$** were determined to be $\nu_0 = 1825.5610(13)\ cm^{-1}$, $B_0 = 1.30839(12)\ cm^{-1}$, and $B_3 = 1.29973(10)\ cm^{-1}$; the centrifugal distortion constant at the $v_3 = 1$ level is 2.66×10^{-6} [2].

The rotational transitions in the excited vibrational levels v_1, v_2, $v_3 = 1$ and $v_2 = 2$ were also observed for $H^{11}BO$ and **$H^{10}BO$** by microwave spectroscopy. The structure was determined as $r_e(BH) = 116.81(10)$ pm and $r_e(BO) = 120.04(3)$ pm with estimated incertainties in parentheses; v_1 is located nearly 2900 cm^{-1} above the ground state. Microwave studies of such highly excited vibrational states are uncommon. The rotational constants are B = 39224.247(5) for $H^{11}BO$ and B = 40575.396(18) for $H^{10}BO$; the centrifugal distortion constants are D = 0.08024(15) for $H^{11}BO$ and D = 0.08576(46) for $H^{10}BO$, and the nuclear quadrupole coupling constants in the vibrationally ground state are eQq = –3.80(10) for $H^{11}BO$ and eQq = –8.20(44) for $H^{10}BO$ (all in MHz) [3].

The HBO molecule was generated as a transient intermediate in the reaction of diborane(6) with oxygen in helium, induced by a d.c. discharge. The $J = 3 \leftarrow 2$ transition was observed for $H^{10}BO$ and was combined with previously reported spectra in order to derive the rotational constants, centrifugal distortion constants, and the nuclear quadrupole coupling constants for this species. The rotational and centrifugal distortion constants were determined for **$D^{11}BO$**, **$D^{10}BO$**, **$H^{11}B^{18}O$**, and **$H^{10}B^{18}O$**. The measurements were extended to rotational transitions of $H^{11}BO$ and $H^{10}BO$ in the v_1, v_2, $v_3 = 1$ and $v_2 = 2$ levels and to those of $D^{11}BO$ and $D^{10}BO$ in the v_2, $v_3 = 1$ and $v_2 = 2$ levels. The vibrational frequencies and the rotational, vibration-rotational *l*-type doubling, and centrifugal distortion constants, determined for the six isotopic species were simultaneously analyzed in order to derive the equilibrium molecular structure and the harmonic and third-order anharmonic potential constants of the HBO molecule. The resultant equilibrium bond lengths are $r_e(BH) = 116.667(41)$ pm and $r_e(BO) = 120.068(10)$ pm. The substitution structure was also calculated by choosing various isotopic species as the parent species. The results were simply averaged to $r_s(BH) = 116.732(16)$ pm and $r_s(BO) = 120.211(2)$ pm, where the values in parentheses denote the range of discrepancies among the various sets of isotope combinations [4].

The reactants, transition state, and products for the collinear abstraction reaction $H_2 + BO \rightarrow H + HBO$ were studied by using ab initio multiconfiguration self-consistent field (MCSCF) and multireference configuration interaction (CI) techniques with basis sets up to triple zeta plus double polarization quality. At the best level of theory, the reaction is calculated to be exothermic by 6.4 kcal/mol and has a zero-point corrected barrier of 9.5 kcal/mol. Conventional transition-state-theory calculations of the rate coefficient with a tunneling correction through an Eckart barrier predict notable curvature in the Arrhenius plot over the temperature range 300 to 3000 K. At low temperatures (e.g., ca. 1700 °C), this curvature is attributed to tunneling. At high temperatures, the curvature is attributed to the temperature dependence of the vibrational partition function caused by anomalously low doubly degenerate bending frequencies at the saddle point. The computed rate coefficient for this abstraction reaction is well represented over a wide temperature range by the three-parameter expression $k(T)=2.96 \times 10^{-22} \cdot T^{3.29} \cdot e^{-2200/T}$. A new and considerably more negative value of the standard heat of formation of HBO of ca. −60 kcal/mol (previous estimate = −47 kcal/mol) is recommended [5].

Ab initio calculations for HBO with STO-3G and 4-31G basis sets gave optimized total energies of E = −98.826900 and −100.018673 a.u., respectively. The bond lengths for $C_{\infty v}$ symmetry are r(BH) = 114.2 and 115.2 pm, r(BO) = 117.6 and 119.9 pm, respectively. The Mulliken charges are q(H) = −0.0077 and 0.0655, q(B) = 0.2979 and 0.5284, q(O) = −0.2902 and −0.5940, q_π(B) = −0.1278 and 0.1580, and q_σ(BO) = 0.0077 and −0.0655, with STO-3G and 4-31G basis sets, respectively. The calculated and experimental (in parentheses) vibrational frequencies are for $H^{10}BO$ 3116.8 (2874) cm^{-1}, 1974.5 (1855) cm^{-1}, 872.6 (764) cm^{-1}; for $H^{11}BO$, they are 3090.7 (2849) cm^{-1}, 1932.7 (1817) cm^{-1}, 861.8 (754) cm^{-1}; for $D^{10}BO$, they are 2515.5 (2303) cm^{-1}, 1767.0 (1663) cm^{-1}, 707.7 (617) cm^{-1}; and for $D^{11}BO$, they are 2460.4 (2259) cm^{-1}, 1752.5 (1648) cm^{-1}, 694.4 (606) cm^{-1}. The B–O stretching force constant is 15.5 mdyn/Å. For a comparison with H_3C–B=O, see p. 147 [6].

The relative stabilities and barriers of isomerization (h_{iso}) in the system HBO-BOH have been calculated ab initio by using the basis double-zeta Huzinaga-Dunning (DZHD) with polarization and 6-31G** basis sets within the self-consistent electron pair approximation with consideration of configuration interaction (CI) calculations. The total energies are for HBO ($C_{\infty v}$) with self-consistent field (SCF) −100.1660 and self-consistent electron pairs (SCEP) −100.4198 a.u., for bridged **(-B-H-O-)** (C_s) −100.0179 and −100.2893 a.u., for **BOH** (C_s) −100.0983 and −100.3426 a.u., and for BOH($C_{\infty v}$) −100.0905 and −100.3362 a.u., respectively. The calculated isomerization energies for HOB-HBO are for the DZHD basis set E_{iso} = 38 kcal/mol; for the DZHD + P basis set, E_{iso} = 50 kcal/mol; for the 6-31G** basis set, E_{iso} = 42 kcal/mol and corrected by SCEP, E_{iso} = 49 kcal/mol. The calculated isomerization barriers for angular BOH are for DZHD h_{iso} = 50 kcal/mol; for DZHD + P, h_{iso} = 50 kcal/mol; for 6-31G**, h_{iso} = 51 kcal/mol; and after calculation with allowance for electron correlation, h_{iso} = 34 kcal/mol. The isomerization energy of the rearrangements by 1,2-hydrogen shift, calculated with SCEP/6-31G** without and with allowance for electron correlation, are E_{iso} = 42.5 and 48.5 kcal/mol; the h_{iso} values are 50.5 and 33.5 kcal/mol, respectively [7].

Approximations on the basis Hartree-Fock-double-zeta-Huzinaga-Dunning (HF/DZHD + P) show for the transition of the ground state HBO to the excited state BOH an energy of isomerization of 40 kcal/mol, and a barrier energy of isomerization of ca. 50 kcal/mol; on the basis Hartree-Fock-double-zeta-Roos-Siegbahn (HF/DZRS) for the transition of the ground state $[BOH]^+$ to the excited state $[HBO]^+$, an energy of isomerization of 16 kcal/mol and a barrier energy of isomerization of 50 kcal/mol are obtained [8].

3.5.2.2 Methyloxoborane, H_3C–B=O

The electronic structure of **H_3C–B=O** is elucidated employing one determinant ab initio calculation with STO-3G and 4-31G basis sets and the semiempirical modified neglect of diatomic overlap (MNDO) theory. The optimized total energies of H_3C–B=O (C_{3v}) are with

STO-3G, E = −137.425791 and with 4-31G, E = −139.026576 a.u.. The bond lengths are (first value with STO-3G, second value with 4-31G, third value with MNDO) r(CH) = 108.5, 108.3, and 110.9 pm; r(CB) = 154.6, 152.8, and 152.5 pm; r(B=O) = 117.8, 120.4, and 117.6 pm; the bond angles are ∢(HCH) = 108.2°, 108.1°, and 109.2°; ∢(HCB) = 110.7°, 110.8°, and 109.7°; the Mulliken charges are q(H) = 0.0798, 0.2120, and 0.0210; q(C) = −0.2927, −0.7732, and 0.1331; q(B) = 0.3661, 0.7289, and −0.0618; q(O) = −0.3128, −0.5916, and −0.1343; $q(CH_3)$ = −0.0534, −0.1373, and 0.1961; q_π(B) = −0.1381, 0.1245, and −0.0689; q_π(O) = 0.0945, −0.1802, and 0.0144; q_π(BO) = −0.0436, −0.0557, and −0.0536; q_σ(BO) = 0.0970, 0.1931, and −0.1425, respectively [6].

The 4-31G calculated vibrational frequencies are for **H_3C–^{11}B=O** with A_1 symmetry 3173.1(2), 2112.5(3), 1541.2(5), and 860.9(7) cm^{-1}; for E symmetry 3251.7(1), 1617.3(4), 1052.6(6), and 384.0(8) cm^{-1}. The 4-31G calculated vibrational frequencies are for **H_3C–^{10}B=O** with A_1 symmetry 3173.2(2), 2183.2(3), 1541.2(5), and 863.2(7) cm^{-1}; for E symmetry 3251.7(1), 1617.6(4), 1061.2(6), and 394.6(8) cm^{-1}. The B–O stretching force constant is 14.9 mdyn/Å. From a comparison of the force constants for stretching the XY bonds in H–XY and H_3C–XY molecules, the HY bond is softened upon CH_3 substitution. This effect is found to decrease in the order H_3C–BO > H_3C–CN > H_3C–NC. For XY = BO, the difference between the first ionization potentials of H–XY and H_3C–XY is somewhat smaller than for XY = CN, but slightly higher than for XY = NC [6].

H_3C–B=O is kinetically unstable in comparison to its trimer. By pyrolysis of 2-methyl-1,3,2-dioxaborolane-4,5-dione at 1100 K, optimized by photoelectron spectroscopic gas analysis and digital substraction of the ionization patterns of CO and CO_2, it has been possible to obtain the He(I)-photoelectron spectrum of the "pure" H_3C–B=O (see **Fig. 3-12**). The six radical cation conditions were assigned based on the overall energy differences, calculated by the "algebraic diagrammatic construction(3)" method (ADC(3) method) [9].

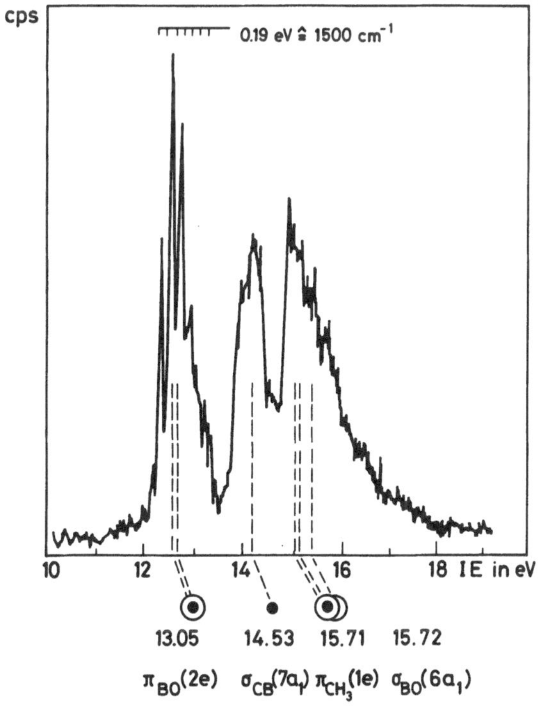

Fig. 3-12. He(I) photoelectron spectrum of H_3C–B=O [9].

References on p. 148

3.5.2.3 2,4,6-Tris(*t*-butyl)phenyloxoborane, 2,4,6-(t-C_4H_9)$_3$$C_6H_2$–B=O

Hydrolysis and dehydration of 2,4,6-(t-C_4H_9)$_3$$C_6H_2$–B($OCH_3$)$_2$, prepared from 2,4,6-($t$-$C_4H_9$)$_3$$C_6H_2$Li and B($OCH_3$)$_3$, led to 1,3-bis[2,4,6-tris(*t*-butyl)phenyl]-1,3-dibora-2,4-dioxetane, which, on photolysis, gave trapping products consistent with intermediate formation of the oxoborane **2,4,6-(t-C_4H_9)$_3$$C_6H_2$–B=O** with a weak UV absorption band at 314 nm [10].

2,4,6-(t-C_4H_9)$_3$$C_6H_2$–B($OCH_3$)$_2$ was reacted with Li[AlH_4] and tetramethylethylenediamine in diethyl ether to give a lithium intermediate, which was hydrolyzed to yield 2,4,6-(t-C_4H_9)$_3$-C_6H_2–BH(OH) (**1**); in strongly acidic media, hydrolysis involved an attack of a neighboring CH_3 group to give a 1-hydroxo-1-boraindane derivative (**2**) with 2,4,6-(t-C_4H_9)$_3$$C_6H_2$–B=O as an intermediate [11].

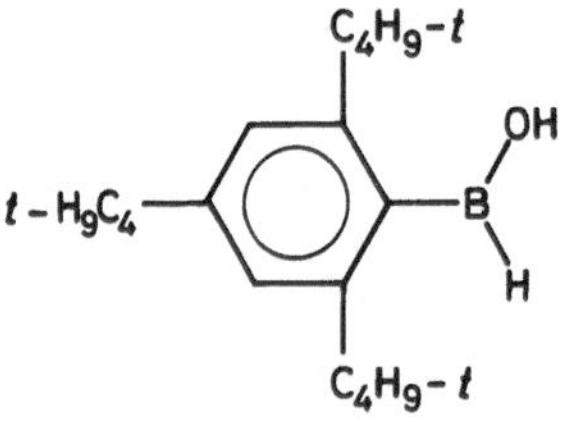

1

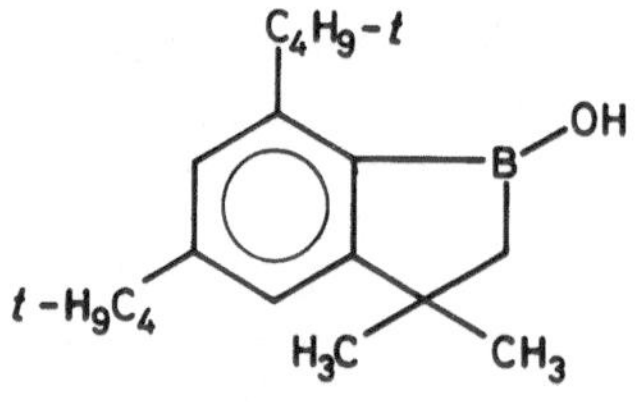

2

References for 3.5.2:

[1] Peterson, K. A.; Woods, R. C. (J. Chem. Phys. **90** [1989] 7239/50).
[2] Kawashima, Y.; Kawaguchi, K.; Hirota, E. (Chem. Phys. Lett. **131** [1986] 205/8).
[3] Kawashima, Y.; Endo, Y.; Kawaguchi, K.; Hirota, E. (Chem. Phys. Lett. **135** [1987] 441/5).
[4] Kawashima, Y.; Endo, Y.; Hirota, E. (J. Mol. Spectrosc. **133** [1989] 116/27).
[5] Page, M. (J. Phys. Chem. **93** [1989] 3639/43; AD-A 201395 [1988] 1/22).
[6] Raabe, G.; Schleker, W.; Straßburger, W.; Heyne, E.; Fleischhauer, J.; Bachler, V. (Z. Naturforsch. **42a** [1987] 1027/36).
[7] Zyubina, T. S.; Zyubin, A. S.; Gorbik, A. A.; Charkin, O. P. (Zh. Neorg. Khim. **30** [1985] 2739/44; Russ. J. Inorg. Chem. **30** [1985/86] 1559/62).
[8] Charkin, O. P.; Zyubina, T. S. (Koord. Khim. **12** [1986] 1011/37; Sov. J. Coord. Chem. **12** [1987] 583/97).
[9] Bock, H.; Cederbaum, L.; Von Niessen, W.; Paetzold, P.; Rosmus, P.; Solouki, B. (Angew. Chem. **101** [1989] 77/8).
[10] Pachaly, B.; West, R. (J. Am. Chem. Soc. **107** [1985] 2987/8).

[11] Groteklās, M.; Paetzold, P. (Chem. Ber. **121** [1988] 809/10).

3.5.3 The Radical Species [HBOH]$^•$ and [B(OH)$_2$]$^•$

The structure and vibrational frequencies of **[HBOH]$^•$** were determined by the self-consistent field (SCF) procedure. It exists as *cis* and *trans* isomers. The theoretical energies for the B + H_2O → [HBOH]$^•$ process with the 6-31G* basis sets are close to those obtained with the 3-21G basis sets. In the following, the first value given for the theoretical geometries is from 3-21G and the second value is from the 6-31G* basis set. For ***cis*-[HBOH]$^•$**, r(BH) = 119.45 and 119.78 pm; r(BO) = 135.96 and 133.33 pm; r(OH) = 96.69 and 94.96 pm. The angle ∢(HBO) is

125.94° and 123.89°, and the angle ∢(BOH) is 122.12° and 115.11°. For ***trans*-[HBOH]•**, r(BH) = 118.70 and 119.21 pm; r(BO) = 136.17 and 133.49 pm; r(OH) = 96.34 and 94.74 pm. The angle ∢(HBO) is 121.21° and 120.73°, and the angle ∢(BOH) is 120.60° and 114.42°. The reaction energy is for *cis*-[HBOH]• −92.48 kcal/mol from 3-21G and −91.14 kcal/mol from 6-31G*, and for *trans*-[HBOH]• −93.66 and −92.54 kcal/mol, respectively. The calculated normal mode vibration frequencies and displacements for [HBOH]• are given in **Fig. 3-13**. The upper values are the 6-31G* frequencies, while the lower values, reported in parentheses, give the 3-21G results. The barrier of isomerization in the MP2 approximation is 11.5 kcal/mol [1].

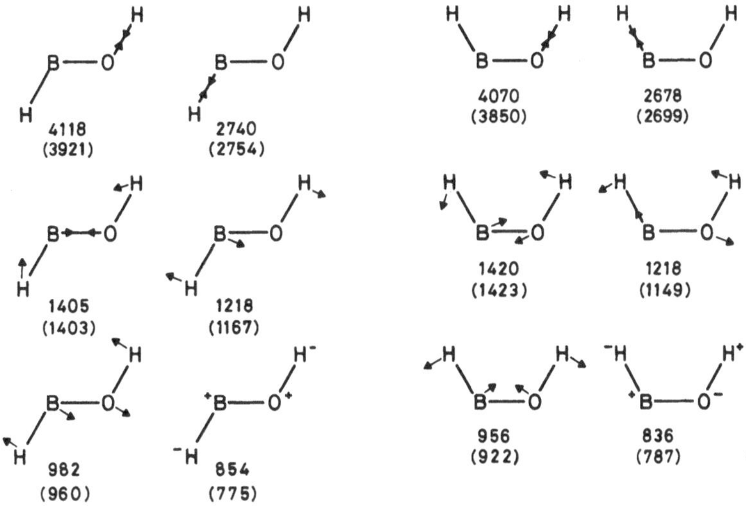

Fig. 3-13. Calculated normal mode frequencies and displacement for [HBOH]•. Upper values 6-31G* frequencies, 3-21G results in parentheses; all in cm^{-1} [1].

The partition functions of polyatomic molecules such as **$[B(OH)_2]^•$** affect significantly the stellar atmospheric equation of state. Least squares polynomal fits are presented for the polyatomic partition functions mostly derived from the revised JANAF thermochemical tables [2].

References for 3.5.3:

[1] Sakai, Sh.; Jordan, K. D. (Chem. Phys. Lett. **130** [1986] 103/10).
[2] Irwin, A. W. (Astron. Astrophys. Suppl. Ser. **74** [1988] 145/60 from C.A. **109** [1988] No. 100938).

3.5.4 The $[H_2BO]^-$ Ion

Ab initio Hartree-Fock-Roothaan calculations with large polarized basis sets gave ^{17}O nuclear quadrupole constants e^2qQ/h. The calculated ^{17}O NMR shielding constant changed from $\sigma = -91$ ppm for the $[H_2BO]^-$ ion to $\sigma = +87$ ppm for the planar $H_2B–O–BH_2$, see Section 3.5.9, p. 250.

Reference on p. 150

Reference for 3.5.4:

Tossell, J. A.; Lazzeretti, P. (J. Non-Cryst. Solids **99** [1988] 267/75).

3.5.5 Borinic Acid, H_2B–OH, and Its Derivatives

3.5.5.1 General Remarks

A review on organic derivatives of hydroxoborane for the year 1983 has been compiled [1], and a book on organometallic compounds of boron including organyloxyboranes has been published [2]. Acyclic trialkylboranes are oxidized with anhydrous trimethylamine N-oxide in order to give, initially, borinate esters R_2B–OR′, then boronate esters $RB(OR')_2$, and finally boric acid esters $B(OR')_3$ [3]. The effect of OH substituents on the structures and stabilities of borirenes, borinanes, and boronanes has been studied by ab initio methods. Due to σ effects and π effects, the bonds to boron are stronger than those to carbon. In the borirenes, π bonding with the coplanar OH groups competes with the aromatic 2π delocalization resulting in a reduced resonance energy of 40 kcal/mol [4]. For new syntheses with organyl-oxyboranes, see [5]. Crystalline chelates of 100% optical purity were synthesized from borinic acid esters [38].

3.5.5.2 Borinic Acid, H_2B–OH

The heat of formation (for C_s symmetry) of H_2B–OH was calculated by AM1 to be −73.4 kcal/mol and experimentally determined to be −69.4 kcal/mol; the dipole moment was calculated to be 1.34 D, and the ionization potential was found to be 11.17 eV. The following bond lengths were calculated: r(BH) = 119.4 pm (observed 119.2 pm), and r(BO) = 133.1 pm; the bond angles have been calculated to be ∢(HBO) = 121.4° (observed 114.0°) and ∢(HOB) = 109.0° (observed 107.4°) [7].

3.5.5.3 Borinic Acid Esters, H_2B–OR

In the reaction of B_2H_6 with CH_3OH or HCHO, the short-lived **H_2B–OCH_3** was identified by microwave spectroscopy, which shows large splittings due to the internal rotation of the CH_3 top. An analysis of the spectrum leads to the following parameters for the normal species: A = 50684.2(13) MHz; B = 10284.41(11) MHz; C = 9024.80(11) MHz; θ = 26.06(18)°; s = 15.983(5); I_α = 3.226(11) u·Å^2; and V_3 = 740(5) cal/mol. The total electric dipole moment is μ = 1.61(10) D. The molecular structure was determined from the rotational constants of ^{10}B, ^{13}C, the deuterated species **H_2B–OCD_3**, **H_2B–$OCHD_2$**, and **D_2B–OCH_3** with the aid of an ab initio molecular orbital (MO) calculation [9].

When monosolvated alkoxide ions $[RO–H–OR']^-$ react with borinic acid esters, H_2B–OCH_3 or R_2B–OC_2H_5, the major displacement reaction is always that the smaller ion RO^- reacts with the central atom, while the larger alcohol R′OH becomes part of the solvated alkoxide product ion. Such reactions are exothermic and very fast. The presence of initial orientation effects controlling the course of reactions was suggested by ab initio 6-31G studies for the model systems (F^-–HOH)–H_2BOCH_3 and ($[CH_3O]^-$–HOH)–H_2BOCH_3. In the former case, F–H bonds to the methoxy hydrogen atom are formed, thereby effecting subsequent attack of $[HO]^-$ at the boron atom. Internal barriers are small [10].

Empirical boron enolate parameters for incorporation into Allinger's MM2 force field were derived from ab initio studies of model systems of enolic alkenoxyboranes **H_2B–OR** (R = CH_2=CH, (Z)-CH_3CH=CH, (E)-CH_3CH=CH, and $(CH_2{=}CH)_3C$) [11].

A theoretical molecular orbital (MO) study of the migration of H_2B between the oxygen atoms of OCH=C(H)O (migration of free valences in the $[H_2B]^\bullet$ radical between oxygen atoms in the **H_2B–O–CH=C(H)$O^\bullet$** radical) has been described [12].

3.5.5.4 2,4,6-Tris(*t*-butyl)phenylborinic Acid, 2,4,6-(*t*-C_4H_9)$_3$$C_6H_2$–BH–OH

Li[AlH_4] and tetramethylethylenediamine react with 2,4,6-(*t*-C_4H_9)$_3$$C_6H_2$–B$(OCH_3)_2$ in diethyl ether to give a lithium product, which was hydrolyzed to yield the compound **2,4,6-(*t*-C_4H_9)$_3$-C_6H_2–BH–OH** (see Formula **1** in Section 3.5.2.3, p. 148). This compound melts at 153°C. ^{1}H NMR data (in CCl_4): δ = 1.32 (s, 9H, 4-C_4H_9-*t*), 1.36 (s, 18H, 2,6-C_4H_9-*t*), 5.07 (d, 1H, OH with J(H,H) = 10.1 Hz), and 7.24 ppm (s, 2H, H-3). ^{13}C NMR data (in $CDCl_3$): δ = 31.3, 33.3, 34.9, 37.5, 120.7, 134.3, 149.2, and 153.2 ppm. IR data: ν(OH) = 3640, ν(BH) = 2500 cm^{-1}; mass spectrum: $[M]^+$ (48%) [13].

3.5.5.5 Diorganylborinic Acids, R_2B–OH and RB(R′)–OH

Data on **$(CH_3)_2BOH$**, dimethylborinic acid (or "hydroxy-dimethylborane"), have been calculated by AM1; the heat of formation, ΔH_f, is −101.0 kcal/mol, the dipole moment is 1.34 D (observed 1.53 D), and the ionization potential is 11.05 eV [7].

$(C_2H_5)_2BOH$ reacts with R_3Al (R = CH_3, C_2H_5, or *i*-C_3H_7) with evolution of RH and formation of the intermediate $(C_2H_5)_2B$–O–AlR_2, which gave $(C_2H_5)_2BR$ on distillation. In the case of (*i*-C_3H_7)$_3$Al, the compound $(C_2H_5)_2B$–O–Al(C_3H_7-*i*)–O–Al(C_3H_7-*i*)$_2$ was also obtained [14]. Bicyclic boroxazolidones, on hydrolysis in dimethylformamide, also formed $(C_2H_5)_2BOH$ [15].

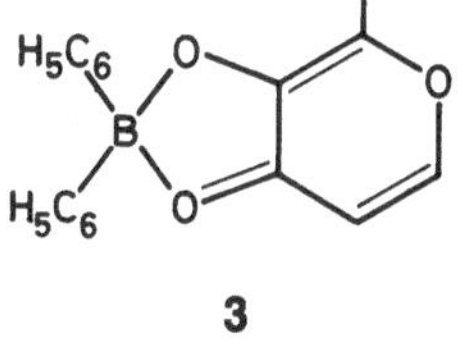

3

$(C_6H_5)_2BOH$, prepared by acid hydrolysis of (2-aminoethanolato)-diphenylborane, reacts with 3-hydroxy-2-methyl-4-pyrone in order to give "maltolato-diphenylboron" (**3**), melting point 195 to 196.5°C. IR data: ν(C=O) = 1640 cm^{-1}, ν(C=C) = 1570 and 1560 cm^{-1}; ^{1}H NMR data (CD_3COCD_3; δ in ppm): 2.63 (s, 3H, CH_3), 7.25 (m, 11H), and 8.62 (d, 1H, OCHCH). The compound crystallizes in the monoclinic space group $P2_1/c-C^5_{2h}$ (No. 14). It contains a five-membered C_2O_2B ring [17]. With hydroxypyridinone derivatives, nitrogen analogs of **3** are obtained [16].

The reaction between $(C_6H_5)_2BOH$ and an α-amino acid proceeds with the elimination of water and donation of an electron pair from N to B to form a kinetically stable boroxazolidone (**4**) at pH 2 to 5; each boroxazolidone, formed from each variety of α-amino acids in a protein, can be qualitatively and quantitatively identified to provide valuable information about the nature of a protein [18].

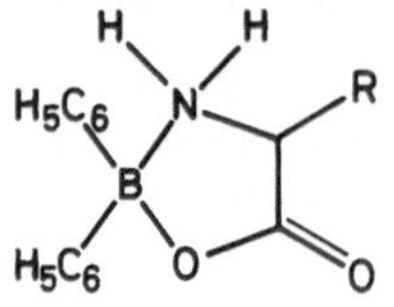

4

$(C_6H_5)_2BOH$ was recovered in an 83% yield from aqueous solutions, using C_{18} reverse-phase liquid chromatography [19]. A mixture of arylborinic acids and arylboronic acids is recovered by neutralization in two stages [20]. Catecholamines have been selectively extracted from body fluids with $(C_6H_5)_2BOH$ [21].

C_6H_5–B(OH)–C_4H_9-*n* inhibits cholesterol esterase with $K_i = 2.9$ nM and lipoprotein lipase-catalyzed hydrolysis with $K_i = 1.7$ μM [22].

The reaction of diarylborinic acids with benzimidazoles yields organyloxydiarylborane chelates containing benzimidazoles as ligands [23].

(4-CH_3–O–C_6H_4)$_2$BOH was obtained from a Grignard reaction of 4-CH_3–O–C_6H_4Br with $B(OC_4H_9\text{-}n)_3$; it is used for the preparation of antitumor compounds [24]. **(3-FC_6H_4)$_2$BOH · D,L-tryptophane** forms tetragonal crystals, space group $I4_1/a-C_{4h}^6$ (No. 88) with a = 1950.4(3) and c = 2140.5(4) pm; Z = 16; $D_m = 1.304$ g/cm³ and $D_c = 1.319$ g/cm³; R = 9.1%. The boron atom bonds through sp^3 hybrid orbitals in forming a five-membered ring [25].

3.5.5.6 Diorganylborinic Acid Esters, R_2B–OR′ and RR′B–OR″

$(C_2H_5)_2B$–OCH_3 was obtained by the reaction of C_2H_5–$B(OCH_3)_2$ with octakis-*O*-(diethylboryl)sucrose [30] or in situ from $(C_2H_5)_3B$ and CH_3OH at −78 °C in tetrahydrofuran without using any other catalyst. The ^{11}B NMR spectra of the product and authentic $(C_2H_5)_2B$–OCH_3 are essentially identical. Using the product as a chelating agent, *syn*-1,3-diols are prepared in over 98% stereochemical purity by reducing β-hydroxyketones with $Na[BH_4]$ [31, 32]. The following NMR value (versus $(C_2H_5)_2O$–BF_3, no solvent given) is reported: $\delta^{11}B = 53.6$ ppm [46].

Compounds of the type R_2B–OCH_3 were prepared by dropping CH_3OH in pentane to a solution of R_2B–$N(CH_3)_2$ in pentane, boiling 2 h under reflux, distilling off pentane at 25 °C/80 Torr, and distillation of the product at 25 °C/10 Torr; **(*i*-C_3H_7)$_2$B–OCH_3**, 83% yield, boiling point 45 to 47 °C/41 Torr; **(*t*-C_4H_9)$_2$B–OCH_3**, 88% yield, boiling point 43 to 44 °C/15 Torr [47]. The following NMR data (in ppm, versus $(C_2H_5)_2O$–BF_3, no solvent given) are reported for (*t*-C_4H_9)$_2$B–OCH_3: $\delta^{11}B = 51.0$, $\delta^{13}C = 24.4$, respectively for (*i*-C_3H_7)$_2$B–OCH_3: $\delta^{11}B = 53.3$, $\delta^{13}C = 16.0$ [46].

Alkenyldicyclohexylboranes are readily cleaved by 1 mol CH_3OH to give the corresponding (Z)-alkene and **(*cyclo*-C_6H_{11})$_2$B–OCH_3** [34]. Boron heterocycles have been synthesized from (*cyclo*-C_6H_{11})$_2$B–OCH_3 and α-aminoacids [37].

1,1,2-Trimethylpropyl-alkyl borinic acid methyl esters, ***i*-C_3H_7–C(CH_3)$_2$–BR–OCH_3**, intermediates for unsymmetrical alkynes, were prepared from 1,1,2-trimethylpropyl-alkylchloroborane [39].

The effectiveness of $LiCHCl_2$ as a homologating agent for dialkylborinic acid esters such as **$R_2B{-}OCH_3$** has been investigated with and without the addition of $HgCl_2$ [35]. A more convenient route to achieve homologation of borinic acid esters utilizes the in situ formation of $LiCHCl_2$ at −78°C [36].

The dimethylborylation of aryllithium reagents with **$(CH_3)_2B{-}O{-}C_2H_5$** was generalized [26]. $(CH_3)_2B{-}O{-}C_2H_5$ has been used in synthesis of 1,8-naphthalenediylbis(dimethylborane) [27] and 8-(ethoxydimethylsilanyl)-1-naphthyl-dimethylborane [28].

Volatile mixed esters of the type **$RB(R')$–O–C_3H_7-*i*** have been obtained in high purity by pyrolysis of $Li[RB(R')(OC_3H_7\text{-}i)_2]$, leaving a residue of $LiOC_3H_7$-*i* [43]. Mass spectra of mixed borinate esters (with R = CH_3, C_6H_5, *i*-C_3H_7, *n*-C_4H_9, *t*-C_4H_9, or *cyclo*-C_6H_{11}, and R′ = C_6H_5, *t*-C_4H_9, or *cyclo*-C_6H_{11}) were recorded and the fragmentation patterns discussed [44].

β-Hydroxyketones are always reduced 1,3-*syn*-diastereoselectively by $Na[BH_4]$ if they are complexed with **$(C_2H_5)_2B{-}O{-}R$** (with R = CH_3, C_2H_5, *i*-C_3H_7, CH_2=$CHCH_2$, *n*-C_4H_9, or *t*-C_4H_9), or **(*n*-C_4H_9)$_2$B–O–CH_3** [32].

$(C_6H_5)_2B{-}O{-}C_4H_9$-*s* has been used to synthesize ammonium 1,3-dioxa-2-borata-5-phosphorinanes [49].

The compounds $R_2B{-}O{-}C_4H_9$-*t* were prepared by addition of ethereal HCl with stirring to a solution of $R_2B{-}N(CH_3)_2$ and *t*-C_4H_9OH in hexane; **(*i*-C_3H_7)$_2$B–O–C_4H_9-*t***, boiling point 25°C/10^{-2} Torr; **(*t*-C_4H_9)$_2$B–O–C_4H_9-*t***, 58% yield, boiling point 57°C/4 Torr [47]. The following NMR data (in ppm, versus $(C_2H_5)_2O{-}BF_3$, no solvent given) are reported for (*i*-C_3H_7)$_2$-B–O–C_4H_9-*t*: $\delta^{11}B$ = 52.0, respectively (*t*-C_4H_9)$_2$B–O–C_4H_9-*t*: $\delta^{11}B$ = 49.9, $\delta^{13}C$ = 25.0 [46].

NMR data are also available for **$(C_2H_5)_2B{-}O{-}C_4H_9$-*t***: $\delta^{11}B$ = 52.0 ppm (versus $(C_2H_5)_2O{-}BF_3$, no solvent given) [46].

The compounds **$(C_2H_5)_2B{-}O{-}C_6H_{11}$-*cyclo*** and **$(C_2H_5)_2B{-}O{-}C_4H_9$-*i*** have been prepared from *cyclo*-$C_6H_{11}OH$, or *i*-C_4H_9OH and $B(C_2H_5)_3$ and were treated with *t*-$C_4H_9{-}Si(CH_3)_2{-}O{-}C(CH_3)$=$CH{-}COCH_3$ in heptane solution in the presence of $CF_3{-}S(=O)_2O{-}Si(CH_3)_3$ in order to introduce *t*-butyl-dimethylsilanyl groups [33].

(*n*-C_3H_7)$_2$B–O–C_6H_{13}-*n* reacts with 2-bromomethyl-1,3-butadiene in the presence of aluminium turnings, activated with $HgCl_2$, to give 2-di-*n*-propylborylmethyl-1,3-butadiene (introduction of isoprene units) [45].

The mixed anhydride **$(C_2H_5)_2B{-}O{-}SO_2{-}C_6H_4CH_3$-4** reacts with *B*-pyrazol-1′-yl-pyrazaboles to form polyboron spiro cations [51].

Interaction between **$(C_6H_5)_2B{-}OR$** and R′OH (R, R′ = CH_3, C_2H_5, *i*-C_3H_7, or *t*-C_4H_9; except for R = R′ = *t*-C_4H_9) results in an exchange process yielding $(C_6H_5)_2B{-}OR'$ and ROH [48].

Treating diazoketones, RC(=O)–CHN_2, with dicyclohexylborane produces enol borinates of methylketones, **(*cyclo*-C_6H_{11})$_2$B–OC(R)=CH_2** [50].

Organyl-1-alkynylborinic acid esters, **R–C≡C–B(R′)–OR″**, are obtained by adding RC≡CLi to $R'B(OR'')_2$ [40].

Optically pure alkynylborinic acid esters such as **(R)-(−)-*s*-C_4H_9–B(OC_3H_7-*i*)–C≡C–CH_2-CH_2CH_2Cl** are obtained at low temperatures from optically pure boronic acid esters, e.g., (R)-(−)-*s*-C_4H_9–B(OC_3H_7-*i*)$_2$ and lithium acetylides, e.g., LiC≡C–$CH_2CH_2CH_2Cl$, followed by treatment of the "ate" complex with ethereal HCl; they were converted to α-chiral acyclic ketones of exceptionally high enantiomeric excess [41, 42].

References on pp. 154/6

3.5.5.7 Cyclic Esters of the Type (–ZO–)B–R and (–Z–)B–OR

The oxaborolane **5** (R = CH_2=$CHCH_2$–) can be viewed as a cyclic borinic acid ester. Its Raman spectrum was studied [6]. For alkoxy substituted borolanes (cyclic esters of boronic acid) with R = OR^1, see Section 3.5.6.4.3, p. 176.

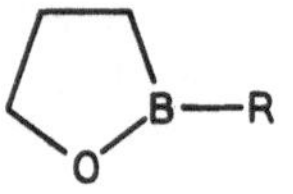

5

Sequential one-carbon homologation of **1-methoxyboracyclanes** (**6**) has been achieved in 75 to 85% yields utilizing the successive reaction of with in situ generated $LiCH_2Cl$. The reaction originates with *B*-methoxyborinane (n = 3) and the highest homolog obtained is *B*-methoxyboracyclododecane (n = 9) [8].

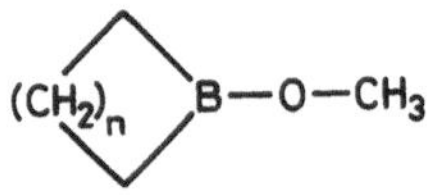

6

The reaction of **7** with ethereal HCl and CH_3OH yields a 47 : 53 *cis-trans* mixture of **1-methoxy-2,5-dimethylborolane** (**7**, OCH_3 instead of $N(C_2H_5)_2$) in 86% yield. The reaction of this isomeric mixture with N,N-dimethylethanolamine in pentane at room temperature, separation of an amine adduct, and vacuum distillation of the uncomplexed fraction gives the *trans* compound **(S,S)-1-methoxy-2,5-dimethylborolane** (**8**), which is a highly enantioselective hydroborating reagent [29].

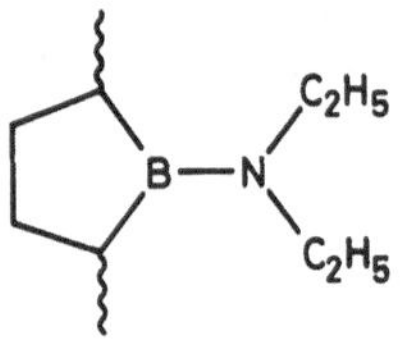

7

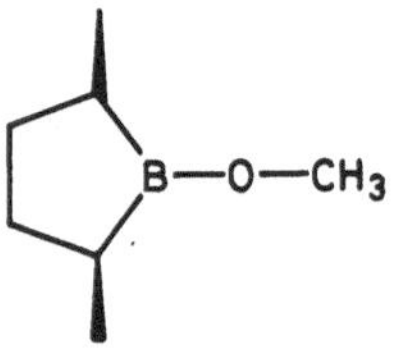

8

3.5.5.8 Dimethylperoxoborinic Acid Methyl Ester, $(CH_3)_2B$–OO–CH_3

The molecular structure and the mechanism of the intramolecular rearrangement of **$(CH_3)_2$-B–$OOCH_3$** has been studied by the SCF-MO-LCAO MINDO/3 method of the molecular orbital theory; the open form of organic peroxides is 27.6 kJ/mol higher in energy than a cyclic form with intramolecular coordination of the B–O bond [52], cf. also [53].

References for 3.5.5:

[1] Wilson, J. W. (Organomet. Chem. **13** [1985] 24/35).
[2] Smith, K. (in: Organometallic Compounds of Boron, Chapman & Hall, London 1985, 304 pp.).
[3] Soderquist, J. A.; Najati, M. R. (J. Org. Chem. **51** [1986] 1330/6).

[4] Budzelaar, P. H.; Kos, A. J.; Clark, T.; Schleyer, P. v. R. (Organometallics **4** [1985] 429/31).
[5] Braun, M. (Nachr. Chem. Tech. Lab. **33** [1985] 504/6).
[6] Zhu, Ziying; Fang, Yixing; He, Dajun; Zhou, Weike; Zhang, Gaoyi; Ding, Hongxun (Huaxue Xuebao **43** [1985] 640/4 from C.A. **105** [1986] No. 24299).
[7] Dewar, M. J. S.; Jie, Caoxian; Zoebisch, E. G. (Organometallics **7** [1988] 513/21).
[8] Brown, H. C.; Phadke, A. S.; Rangaishenvi, M. V. (J. Am. Chem. Soc. **110** [1988] 6263/4).
[9] Kawashima, Y.; Takeo, H.; Matsumura, Ch. (J. Mol. Spectrosc. **116** [1986] 23/32).
[10] Van der Wel, H.; Nibbering, N. M. M.; Sheldon, J. C.; Hayes, R. N.; Bowie, J. H. (J. Am. Chem. Soc. **109** [1987] 5823/8).

[11] Goodman, J. M.; Paterson, I.; Kahn, S. D. (Tetrahedron Lett. **28** [1987] 5209/12).
[12] Raevskii, N. I.; Borisov, Yu. A. (Dokl. Akad. Nauk SSSR **288** [1986] 664/8).
[13] Grotekläs, M.; Paetzold, P. (Chem. Ber. **121** [1988] 809/10).
[14] Synoradzki, L.; Boleslawski, M.; Lewinski, J. (J. Organomet. Chem. **284** [1985] 1/4).
[15] Garrigues, B.; Mulliez, M. (J. Organomet. Chem. **314** [1986] 19/24).
[16] Nelson, W. O.; Orvig, C.; Rettig, S. J.; Trotter, J. (Can. J. Chem. **66** [1988] 132/8).
[17] Orvig, C.; Rettig, S. J.; Trotter, J. (Can. J. Chem. **65** [1987] 590/4).
[18] Strang, C. J.; Henson, E.; Okamoto, Y.; Paz, M. A.; Gallop, P. M. (Anal. Biochem. **178** [1989] 276/86).
[19] Mills, G. L.; Schwind, D.; Adriano, D. C. (Chemosphere **17** [1988] 937/42 from C.A. **109** [1988] No. 27375).
[20] Ostermaier, J. J. (Eur. Appl. 131307 [1983/85] 1/10 from C.A. **102** [1985] No. 204099).

[21] MacDonald, J. A.; Lake, D. M. (J. Neurosci. Methods **13** [1985] 239/48 from C.A. **103** [1985] No. 116397).
[22] Sutton, L. D.; Stout, J. S.; Hosie, L.; Spencer, P. S.; Quinn, D. M. (Biochem. Biophys. Res. Commun. **134** [1986] 386/92 from C.A. **104** [1986] No. 104930).
[23] Yuan, Guozheng; Li, Guiying; Zhang, Guomin (Gaodeng Xuexiao Huaxue Xuebao **8** No. 5 [1987] 398/402 from C.A. **109** [1988] No. 54817).
[24] Lin, Kai; Zhang, Guomin; Fu, Naiwu (Youji Huaxue **1985** No. 3, pp. 228/32 from C.A. **104** [1986] No. 207334).
[25] Wu, Ziwu; Zhang, Guomin; Dong, Shihua (Jiegou Huaxue **4** [1985] 5/8 from C.A. **104** [1986] No. 13286).
[26] Katz, H. E. (Organometallics **5** [1986] 2308/11).
[27] Katz, H. E. (J. Am. Chem. Soc. **107** [1985] 1420/1).
[28] Katz, H. E. (J. Am. Chem. Soc. **108** [1986] 7640/5).
[29] Masamune, S.; Kim, B. M.; Petersen, J. S.; Sato, T.; Veenstra, S. J.; Imai, T. (J. Am. Chem. Soc. **107** [1985] 4549/51).
[30] Taba, K. M.; Köster, R.; Dahlhoff, W. V. (Liebigs Ann. Chem. **1987** 463/6).

[31] Chen, K. M.; Gunderson, K. G.; Hardtmann, G. E.; Prasad, K.; Repic, O.; Shapiro, M. J. (Chem. Lett. **1987** 1923/6).
[32] Chen, K. M.; Hardtmann, G. E.; Prasad, K.; Repic, O.; Shapiro, M. J. (Tetrahedron Lett. **28** [1987] 155/8).
[33] Dahlhoff, W. V.; Taba, K. M. (Synthesis **1986** 561/2).
[34] Brown, H. C.; Molander, G. A. (J. Org. Chem. **51** [1986] 4512/4).
[35] Brown, H. C.; Singh, S. M. (Organometallics **5** [1986] 998/1001).
[36] Brown, H. C.; Singh, S. M. (Organometallics **5** [1986] 994/7).
[37] Brown, H. C.; Gupta, A. K. (J. Organomet. Chem. **341** [1988] 73/81).
[38] Brown, H. C.; Vara Prasad, J. V. N. (J. Org. Chem. **51** [1986] 4526/30).

[39] Sikorski, J. A.; Bhat, N. G.; Cole, T. E.; Wang, K. H.; Brown, H. C. (J. Org. Chem. **51** [1986] 4521/5).
[40] Brown, H. C.; Srebnik, M. (Organometallics **6** [1987] 629/31).

[41] Brown, H. C.; Srebnik, M.; Bakshi, R. K.; Cole, T. E. (J. Am. Chem. Soc. **109** [1987] 5420/6).
[42] Brown, H. C.; Gupta, A. K.; Vara Prasad, J. V. N.; Srebnik, M. (J. Org. Chem. **53** [1988] 1391/4).
[43] Brown, H. C.; Srebnik, M.; Cole, T. E. (Organometallics **5** [1986] 2300/3).
[44] Rothwell, A. P.; Wood, K. V.; Srebnik, M.; Cole, T. E. (Org. Mass. Spectrom. **21** [1986] 165/7).
[45] Bubnov, Yu. N.; Etinger, M. Yu. (Tetrahedron Lett. **26** [1985] 2797/800).
[46] Nöth, H.; Prigge, H. (Chem. Ber. **119** [1986] 338/48).
[47] Höbel, U.; Nöth, H.; Prigge, H. (Chem. Ber. **119** [1986] 325/37).
[48] Domaille, P. J.; Druliner, J. D.; Gosser, L. W.; Read, J. M.; Schmelzer, E. R.; Stevens, W. R. (J. Org. Chem. **50** [1985] 189/94).
[49] Arbuzov, B. A.; Erastov, O. A.; Nikonov, G. N.; Karasik, A. A. (Izv. Akad. Nauk SSSR Ser. Khim. **1986** 1641/4; Bull. Acad. Sci. USSR Div. Chem. Sci. **35** [1986] 1490/3).
[50] Hooz, J.; Oudenes, J.; Roberts, J. L.; Benderly, A. (J. Org. Chem. **52** [1987] 1347/9).

[51] Clarke, C. M.; Niedenzu, K.; Niedenzu, P. M.; Trofimenko, S. (Inorg. Chem. **24** [1985] 2648/51).
[52] Kokorev, V. N.; Aleksandrov, Yu. A.; Vyshinskii, N. N.; Maslennikov, V. P.; Abronin, I. A. (Izv. Akad. Nauk SSSR Ser. Khim. **1985** 1788/94; Bull. Acad. Sci. USSR Div. Chem. Sci. **34** [1985] 1637/43).
[53] Aleksandrov, Yu. A.; Vyshinskii, N. N.; Kokorev, V. N.; Al'ferov, V. A.; Chikinova, N. V.; Makin, G. I. (J. Organomet. Chem. **332** [1987] 259/69).

3.5.6 HBO_2, $HB(OH)_2$, and Derivatives Thereof

For HBO_2, see also $H_3B_3O_6$, Section 3.5.11, p. 256.

3.5.6.1 "Monomeric" Metaboric Acid, HBO_2

Analysis of currently available data for the system containing 7.5 wt% B_2O_3 and 92.5 wt% H_2O indicate that the gas phase between 1350 and 1500 K consists mainly of H_2O and **HBO_2** vapor, and between 1200 and 1350 K of H_2O, HBO_2, H_3BO_3, and $H_3B_3O_6$ vapor [1]. A review is given in [2]. For the molecule O=B–OH with C_s symmetry and r(B=O) = 117.5 pm, r(BO) = 132.2 pm, and ∢(OBO) = 175.8°, the following values have been calculated by AM1: heat of formation, $\Delta H_f = -138.4$ kcal/mol (observed −134.2 kcal/mol), dipole moment, $\mu = 2.78$ D, and ionization potential, I = 12.95 eV [9].

The hot oxidizing atmosphere provided by a propane-air flame contains, after feeding boron particles into it, HBO_2 besides BO, BO_2, CO, and CO_2 [3]. The presence of gaseous H_2O in $CH_4/B(OCH_3)_3$ flames (in air) contributed to the formation of gaseous HBO_2 in preference to condensed-phase B_2O_3 [4].

Boron doping with a high carrier concentration for silicon molecular beam epitaxy has been realized using an HBO_2 gas source with the usual Knudsen cell [5]. Maximal carrier concentration reached 6×10^{20} cm^{-3} at crucible temperatures of 900°C. A comparison between the activation energy for vapor pressure and the carrier concentration dependence on crucible temperatures shows that boron evaporated in the form of HBO_2 molecules from the HBO_2 gas source. The boron profile, achieved by Knudsen cell shutter opening and closing, was sufficiently deep. Boron can be used as the p-type dopant in silicon molecular beam epitaxy without using ion imbedding or a very high temperature crucible [6]. When a 1.0 monolayer coverage of boron atoms was evaporated on a clean silicon (100) surface, followed by a 200 Å n-type silicon epitaxial layer, ca. 70% of the boron atoms were activated. Surface formation of HBO_2 plays a role in this process [7].

Boron doping in silicon layers grown by molecular beam epitaxy using an HBO_2 source was also studied between 500 and 700°C. The maximal boron concentration (without detectable oxygen incorporation) for a given substrate temperature and the silicon growth rate were determined using secondary-ion mass spectrometry analysis. Boron present in the silicon layers grown between 550 and 700°C was electrically active, independent of the amount of oxygen incorporation. By reducing the silicon growth rate, highly boron-doped layers were grown at 600°C without detectable oxygen incorporation [8].

References for 3.5.6.1:

[1] Zav'yalov, V. V.; Tarasevich, B. P.; Sirotkin, O. S.; Khitrov, M. Yu. (Zh. Obshch. Khim. **55** [1985] 2157/61; J. Gen. Chem. [USSR] **55** [1985] 1915/9).
[2] Tarasevich, B. P.; Sirotkina, N. Z.; Il'in, A. S.; Kuznetsov, E. V. (Vysokomol. Soedin. A **27** [1985] 451/63; Polym. Sci. [USSR] **27** [1985] 503/17).
[3] Eisenreich, N.; Liehmann, W. (Propellants Explos. Pyrotech. **12** No. 3 [1987] 88/91 from C.A. **107** [1987] No. 80591).
[4] Turns, S. R.; Funari, M. J.; Khan, A. (Combust. Flame **75** [1989] 183/95 from C.A. **110** [1989] No. 98254).
[5] Tatsumi, T.; Hirayama, H.; Aizaki, N. (Appl. Phys. Lett. **50** [1987] 1234/6).
[6] Tatsumi, T.; Hirayama, H.; Aizaki, N. (Proc. Electrochem. Soc. **88-8** [1988] 430/7 from C.A. **109** [1988] No. 15010).
[7] Tatsumi, T.; Hirayama, H.; Aizaki, N. (Jpn. J. Appl. Phys. **27** Pt. 2 [1988] L954/L956 from C.A. **109** [1988] No. 139950).
[8] Lin, T. L.; Fathauer, R. W.; Grunthaner, P. J. (Appl. Phys. Lett. **55** [1989] 795/7).
[9] Dewar, M. J. S.; Jie, Caoxian; Zoebisch, E. G. (Organometallics **7** [1988] 513/21).

3.5.6.2 Boronic Acid, $HB(OH)_2$, and Organylboronic Acids, $RB(OH)_2$

For some general remarks on dihydroxy(organyl)boranes (= organylboronic acids), see a section in a recently published monograph [2].

3.5.6.2.1 Boronic Acid, $HB(OH)_2$

B_2H_6 reacts with H_2O to form **$HB(OH)_2$** as an intermediate on the way to $B(OH)_3$ and H_2. **Fig. 3-14**, p. 158, shows the possible conformations and the determined orientation of the dipole moments. For ***cis,trans*-$H^{11}B(OH)_2$** and ***cis,trans*-$H^{10}B(OH)_2$**, the observed transition

References on pp. 164/7

frequencies and their deviations from the calculated values, as well as the derived molecular constants of additional 16 isotopically labeled species (containing H, D, ^{10}B, ^{11}B, ^{16}O, and ^{18}O), are tabulated. The microwave spectra of various isotopic species such as **$HB(OD)_2$**, **$DB(OH)_2$**, or **$DB(OD)_2$** were observed. The r_s and r_0 structures were derived from the rotational constants obtained. Since Kraitchman's equation gives imaginary values for some coordinates, the double-substitution method was applied to the determination of these coordinates. The determined r_s structural parameters are in agreement with the r_0 parameters determined by least squares fitting of all the observed rotational constants. The structural parameters (r_s) for *cis,trans*-$HB(OH)_2$ are (in pm): r(BH)=119.72(3), $r(B–O_t)$=135.9(9), and $r(B–O_c)$=136.5(9), $r(O_t–H_t)$=95.90(8) and $r(O_c–H_c)$=94.98(4); the angles are (in degrees): $\sphericalangle(H–B–O_t)$=118.2(12), $\sphericalangle(H–B–O_c)$=122.8(12), $\sphericalangle(O_t–B–O_c)$=119.1(13), $\sphericalangle(B–O_t–H_t)$=111.8(13), and $\sphericalangle(B–O_c–H_c)$= 113.3(17). The dipole moments and their directions were obtained from the measurements of the Stark effect. These molecular parameters show agreement with those predicted by ab initio molecular orbital calculations [3].

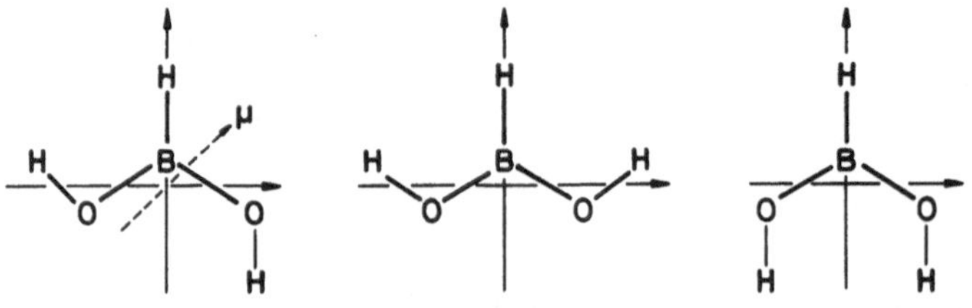

Fig. 3-14. The possible conformations of $HB(OH)_2$ and the determined orientation of the dipole moment. The dotted-line arrow indicates the direction of plus to minus charge [3].

For $HB(OH)_2$, symmetry C_s, with r(BH)=118.7 pm, r(BO)=134.1 and 135.6 pm, and $\sphericalangle$(HOB)=119.6° and 127.2°, the following values have been calculated by AM1: heat of formation, ΔH_f=−166.2 kcal/mol (observed −153.1 kcal/mol), dipole moment, μ=1.54 D, and ionization potential, I=12.54 eV [4].

3.5.6.2.2 Organylboronic Acids, $RB(OH)_2$

3.5.6.2.2.1 Alkylboronic Acids

For cycloalkyl-, alkenyl-, allenyl-, and aralkylboronic acids, see Section 3.5.6.2.2.3 on pp. 162/3.

CH_3–$B(OH)_2$, dihydroxy-methylborane or methylboronic acid, has been prepared in yields between 73 and 94% by three routes: (a) treating $B(OC_3H_7\text{-}i)_3$ with CH_3Li in ether and subsequent hydrolysis by HCl, (b) carbonylation of $(CH_3)_2S–BH_3$ by CO at atmospheric and higher pressures, or (c) hydrolysis of trimethylboroxin. The compound has a melting point of 88 to 90°C; NMR data (in ppm) are given: δ^1H=0.26 (br s, 3H), 4.5 (br s, 2H), both in $CDCl_3$; $\delta^{11}B$= 31.9; $\delta^{13}C$=−3 (br s), both in D_2O [5]. The following values have been calculated by AM1: heat of formation, ΔH_f=−177.7 kcal/mol, dipole moment, μ=1.60 D (observed 1.16 D), and ionization potential, I=12.31 eV [4]. A stable oxazaborolidine was prepared from CH_3–$B(OH)_2$ which is a catalyst for the enantioselective reduction of ketones [7]. $CH_3B(OH)_2$ has been used for gas chromatographic determinations of racemic modifications present in labetalol hydrochloride [8]. In the presence of CH_3–B(OH)$_2$ and C_2H_5–B(OH)$_2$, the acetylcholinesterase alkylation with the N,N-dimethyl-2-phenylaziridinium ion was accelerated [9].

C_2H_5–B(OH)$_2$ was prepared in quantitative yield by the hydrolysis of triethylboroxin, [–B(C_2H_5)–O–]$_3$. C_2H_5–B(OH)$_2$ was treated with $(C_2H_5)_3B$ in order to give a mixture of tetraethyldiboroxane and triethylboroxin, which is suitable for the introduction of *O*-diethylboryl and *O*-ethylboranediyl groups into hydroxy compounds [10]. C_2H_5–B(OH)$_2$ is formed by hydrolysis of bicyclic boroxazolidones in dimethylformamide [12]. C_2H_5–B(OH)$_2$ was used to prepare bis-*O*-ethylboranediyl-α-D-mannofuranose in order to get stereoselective glycosylate ions [11].

The mutual exclusivity of binding of a Co aminopeptidase for hydroxamate inhibitors and ***n*-C_4H_9–B(OH)$_2$** has been reflected in the characteristic absorption spectra [95]. *n*-C_4H_9–B-(OH)$_2$ is shown to inhibit urease completely [96] and it was used as a catechol-specific HPLC mobile phase pairing agent [97], as a group-specific pairing agent in the mobile phase for catecholamine separations [98], and to determine ethylene glycol in biological materials like blood by gas chromatography after esterification [99]. For determination of serum glucose by isotope dilution mass spectrometry, uniformely labeled [$^{13}C_6$] α-D-glucose was added to the serum and glucose was separated from the serum matrix and converted into α-D-glucofuranose cyclic 1,2:3,5-bis(*n*-butylboronate)-6-acetate by *n*-C_4H_9–B(OH)$_2$ [100].

The inhibitory effect of organoboronic acids of the type **RB(OH)$_2$** (R = C_2H_5, *n*-C_3H_7, *n*-C_4H_9, *n*-C_5H_{11}, *n*-C_6H_{13}, *n*-C_7H_{15}) on the enzymatic activity of carboxypeptidase Y depends on the length of n and of the structure of the hydrocarbon chain [89].

Hydroboration of enamines with one equivalent of H_3B–S$(CH_3)_2$, methanolysis, and subsequent hydrolysis with water leads to the formation of **[2-(dialkylamino)alkyl]boronic acids** [6].

3.5.6.2.2.2 Phenylboronic Acid, C_6H_5–B(OH)$_2$

C_6H_5–B(OH)$_2$. The following values have been calculated for C_6H_5–B(OH)$_2$ by AM1: heat of formation, $\Delta H_f = -135.5$ kcal/mol (observed −151.8 kcal/mol), dipole moment, $\mu = 3.29$ D, and ionization potential, I=12.95 eV [4].

C_6H_5–B(OH)$_2$ was used to prepare heterocycles, e.g., 3-(phenylmethylidene)-4-methyl-1-phenyl-2,6,7-trioxa-3-azonia-1-boratabicyclo[2.2.2]octane [13], 1,9-dimethyl-3,5,7-triphenyl-2,4,6,8-tetraoxo-1,9-diazonia-5-borata-3,7-diboratricyclo[5.4.0.0^{3,9}]undecane [14], or 3,3-dimethyl-2-phenoxy-2-phenyl-1-oxa-3-azonia-2-boratacyclopentane [15].

Stable liquid crystal phases for use in preparing electro-optical display devices contain boron heterocycles such as 2-(4-propoxyphenyl)-5-pentyl-1,3,2-dioxaborinane, which is prepared from C_6H_5–B(OH)$_2$ [16].

C_6H_5–B(OH)$_2$ and its derivatives react with bromophenyldiamines with cyclization to form bromobenzodiazaboroles [17]. C_6H_5–B(OH)$_2$ is the precursor of new diastereomeric complicated heterocycles [18]. It reacts with iminodiacetic acids or N-methyliminodiacetic acids to form air-stable bicyclic esters [19], with N-alkyl-N-(2-hydroxyethyl)aminoacetic acids stereoselectively to yield new bicyclic phenylboronic acid esters containing stable chiral nitrogen and boron atoms [20], and with bis-(2-hydroxyphenyl)amines to give new dibenzobicyclic phenylboronates [21].

References on pp. 164/7

C_6H_5–$B(OH)_2$ reacts with hydroxamic acids, RCONR′OH, in nitromethane to form C_6H_5–B-(OH)ON(R′)C(O)R nonelectrolytes [22], and with a mixture of 3-(1′-hydroxyethyl)-5,8-dimethoxy-1,2-dihydronaphthalen-1-ols to yield cyclic phenylboronates [23]. C_6H_5–$B(OH)_2$ was used to convert *cis*-diol compounds into their cyclic phenylboronates, while the *trans*-diols do not react [24]. C_6H_5–$B(OH)_2$ protecting groups can serve to moderate the basicity of 1,2-diols [25].

In a solution of C_6H_5–$B(OH)_2$ and either 1,2-$(OH)_2C_6H_4$ (pyrocatechol), or L-dopa (3-(3,4-dihydroxyphenyl)-alanine) at various pH values, the equilibrium between phenylboronate anion and pyrocatechol or L-dopa to form the anionic complexes has been demonstrated by the observation of ^{11}B NMR signals for the complex and either phenylboronate anion or phenylboronic acid. The ionization constant of C_6H_5–$B(OH)_2$ has been estimated by the pH dependence of the ^{11}B NMR chemical shift of phenylboronate-phenylboronic acid solution to be $pK_a = 8.90$. The ^{11}B NMR spectra of solutions with pH values below 7 were used to deduce the complex formation constant with log K = 4.5 for pyrocatechol and with log K = 4.6 for L-dopa. The ^{13}C NMR spectra also indicate the complex formations, but in solutions of phenylboronic acid at pH 8 and at pH 9 with L-dopa, the ^{13}C NMR signal of the carbonyl carbon of L-dopa disappears [26]. ^{11}B NMR studies were also performed on phenylboronic acid in the absence or presence of chymotrypsin and subtilisin [28].

The conformation of the phenylboronates of β-5-lignin gave rise to unusual spectral dispersion of the aromatic protons in the ^1H NMR spectrum [27].

Stability constants of a number of pentoses and pentitols with C_6H_5–$B(OH)_2$ were determined in solution at 25°C and at an ionic strength of 0.1M (NaCl) by potentiometric titration. The solution structure of the complexes is discussed [29]. C_6H_5–$B(OH)_2$ mediates the uphill transport of monosaccharides, when a pH difference is imposed [30]. C_6H_5–$B(OH)_2$ was used for the conversion of completely constructed *cis,trans*-anthracyclinone mixtures into pure *cis*-isomers [33].

The combined use of C_6H_5–$B(OH)_2$ and OsO_4 is a convenient method for the *cis*-dihydroxylation of olefins, giving the corresponding 2-phenyl-1,3,2-dioxaborolane derivatives (cyclic phenylboronic acid esters) in good yield [31]. C_6H_5–$B(OH)_2$ and 1,2-cyclohexanedione dioxime are used to form a cage ligand for Ru^{2+} ions [32]. Thallated indole reacts with C_6H_5–$B(OH)_2$ in the presence of $Pd[OC(O)CH_3]_2$ to give 4-phenylindole [34]. C_6H_5–$B(OH)_2$ can transfer its C_6H_5 group to Pt in cationic *cis*-$[Pt(CH_3OH)_2(PR_3)_2]^{2+}$ [35]. C_6H_5–$B(OH)_2$, $(C_4H_9)_2SnO$, and Schiff bases were used to synthesize Schiff base stannoboroxanes [36].

C_6H_5–$B(OH)_2$ is a corrosion inhibitor for the local dissolution of aluminium in buffered borate solutions [37]. Addition of C_6H_5–$B(OH)_2$ to a model metal rolling lubricant reduces the crack propagation rate in the precracked die steel under stress [38]. C_6H_5–$B(OH)_2$ is effective in eliminating almost totally the increase in the rate of crack propagation in steels by water-based lubricants at high temperatures [39].

Reaction of alkenes with C_6H_5–$B(OH)_2$ in the presence of $AlHCl_2$ followed by air oxidation gives *anti*-Markovnikov alcohols [40]. A C_6H_5–$B(OH)_2/B(C_6H_5)_3/AlHCl_2$ catalyst is highly efficient for regio- and chemoselective hydroalumination of substituted olefins [41].

For the carbonylation of C_6H_5–$B(OH)_2$ with CO in methanol in the presence of catalytic amounts of Pd compounds, see Section 3.5.6.2.2.3 on p. 162 [79].

The $C_6H_5B(OH)_2$–H_2O_2 system, as studied by pH titration methods, is characterized by larger equilibrium constants than the $B(OH)_3$–H_2O_2 system; after proton release, C_6H_5OH and $B(OH)_3$ are formed [42]. The oxidation of a methionine residue in subtilisine-type proteinases by the alkaline $C_6H_5B(OH)_2$–H_2O_2 system is an active site-directed reaction [43]. The kinetics of the reaction of methionine sulfoxide thermitase with the reversible inhibitor $C_6H_5B(OH)_2$ was investigated [44]. C_6H_5–$B(OH)_2$ is apparently oxygenated by the hydroperoxide of liver microsomal flavin-containing monooxygenase (4a-flavin hydroperoxide) [45].

C_6H_5–$B(OH)_2$ inhibits the lipoprotein lipase-catalyzed hydrolysis of 4-nitrophenylbutyrate with K = 6.9 μM at pH 7.36 and 25 °C [46]. The activity of cholesterol esterase in the presence of carbamates can be protected by the competitive inhibitor C_6H_5–$B(OH)_2$ [47]. C_6H_5–$B(OH)_2$ binds 15-fold as tightly to bile-salt (e.g., sodium taurocholate) dependent lipase as does 4-nitrophenyl acetate [48]. C_6H_5–$B(OH)_2$ is used as a probe for the catalytic site of human plasma lecithine-cholesterol acetyltransferase [49].

Lipophilic salts of phenylboronic acid facilate the transport of ribonucleosides across a liquid membrane [50]. C_6H_5–$B(OH)_2$ reacts with a substituted bromoisoquinoline in the ultimate regiospecific synthesis of ellipticine [51] and may form histidine adducts with subtilisin and chymotrypsin, as indicated by low-field ^1H NMR [87] and ^{15}N NMR [88] studies. The inhibitory effect of C_6H_5–$B(OH)_2$ on the enzymatic activity of carboxypeptidase Y is investigated in [89].

Ethylene glycol is analyzed by mixing C_6H_5–$B(OH)_2$ with CH_3CN and acidified 2,2-dimethoxypropane [52]. Water acts to promote phenylboronate formation of 1,3-diols in a CH_3OH/H_2O mixture (2:1) for analysis by gas chromatography [53].

Treatment of C_6H_5–$B(OH)_2$ with bromonitrotoluenes afforded methylnitrobiphenyl in high yield [54]. Treatment of substituted biphenylboronic acids with N,N-diethyl-2-bromophenyl carbamate gave substituted *m*-terphenyls [55].

C_6H_5–$B(OH)_2$ reacts with diols in order to give the corresponding, previously unknown 2-phenyl-1,3,2-dioxaboroles [72].

In order to form cyclic boronates for a new type of gas chromatographic analysis of glycerol-1-nitrate, the latter compound is first treated with C_6H_5–$B(OH)_2$ [56]. The use of phenylboronic acid-substituted, amine-modified silica gel stationary phases for the high-performance liquid chromatography (HPLC) separation of saccharides and nucleosides under neutral conditions was studied; the apparent pK_a of the acid group is lowered. This surface buffering effect permits boronate-saccharide complexation to occur at much lower pH values than is typically the case [57]. Nonenzymically glycosylated proteins have been separated by phenylboronate affinity chromatography [58]. The boronate affinity chromatographical method of Little et al. [59] was evaluated in order to determine glycated hemoglobin in dried blood on filter paper [60]. Boronate chromatography of adenosine-5′-diphosphate (ADP) and adenosine-5′-triphosphate (ATP) has been summarized [61].

Cl^-, Br^-, and I^- ions were simultaneously determined by conversion to RHgX compounds by treatment with C_6H_5–$B(OH)_2$ and $Hg[NO_3]_2$ or $Hg[ClO_4]_2$, followed by HPLC [62].

C_6H_5–$B(OH)_2$ and 4-substituted derivatives thereof were used to prepare no-carrier-added ^{123}I-labeled RC_6H_4I [63]. C_6H_5–$B(OH)_2$ is a labeling substrate for the generator-produced positron emitter ^{122}I [64].

3.5.6.2.2.3 Additional Organylboronic Acids

2-$CH_3C_6H_4$–$B(OH)_2$, 2,6-$(CH_3)_2C_6H_3$–$B(OH)_2$, and other organylboronic acids react with diols in order to give the corresponding previously unknown 2-organyl-1,3,2-dioxaboroles [72].

References on pp. 164/7

4-Methylphenylboronic acid, **4-$CH_3C_6H_4$–$B(OH)_2$**, melting point 245 to 247 °C, was prepared by dropwise addition of *p*-tolylmagnesium bromide to an excess of $B(OCH_3)_3$ in dry tetrahydrofuran at −70 °C, 1 h stirring of the reaction mixture below −60 °C, and subsequent hydrolysis with 10% H_2SO_4. After filtering off the inorganic salt, separation of the tetrahydrofuran layer, extracting the aqueous layer with ether, washing the combined organic layers with water, drying over $Na_2[SO_4]$, and concentration of the solution, an 86% yield of product is obtained. IR data in Nujol: ν(OH) = 3480 and 3400 cm^{-1} (OH), ν(BO) = 1365 cm^{-1} [70].

4-Boronobenzenecarboxylic acid, **4-HOOC–C_6H_4–$B(OH)_2$**, melting point 232 to 234 °C, was prepared by dissolving 4-$CH_3C_6H_4$–$B(OH)_2$ in NaOH/H_2O, dilution of the solution, adding an aqueous solution of excess $K[MnO_4]$ (divided into eight parts) into the agitate mixture (one part every one hour), stirring the mixture overnight, destroying the excess of $K[MnO_4]$ by addition of ethanol at 50 °C, and filtering off the MnO_2. The filtrate was concentrated to ca. 200 mL below 40 °C, and then carefully acidified with HCl, and the crystals washed with and recrystallized from water (58% yield); IR data in Nujol: ν(OH) = 3500 to 2500 cm^{-1}, ν(C=O) = 1700 cm^{-1}, ν(BO) = 1355 cm^{-1} [70]. Resolution of the racemic 2,3-*trans*-3,4-*cis*-3′,4′,5,7-tetramethoxyflavan-3,4-diol by crystallization and decomposition of the ephedrine salts formed by its ester with 4-HOOC–C_6H_4–$B(OH)_2$ was reported [75].

Several **arylboronic acids**, used for affinity chromatography, are synthesized with ionization constants nearly neutral, which was determined by a new spectral-difference method. Arylboronic acids attached to solid matrixes have proved useful for the diol-specific chromatography of biomolecules and affinity purification of enzymes by exchangeable ligand chromatography [65]. Isoflavones are synthesized by the cross-coupling reaction of arylboronic acids and 3-iodochromone [66]. The synthesis of new triarylmesitylenes was performed by using the Pd-catalyzed arylboronic coupling method of Suzuki [76], e.g., with naphth-1-yl, 2-methylphenyl, or 1-dibenzofuranyl as the aryl group [77]. The coupling of halogenated pyrazines and pyridines with arylboronic acids in the presence of Pd(0) catalysts is described [78].

Principles, procedures, and applications of boronate-containing gels for affinity electrophoretic analyses of ribonucleic acid (RNA) molecules are described [67].

Difluoroboranes R–BF_2, produced from boronic acids by treatment with HF, are more easily purified than boronic acids [68]. A mixture of arylboronic acids and borinic acids is recovered by neutralization in two stages [69].

Using substituted phenylboronic acids as the binding sites, the influence of the structure of the binding sites in template-imprinted polymers on the selectivity for racemic resolution was investigated [71]. An *m*-phenylboronate agarose column has been used to selectively bind DOPA-containing proteins [73]. Dihydroxyboryl-substituted resins, e.g., **$\{-CH_2CH[C_6H_4-B(OH)_2\text{-}4]-\}_n$** containing Mo are thermally stable [74].

Hydrogenation of C_6H_5–C≡C–$B(OC_3H_7\text{-}i)_2$ over Lindlar catalyst in pure 1,4-dioxane (or better yet in mixtures of 1,4-dioxane with small amounts of pyridine or quinoline) followed by hydrolysis forms high stereochemically pure **(Z)-C_6H_5–CH=CH–$B(OH)_2$** in good yields. It was characterized as the 2-phenylethenyl dioxaborinane after cyclocondensation with $HOCH_2CH_2CH_2OH$ [94]. Carbonylation of alkenyl- and arylboronic acids such as **(E)-C_6H_{13}CH=CH–$B(OH)_2$**, **(E)-C_6H_5CH=CH–$B(OH)_2$**, or **4-RC_6H_4–$B(OH)_2$** (R = H, CH_3, or OCH_3) with CO in methanol in the presence of catalytic amounts of $Pd[O(CO)CH_3]_4$ or $Pd[P(C_6H_5)_3]_4$ gave, besides ketones, the corresponding methyl carboxylates, e.g., C_6H_5CH=CH–$COOCH_3$ or RC_6H_4–$COOCH_3$, as the major products in less than 31% yield [79].

For the use of 1-alkenylboronic acids in the formation of dienes, see [80], and for the preparation of stereospecific conjugated dienoates and dienones via alkenylboronic acids, see [81].

High stereochemically pure **(Z)-R–CH=CH–B(OH)$_2$** (R = i-C_3H_7, n-C_4H_9, t-C_4H_9, *cyclo*-C_5H_{11}, n-C_6H_{13}) were prepared from R–C≡C–B(OC_3H_7-i)$_2$ by hydrogenation over Lindlar catalyst in pure 1,4-dioxane (or better yet in mixtures with small amounts of pyridine or quinoline) followed by hydrolysis. The compounds were characterized as the corresponding alkenyldioxaborinanes after cyclocondensation with $HOCH_2CH_2CH_2OH$ [94].

The R–B(OH)$_2$ moiety was formed after boronation of an organic lithium compound (prepared from the analogous bromo compound and n-C_4H_9Li) with B(OC_4H_9-n)$_3$ at −90°C under Ar and hydrolysis with HCl [82].

A review describes the Pd^0-catalyzed coupling of aromatic boronic acids such as **2-(O=CH)–C_6H_4–B(OH)$_2$**, **2-(O=CH)–SC_4H_2–B(OH)$_2$-3**, and **4-(O=CH)–SC_4H_2–B(OH)$_2$-3** with aromatic halides for the preparation of unsymmetrical heterocyclic condensed ring systems (C_4H_4S is thiophene, C_4H_4Se is selenophene) [83]. **SC_4H_3–B(OH)$_2$-2**, **SC_4H_3–B(OH)$_2$-3**, **Se-C_4H_3–B(OH)$_2$-2**, and **SeC_4H_3–B(OH)$_2$-3** couple under Pd^0-catalysis with halopyrimidines to yield thienyl- or selenienyl-pyrimidines [83, 84]. **2-Formylaryl-boronic acids** react with 2-nitroaryl-halides to give unsymmetric biaryls [85].

***Cyclo*-C_6H_{11}–B(OH)$_2$** and other organylboronic acids react with diols in order to give the previously unknown corresponding 2-organyl-1,3,2-dioxaboroles [72].

A modified neglect of diatomic overlap (MNDO) calculation of the values of the 1,5-sigmatropic rearrangement of cyclopentadienylboronic acid was given. For **η^1-cyclopentadienylboronic acid** (σ structure, C_s symmetry; in pm): r(BC) = 160.2, r(C–C) = 152.8, r(C=C) = 136.2, r(BO) = 136.4; the angles O–B–O are 111.1° (with an inversion of configuration) or 110.2° (with the retention of configuration); ΔH_f = −123.78 kcal/mol (inversion) or ΔH_f = −122.33 kcal/mol (retention). For **η^2-cyclopentadienylboronic acid** (C_s symmetry imposed) with inversion of configuration (in pm): r(BC) = 180.0, r(C–C) = 143.6, r(C=C) = 141.4, r(BO) = 137.2; ∢(OBO) = 110.6°; ΔH_f = −86.00 kcal/mol. For η^2-cyclopentadienylboronic acid with retention of configuration (in pm): r(BC) = 194.3, r(C–C) = 143.0, r(C=C) = 141.4, r(BO) = 136.3; ∢(OBO) = 108.4°; ΔH_f = −66.44 kcal/mol [86].

The inhibitory effect of **$C_6H_5CH_2CH_2$–B(OH)$_2$**, **$C_6H_5CH_2CH_2CH_2$–B(OH)$_2$**, and **$C_6H_5(CH_2)_4$–B(OH)$_2$** on the enzymatic activity of carboxypeptidase Y depends on the length and of the structure of the hydrocarbon chain [89]. $C_6H_5CH_2CH_2$–B(OH)$_2$ may form histidine adducts with subtilisin and chymotrypsin, as indicated by low-field ^{1}H NMR [87] and ^{15}N NMR studies [88].

A reaction sequence beginning with the mercuration or thallation of vinylboronic acids leads to vinyl iodides [90]. **CH_2=C=CH–B(OH)$_2$** was enantioselectively condensed with $(CH_3)_2CH$-CH_2CHO [91] and underwent stereoselective addition with 3-hydroxyketones to give propargyl diols [92]. **CH_2=$CHCH_2$–B(OH)$_2$** reacts with formaldehyde with the formation of chair and twist-boat transition structures according to ab initio molecular orbital (MO) calculations with the 3-21G basis set [93].

Specific boronic acids have been prepared which are inhibitors for serine β-lactamases [101]. Enzyme models based on boronic acids and on monoclonal antibodies are described [102].

For the hydrolysis of cyclic α-chloromethaneboronic acid esters to give α-chloromethaneboronic acids, see [103]; for its reaction with alkyl lithium compounds, e.g., dichloromethyl lithium, see [104].

Aryltin compounds of the types **4-R_3Sn–C_6H_4–B(OH)$_2$** (and also **4-(O=CH)–C_6H_4–B(OH)$_2$**) were reacted with halogenated nucleosides using a Pd[P(C_6H_5)$_3$]$_4$-catalyzed coupling reaction to form ^{10}B-containing nucleoside derivatives for neutron-capture therapy [1, 105].

Cross-linked polymers from acenaphthylene metalated by Hg or Tl were reacted with boranes in order to give, after hydrolysis, polymers containing boronic acid groups; these materials can serve as polymer-supported projecting groups for diols [106].

References for 3.5.6.2:

[1] Yamamoto, Y.; Seko, T.; Rong, Feng Guang; Nemoto, H. (Tetrahedron Lett. **30** [1989] 7191/4).
[2] Matteson, D. S. (in: Liebmann, J. F.; Advances in Boron and in Boranes, Verlag Chemie, Weinheim 1988, pp. 343/56).
[3] Kawashima, Y.; Takeo, H.; Matsumura, Ch. (Nippon Kagaku Kaishi **1986** 1465/75).
[4] Dewar, M. J. S.; Jie, Caoxin; Zoebisch, E. G. (Organometallics **7** [1988] 513/21).
[5] Brown, H. C.; Cole, T. E. (Organometallics **4** [1985] 816/21).
[6] Goralski, C. T.; Singaram, B.; Brown, H. C. (J. Org. Chem. **52** [1987] 4014/9).
[7] Corey, E. J.; Bakshi, R. K.; Shibata, S.; Chen, Ch. P.; Singh, V. K. (J. Am. Chem. Soc. **109** [1987] 7925/6).
[8] Cholerton, T. J.; Hunt, J. H.; Martin-Smith, M. (J. Chromatogr. **333** [1985] 175/85).
[9] Palumaa, P.; Jaro, J. (Bioorg. Khim. **11** [1985] 1210/6 from C.A. **103** [1985] No. 174546).
[10] Köster, R.; Idelman, P. (Inorg. Synth. **24** [1986] 83/7 from C.A. **105** [1986] No. 153107).

[11] Dahlhoff, W. V.; Geisheimer, A. (Z. Naturforsch. **40b** [1985] 141/9).
[12] Garrigues, B.; Mulliez, M. (J. Organomet. Chem. **314** [1986] 19/24).
[13] Kliegel, W.; Preu, L.; Rettig, S. J.; Trotter, J. (Can. J. Chem. **63** [1985] 509/15).
[14] Arnt, H.; Kliegel, W.; Rettig, S. J.; Trotter, J. (Can. J. Chem. **66** [1988] 1117/22).
[15] Ebeling, E.; Kliegel, W.; Rettig, S. J.; Trotter, J. (Can. J. Chem. **67** [1989] 933/40).
[16] Waechtler, A.; Krause, J.; Eidenschink, R.; Eichler, J.; Scheuble, B. (Ger. Offen. 3608714 [1985/86] 1/64 from C.A. **106** [1987] No. 93757).
[17] Kanabur, V. V.; Dandegaonkar, S. H. (Curr. Sci. **54** [1985] 1199/200 from C.A. **105** [1986] No. 78 988).
[18] Farfán, N.; Contreras, R. (Heterocycles **23** [1985] 2989/93 from C.A. **105** [1986] No. 226697).
[19] Mancilla, T.; Contreras, R.; Wrackmeyer, B. (J. Organomet. Chem. **307** [1986] 1/6).
[20] Mancilla, T.; Contreras, R. (J. Organomet. Chem. **321** [1987] 191/8).

[21] Farfán, N.; Joseph-Nathan, P.; Chiquete, L. M.; Contreras, R. (J. Organomet. Chem. **348** [1988] 149/56).
[22] Das, M. K.; Biswas, B.; Chakraborty, S. (Synth. React. Inorg. Met.-Org. Chem. **16** [1986] 77/84 from C.A. **105** [1986] No. 226701).
[23] Irvine, R. W.; Russell, R. A.; Warrener, R. N. (Tetrahedron Lett. **26** [1985] 6117/20).
[24] Machida, M.; Oda, K.; Kanaoka, Y. (Tetrahedron **41** [1985] 4995/5001).
[25] McMurry, J.; Erim, M. D. (J. Am. Chem. Soc. **107** [1985] 2712/20).
[26] Yoshino, K.; Oguchi, S.; Shimakata, Y.; Mori, Y.; Okamoto, M.; Kotaka, M.; Kakihana, H. (Kyoto Daigaku Genshiro Jikkensho Tech. Rep. KURRI-TR-260 [1985] 221/32 from C.A. **104** [1986] No. 109729).
[27] Ede, R. M.; Ralph, J.; Wilkins, A. L. (Holzforschung **41** [1987] 239/45 from C.A. **107** [1987] No. 178376).
[28] Adebodun, F.; Jordan, F. (J. Cell. Biochem. **40** [1989] 249/60 from C.A. **111** [1989] No. 92770).
[29] Huttunen, E. (Ann. Acad. Sci. Fenn. A II No. 201 [1984] 1/45 from C.A. **102** [1985] No. 120917).
[30] Shinbo, T.; Nishimura, K.; Yamaguchi, T.; Sugiara, M. (J. Chem. Soc. Chem. Commun. **1986** 349/51).

[31] Iwasawa, N.; Kato, T.; Narasaka, K. (Chem. Lett. **1988** 1721/4).
[32] Muller, J. G.; Grzybowski, J. J.; Takeuchi, K. J. (Inorg. Chem. **25** [1986] 2665/7).
[33] Ravichandron, K.; Francis, F. A. J.; Cava, M. P. (J. Org. Chem. **51** [1986] 2044/6).
[34] Somei, M.; Amari, H.; Makita, Y. (Chem. Pharm. Bull. **34** [1986] 3971/3 from C.A. **106** [1987] No. 213695).
[35] Siegmann, K.; Pregosin, P. S.; Venanzi, L. M. (Organometallics **8** [1989] 2659/64).
[36] Chaturvedi, V.; Bhal, L.; Tandon, J. P. (Indian J. Chem. **24A** [1985] 1039/41).
[37] Kusnetsov, Yu. L.; Andreev, N. N. (Zashch. Met. **21** [1985] 484/7 from C.A. **103** [1985] No. 25902).
[38] Lockwood, F. E.; Christodoulou, L.; Langan, T. J. (Addit. Schmierst. Arbeitsfluessigkeiten 5th Int. Kolloq., Ostfildern, FRG, 1986, Vol. 2, pp. 11/6/1-11/6/10, from C.A. **105** [1986] No. 175537).
[39] Langan, T. J.; Christodoulou, L.; Lockwood, F. E. (ASLE Trans. **30** [1987] 105/10 from C.A. **106** [1987] No. 88417).
[40] Maruoka, K.; Sano, H.; Shinoda, K.; Yamamoto, H. (Chem. Lett. **1987** 73/7).

[41] Maruoka, K.; Sano, H.; Shinoda, K.; Nakai, S.; Yamamoto, H. (J. Am. Chem. Soc. **108** [1986] 6036/7).
[42] Pizer, R.; Tihal, C. (Inorg. Chem. **26** [1987] 3639/42).
[43] Hausdorf, G.; Krueger, K.; Kuettner, G.; Holzhuetter, H. G.; Froemmel, C.; Hoehne, W. E. (Biochim. Biophys. Acta **952** No. 1 [1988] 20/6).
[44] Hausdorf, G.; Krueger, K.; Hoehne, W. E. (Biomed. Biochim. Acta **48** [1989] 577/81 from C.A. **111** [1989] No. 190201).
[45] Jones, C. K.; Ballou, D. P. (J. Biol. Chem. **261** [1986] 2553/9 from C.A. **104** [1986] No. 164167).
[46] Quinn, D. M. (Biochemistry **24** [1985] 3144/9).
[47] Hosie, L.; Sutton, L. D.; Quinn, D. M. (J. Biol. Chem. **262** [1987] 260/4).
[48] Abouakil, N.; Lombardo, D. (Biochim. Biophys. Acta **1004** [1989] 215/20 from C.A. **111** [1989] No. 149363).
[49] Jauhiainen, M.; Ridgway, N. D.; Dolphin, P. J. (Biochim. Biophys. Acta **918** [1987] 175/88).
[50] Grotjohn, B. F.; Czarnik, A. W. (Tetrahedron Lett. **30** [1989] 2325/8).

[51] Miller, R. B.; Dugar, S. (Tetrahedron Lett. **30** [1989] 297/300).
[52] Flanagan, R. J.; Dawling, S.; Buckley, B. M. (Ann. Clin. Biochem. **24** [1987] 80/4 from C.A. **106** [1987] No. 208884).
[53] Flanagan, R. J.; Chan, M. W. J. (Analyst [London] **114** [1989] 703/6).
[54] Iihama, T.; Fu, J.-M.; Bourguignon, M.; Snieckus, V. (Synthesis **1989** 184/7).
[55] Cheng, W.; Snieckus, V. (Tetrahedron Lett. **28** [1987] 5097/8).
[56] Scharpf, F.; Yeates, R. A.; Laufen, H.; Eibel, G. (J. Chromatogr. **143** [1987] 91/9).
[57] Lochmuller, C. H.; Hill, W. B. (ACS Symp. Ser. No. 297 [1986] 210/25 from C.A. **105** [1986] No. 17540).
[58] Ducrocq, R.; Cahour, A.; Berriche, S.; Intrator, S.; Elion, J. (Protides Biol. Fluids **33** [1985] 651/4 from C.A. **104** [1986] No. 65099).
[59] Little, R. R.; McKenzie, E. M.; Wiedmeyer, H. M.; England, J. D. (Clin. Chem. [Winston-Salem, N. C.] **32** [1986] 869/71 from C.A. **105** [1986] No. 2963).
[60] Eross, J.; Kreutzmann, D.; Crowell, C.; Silink, M. (Clin. Chem. [Winston-Salem, N. C.] **32** [1986] 2222 from C.A. **106** [1987] No. 64017).

[61] Barnes, L. D.; Robinson, A. K.; Mumford, C. H.; Garrison, P. N. (Anal. Biochem. **144** [1985] 296/304 from C.A. **102** [1985] No. 91765).

[62] Moss, P. E.; Stephen, W. I. (Anal. Proc. [London] **22** [1985] 5/6 from C.A. **102** [1985] No. 124757).
[63] Kabalka, G. W. (DOE Symp. Ser. No. 56 [1985] 377/83 from C.A. **105** [1986] No. 6240).
[64] Moerlein, S. M.; Mathis, C. A.; Yano, Y. (Appl. Radiat. Isot. **38** [1987] 85/90 from C.A. **107** [1987] No. 23424).
[65] Soundararajan, S.; Badawi, M.; Kohlrust, C. M.; Hagemann, J. H. (Anal. Biochem. **178** [1989] 125/34 from C.A. **111** [1989] No. 20318).
[66] Yokoe, I.; Sugita, Y.; Shirataki, Y. (Chem. Pharm. Bull. **37** [1989] 529/30 from C.A. **111** [1989] No. 133838).
[67] Igloi, G. L.; Koessel, H. (Methods Enzymol. **155** [1987] 433/8 from C.A. **109** [1988] No. 3313).
[68] Kinder, D. H.; Katzenellenbogen, J. A. (J. Med. Chem. **28** [1985] 1917/25 from C.A. **103** [1985] No. 196378).
[69] Ostermaier, J. J. (Eur. Appl. 131 307 [1983/85] 1/10 from C.A. **102** [1985] No. 204099).
[70] Matsubara, H.; Seto, K.; Tahara, T.; Takahashi, S. (Bull. Chem. Soc. Jpn. **62** [1989] 3896/901).

[71] Wulff, G.; Poll, H. G. (Makromol. Chem. **188** [1987] 741/8).
[72] Wulff, G.; Hansen, A. (Carbohydr. Res. **164** [1987] 123/40).
[73] Hawkins, C. J.; Lavin, M. F.; Parry, D. L.; Ross, I. L. (Anal. Biochem. **159** [1986] 187/90 from C.A. **106** [1987] No. 98743).
[74] Tempesti, E.; La Ginestra, A.; Pelino, M.; Di Renzo, F.; Mazzocchia, C. (Thermochim. Acta **137** [1988] 71/8).
[75] Brown, B. R.; Fuller, M. J. (J. Chem. Res. Synop. **1986** No. 4, pp. 140/1 from C.A. **105** [1986] No. 133581).
[76] Miyaura, Norio; Yanagi, T.; Suzuki, Akira (Synth. Commun. **11** [1981] 513/9).
[77] Katz, H. E. (J. Org. Chem. **52** [1987] 3932/4).
[78] Thompson, W. J.; Jones, J. H.; Lyle, P. A.; Thies, J. E. (J. Org. Chem. **53** [1988] 2052/5).
[79] Ohe, T.; Ohe, K.; Uemara, S.; Sugita, N. (J. Organomet. Chem. **344** [1988] C5/C7).
[80] Kim, J. I.; Lee, J. T.; Yeo, K. D. (Bull. Korean Chem. Soc. **6** [1985] 366/9 from C.A. **106** [1987] No. 49558).

[81] Mavrov, M. V.; Urdaneta, N. A.; Nguyen Cong Hao; Serebryakov, E. P. (Izv. Akad. Nauk SSSR Ser. Khim. **1987** 2633/4; Bull. Acad. Sci. USSR Div. Chem. Sci. **36** [1987] 2447/8).
[82] Schinazi, R. F.; Prusoff, W. H. (J. Org. Chem. **50** [1985] 841/7).
[83] Gronowitz, S. (Chem. Scr. **27** [1987] 535/7).
[84] Gronowitz, S.; Hoernfeldt, A.-B.; Kristjansson, V.; Musil, T. (Chem. Scr. **26** [1986] 305/9).
[85] Gronowitz, S.; Hoernfeldt, A.-B.; Yang, Youhua (Org. Synth. Mod. Trends Proc. 6th IUPAC Symp., Moscow 1986 [1987], pp. 253/62 from C.A. **110** [1989] No. 173287).
[86] Schoeller, W. W. (J. Chem. Soc. Dalton Trans. **1984** 2233/6).
[87] Bachovdin, W. W.; Wang, W. Y. L.; Farr-Jones, S.; Kettner, C. A.; Shenvi, A. B. (Biochemistry **27** [1988] 7689/97).
[88] Farr-Jones, S.; Smith, S. O.; Kettner, C. A.; Griffin, R. G.; Bachovdin, W. W. (Proc. Natl. Acad. Sci. USA **86** [1989] 6922/4).
[89] Popov, A. A.; Rotanova, T. V.; Rumsh, L. D. (Bioorg. Khim. **14** [1988] 1650/5 from C.A. **110** [1989] No. 150419).
[90] Srivastava, P. C.; Knapp, F. F.; Kabalka, G. W.; Kunda, S. A. (Synth. Commun. **15** [1985] 355/64 from C.A. **103** [1985] No. 104502).

[91] Ikeda, N.; Arai, J.; Yamamoto, H. (J. Am. Chem. Soc. **108** [1986] 483/6).
[92] Ikeda, N.; Omori, K.; Yamamoto, H. (Tetrahedron Lett. **27** [1986] 1175/8).

[93] Li, Yi; Houk, K. N. (J. Am. Chem. Soc. **111** [1989] 1236/40).
[94] Srebnik, M.; Bhat, N. G.; Brown, H. C. (Tetrahedron Lett. **29** [1988] 2635).
[95] Wilkes, S. H.; Prescott, J. M. (J. Biol. Chem. **262** [1987] 8621/5 from C.A. **107** [1987] No. 111685).
[96] Breitenbach, J. M.; Hausinger, R. P. (Biochem. J. **250** [1988] 917/20 from C.A. **108** [1988] No. 182602).
[97] Joseph, M. H. (J. Chromatogr. **342** [1985] 370/5).
[98] Joseph, M. H. (Curr. Separ. **7** No. 1 [1985] 2/4 from C.A. **105** [1986] No. 165150).
[99] Bogusz, M.; Bialka, J.; Gierz, J.; Klys, M. (Z. Rechtsmed. **96** [1986] 23/6 from C.A. **105** [1986] No. 73804).
[100] Takatsu, A.; Nishi, S. (Anal. Sci. **4** [1988] 487/91 from C.A. **110** [1989] No. 4061).

[101] Crompton, I. E.; Cuthbert, B. K.; Lowe, G.; Waley, S. G. (Biochem. J. **251** [1988] 453/9 from C.A. **108** [1988] No. 218119).
[102] Rao, G. C. (Diss. City Univ. New York 1987, pp. 1/157 from C.A. **110** [1989] No. 3619).
[103] Sadhu, K. M.; Matteson, D. S.; Hurst, G. D.; Kurosky, J. M. (Organometallics **3** [1984] 804/6).
[104] Matteson, D. S.; Kandil, A. A. (Tetrahedron Lett. **27** [1986] 3831/4).
[105] Yamamoto, Y.; Seko, T.; Nemoto, H. (J. Org. Chem. **54** [1989] 4734/6).
[106] Al-Kadhumi, A. A. H. A.; Hodge, P.; Hunt, B. J.; Thorpe, F. G. (React. Polym. Ion Exch. Sorbents **7** [1987] 15/23 from C.A. **107** [1987] No. 176611).

3.5.6.3 Boronic Acid Esters

3.5.6.3.1 Compounds of the Type $HB(OR)_2$

$HB(OCH_3)_2$ shows the symmetry C_s, the distances r(BH)=118.7 pm, r(BO)=134.9 and 136.7 pm, and the angles ∢(HOB)=118.2° and 127.4°. The following values are calculated at the AM1 level: heat of formation, ΔH_f=−144.3 kcal/mol (observed −138.4), dipole moment, μ=1.50 D, and ionization potential, I=11.13 eV [1].

The reduction of nitriles, aldehydes, and ketones with **$(CH_3O)_2BH \cdot CoCl_2$** or **$(CH_3O)_2BH \cdot NiCl_2$** occurs under mild conditions in good yields [2].

$HB[OC(O){-}C_6H_5]_2$, rather a mixed anhydride, has been mentioned as a precursor for organic syntheses [3].

3.5.6.3.2 Cyclic Esters of the Type HB(−OZO−)

For 1,3,2-benzodioxaboroles **9** with B–R instead of B–H, see Section 3.5.6.4.2.3, p. 173, and for 1,3,2-benzodioxaboroles **9** with B–OR instead of B–H, see Section 3.5.8.6, p. 247.

$HB({-}OCH_2CH_2O{-})$ has the following values calculated by AM1: heat of formation, ΔH_f= −138.0 kcal/mol, dipole moment, μ=1.71 D (observed 2.28 D), and ionization potential, I= 10.95 eV [1].

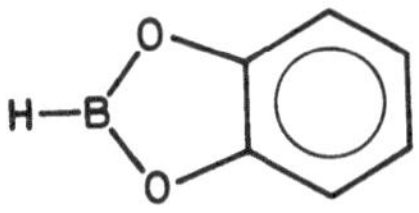

9

References on p. 168

$HB[-OC_6H_4O-]$, "catecholborane" (**9**), is used in selective hydroboration of a series of organic compounds. A survey with 226 references was published [3]. Compound **9** permits the monohydroboration of terminal alkynes with high stereoselectivity, thus providing access to the corresponding (E)-1-alkenyldioxaborolanes [4]. For additional uses as a hydroboration reagent, see [6 to 11].

Compound **9** is easily prepared by refluxing a solution of 1,2-$(OH)_2C_6H_4$ (pyrocatechol) and B_2O_3 (3:1) in benzene for 20 h in order to give 2,2′-phenylenebis-(2-oxy-1,3,2-benzodioxaborole) (**10**) in 93% yield, which was treated with $Na[BH_4]$ (2:1) in diglyme (containing small amounts of LiCl) under N_2; the yield of this step was 97% [5].

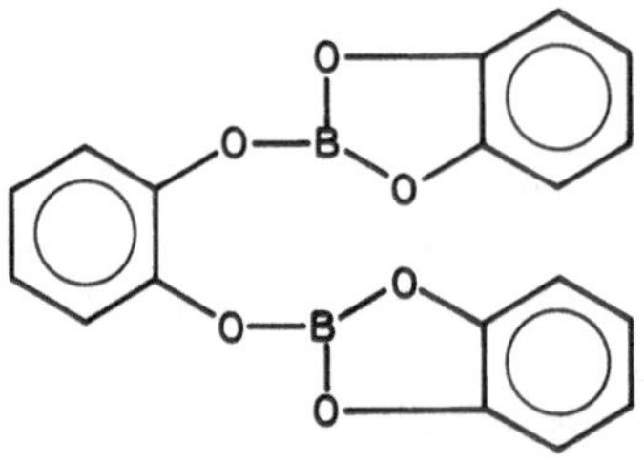

10

$HB[-OCH_2CH(C_6H_5)CH_2O-]$ has been prepared as the ultimate product of the exceptionally slow reduction of phenylmalonic acid with the tetrahydrofuran adduct of BH_3 [12].

References for 3.5.6.3:

[1] Dewar, M. J. S.; Jie, Caoxian; Zoebisch, E. G. (Organometallics **7** [1988] 513/21).
[2] Nose, A.; Kudo, T. (Chem. Pharm. Bull. **37** [1989] 808/10 from C.A. **111** [1989] No. 114788).
[3] Suzuki, A.; Dhillon, R. S. (Top. Curr. Chem. **130** [1986] 23/88, 82).
[4] Suzuki, A. (Pure Appl. Chem. **57** [1985] 1749/58).
[5] Männig, D.; Nöth, H. (J. Chem. Soc. Dalton Trans. **1985** 1689/92; Ger. Offen. 3528321 [1985/87] 1/3 from C.A. **106** [1987] No. 156662).
[6] Männig, D.; Nöth, H. (Angew. Chem. **97** [1985] 854/5; Angew. Chem. Int. Ed. Engl. **24** [1985] 878/9; Ger. Offen. 3528320 [1985/87] 1/5 from C.A. **106** [1987] No. 176654).
[7] Evans, D. A.; Fu, G. C.; Hoveyda, A. H. (J. Am. Chem. Soc. **110** [1988] 6917/8).
[8] Burgess, K.; Ohlmeyer, M. J. (J. Org. Chem. **53** [1988] 5178/9; Tetrahedron Lett. **30** [1989] 395/8).
[9] Satoh, Makoto; Miyaura, Norio; Suzuki, Akira (Synthesis **1987** 373/7).
[10] Satoh, Naoyuki; Ishiyama, Tatsuo; Miyaura, Norio; Suzuki, Akira (Bull. Chem. Soc. Jpn. **60** [1987] 3471/3).
[11] Yanagi, Teiji; Oh-e, Takayuki; Miyaura, Norio; Suzuki, Akira (Bull. Chem. Soc. Jpn. **62** [1989] 3892/5).
[12] Choi, Y. M.; Emblidge, R. W.; Kucharczyk, N.; Sofia, R. D. (J. Org. Chem. **52** [1987] 3925/7).

3.5.6.4 Organylboronic Acid Esters

In this Gmelin Handbook of Inorganic Chemistry, the "full organic derivatives" of boranes etc. are only given as a general survey.

3.5.6.4.1 Esters of the Type $RB(OR')_2$ and $RB(OR')$–OR"

The homologation of chiral bis(organyloxy)organylboranes (boronic acid esters) [2], the use of chiral organoboranes in organic syntheses [3], and directed asymmetric syntheses with boronic acid esters [4] have been reviewed. Also a review about asymmetric syntheses with alkylboronic acid esters, 1-haloalkylboronic acids and their esters [5], and a summary on enantioselective syntheses of natural products utilizing the method of insertion of asymmetric carbon into an existing B–C bond of boronic acid esters [6] have been compiled. Another review describes asymmetric syntheses with boronic acid esters, where synthetic applications include amidoboronic esters and insect pheromones [7].

One of the most often used procedures for the preparation of bis(organyloxy)organylboranes (boronic acid esters), $RB(OR')_2$, especially for the preparation of ^{10}B-labeled species, is the direct reaction of carbanions from R^-Li^+ with $B(OR')_3$ [1].

Chirally selective syntheses of diols are achieved via (+)-pinanediol alkylboronates [8].

$CHCl_2$–$B(OC_3H_7\text{-}i)_2$ and **CH_3CCl_2–$B(OCH_3)_2$** have been prepared from a mixture of $B(OC_3H_7\text{-}i)_3$ and CH_2Cl_2 (respectively $B(OCH_3)_3$ and CH_3CHCl_2) in tetrahydrofuran by the addition of lithium diisopropylamide or lithium dimethylamide [10].

A mixture of boronic acid esters and CH_2Cl_2 has been reacted with $n\text{-}C_4H_9Li$; for details, see [9]. A series of new chiral syntheses [11, 12] and asymmetric syntheses [13] with boronic acid esters was described. For the preparation of homoallylic alcohols from crotylboronates, see [14].

The cross- and homocoupling of $4\text{-}CH_3OC_6H_4I$ with **C_6H_5–$B(OCH_3)_2$** is catalyzed by $PdCl_2\text{-}(CH_3CN)_2$ [15]. Bisalkylboronate derivates, formed from labetalol and methyl- or *n*-butylboronic acid, have been used for the gas chromatographic determination of the properties of the two racemic modifications present in labetalol hydrochloride, an important constituent of various pharmaceuticals [16].

The compound **$2,4,6\text{-}(t\text{-}C_4H_9)_3C_6H_2$–$B(OCH_3)_2$** reacts with $Li[AlH_4]$ and tetramethylethylenediamine in ether to give a lithium product, which hydrolyzes to give $2,4,6\text{-}(t\text{-}C_4H_9)_3C_6H_2$–BH–OH (see Formula 1 in Section 3.5.2.3, p. 148); in strongly acidic media, the hydrolysis involves the attack of a neighboring CH_3 group to give a 1-hydroxy-1-boraindane derivative [25].

Allylic boronic acid esters react with aldehydes [17].

Vinyl di-*n*-octylboronate, **CH_2=CH–$B(OC_8H_{17}\text{-}n)_2$**, boiling point 105 to 108°C/0.01 Torr, has been prepared from a solution of $B(OCH_3)_3$ in ether and vinylmagnesium chloride in tetrahydrofuran, by addition of concentrated HCl, then phenothiazine, extraction with *n*-octanol and concentration under vacuum [18]. For the preparation of cyclic vinylboronates from vinylbromides and Grignard reagents, see [19].

(Z)-1-pentenylboronate and (Z)-1-hexenylboronate react with 1-alkenylhalides by cross-coupling [20]. For the preparation of the diisopropylesters of substituted (Z)-1-alkenylboronic acids originating from 1-bromoalkynes and $HBBr_2$–$S(CH_3)_2$, see [21]. For modified neglect of diatomic overlap (MNDO) calculations of the activation energies for the aldol addition of cyclopentanone enol boron compounds to aldehydes, see [22]. For the oxidation of a 2-methylcyclohexylboronic acid ester to (S,S)-2-methylcyclohexanol, see [23]. Hydrolysis and subsequent dehydration of tris(2,4,6-*t*-butyl)phenyldialkoxyboranes, prepared from tris(2,4,6-*t*-butyl)-phenyllithium and $B(OCH_3)_3$ in hexane, yield 2,4-dioxa-1,3-diboretanes [24].

One-carbon homologated boronic acid esters have been prepared from alkylboronate esters [26]. Boronic acid esters were used to prepare optically pure 2-alkyl-1,3,2-dioxaborinanes by asymmetric hydroboration [27]. Alkylboronic acid esters react with $LiCHCl_2$ to give

References on pp. 170/1

homologated primary and secondary alcohols [28]. Boronic acid esters react with $Li[AlH_4]$ in ether/pentane to form the corresponding optically pure lithium monoalkylborohydrides [29]. Monoorganyldiisopropoxyboranes react cleanly with organolithium compounds at −78 °C [30]. The pyrolysis of $Li[RB(OC_3H_7\text{-}i)_3]$ yields the relatively volatile boronic acid esters in high purity (leaving a residue of $LiOC_3H_7\text{-}i$). Treatment of the lithium organylborates with acetyl or benzoyl chloride cleanly generates boronic acid esters [31]. The effectiveness of $LiCHCl_2$ and $LiCHClSi(CH_3)_3$ for the homologation of boronate esters has been investigated with and without the addition of $HgCl_2$ [32]. A more convenient route for the homologation of boronic acid esters utilizes the in situ formation of $LiCHCl_2$ at −78 °C [33]. The available procedures for convenient homologation of boronic acid esters have been evaluated [34].

Heterocyclic boronates such as diethyl 3-tetrahydrofuranylboronate, $\mathbf{OC_4H_7{-}B(OC_2H_5)_2\text{-}3}$, of very high enantiomeric purity have been obtained by treatment of diisopinocampheyl-3-tetrahydrofuranylborane with acetaldehyde [35]. Crystalline chelates of 100% optical purity have been synthesized from boronate esters [36].

Boronate esters react with excess KH in tetrahydrofuran under N_2 to give the corresponding borohydrides [37]. Treatment of vinyl(dimethyloxy)boranes with Br_2 and then with a base yields the corresponding vinyl bromides [38]. Optically pure alkynylborinic acid esters (see Section 3.5.5.6, p. 153) have been prepared from optically pure boronic acid esters such as $\mathbf{(R)\text{-}(-)\text{-}s\text{-}C_4H_9{-}B(OC_3H_7\text{-}\mathit{i})_2}$ [39]. For the syntheses of chiral boron heterocycles from boronic acids and boronic acid esters with iminodiacetic acids or α-amino acids, see [40]. Low-temperature autoxidation of R_3B (R = cyclopentyl, methylpentyl, etc.) in tetrahydrofuran gave $RB(OR)_2$ compounds; the reaction is inhibited by iodine [41]. 1-Alkynyldi(isopropoxy)boranes, $\mathbf{RC{\equiv}CB(OC_3H_7\text{-}\mathit{i})_2}$, have been prepared by reaction of 1-lithio-1-alkynes with tris(isopropoxy)-borane at −78 °C, followed by treatment with HCl [42].

Arylboronic acid derivatives are good antioxidants. Alkyl- and arylboronates decompose hydroperoxides in a nonradical way. Phenylboronates initially give phenylborates by oxygen insertion. Aryl esters of phenylboronic acid are very effective antioxidants; they considerably surpass the antioxidative activity of the corresponding phenols [43]. The rate constants of the reaction of aryl phenylboronates with peroxy radicals are approximately 10 times larger than that for structurally analogous phosphites [44]. As antioxidants for polymers, arylboronic acid esters were found to destroy hydroperoxides by three mechanisms: (a) catalytic decomposition after an induction period, (b) reduction of the hydroperoxides by oxidizable parts of the organoboron compounds, and (c) reduction of the hydroperoxides by phenylboronic acid esters to give alcohols and the corresponding boric acid esters [45].

References for 3.5.6.4.1:

[1] Schinazi, R. F.; Prusoff, W. H. (J. Org. Chem. **50** [1985] 841/7).
[2] Matteson, D. S. (Kagaku Zokan [Kyoto] No. 105 [1985] 25/36 from C.A. **103** [1985] No. 5638).
[3] Matteson, D. S. (Synthesis **1986** 973/85).
[4] Matteson, D. S.; Sadhu, K. M.; Ray, R.; Pradipta, P. K.; Majumdar, D.; Tsai, D. J. S.; Hurst, G. D.; Erdik, E. (J. Organomet. Chem. **281** [1985] 15/23).
[5] Matteson, D. S.; Sadhu, K. M.; Ray, R.; Peterson, M. L.; Majumdar, D.; Hurst, G. D.; Jesthi, P. K.; Tsai, D. J. S.; Erdik, E. (Pure Appl. Chem. **57** [1985] 1741/8).
[6] Matteson, D. S. (Acc. Chem. Res. **21** [1988] 294/300 from C. A. **109** [1988] No. 73499).
[7] Matteson, D. S. (Mol. Struct. Energ. **5** [1988] 343/56 from C.A. **110** [1989] No. 153395).
[8] Matteson, D. S.; Sadhu, K. M.; Peterson, M. L. (J. Am. Chem. Soc. **108** [1986] 810/9).
[9] Sadhu, K. M.; Matteson, D. S. (Organometallics **4** [1985] 1687/9).
[10] Matteson, D. S.; Hurst, G. D. (Organometallics **5** [1986] 1465/7).

[11] Matteson, D. S.; Kandil, A. A. (Tetrahedron Lett. **27** [1986] 3831/4).
[12] Hurst, G. D. (Diss. Washington State Univ. Pullman, Wash., 1987, pp. 1/193 from C.A. **109** [1988] No. 129092).
[13] Peterson, M. L. (Diss. Washington State Univ., Pullman, Wash., 1987, pp. 1/190 from C.A. **109** [1988] No. 73782).
[14] Roush, W. R.; Ando, K.; Powers, D. B.; Haltermann, R. L.; Palkowitz, A. D. (Tetrahedron Lett. **29** [1988] 5579/82).
[15] Bumagin, N. A.; Ponomarev, A. B.; Beletskaya, I. P. (J. Organomet. Chem. **291** [1985] 129/32).
[16] Cholerton, T. J.; Hunt, J. H.; Martin-Smith, M. (J. Chromatogr. **333** [1985] 178/85).
[17] Hoffmann, R. W. (in: Bartmann, W.; Trost, B. M.; Selectivity, a Goal for Synthetic Efficiency, Verlag Chemie, Weinheim 1984, pp. 87/100).
[18] Hoffmann, R. W.; Landmann, B. (Chem. Ber. **119** [1986] 2013/24).
[19] Hoffmann, R. W.; Ditrich, K.; Fröch, S. (Liebigs Ann. Chem. **1987** 977/85).
[20] Miyaura, N.; Satoh, M.; Suzuki, A. (Tetrahedron Lett. **32** [1986] 3745/8).

[21] Satoh, M.; Miyaura, N.; Suzuki, A. (Chem. Lett. **1986** 1329/32).
[22] Gennari, C.; Todeschini, R.; Beretta, M. G.; Favini, G.; Scolastico, C. (J. Org. Chem. **51** [1986] 612/6).
[23] Latham, J. A.; Walsh, C. (J. Chem. Soc. Chem. Commun. **1986** 527/8).
[24] Pachaly, B.; West, R. (J. Am. Chem. Soc. **107** [1985] 2987/8).
[25] Groteklaes, M.; Paetzold, P. (Chem. Ber. **121** [1988] 809/10).
[26] Brown, H. C.; Naik, R. G.; Singaram, B.; Pyun, C. (Organometallics **4** [1985] 1925/9).
[27] Brown, H. C.; Naik, R. G.; Bakshi, R. K.; Pyun, C.; Singaram, B. (J. Org. Chem. **50** [1985] 5586/92).
[28] Brown, H. C.; Imai, T.; Perumal, P. T.; Singaram, B. (J. Org. Chem. **50** [1985] 4032/6).
[29] Brown, H. C.; Singaram, B.; Cole, T. E. (J. Am. Chem. Soc. **107** [1985] 460/4).
[30] Brown, H. C.; Cole, T. E.; Srebnik, M. (Organometallics **4** [1985] 1788/92).

[31] Brown, H. C.; Srebnik, M.; Cole, T. E. (Organometallics **5** [1986] 2300/3).
[32] Brown, H. C.; Singh, S. M. (Organometallics **5** [1986] 998/1001).
[33] Brown, H. C.; Singh, S. M. (Organometallics **5** [1986] 994/7).
[34] Brown, H. C.; Singh, S. M.; Rangaishevi, M. V. (J. Org. Chem. **51** [1986] 3150/5).
[35] Brown, H. C.; Vara Prasad, J. V. N. (J. Am. Chem. Soc. **108** [1986] 2049/54).
[36] Brown, H. C.; Vara Prasad, J. V. N. (J. Org. Chem. **51** [1986] 4526/30).
[37] Brown, H. C.; Park, W. S.; Cha, J. S.; Cho, B. T.; Brown, C. A. (J. Org. Chem. **51** [1986] 337/42).
[38] Brown, H. C.; Bhat, N. G.; Rajagopalan. S. (Synthesis **1986** 480/2).
[39] Brown, H. C.; Gupta, A. K.; Vara Prasad, J. V. N.; Srebnik, M. (J. Org. Chem. **53** [1988] 1391/4).
[40] Brown, H. C.; Gupta, A. K. (J. Organomet. Chem. **341** [1988] 73/81).

[41] Brown, H. C.; Midland, M. M. (Tetrahedron **43** [1987] 4059/70).
[42] Brown, H. C.; Bhat, N. G.; Srebnik, M. (Tetrahedron Lett. **29** [1988] 2631/4).
[43] Schwetlick, K.; Koenig, T. (Polym. Degrad. Stab. **24** [1989] 279/87 from C.A. **111** [1989] No. 116240).
[44] Koenig, T.; Maennel, D.; Habicher, W. D.; Schwetlick, K. (Polym. Degrad. Stab. **22** [1988] 137/45 from C.A. **110** [1989] No. 96437).
[45] Koenig, T.; Maennel, D.; Schwetlick, K. (Makromol. Chem. Macromol. Symp. **27** [1989] 223/30 from C.A. **111** [1989] No. 175283).

3.5.6.4.2 Cyclic Esters of the Type RB(–OZO–)

3.5.6.4.2.1 General Remarks

Cyclic boronic esters have been used in new chiral syntheses [1, 3] and in asymmetric syntheses [2]. A review with 40 references describes asymmetric syntheses with cyclic boronic acid esters and their applications [4]. Stereospecific cyclic boronic acid esters have been used to prepare (2S,3S)-phenylalanine-3-D [5]. The application of allylboronates in the synthesis of carbohydrates and polyhydroxylated natural products has been reviewed [6]. A review elaborates on general and convenient methods for the stereo- and regioselective synthesis of conjugated alkanes, alkynes, arylated alkenes, and heterobiaryls via Pd(0)-catalyzed cross-coupling reactions of aryl and vinyl boron compounds with organic halides [7].

References for 3.5.6.4.2.1:

[1] Braun, M. (Nachr. Chem. Tech. Lab. **33** [1985] 504/6).
[2] Matteson, D. S.; Sadhu, K. M.; Ray, R.; Peterson, M. L.; Majumdar, D.; Hurst, G. D.; Jesthi, P. K.; Tsai, D. J. S.; Erdik, E. (Pure Appl. Chem. **57** [1985] 1741/8).
[3] Matteson, D. S.; Sadhu, K. M.; Peterson, M. L. (J. Am. Chem. Soc. **108** [1986] 810/9).
[4] Matteson, D. S. (Mol. Struct. Energ. **5** [1988] 343/56 from C.A. **110** [1989] No. 153395).
[5] Matteson, D. S.; Beedle, E. C.; Christenson, E.; Dewey, M. A.; Peterson, M. L. (J. Labelled Compd. Radiopharm. **25** [1988] 675/83 from C.A. **110** [1989] No. 39339).
[6] Roush, W. R. (ACS Symp. Ser. **386** [1989] 242/77 from C.A. **111** [1989] No. 58162).
[7] Miyaura, Norio; Suzuki, Akira (Yuki Gosei Kagaku Kyokaishi **46** [1988] 848/60 from C.A. **111** [1989] No. 153 263).

3.5.6.4.2.2 1,3,2-Dioxaborolanes

A review substantiating allylic strain as a controlling factor in stereospecific transformations involving 1,3,2-dioxaborolanes has been compiled [2], and the preparation and testing of some α-chiral allyl- and crotylboronates has been reviewed [3]. Newer approaches to the syntheses of stereoselective parts of sugars or other natural products utilize allylboronic or crotylboronic acid esters, which add to aldehydes and prochiral ketones and give the homoallyl alcohols [1, 2]; see also [5, 6]. The selectivity for the addition of allylboronic acid esters to α-chiral aldehydes has been discussed [1]; see also [4].

By homologation of **CH_2=CH–B[–OC(CH_3)$_2$C(CH_3)$_2$O–]**, several α-hetero-substituted allylboronates have been prepared [7]. Ketone-derived enol borates having a 4,4,5,5-tetramethyl-1,3,2-dioxaborolane unit are stable compounds [8]. For other such species containing the same dioxaborolane ring system, see [9 to 25].

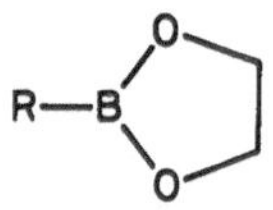

11

2-Phenyl-1,3,2-dioxaborolane (**11**) (R = C_6H_5) and alkylnaphthalene are components of an insulating oil for a capacitor [26]. The species **11** (R = 4-vinylphenyl) has been used to polymerize divinylbenzene to macroporous polymers [27]. Complex formation occurs between **11** (R = C_6H_5 or 2-$CH_3C_6H_4$) and nitrogen bases such as piperidine [29].

Compound **11** (R = 4-hydroxyphenyl) has a melting point of 108 °C and was prepared from 4-hydroxyphenylboronic acid and ethylene glycol in toluene by azeotropic distillation of the formed water and concentration of the solution [28].

The reaction of 2-diethylamino-1,3,2-dioxaborolane with *t*-butyllithium at −78 °C yields **11** with R = t-C_4H_9 [30].

Compound **11** (R = $C_5(CH_3)_5$), boiling point 60 °C/0.5 Torr, melting point 62 to 65 °C, has been prepared from **11**, R = Cl or Br, and (pentamethylcyclopentadienyl)lithium in petroleum ether. ^{1}H NMR data (in C_7D_8): δ^1H = 1.22 (s, 3H), 1.77 (s, 6H), 1.93 (s, 6H), 3.61 ppm (s, 4H). ^{11}B NMR data (in CH_2Cl_2, versus external $(C_2H_5)_2O{-}BF_3$): 34.4 ppm, at higher concentrations δ = 5.3 ppm. ^{13}C NMR data (in $CDCl_3$): 16.0, 45.2, 134.3, and 139.0 ppm [31].

2-(2′-Propenyl)-1,3,2-dioxaborolane (**11**, R = $CH_2CH{=}CH_2$), boiling point 77 to 78 °C/95 Torr [32] (44 to 46 °C/15 Torr [17]), and various other 2-substituted 1,3,2-dioxaborolanes including derivatives of substituted glycols have been prepared and their reaction chemistry has been studied [33 to 48]. For the compound with R = OH , see **22** in Section 3.5.8.6, p. 245.

3.5.6.4.2.3 1,3,2-Benzodioxaboroles

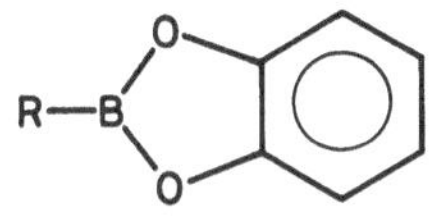

12

$C_6H_5B(OH)_2$ and OsO_4 react with olefins or benzene to yield the corresponding 2-phenyl-1,3,2-dioxaborolane derivatives such as **12** (R = C_6H_5) in good yields [49]. ^{1}H, ^{11}B, and ^{13}C NMR chemical shifts are reported for pyridine and 2-ethylpyridine complexes of **12** (R = C_6H_5) [50].

The preparation and the chemistry of compounds **12** with various R substituents has been published [15, 21, 51 to 62]. For compounds **12** with OR substituents like the "pyrocatechol ester" **12** (R = OC_6H_5), which is prepared by heating H_3BO_3 with 1,2-$(OH)_2C_6H_4$ (pyrocatechol) and phenol, see Section 3.5.8.6, p. 247 [35]. For 1,3,2-benzodioxaborole **12** with R = H, see Section 3.5.6.3.2., p. 167.

References for 3.5.6.4.2.2 and 3.5.6.4.2.3:

[1] Hoffmann, R. W. (Angew. Chem. **99** [1987] 503/17; Angew. Chem. Int. Ed. Engl. **99** [1987] 489/503).
[2] Hoffmann, R. W. (Chem. Rev. **89** [1989] 1841/60).
[3] Hoffmann, R. W. (Chem. Scr. **25** [1985] 53/60).
[4] Hoffmann, R. W.; Weidmann, U. (Chem. Ber. **118** [1985] 3966/79).
[5] Hoffmann, R. W.; Kemper, B.; Metternich, R.; Lehmeier, T. (Liebigs Ann. Chem. **1985** 2246/60).
[6] Hoffmann, R. W.; Metternich, R. (Liebigs Ann. Chem. **1985** 2390/402).
[7] Hoffmann, R. W.; Landmann, B. (Chem. Ber. **119** [1986] 1039/53).
[8] Hoffmann, R. W.; Ditrich, K.; Fröch, S. (Liebigs Ann. Chem. **1987** 977/85).
[9] Andersen, M.; Hildebrandt, B.; Köster, G.; Hoffmann, R. W. (Chem. Ber. **122** [1989] 1777/82).

[10] Hoffmann, R. W.; Dresely, S. (Angew. Chem. **98** [1986] 186/7; Angew. Chem. Int. Ed. Engl. **98** [1986] 189/90).

[11] Hoffmann, R. W.; Dresely, S. (Synthesis **1988** 103/6).
[12] Hoffmann, R. W.; Dresely, S. (Tetrahedron Lett. **28** [1987] 5303/6).
[13] Hoffmann, R. W.; Dresely, S.; Lanz, J. W. (Chem. Ber. **121** [1988] 1501/7).
[14] Hoffmann, R. W.; Dresely, S. (Chem. Ber. **122** [1989] 903/9).
[15] Vaulthier, M.; Truchet, F.; Carboni, B.; Hoffmann, R. W.; Denne, I. (Tetrahedron Lett. **28** [1987] 4169/72).
[16] Roush, W. R.; Adam, M. A.; Harris, D. J. (J. Org. Chem. **50** [1985] 2000/3).
[17] Roush, W. R.; Adam, M. A.; Walts, A. E.; Harris, D. J. (J. Am. Chem. Soc. **108** [1986] 3422/34).
[18] Roush, W. R.; Walts, A. E. (Tetrahedron Lett. **26** [1985] 3427/30).
[19] Wuts, P. G. M.; Bigelow, S. S. (J. Org. Chem. **53** [1988] 5023/34).
[20] Wuts, P. G. M.; Jung, Yong Woon (Tetrahedron Lett. **27** [1986] 2079/82).

[21] Yamamoto, Y.; Seko, T.; Rong, Feng Guang; Nemoto, H. (Tetrahedron Lett. **30** [1989] 7191/4).
[22] Matteson, D. S.; Wilson, J. W. (Organometallics **4** [1985] 1690/2).
[23] Sadhu, K. M.; Matteson, D. S. (Organometallics **4** [1985] 1687/9).
[24] Shenvi, A. B. (U.S. 4537773 [1983/85] 1/14 from C.A. **104** [1986] No. 19668).
[25] Phillion, D. P.; Neubauer, R.; Andrew, S. S. (J. Org. Chem. **51** [1986] 1610/2).
[26] Matsushita Electric Industrial Co., Ltd. (Jpn. Tokkyo Koho 85-22454 [1976/85] 1/5 from C.A. **103** [1985] No. 170776).
[27] Sarhan, A.; Ali, M. M.; Abd El-Al, M. Y. (J. Prakt. Chem. **330** [1988] 71/8 from C.A. **110** [1989] No. 23927).
[28] Wulff, G.; Poll, H. G. (Makromol. Chem. **188** [1987] 741/8).
[29] Lauer, M.; Wulff, G. (J. Chem. Soc. Perkin Trans. II **1987** 745/9).
[30] Höbel, U.; Nöth, H.; Prigge, H. (Chem. Ber. **119** [1986] 325/37).

[31] Jutzi, P.; Krato, B.; Hursthouse, M.; Howes, A. J. (Chem. Ber. **120** [1987] 565/74).
[32] Hoffmann, R. W.; Endesfelder, A. (Liebigs Ann. Chem. **1986** 1823/36).
[33] Hoffmann, R. W.; Endesfelder, A. (Liebigs Ann. Chem. **1987** 215/9).
[34] Hoffmann, R. W.; Ladner, W.; Ditrich, K. (Liebigs Ann. Chem. **1989** 883/9).
[35] Grachek, V. I.; Naumova, S. F.; Motol'ko, G. R.; Smolyakov, A. V.; Bukanova, N. N.; Kozlov, N. S. (Dokl. Akad. Nauk BSSR **29** [1985] 536/9 from C.A. **104** [1986] No. 148952).
[36] Nilsson, K.; Hallberg, A. (Acta Chem. Scand. B **41** [1987] 569/76).
[37] Hoffmann, R. W.; Landmann, B. (Chem. Ber. **119** [1986] 2013/24).
[38] Matteson, D. S.; Tripathy, P. B.; Sarkar, A.; Sadhu, K. M. (J. Am. Chem. Soc. **111** [1989] 4399/402).
[39] Matteson, D. S.; Sadhu, K. M.; Ray, R.; Peterson, M. L.; Majumdar, D.; Hurst, G. D.; Jesthi, P. K.; Tsai, D. J. S.; Erdik, E. (Pure Appl. Chem. **57** [1985] 1741/8).
[40] Ditrich, K.; Bube, T.; Stürmer, R.; Hoffmann, R. W. (Angew. Chem. **98** [1986] 1016/8; Angew. Chem. Int. Ed. Engl. **98** [1986] 1028/30).

[41] Hoffmann, R. W.; Ditrich, K.; Köster, G.; Stürmer, R. (Chem. Ber. **122** [1989] 1783/9).
[42] Roush, W. R.; Ando, K.; Powres, D. B.; Halterman, R. L.; Palkowitz, A. D. (Tetrahedron Lett. **29** [1988] 5579/82).
[43] Roush, W. R.; Walts, A. E.; Hoong, L. K. (J. Am. Chem. Soc. **107** [1985] 8186/90).
[44] Roush, W. R.; Halterman, R. L. (J. Am. Chem. Soc. **108** [1986] 294/6).

[45] Roush, W. R.; Balkowitz, A. D.; Palmer, H. J. (J. Org. Chem. **52** [1987] 316/8).
[46] Lauer, M.; Wulff, G. (Chem. Ber. **118** [1985] 246/60).
[47] Wulff, G.; Hansen, A. (Carbohydr. Res. **164** [1987] 123/40).
[48] Wulff, G.; Hansen, A. (Angew. Chem. **98** [1986] 552/3).
[49] Iwasawa, N.; Kato, T.; Narasaka, K. (Chem. Lett. **1988** 1721/4).
[50] Farfán, N.; Contreras, R. (J. Chem. Soc. Perkin Trans. II **1987** 771/3).

[51] Baban, J. A.; Goodchild, N. J.; Roberts, B. P. (J. Chem. Soc. Perkin Trans. II **1986** 157/61).
[52] Miyaura, N.; Yamada, K.; Suginome, H.; Suzuki, A. (J. Am. Chem. **107** [1985] 972/80).
[53] Satoh, M.; Ishiyama, T.; Miyaura, N.; Suzuki, A. (Bull. Chem. Soc. Jpn. **60** [1987] 3471/3).
[54] Ohe, T.; Ohe, K.; Uemara, S.; Sugita, N. (J. Organomet. Chem. **344** [1988] C5-C7).
[55] Suzuki, A. (Pure Appl. Chem. **57** [1985] 1749/58).
[56] Yanagi, T.; Oh-e, T.; Miyaura, N.; Suzuki, A. (Bull. Chem. Soc. Jpn. **62** [1989] 3892/5).
[57] Satoh, M.; Miyaura, N.; Suzuki, A. (Synthesis **1987** 373/7).
[58] Miyaura, N.; Satoh, M.; Suzuki, A. (Tetrahedron Lett. **27** [1986] 3745/8).
[59] Suzuki, A.; Dhillon, R. S. (Top. Curr. Chem. **130** [1986] 23/88; 62/3; 70).
[60] Miyaura, N.; Ishiyama, T.; Sasaki, H.; Ishikawa, M.; Satoh, M.; Suzuki, A. (J. Am. Chem. Soc. **111** [1989] 314/9).

[61] Miyaura, N.; Suzuki, A. (Chem. Lett. **1986** 1329/39).
[62] Oki, M.; Yamada, Y. (Bull. Chem. Soc. Jpn. **61** [1988] 1191/4).

3.5.6.4.2.4 1,3,2-Dioxaborinanes

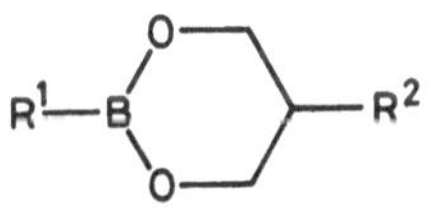

13

Compound **13** ($R^1 = CH_3$, $R^2 = H$) was prepared by methylation of 2-chloro-1,3,2-dioxaborinane with CH_3MgBr in 31% yield [1]. Compound **13** ($R^1 = C_2H_5$, $R^2 = H$) and ring substituted species thereof were obtained by treatment of 1,3-dioxanes with C_2H_5–BCl_2 at room temperature [2].

2,5-Diaryl-1,3,2-dioxaborinanes, a new series of liquid crystals, were synthesized and found to form mesomorphic phases over a wide temperature range [19]. The following compounds are of special interest:

13 ($R^1 = 4\text{-}CH_3C_6H_4$, $R^2 = 4\text{-}n\text{-}C_4H_9\text{–}OC_6H_4$), transition temperature 126.4 °C, was prepared by stirring a mixture of $4\text{-}CH_3C_6H_4\text{–}B(OH)_2$ and $4\text{-}n\text{-}C_4H_9\text{–O–}C_6H_4\text{–}CH(CH_2OH)_2$ in toluene under reflux for 2 h. The water formed during the reaction was removed azeotropically using a Dean-Stark apparatus. When the reaction was completed, toluene was distilled off, and the resulting residue was purified by chromatography on silica gel using CH_2Cl_2 as an eluent and recrystallized from *n*-hexane. IR (Nujol): 1315 cm^{-1} (BO); ^1H NMR [19].

13 ($R^1 = 4\text{-}CH_3O\text{–}C_6H_4$, $R^2 = 4\text{-}n\text{-}C_4H_9\text{–}OC_6H_4$), transition temperature 133.9 °C; IR: 1315 cm^{-1} (BO); ^1H NMR; mass spectrum: m/z = 340 ($[M]^+$) [19].

References on pp. 177/9

13 (R^1=4-CH_3O–C_6H_4, R^2=*n*-C_6H_{13}), transition temperature 103.0°C; IR: 1320 cm^{-1} (BO) and 1240 cm^{-1} (ArO); ^{1}H NMR; mass spectrum: m/z=352 ($[M]^+$) [19].

13 (R^1=4-$CH_3OC(O)$–C_6H_4, R^2=4-*n*-C_4H_9–OC_6H_4), transition temperature 156.2°C, was prepared by reaction of 2-(4-carboxyphenyl)-5-(4-*n*-butoxyphenyl)-1,3,2-dioxaborinane and excess methanol in the presence of H_2SO_4 under reflux for 6 h. After cooling of the reaction mixture, excess methanol was removed, and the residue was purified by chromatography on silica gel using CH_2Cl_2 as an eluent; recrystallization from hexane gave a 67% product yield. IR data (Nujol; in cm^{-1}): 1735 (C=O), 1320 (BO), and 1280 (COC); ^{1}H NMR; mass spectrum: m/z= 368 ($[M]^+$) [19].

13 (R^1=4-$CH_3OC(O)$–C_6H_4, R^2=*n*-C_6H_{13}), transition temperature 140.5°C, was obtained by the same method as outlined above; IR data (Nujol; in cm^{-1}): 1735, 1320, and 1275; ^{1}H NMR [19].

Various additional 1,3,2-dioxaborinanes have been reported. They include the species **13** with R^1=*i*-C_3H_7 or *t*-C_4H_9 [3] and numerous others [4 to 37]. For *B*-alkoxy substituted dioxaborinanes (=cyclic esters of boric acid), see **23** in Section 3.5.8.6, p. 245).

3.5.6.4.3 Cyclic Esters of the Type RO–B(–ZO–)

The following 2-alkoxy-1,2-oxaborolanes and 1,3-dihydro-1-hydroxy-2,1-benzoxaborole can be viewed as cyclic esters of boronic acid.

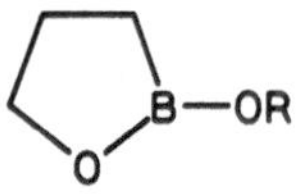

14

Compound **14** (R=CH_2=$CHCH_2$) has been prepared in a thermal exchange reaction of 2-allyloxy-1,2-oxaborolane, which is easily obtained from allyl alcohol via hydroboration [38]. 3-Alkyl-2-alkoxy-1,2-oxaborinanes have been prepared by the reaction of 2-alkoxy-1,2-oxoborolanes with $LiCHCl_2$ and subsequent treatment with RLi or RMgCl. On reaction with KOH/H_2O_2, they are converted to 1,4-alkanediols [39].

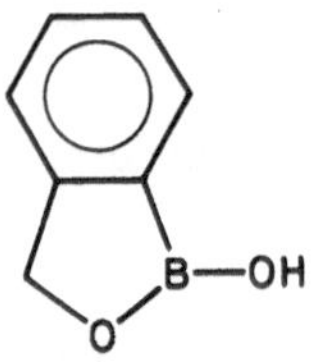

15

1,3-Dihydro-1-hydroxy-2,1-benzoxaborole (**15**), melting point 90 to 97°C, has been prepared by treatment of 1,2-dihydro-1-hydroxy-2-(4-methylphenylsulfonyl)-2,3,1-benzodiazaborine with aqueous 2N NaOH solution at 100°C [41].

3.5.6.4.4 Cyclic Esters Containing Four-Coordinate Boron, $R_2B(-OZO-)$ and $(-Z'-)B(-OZO-)$

The molecular structures of the following boron acetylacetonates (**16**) were determined [40].

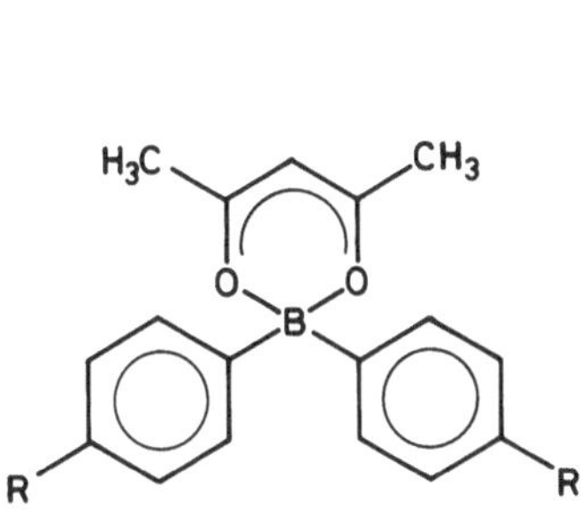

16a

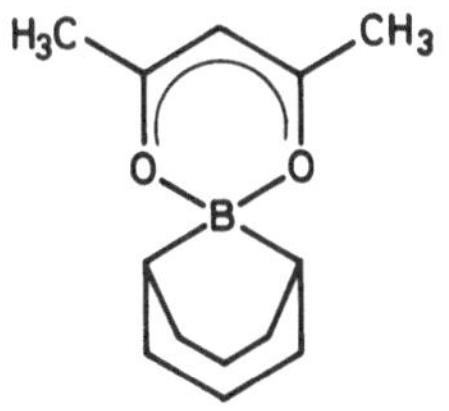

16b

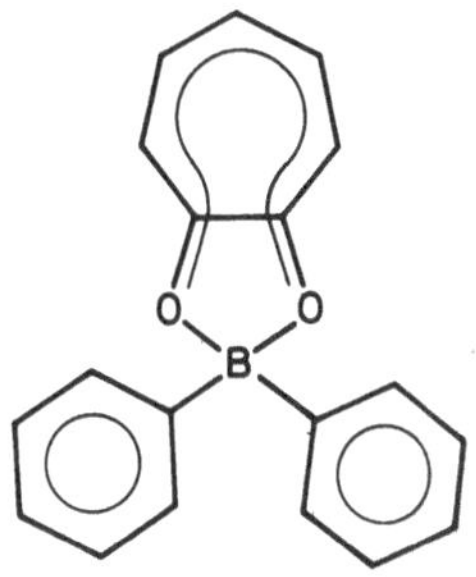

16c

The compound **16a**, R = H, was prepared by reacting (2-aminoethanolato)diphenylboron with acetylacetone in ethanol/acetone. (Acetylacetonato)diphenylboron is triclinic, space group $P\bar{1}-C_i^1$ (No. 2) with a = 8.3557(8), b = 9.3519(8), and c = 9.6103(9) Å; α = 96.094(5)°, β = 94.904(6)°, and γ = 93.489(6)°; Z = 2; D_c = 1.182 g/cm³ [43].

The compound **16a**, R = CH_3 [40], is monoclinic, space group $P2_1/c-C_{2h}^5$ (No. 14) with a = 9.515(2), b = 13.164(5), and c = 14.281(4) Å; β = 102.16(2)°; D_c = 1.11 g/cm³ [42].

The compound **16b** [40], containing a bicyclic boron moiety, is triclinic, space group $P\bar{1}-C_i^1$ (No. 2) with a = 7.327(2), b = 8.833(2), and c = 10.524(3) Å; α = 93.09(2)°, β = 108.61(2)°, and γ = 97.89(2)°; Z = 2; D_c = 1.14 g/cm³ [42].

(Tropolonato)diphenylboron (**16c**) was prepared by reacting (2-aminoethanolato)diphenylboron with tropolone in ethanol/acetone. The compound is monoclinic, space group $P2_1/c-C_{2h}^5$ (No. 14) with a = 12.1596(5), b = 10.2614(4), and c = 12.4883(6) Å; β = 100.965(2)°; Z = 4; D_c = 1.242 g/cm³ [43].

References for 3.5.6.4.2.4 to 3.5.6.4.4:

[1] Brown, H. C.; Cole, T. E. (Organometallics **4** [1985] 816/21).
[2] Kuznetsov, V. V.; Gren, A. I. (Zh. Org. Khim. **22** [1986] 2237; J. Org. Chem. [USSR] **22** [1986] 2009/10; Zh. Obshch. Khim. **55** [1985] 1651/2; J. Gen. Chem. [USSR] **55** [1985] 1469/70 from C.A. **104** [1986] No. 129954).
[3] Hoebel, U.; Nöth, H.; Prigge, H. (Chem. Ber. **119** [1986] 325/37).
[4] Brown, H. C.; Naik, R. G.; Bakshi, R. K.; Pyun, C.; Singaram, B. (J. Org. Chem. **50** [1985] 5586/92).
[5] Brown, H. C. (Eur. Appl. 188235 [1985/86] 1/41 from C.A. **106** [1987] No. 214124).
[6] Brown, H. C.; Park, W. S.; Cha, J. S.; Cho, B. T.; Brown, C. A. (J. Org. Chem. **51** [1986] 337/42).
[7] Brown, H. C.; Cho, B. T.; Park, W. S. (J. Org. Chem. **52** [1987] 4020/4).
[8] Brown, H. C.; Singh, S. M.; Rangaishenvi, M. V. (J. Org. Chem. **51** [1986] 3150/5).
[9] Lauer, M.; Wulff, G. (J. Chem. Soc. Perkin Trans. II **1987** 745/9).
[10] Brown, H. C.; Bakshi, R. K.; Singaram, B. (J. Am. Chem. Soc. **110** [1988] 1529/34).

[11] Srebnik, M.; Bhat, N. G.; Brown, H. C. (Tetrahedron Lett. **29** [1988] 2635/6).
[12] Brown, H. C.; Imai, T.; Desai, M. C.; Singaram, B. (J. Am. Chem. Soc. **107** [1986] 4980/3).
[13] Brown, H. C.; Kim, K. W.; Cole, T. E.; Singaram, B. (J. Am. Chem. Soc. **108** [1986] 6761/4).
[14] Roush, W. R.; Walts, A. E.; Hoong, L. K. (J. Am. Chem. Soc. **107** [1985] 8186/90).
[15] Brown, H. C.; Imai, T.; Bhat, N. G. (J. Org. Chem. **51** [1986] 5277/82).
[16] Brown, H. C.; Bhat, N. G. (Tetrahedron Lett. **29** [1988] 21/4).
[17] Yamamoto, Y.; Seko, T.; Nemoto, H. (J. Org. Chem. **54** [1989] 4734/6).
[18] Yamamoto, Y.; Seko, T.; Rong, Feng Guang; Nemoto, H. (Tetrahedron Lett. **30** [1989] 7191/4).
[19] Matsubara, H.; Seto, K.; Tahara, T.; Takahashi, S. (Bull. Chem. Soc. Jpn. **62** [1989] 3896/901).
[20] Takahashi, S. (Mem. Inst. Sci. Ind. Res. Osaka Univ. **42** [1985] 1/11 from C.A. **103** [1985] No. 46329).

[21] Seto, K.; Takahashi, S.; Tahara, T. (J. Chem. Soc. Chem. Commun. **1985** 122/3).
[22] Kuznetsov, V. V.; Gren, A. I. (Khim. Geterotsikl. Soedin **1985** 710 from C.A. **104** [1986] No. 68908).
[23] Kuznetsov, V. V.; Gren, A. I. (Ukr. Khim. Zh. [Russ. Ed.] **51** [1985] 535 from C.A. **104** [1986] No. 148953).
[24] Kuznetsov, V. V.; Gren, A. I. (Zh. Org. Khim. **21** [1985] 2016/7; J. Org. Chem. [USSR] **21** [1985] 1846/7).
[25] Gren, A. I.; Kuznetsov, V. V.; Zakharov, K. S. (Khim. Geterotsikl. Soedin. **1986** No. 4, pp. 558/61 from C.A. **105** [1986] No. 97533).
[26] Kuznetsov, V. V.; Gren, A. I. (Zh. Obshch. Khim. **56** [1986] 613/6; J. Gen. Chem. [USSR] **56** [1986] 542/5 from C.A. **106** [1987] No. 18649).
[27] Kuznetsov, V. V. (Deposited Doc. VINITI-5646-83 [1983] 1/13 from C.A. **102** [1985] No. 113556).
[28] Kuznetsov, V. V.; Gren, A. I.; Timofeev, O. S. (Deposited Doc. VINITI-5648-83 [1983] 1/14 from C.A. **102** [1985] No. 113555).
[29] Kuznetsov, V. V.; Gren, A. I.; Staikov, A. I.; Alekseenko, L. I.; Mazepa, A. V. (Deposited Doc. VINITI-5649-83 [1983] 1/21 from C.A. **102** [1985] No. 113554).
[30] Kuznetsov, V. V.; Gren, A. I.; Selyanin, V. N.; Trigub, L. P. (Deposited Doc. VINITI-5647-83 [1983] 1/12 from C.A. **102** [1985] No. 113553).

[31] Kuznetsov, V. V.; Gren, A. I. (Ukr. Khim. Zh. [Russ. Ed.] **52** [1986] 882/4 from C.A. **107** [1987] No. 176080).
[32] Kuznetsov, V. V.; Gren, A. I. (Zh. Org. Khim. **22** [1986] 2237; J. Org. Chem. [USSR] **22** [1986] 2009/10).
[33] Kuznetsov, V. V.; Gren, A. I.; Alekseenko, L. I.; Novikova, E. D. (Ukr. Khim. Zh. [Russ. Ed.] **53** [1987] 535/7 from C.A. **108** [1988] No. 75455).
[34] Kuznetsov, V. V.; Alekseenko, L. I.; Staikov, A. I.; Gren, A. I. (Ukr. Khim. Zh. [Russ. Ed.] **54** [1988] 1315/9 from C.A. **110** [1989] No. 207795).
[35] Wächtler, A.; Krause, J.; Eidenschink, R.; Eichler, J.; Scheuble, B. (Ger. Offen. 3608714 [1985/86] 1/64 from C.A. **106** [1987] No. 93757).
[36] Hoffmann, R. W.; Ditrich, K.; Fröch, S. (Liebigs Ann. Chem. **1987** 977/85).
[37] Kölle, P.; Nöth, H. (Chem. Rev. **85** [1985] 399/418).
[38] Zhou, Weike; Zhang, Gaoyi; Ding, Hongxun (Youji Huaxue **1982** No. 1, pp. 19/25 from C.A. **96** [1982] No. 181336).
[39] Ding, Hongxun; Zhou, Weike; Bai, Junchai (Tetrahedron Lett. **28** [1987] 2599/602).
[40] Köster, R.; Rotermund, G. W. (Liebigs Ann. Chem. **689** [1965] 40/64).

[41] Grassberger, M. A. (Liebigs Ann. Chem. **1985** 683/8).
[42] Boese, R.; Köster, R.; Yalpani, M. (Chem. Ber. **118** [1985] 670/5).
[43] Rettig, S. J.; Trotter, J. (Can. J. Chem. **60** [1982] 2957/64).

3.5.7 Orthoboric Acid, H_3BO_3

For a previous review, see "Boron Compounds" 3rd Suppl. Vol. 2, 1987, pp. 96/132.

3.5.7.1 Preparation and Recovery

The technology and expediency of using different methods for preparing raw materials in the production of $\mathbf{H_3BO_3}$ from eluvial ores [1] as well as the improvement by utilizing mother liquors in the processing of eluvial ores were described [2]. Based on the optimization of the filtration of slurries from H_2SO_4-decomposition of an eluvial borate ore and the optimization of centrifugation, the yield of H_3BO_3 from the ores was increased by 8.2% [3]. The effect of $Mg[SO_4]$ and silicic acid on the recovery of H_3BO_3 from borate ores by their decomposition with H_2SO_4 has been studied. The filtering rate of the pulp and the impurity content (SiO_2 and other insoluble residue) increased with increasing silicic acid content. The recirculating leach liquors were treated with MgO or $Ca[CO_3]$ in order to decrease the silicic acid and metal oxide contents in the pulp [4]. The minimal solubility of silicic acid in an H_3BO_3–SiO_2–H_2O system occurs at 60°C at B_2O_3=3.99 and SiO_2=0.0209 mol/L, at 90°C at B_2O_3=0.39 and SiO_2= 0.0326 mol/L [5]. In suspensions from H_2SO_4 treatment of eluvial borate ores, silicic acid ($\leq$0.13% SiO_2) affects the crystallization and filterability of H_3BO_3, for it is not only adsorbed by the H_3BO_3 crystals, hampering their growth and leading to lattice distortions, but also forms solid solutions with H_3BO_3 [6]. The solubility of hydrated and dehydrated SiO_2 in the H_3BO_3–SiO_2–H_2O system was studied at 30°C. H_3BO_3 has a salting-out effect on the dissolved SiO_2. The solubility of SiO_2 in H_3BO_3 solutions decreases with the decreasing degree of hydration of the acid [7].

The effect of the mineralogical composition of borate ores, the preliminary heat-treatment temperature, and the degree of decomposition using the minimal amount of H_2SO_4 for the highest boron recovery in the manufacture of H_3BO_3 has been studied [8]. The effect of $Fe[SO_4]$, $Fe_2[SO_4]_3$, $Al_2[SO_4]_3$, and SiO_2 on the crystallization of H_3BO_3 and $Mg[SO_4]\cdot 7H_2O$ at 25°C and their subsequent separation by flotation depends on the particle-size distribution and their concentration ratios [9]. The metal sulfates and H_2SiO_3 act as depressants on $Mg[SO_4]\cdot 7H_2O$, thus increasing selectivity and effectiveness of the flotation separation of H_3BO_3 from $Mg[SO_4]\cdot 7H_2O$ [10].

The recovery of boron from stratal water was optimized by sorption with a special ion exchanger (CS 3) between 50 and 60°C, elution, and conversion into H_3BO_3 and its salts; the loss of boron was reduced to 8% [11, 12].

References for 3.5.7.1:

[1] Abakumov, V. I.; Tkachev, K. V.; Savinykh, Yu. G.; Nedopekina, V. A. (Tr. Ural. Nauchno Issled. Khim. Inst. No. 57 [1984] 19/22 from C.A. **105** [1986] No. 63099).
[2] Abakumov, V. I.; Savinykh, Yu. G.; Shubin, A. S.; Nedopekina, V. A. (Tr. Ural. Nauchno Issled. Khim. Inst. No. 57 [1984] 26/9 from C.A. **105** [1986] No. 63098).

[3] Abakumov, V. I.; Kardashina, L. F. (Khim. Promst. [Moscow] **1987** No. 9, pp. 534/6 from C.A. **107** [1987] No. 179352).
[4] Abakumov, V. I.; Shubin, A. S.; Kardashina, L. F. (Zh. Prikl. Khim. **60** [1987] 605/7, 1164/6; J. Appl. Chem. [USSR] **60** [1987] 568/70, 1097/9).
[5] Abakumov, V. I.; Diarov, M. D.; Korznikov, V. A.; Kalacheva, V. G. (Izv. Akad. Nauk Kaz. SSR Ser. Khim. **1987** No. 6, pp. 15/8 from C.A. **108** [1988] No. 82985).
[6] Abakumov, V. I.; Diarov, M. D.; Kardashina, L. F.; Kalacheva, V. G. (Izv. Akad. Nauk Kaz. SSR Ser. Khim. **1988** No. 4, pp. 3/6 from C.A. **109** [1988] No. 152376).
[7] Abakumov, V. I.; Shubin, A. S.; Kardashina, L. F.; Korznikov, V. A.; Vasina, Z. I. (Zh. Prikl. Khim. **61** [1988] 2111/3; J. Appl. Chem. [USSR] **61** [1988] 1909/11 from C.A. **109** [1988] No. 238232).
[8] Abakumov, V. I.; Bessonov, N. N.; Kardashina, L. F.; Shubin, A. S. (Zh. Prikl. Khim. **61** [1988] 640/3; J. Appl. Chem. [USSR] **61** [1988] 576/9 from C.A. **108** [1988] No. 189295).
[9] Abakumov, V. I.; Kardashina, L. F.; Nedopekina, V. A. (Obogashch. Rud [Leningrad] **1989** No. 5, pp. 12/5 and 16/9 from C.A. **113** [1990] No. 8916 and No. 62218).
[10] Abakumov, V. I.; Kardashina, L. F.; Nedopekina, V. A.; Kotikova, L. I. (Zh. Prikl. Khim. **61** [1988] 1896/8; J. Appl. Chem. [USSR] **61** [1988] 1711/3 from C.A. **111** [1989] No. 60425).

[11] Kuzman-Anton, R.; Parlea, M.; Parlea, G. (Rev. Chim. [Bucharest] **39** [1988] 284/5 from C.A. **109** [1988] No. 95528).
[12] Parlea, M.; Kuzman-Anton, R.; Parlea, G. (Rev. Chim. [Bucharest] **39** [1988] 370 from C.A. **109** [1988] No. 95527).

3.5.7.2 Properties

The nature of the unoccupied molecular orbitals associated with the B–O bond in gas-phase molecules and solids was determined by performing molecular orbital calculations using the multiple scattering X_α molecular orbital method, and by comparison of the calculated properties with those obtained from X-ray absorption near-edge spectroscopy (XANES) and electron-transmission spectroscopy (ETS). For three-coordinate boron, the lowest-energy unoccupied molecular orbital is essentially a B $2p_\pi$ nonbonding or weakly antibonding orbital of A_2 symmetry within the D_{3h} point group. For H_3BO_3, this orbital is less stable by about 3 eV than its analog in the isoelectronic gas-phase molecule BF_3, and its energy changes only slowly with the B–O distance. It generates an absorption of ca. 4 eV below the B 1s ionization potential in XANES and a scattering resonance of ca. 4 eV above the threshold in ETS. It is the final state for the lowest-energy UV absorptions of the $[BO_3]^{3-}$ group, and excitations to it from the e′ B–O bonding molecular orbital dominate the paramagnetic contribution to the ^{11}B NMR chemical shift. The only other unoccupied orbital observed for three-coordinate boron is the e′ B–O antibonding orbital that appears above the threshold both in XANES and ETS. The energy of this e′ antibonding orbital is highly distance dependent, and thus may provide useful information on B–O distances in amorphous materials [1].

Ab initio Hartree-Fock calculations by SCF/DZ, CI/DZ, SCF/DZP, or CI/DZP methods of the electrical field gradient tensor are reported for H_3BO_3 and $(H_3BO_3)_2$ (CI = configuration interaction, DZ = double-zeta level, DZP = double-zeta plus polarization level with an extended basis set) in order to investigate polarization and correlation effects. Calculations at the double-zeta level show the best agreement with the experimental data for the nuclear quadrupole coupling constant (calculated = 2.55 Hz, experimental = 2.57 Hz). The following parameters are found for H_3BO_3 with symmetry C_{3h}: r(BO)=2.604 a.u., r(OH)=1.795 a.u., ∢(BOH)=111.36°; for $(H_3BO_3)_2$ with the symmetry C_{2h}, the following values were obtained: r(O··H)=3.345 a.u., ∢(O–H··O)=180.0° [2].

The charge distribution in H_3BO_3 was investigated experimentally using low-temperature X-ray diffraction data as well as theoretically by using ab initio calculations. H_3BO_3 is triclinic, space group $P\bar{1}-C_i^1$ (No. 2) with a=701.87(14), b=703.50(20), and c=634.72(12) pm; α=92.49(12)°, β=101.46(2)°, and γ=119.76(2)°; Z=4; D_c=1.562(1) g/cm³ at 105 K; the final R value was 3.8% for 3370 reflections. The two independent boron atoms are planar, each having threefold symmetry. The structural parameters for boron and oxygen, determined from high-order refinement, are virtually identical to those obtained from a multipole refinement using all these data. The static deformation density in the B–O bonds (as determined from a multipole refinement) agrees with the theoretical deformation density obtained from an ab initio calculation at the Hartree-Fock level, which had peaks with a density of 0.3 e/Å³. The features about the oxygen atoms, especially in the lone-pair regions, agree less well [3].

The geometry of H_3BO_3 (assuming C_{3h} symmetry) has been optimized by 6-31G* calculation with r(BO)=135.8 pm, r(OH)=94.7 pm, ∢(OBO)=120°, and E_T=−251.1817 a.u.. The calculated charge density map for an H_3BO_3 monomer is given in **Fig. 3-15**. The bonding density in a B–O bond is 0.5 e/Å³ and in an O–H bond 0.6 e/Å³. In addition, a larger peak of lone pair density (1 e/Å³) is observed on the back side of the B–O–H angle [4].

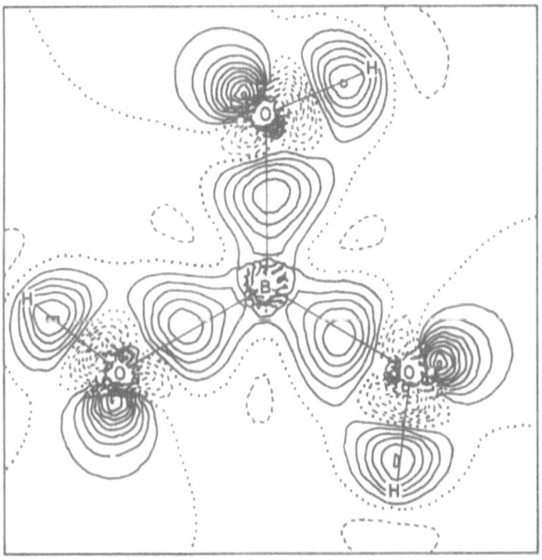

Fig. 3-15. Calculated charge density map for an H_3BO_3 monomer using a 6-31G* basis set. The contour interval is 0.1 e/Å³ with negative contours dashed and the zero level dotted [4].

The heat of formation of H_3BO_3 has been calculated by AM1 approximation to be ΔH_f= −251.6 kcal/mol (observed ΔH_f=−237.4 kcal/mol) for C_{3h} geometry with r(BO)=134.8 pm; the calculated ionization potential was obtained as 12.78 eV [5].

A review on the preparation and properties of the low and high molecular weight forms of H_3BO_3 has been compiled. The preparation was discussed in terms of linear polymers from H_3BO_3 having a functionality of 2, and copolymers with B–O–E bonds (E=Si, P, Al, etc.). The

References on p. 188

system B_2O_3–H_2O shows H_3BO_3 with C_{3h} symmetry and r(BO)=136 pm, r(OH)=96 pm, ∢(OBO)=120°, and ∢(BOH)=105±10° [6]. Thermodynamic calculations, analysis of literature data, and presently obtained data for the B_2O_3–H_2O system indicate that the molar composition forming boron hydroxides depends on the temperature and the water vapor partial pressure. Calculations for the system containing 7.5 wt% B_2O_3 and 92.5 wt% H_2O indicate that H_3BO_3 vapors exist between 1200 and 1350 K [7].

When crystalline H_3BO_3 is gently heated in Pyrex ampules in a vacuum of less than 10^{-6} mbar to 35 to 45°C, it vaporizes to yield molecular H_3BO_3. The IR spectrum of this species (isolated at ca. 12 K in N_2 or Ar matrices) shows characteristic absorptions which could be assigned and which are consistent with the C_{3h} symmetry of molecular H_3BO_3 (**Fig. 3-16**). The assignments are supported by extensive isotope labeling with D, ^{10}B, and ^{18}O, by 20 eV mass spectrometric studies at 40°C (**Fig. 3-17**), as well as by a partial normal coordinate analysis (Table 3/14) [8].

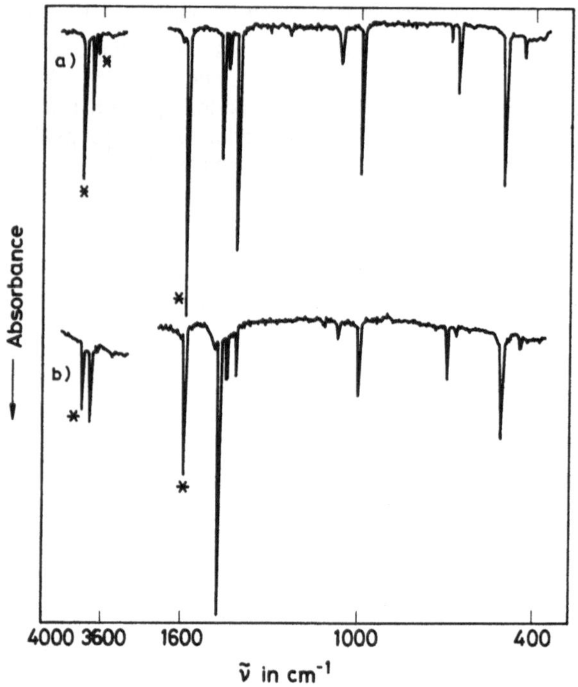

Fig. 3-16. Low-resolution IR spectra obtained from boric acid isolated in nitrogen matrices at ca. 12 K.

a) Spectrum obtained from normal boric acid; and

b) spectrum obtained from sample enriched with 90% ^{10}B.

The bands denoted with an asterisk are due to monomeric H_2O [8].

Fig. 3-17. Mass spectra (20 eV) obtained from samples of boric acid at 40°C.

a) Sample with normal isotopic composition;

b) sample enriched with ^{10}B and ^{18}O; and

c) calculated spectra for $[B(OH)_2]^+$ and $[B(OH)_3]^+$ assuming 90% ^{10}B and 60% ^{18}O with 100% ^1H [8].

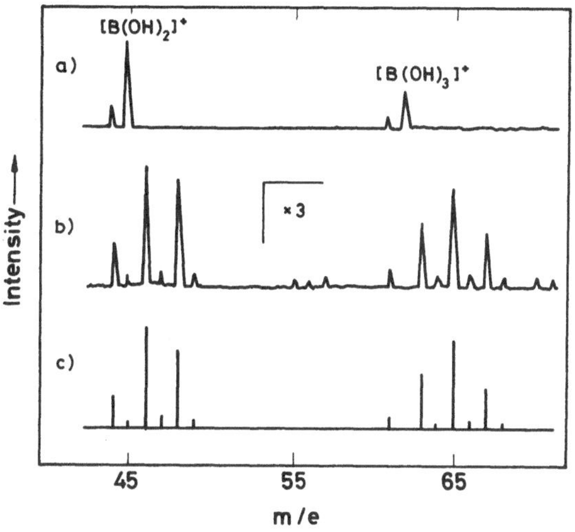

Table 3/14

Observed and Calculated Frequencies (in cm^{-1}) for Selected Fundamentals of Isotopically Labeled Boric Acid [8].

out-of-plane BO_3 bend mode		
observed	calculated[a)]	assignment
675.0	674.4	$^{11}B^{16}O_3$
672.6	672.1	$^{11}B^{16}O_2{}^{18}O$
670.2	669.7	$^{11}B^{16}O^{18}O_2$
668.1	667.4	$^{11}B^{18}O_3$
700.9	701.3	$^{10}B^{16}O_3$
698.6	699.0	$^{10}B^{16}O_2{}^{18}O$
696.4	696.8	$^{10}B^{16}O^{18}O_2$
694.1	694.5	$^{10}B^{18}O_3$

torsion modes			
observed	calculated[b)]	assignment	
–	577.7	H_3BO_3	(E″)
513.8	513.8	H_3BO_3	(A″)
408.4	408.3		
541.8	541.7	H_2DBO_3	(A″)
–	577.7		
561.2	561.4		
390.5	390.3	HD_2BO_3	(A″)
–	434.3		
376.0	375.7	D_3BO_3	(A″)
–	434.3	D_3BO_3	(E″)

References on p. 188

Table 3/14 (continued)

observed	OH stretching modes calculated[c)]	calculated[d)]	assignment	
—	3672.6	3672.7	$H_3B^{16}O_3$	(A′)
3668.5	3668.5	3668.5	$H_3B^{16}O_3$	(E′)
	3668.5	3668.5		
3658.2	3657.6	3658.1	$H_3B^{16}O_2{}^{18}O$	(A′)
3671.8	3671.5	3671.5		
3670.2	3670.2	3670.2		
3659.2	3658.9	3659.4	$H_3B^{16}O^{18}O_2$	(A′)
	3656.5	3657.1		
3657.6	3656.5	3657.1	$H_3B^{18}O_3$	(E′)
—	3660.6	3661.2	$H_3B^{18}O_3$	(A′)
—	3672.6	3672.6	H_3BO_3	(A′)
3668.5	3668.5	3668.5	H_3BO_3	(E′)
	3668.5	3668.5		
3671.1	3671.2	3671.2	H_2DBO_3	(A′)
2705.8	2670.2	2705.9		
3669.7	3669.9	3669.9		
2707.7	2671.2	2706.9	HD_2BO_3	(A′)
	2669.2	2704.9		
2704.6	2669.2	2704.9	D_3BO_3	(E′)
—	2672.4	2707.7	D_3BO_3	(A′)

a) Assuming force constant K=0.2665 mdyn/Å. – b) Assuming F_τ=0.644, $F_{\tau\tau}$=−0.0035 mdyn/Å, and β=131°. – c) Assuming harmonic motion, with F_d= 7.468 mdyn/Å and F_{dd}=0.0056 mdyn/Å. – d) Assuming $\omega_e\chi_e$=90 cm^{-1}, with F_d= 8.219 mdyn/Å and F_{dd}=0.0064 mdyn/Å.

The appearance potentials for **$[B(OH)_2]^+$** and **$[B(OH)_3]^+$** were estimated to be 13.8±0.5 V and 10.8±0.5 V, respectively. The main IR absorptions are at 3668.5 cm^{-1} (the E′ OH stretching mode for $H_3B^{16}O_3$), at 1426.2 cm^{-1} (the E′ BO_3 stretching mode for $H_3{}^{11}BO_3$ which changes to 1478.0 cm^{-1} for $H_3{}^{10}BO_3$), at 1009.9 cm^{-1} (the E′ HOB bending mode which on deuteration shifted to ca. 825 cm^{-1}), at 675.0 cm^{-1} (the A″ out-of-plane BO_3 bending mode which shifted to 700.9 cm^{-1} for $H_3{}^{10}BO_3$), at 513.8 cm^{-1} (the A″ torsion mode), and a weaker feature at 448.9 cm^{-1} (the E′ in-plane OBO bending mode). The weak bands at 3726 and 3633 cm^{-1} are due to matrix isolated H_2O monomer. The weak broad features at ca. 1460 cm^{-1} and ca. 1080 cm^{-1} remain unassigned. The possible structures for molecular H_3BO_3 are given in **Fig. 3-18** [8].

These mass spectrometric and matrix isolation studies on H_3BO_3 not only provided firm evidence for the existence of molecular H_3BO_3 at relatively low temperatures (310 to 400 K), but also established both the frequencies and assignments for the nine fundamentals. Table 3/15 compares these values with other data [9 to 11], and also includes the frequency of the remaining fundamental (A′ BO stretch), which has been estimated from Raman studies on the solid [9]. Based on these most recent spectroscopic data, new estimates (Table 3/16, p. 186) for the thermodynamical functions of molecular H_3BO_3 have been made taking the following geometrical parameters for the monomer: r(BO)=136 pm, r(OH)=96 pm, ∢(OBO)=120°, and ∢(HOB)=130°. Table 3/17, p. 186, shows the free energy changes of the reactions MI(s)+ $H_3BO_3 \rightarrow MBO_2 + HI + H_2O$, where M=Li, Na, K, Rb, or Cs [12].

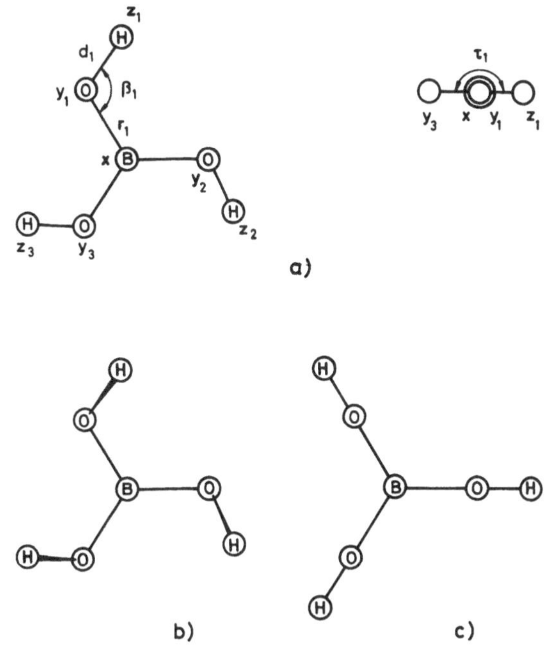

Fig. 3-18. Possible structures for molecular boric acid.
a) C_{3h} model for general $X(YZ)_3$ species with definitions of internal coordinates,
b) C_3 model, and
c) D_{3h} model [8].

Table 3/15
Fundamental Frequencies (in cm^{-1}) of Molecular Boric Acid [12].

most recent[a]	from [9]	from [10]	from [11]	assignment	
3672.6[b]	3250	3250	3600	OH stretch	(A′)
1020.0[b]	1060	1060	1080	HOB bend	(A′)
(880.0)	881	881	880	BO stretch	(A′)
675.0	648	652	625	BO_3 bend	(A″)
3668.5	3150	3150	3600	OH stretch	(E′)
1426.2	1428	1440	1490	BO stretch	(E′)
1009.9	1183	1185	1200	HOB bend	(E′)
448.0	544	544	540	OBO bend	(E′)
513.8	824/303	(rotation)	600	torsion	(A″)
577.7[b]	209	(rotation)	600	torsion	(E″)

[a] Values used to determine thermodynamic functions. – [b] Calculated values.

References on p. 188

Table 3/16
Thermodynamic Functions for Molecular Boric Acid [12].

T (K)	H(T) – H(298) (J/mol)	$C_p(T)$ ($J \cdot mol^{-1} \cdot K^{-1}$)	S(T) ($J \cdot mol^{-1} \cdot K^{-1}$)	[H(298) – G(T)]/T ($J \cdot mol^{-1} \cdot K^{-1}$)
		present work [12]		
100	–10200	34.96	215.37	317.37
200	–5972	51.42	244.01	273.87
298	0	70.02	268.07	268.07
300	140	70.37	268.54	268.07
400	7971	85.54	290.96	271.03
500	17116	96.80	311.32	277.09
600	27233	105.12	329.74	284.35
700	38075	111.46	346.44	292.05
800	49482	116.52	361.67	299.81
900	61350	120.72	375.64	307.47
1000	73607	124.32	388.55	314.94
1100	86200	127.47	400.55	322.19
1200	99090	130.26	411.76	329.19
1300	112241	132.73	422.29	335.95
1400	125627	134.93	432.21	342.47
1500	139220	136.90	441.59	348.77
		selected JANAF data [10]		
298	0	65.27	294.85	249.85
500	15416	86.28	333.87	303.03
1000	66391	113.69	403.45	337.06
1500	126838	126.64	452.29	367.73
		selected data from [11]		
298	0	65.82	264.52	264.52
500	16268	92.90	305.60	273.06
1000	71444	122.76	380.91	309.46
1500	136562	136.31	433.53	342.49

Table 3/17
Free Energy Changes for the Reactions $MI(s) + H_3BO_3 \rightarrow MBO_2(s) + HI(g) + H_2O(g)$ [12].[a)]

M	T (K)	ΔG(T) (kJ/mol) b)	ΔG(T) (kJ/mol) c)	equilibrium constant (atm) b)	equilibrium constant (atm) c)
Li	298	1.06	9.08	6.5×10^{-1}	2.5×10^{-2}
	500	–17.82	–4.64	$7.3 \times 10^{+1}$	3.0
	700	–36.72	–18.08	$5.5 \times 10^{+2}$	$2.5 \times 10^{+1}$

Table 3/17 (continued)

M	T (K)	ΔG(T) (kJ/mol) b)	c)	equilibrium constant (atm) b)	c)
Li	900	−55.75	−33.09	$1.7\times10^{+3}$	$8.7\times10^{+1}$
	1100	−74.88	−48.28	$3.6\times10^{+3}$	$2.0\times10^{+2}$
	1300	−94.11	−63.45	$6.1\times10^{+3}$	$3.5\times10^{+2}$
	1500	−113.04	−78.83	$9.0\times10^{+3}$	$5.6\times10^{+2}$
Cs	298	114.23	112.25	9.0×10^{-21}	3.5×10^{-22}
	500	91.21	104.04	2.9×10^{-10}	1.2×10^{-11}
	700	66.73	84.65	1.0×10^{-5}	4.7×10^{-7}
	900	41.07	63.43	4.0×10^{-3}	2.1×10^{-4}
	1100	14.49	41.09	2.0×10^{-1}	1.1×10^{-2}
	1300	−12.83	17.83	3.3	1.9×10^{-1}
	1500	−40.79	−6.21	$2.6\times10^{+1}$	1.6

[a] Assuming the relationship $\Delta G(T)=\Delta H(298)+\Delta C_p(298)\cdot(T-298)-T[\Delta S(298)+\Delta C_p(298)\cdot \ln(T/298)]$. – [b] With data for H_3BO_3 based on matrix infrared vibrational data. – [c] With data for H_3BO_3 based on JANAF tabulations.

The effects of the conditions of the thermoanalytical examination of the process of dehydration of H_3BO_3 to B_2O_3 were studied; structurally arranged B_2O_3 was obtained directly from H_3BO_3 at 176°C, the melting point of metaboric acid [13].

Differential thermal analysis (DTA) and thermogravimetry (TG) curves of the classical dehydration of H_3BO_3 up to ca. 1273 K were compared with laser-heated decomposition (see Fig. 3-19). The laser irradiation was from a power-stabilized continuous wave (CW) CO_2 laser

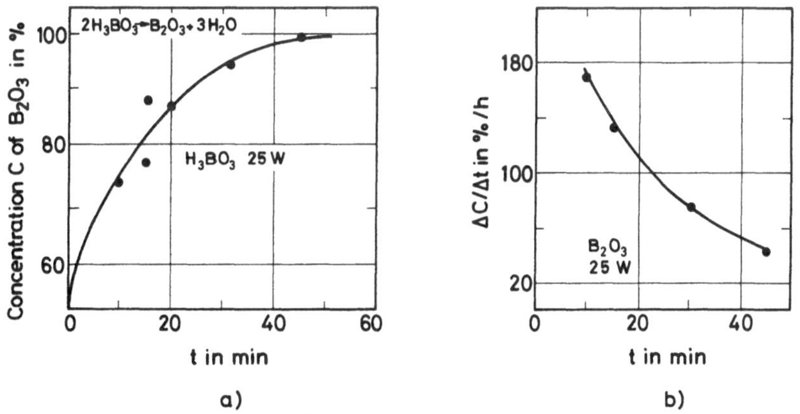

Fig. 3-19. Laser decomposition (25 W) of boric acid: $2H_3BO_3 \rightarrow B_2O_3+3H_2O$.

a) B_2O_3 concentration C versus time;

b) concentration variation rate ($\Delta C/\Delta t$) versus time [14].

References on p. 188

source, which was operated on the 944 cm^{-1} line at an output power of ca. 25 W. Thermal heating is interpreted by the reaction $4H_3BO_3 \rightarrow H_2B_4O_7 + 5H_2O$ (reaction order n = 1; activation energy E_a = 20 kcal/mol), or by the reaction $H_2B_4O_7 \rightarrow 2B_2O_3 + H_2O$ (reaction order n = 3; activation energy E_a = 24.6 kcal/mol). The laser irradiation proceeds by $2H_3BO_3 \rightarrow B_2O_3 + 3H_2O$ (reaction order n = 3/2; ln k = 2.1 min^{-1}, k = reaction rate constant) [14].

Changes in the network structure of H_3BO_3 and materials of the B_2O_3–α-Fe_2O_3 system with the composition $(B_2O_3)_x(Fe_2O_3)_{1-x}$ (x = 0.6, 0.7, or 0.8) occurring as the temperature was varied, were investigated using IR spectroscopy, X-ray analysis, and density and electrical conductivity measurements. Analyses of the network structures were carried out based on a division into three classes: (a) networks of H_3BO_3 treated at various temperatures, (b) Fe_2O_3-modified networks, and (c) networks unmodified by Fe_2O_3. Borate networks are classified into modified and unmodified networks, where the modified networks are formed by the three-dimensional arrangement of simple modified BO_3 and BO_4 groups, while the unmodified networks are made by three-dimensional arrays of clusters with unmodified BO_3 groups, BO_4 groups, B=O bonds, and nonbridging oxygen atoms. The kind of clusters in unmodified networks depends on the stability of the network built up under the given conditions [15].

References for 3.5.7.2:

[1] Tossell, J. A. (Am. Mineral. **71** [1986] 1170/7).
[2] Gajhede, M. (Chem. Phys. Lett. **120** [1985] 266/71).
[3] Gajhede, M.; Larsen, S.; Rettrup, S. (Acta Crystallogr. B **42** [1986] 545/52).
[4] Zhang, Z. G.; Boisen, M. B.; Finger, L. W.; Gibbs, G. V. (Am. Mineral. **70** [1985] 1238/47).
[5] Dewar, M. J. S.; Jie, Caoxian; Zoeblsch, E. G. (Organometallics **7** [1988] 513/21).
[6] Tarasevich, B. P.; Sirotkina, N. Z.; Il'in, A. S.; Kuznetsov, Je. V. (Vysokomol. Soedin. A **27** [1985] 451/63).
[7] Zav'yalov, V. V.; Tarasevich, N. Z.; Sirotkin, O. S.; Khitrov, M. Yu. (Zh. Obshch. Khim. **55** [1985] 2157/61; J. Gen. Chem. [USSR] **55** [1985] 1915/9).
[8] Ogden, J. S.; Young, N. A. (J. Chem. Soc. Dalton Trans. **1988** 1645/52); Young, N. A. (Diss. Univ. Southampton 1988).
[9] Kristiansen, L. A.; Moeney, R. W.; Cyvin, S. J.; Brunvoll, J. (Acta Chem. Scand. **19** [1965] 1749).
[10] JANAF Thermochemical Tables, 2nd. Ed., NSRDS-NBS-37 [1971].

[11] Glushko, V. P.; Gurvich, L. V.; Bergman, G. A.; Veits, I. V.; Medvedev, V. A.; Khachkuruzov, G. A.; Yungman, V. S. (Thermodynamic Properties of Individual Substances, 3rd. Ed., Pt. 1, Pergamon, New York 1988).
[12] Bowsher, B. R.; Dickinson, S.; Ogden, J. S.; Young, N. A. (Thermochim. Acta **141** [1989] 125/30).
[13] Latocha, C.; Uhniat, M. (J. Therm. Anal. **33** [1987/88] 973/5).
[14] Popescu, C.; Jianu, V.; Alexandrescu, R.; Mihailescu, I. N.; Morjan, I.; Pascu, M. L. (Thermochim. Acta **129** [1988] 269/76).
[15] Kim, Yoo Young; Kim, Keu Hong; Choi, Jae Shi (J. Phys. Chem. Solids **50** [1989] 903/8).

3.5.7.3 Solutions of H_3BO_3

The stoichiometric pK* value for the ionization of H_3BO_3, with $K^* = [H^+][B(OH)_4^-]/[B(OH)_3]$, has been determined from electromotive force measurements with a glass electrode and an Ag/AgCl electrode, in pure NaCl solution at 25.0 ± 0.02°C and tabulated (Table 3/18) together with data that were calculated from the equation $-\log K^* = 9.2365 - 0.7927 \cdot I^{1/2} + 0.4013 \cdot I - 0.04554 \cdot I^{3/2}$, and in NaCl solutions with small amounts of added $CaCl_2$ or $MgCl_2$ (Table 3/19) from I = 0.65 to 6.15 M. From the values for pK*, the Pitzer parameters for the interaction with $Ca[B(OH)_4]_2$ have been determined to be $\beta^0 = -1.57$, $\beta^1 = -4.49$, and $C^\phi = -0.17$; for $Mg[B(OH)_4]_2$, $\beta^0 = -0.21$, $\beta^1 = -4.98$, and $C^\phi = -0.36$. Ion-pairing constants for the interaction of $[B(OH)_4]^-$ with Ca^{2+} or Mg^{2+} ions were also determined from I = 0.5 to 6.0 M, and the pK values were extrapolated to pure water to be −1.72 for Ca^{2+} and −1.45 for Mg^{2+} [1].

Table 3/18

Values of pK* for the Ionization of Boric Acid in NaCl Solution at 25°C [1].

m (molarity of NaCl)	pK*	m (molarity of NaCl)	pK*
0	—	0.725	8.824
0.02	9.132	1.0	8.813 a)
0.07	9.054	1.25	8.788
0.36	8.896	2.0	8.790 a)
0.409	8.882	3.0	8.830 a)
0.5	8.856 a)	4.0	8.881 a)
0.621	8.839	5.0	8.952 a)
0.68	8.830	6.0	9.045 a)
0.729	8.824		

a) Experimental data; other values are calculated from the equation given in [1].

Table 3/19

Values of pK* for the Ionization of Boric Acid in $NaCl/CaCl_2$ and $NaCl/MgCl_2$ Solutions at 25°C [1].

m (molarity of Na^+)	I	$NaCl/CaCl_2$ pK*	$NaCl/MgCl_2$ pK*
0.50	0.650	8.675	8.726
1.00	1.15	8.664	8.712
2.00	2.15	8.672	8.710
3.00	3.15	8.706	8.764
4.00	4.15	8.778	8.820
5.00	5.15	8.856	8.897
6.00	6.15	8.942	8.982

Vapor pressures of aqueous solutions of H_3BO_3 over a wide range of acid concentration were measured from 40 to 100°C. The results, together with solubility data taken from the literature, can be described with a thermodynamic model which uses the Wilson equations to express the activity coefficients of water and boric acid. Only three temperature-independent adjustable parameters are required; one of these represents the entropy of fusion of H_3BO_3, data which are not available in the literature [2].

References on pp. 197/201

An UV/VIS spectrometer connected to a high pressure assembly was used to determine the dissociation constant for $B(OH)_3+H_2O \rightleftharpoons H^+(aq)+[B(OH)_4]^-(aq)$ to be at 1 atm $pK_a=9.29 \pm 0.13$, using an indicator technique. The measurements were made at 25°C and for ionic strengths between I=0.1 and 1.0 M over a pressure range of 1.0 to 2040 atm. Extrapolation of pK_a^* to I=0 gave a thermodynamic dissociation constant of $5.16 \pm 0.15 \times 10^{-10}$ at 1.0 atm; the pressure dependence yielded a partial molal volume change of −28.9 and −31.8 cm^3/mol and a compressibility change of −3.1 and -4.8×10^{-3} $cm^3 \cdot mol^{-1} \cdot atm^{-1}$ for the dissociation at I=0.1 and 1.0 M, respectively. The association constant K_A for the formation of the sodium borate ion pair, $Na^+(aq)+[B(OH)_4]^-(aq) \rightleftharpoons Na[B(OH)_4]$, was determined by comparing the acid constants in $[(CH_3)_4N]Cl$ to those in NaCl solutions. Extrapolation to I=0 yielded a K_A for $Na[B(OH)_4]$ of 0.64 ± 0.06 at 1 atm. The pressure dependence of K_A gave the partial molal volume $\Delta\bar{V}^\circ = 8.39 \pm 0.40$ cm^3/mol and the compressibility change $\Delta\bar{\kappa}^\circ = 1.6 \pm 0.1 \times 10^{-3}$ $cm^3 \cdot mol^{-1} \cdot atm^{-1}$ for the formation of the ion pair corrected to I=0 using 1.0 M data [3].

The reactivity of the hydrated electron and the OH radical with H_3BO_3 was investigated by steady-state and pulse radiolytic techniques. No evidence was found for appreciable reactivity with OH radicals by either method, and an upper limit of 5×10^4 $dm^3 \cdot mol^{-1} \cdot s^{-1}$ was estimated for the reaction rate constant. Steady-state irradiations gave no evidence for the reactivity with the hydrated electron, but pulse-radiolysis experiments showed the hydrated electron decay rate to be linearly dependent on the H_3BO_3 concentration, but independent of the pH value. Although superficially suggesting a reaction with H_3BO_3, it seems more likely that the enhanced decay is due to an impurity in the H_3BO_3 (probably nitrate). An upper limit of 5×10^3 $dm^3 \cdot mol^{-1} \cdot s^{-1}$ was estimated for $k(e^-(aq)+H_3BO_3)$ [4].

The variation in volatility with temperature of H_3BO_3 and NH_3 is examined through experimental tests on fluids. The distribution of the two species between liquid and vapor agrees well with known experimental data [5].

Relaxation studies after a temperature jump were performed in dilute aqueous solutions containing H_3BO_3 (A) with different weak acids (B) (silicic acid, phosphoric acid, ammonium chloride). The results are interpreted from a model based on proton transfer between the couples A/$[B(OH)_4]^-$ and A/B [6].

In aqueous solutions similar to seawater and containing H_3BO_3, low concentrations of weak acids such as $[NH_4]^+$, $Si(OH)_4$, $[H_2PO_4]^-$, or $[HCO_3]^-$ significantly modified the relaxation frequency of H_3BO_3, including the equilibrium $B(OH)_3+[OH]^- \rightleftharpoons [B(OH)_4]^-$, which is directly responsible for the attenuation of the absorption coefficient of low-frequency (ca. 1 kHz) sound waves as was measured by using the temperature-jump technique [7].

By studying the effect of the buffer capacity of borate solutions at their different pH values and ionic strengths on polarographical catalytic waves of pachycarpine or lupinine, the surface partial protonation rate constants under the effect of H_3BO_3, H^+, and H_2O were found. The dissociation constant of H_3BO_3 near the electrode at the potentials of the maximal current was also evaluated [8].

Studies of the strength of complexation in the system H^+–Al^{3+}–H_3BO_3 using three independent types of measurement (H^+ potentiometry, ^{27}Al NMR, and ^{11}B NMR) indicate that no (or at most very weak) aluminium borate complexes are formed in seawater [9].

Ag^{3+} has at pH 11 a half-life time of several hundred seconds in aqueous solutions in the presence of 0.1 to 1.0 M concentrations of $[B(OH)_4]^-$; this was compared with a life-time of a few seconds at pH 11 in the absence of borate [10].

Potentials are reported from 5 to 55°C for the cells without liquid junction H_2,Pt|$B(OH)_3(m_1)$, $M[B(OH)_4](m_2)$, $MCl(m_3)$|AgCl,Ag with M=K or Na and $m_1+m_2=0.05$ at an ionic strength up to

3 mol/kg. The total boron concentration $m_1 + m_2$ was maintained at a level sufficiently low to minimize polyborate formation. Ion interaction parameters for mixed sodium and potassium borates and chlorides were calculated for $Na[B(OH)_4]$ with $a_i = -0.0510$ and $b_i = 5.264 \times 10^{-3}$, for $K[B(OH)_4]$ with $a_i = 0.1469$ and $b_i = 2.881 \times 10^{-3}$; the pure electrolyte parameters at 25 °C and their temperature coefficients are also given [11]. Also, for the cells $H_2,Pt|B(OH)_3(m_1)$, $Na[B(OH)_4](m_2)$, $MCl_2(m_3)|AgCl,Ag$ with M = Mg or Ca, the potentials are reported for different ionic strengths between 5 and 55 °C. Parameters of Pitzer's ion interaction treatment were determined for $Ca[B(OH)_4]_2$ and $Mg[B(OH)_4]_2$ by fitting the cell results. The ion interaction parameters β_2 at 25 °C were given in the form $a_i + b_i(T-298.15) + c_i(T-303.15)^2$ and are −11.47 for $Mg[B(OH)_4]_2$ and −15.88 for $Ca[B(OH)_4]_2$ [12].

An unusually high solubility of oxygen in dilute aqueous borate solutions has been reported [13]. The solubility and diffusivity of oxygen in borate solutions were studied on a Pt microelectrode under diffusion-controlled conditions. The solubility of 4.2×10^{-3} mol/L is unusually high, whereas its diffusion coefficient of 5×10^{-6} cm²/s is lower than in other aqueous solutions. The decrease in oxygen solubility at higher ionic strengths is due to the salting-out effect. The results were consistent with the formation of an oxygen-borate complex, whereby the stability of the borate solution containing polymeric anions is enhanced. The borate solutions were prepared by addding 0.1 to 1.0 M $Na_2[SO_4]$ solutions to the original buffer solution, containing 0.0375 M $Na_2[B_4O_7] \cdot 10H_2O$ and 0.15 M H_3BO_3 (pH 8.4) [14]. Similar results are obtained on iron by studying the borate absorption with in situ Fourier transform infrared (FTIR) spectroscopy [15].

In the system H_3BO_3–HOX–H_2O with X = Cl or Br, the stability constant for the hypochloritoborate complex $\beta' = [B(OH)_3OCl^-]/[HOCl] \cdot [B(OH)_4^-]$, where brackets denote activities, was found to be $\log \beta' = 2.25 \pm 0.01$ at 25 °C from a combination of spectrometric measurements at equilibrium and kinetic studies of the oxidation of the bromide ion. The corresponding constants (from equilibrium studies only) for the $[B(OH)_3OBr]^-$ complex are $\log \beta' = 1.83 \pm 0.04$ and $\log \beta = 1.26 \pm 0.04$ over the pH range 8.2 to 9.7 [16].

The activity of water in extremely concentrated aqueous solutions of NaOH and H_3BO_3 at ca. 317 °C was studied. These solutions are relevant to water chemistry and corrosion in the steam generator of a nuclear power plant. **Fig. 3-20** suggests incipient separation into immiscible liquid phases; one aqueous, the other an associated sodium borate liquid [17].

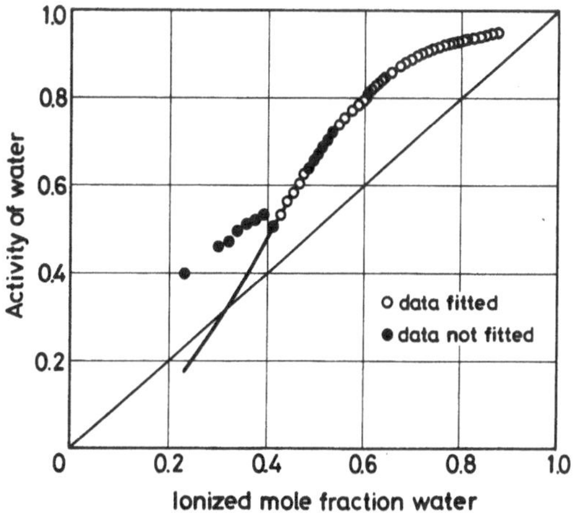

Fig. 3-20. Activity of water in sodium borate solution at 317 °C. Ratio Na:B = 0.75. Borate ions in the melt are assumed to be anhydrous for the purposes of data reduction [17].

References on pp. 197/201

A gravimetric analysis using HBO_2 or DBO_2 as weighting forms was developed for solubility measurements. The method gave satisfactory results for H_3BO_3 in H_2O. By using this method, the solubilities of ^{10}B-enriched D_3BO_3 in D_2O were measured to be (in wt%): 2.67 at 7°C, 3.52 at 15°C, 5.70 at 30°C, 8.87 at 50°C, and 12.92 at 70°C , respectively. These values are ca. 10% lower than those in H_2O. Thermodynamical considerations, based on these data, show that boric acid is the water-structure breaker [18].

A ^1H NMR study of the $(CD_3)_2SO-H_3BO_3-KOH-H_2O$ system (KOH : H_3BO_3 = 13 : 9 mole ratio) shows the formation of $[B(OH)_4]^-$; addition of alcohols (CH_3OH, C_2H_5OH, $C_6H_5CH_2OH$, ethylene glycol, pinacol) leads to the formation of boric acid esters. Pinacol forms cyclic boric acid esters, but no signals for cyclic ethylene glycol esters of boric acid were observed [19]. By ^1H NMR spectroscopy, some lines of alkylborates prepared by reaction of H_3BO_3 with methanol/*i*-propanol mixtures have been assigned. The existence of a linear dependence on chemical shift differences for two different mono- or dialkylborates on the Taft induction constants has been established. The ratio of concentration constants of formation of some (ar)alkylborates (R = C_2H_5, *i*-C_3H_7, *n*-C_4H_9, *i*-C_4H_9, *n*-C_6H_{13}, *n*-C_8H_{17}, $C_6H_5CH_2$) to monomethylborate has been determined [20].

Boron-11 NMR chemical shifts and linewidths have been measured in very dilute aqueous solutions of H_3BO_3, titrated with HCl and NaOH, respectively, and on some salts. **Fig. 3-21** shows the results for concentrations down to 2 millimolal. The curve given in **Fig. 3-22** for the dependence of the chemical shift on the pH is calculated with the result of a least squares fit of $pK_s = 9.3$ by using $\delta = 0$ ppm for $B(OH)_3$ and $\delta = -17.6$ ppm for $[B(OH)_4]^-$. Between pH 8 and 10.5, a strong change in the chemical shift δ is observed. Obviously the observed chemical shift is the weighted mean of the chemical shifts of $B(OH)_3$ and $[B(OH)_4]^-$ due to their different

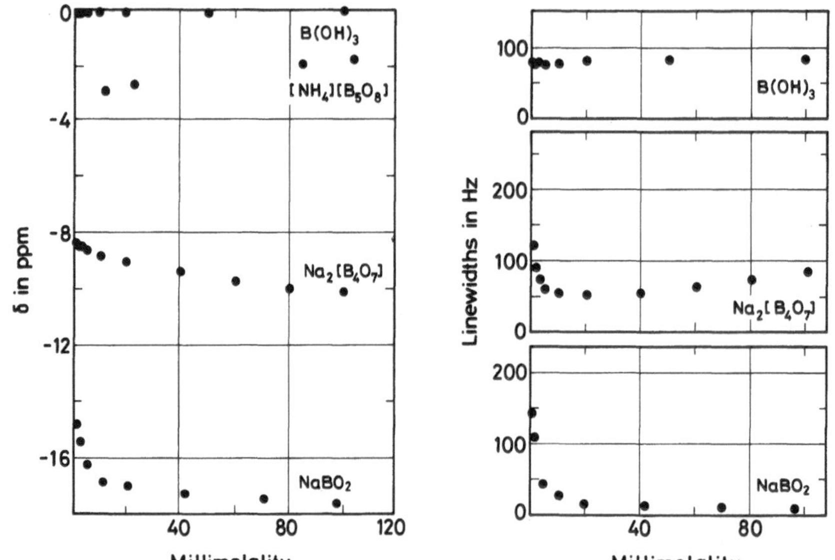

Fig. 3-21. Chemical shifts (left) and linewidths (right) of ^{11}B NMR signals in aqueous solutions of boric acid, sodium tetraborate, sodium borate, and ammonium pentaborate as a function of concentration. The lines, especially those of sodium borate solutions at higher concentrations, are broadened [21].

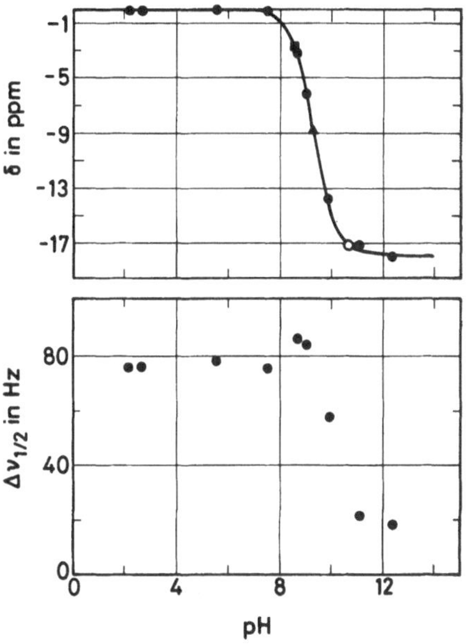

Fig. 3-22. Boron-11 chemical shift and linewidth of about 10 millimolal solutions. Typical borate solutions: 10 mmol $[NH_4][B_5O_8]$ (■), 10 mmol $Na_2[B_4O_7]$ (▲), 10.8 mmol $NaBO_2$ (○). Solutions of $B(OH)_3$ as a function of the pH, titrated with HCl, respectively, NaOH (●) [21].

concentrations as a function of pH. The ^{11}B-^{10}B primary isotope effect on the magnetic shielding is smaller then 3×10^{-8}. The H_2O-D_2O solvent isotope effect on the longitudinal relaxation times T_1 (in ms) of $^{11}B(OH)_3$ is 5.8±0.3, of $^{10}B(OH)_3$ is 9.0±0.3, of $^{11}B(OD)_3$ is 4.8±0.2, and of $^{10}B(OD)_3$ is 7.6±0.4. This has been established for ^{11}B and ^{10}B in the cited species, and from the ratios of T_1 values the quadrupolar origin of the relaxation mechanism has been inferred [21].

Boron-11 NMR studies have been performed in dilute solutions of H_3BO_3 or a borate, respectively, in the presence of hydroxycarboxylic acids (LH_n). The second signals in the spectra for pH≤5.5 appear for L(+)-lactic acid between −9.9 and −10.0 ppm, for L(−)-malic acid near −9.9 ppm, and for D(−)-tartaric acid between −9.4 and −9.7 ppm; these can be assigned to the diester anion $[BL_2]^-$. From about pH 5, an additional signal is observed at −13.3, −13.8, and −12.8 ppm for the cited three acids, which is assigned to the monoester anion $[BL]^-$. No complex formation is observed for succinic acid solution at pH 2.46. Red wine shows a signal near −9.9 ppm, arising from the species $[BL_2]^-$ of various hydroxycarboxylic acids [22].

The anionic complex formation of boric acid (HA) or the tetrahydroxyborate ion with some polyols (L) such as disaccharides, sugar acids, and sugar alcohols was studied potentiometrically in aqueous solution at 25°C and at an ionic strength of 0.1M (KCl). Both 1:1 and 1:2 complexes are formed. The constants were calculated by $HA+L \rightleftharpoons H^+ + [AL]^-$ or $HA+2L \rightleftharpoons H^+ + [AL_2]^-$ with the program SCOGS and/or manually by a graphical method. The overall stability constants are $\beta_1=[AL^-]/[A^-]\cdot[L]$ or $\beta_2=[AL_2^-]/[A^-]\cdot[L]^2$ (see Table 3/20, p. 194) [23].

The pK values of the chemical processes which occur in "tris-borate" buffers (for the existence of a complex "tris-boric acid" in borate buffers, see "Boron Compounds" 3rd Suppl. Vol. 2, 1987, p. 105) are determined with higher accuracy using more precise "tris-ion" and boric acid pK values; the buffer capacities at different pH values have been calculated. The results confirm that "tris-borate" buffers contain a complex compound with a betaine structure, previously described as "tris-boric acid" [24].

Absorption of CO_2 gas into a borate buffer solution is a linear function of time; the stable C-isotope composition is depleted in ^{13}C NMR by 1.95% as compared to the gas phase [25].

Table 3/20
Values of the Overall Stability Constants of Borate Complexes with Selected Carbohydrates in Aqueous Solution at 25°C [23].

carbohydrate	log β_1	log β_2	carbohydrate	log β_1	log β_2
gluconic acid	2.83(2)	4.46(6)	maltose	1.36(8)	
glucuronic acid	1.71(4)	—	melibiose	1.82(2)	2.44(6)
glucaric acid	2.16(2)	3.58(2)	saccharose	0.75(8)	
glucose	2.07(2)	2.80(4)	trehalose	1.04(8)	
galactose	1.97(3)	2.52(5)	turanose	1.91(2)	2.47(5)
cellobiose	1.25(5)		erythritol	1.99(4)	
gentibiose	1.14(7)		*meso*-inositol	1.57(1)	
lactose	1.51(7)		glycerol	1.39(2)	

The strength of hardening structures in the metal carbonate–H_3BO_3–H_2O system decreased in the sequence $Ba[CO_3] > Sr[CO_3] > K_2[CO_3] \approx Y_2[CO_3]_3 \approx$ germanium carbonate $\approx Tl_2[CO_3] > Pb[CO_3] > Na_2[CO_3] > Zn[CO_3] > Li_2[CO_3] > Nd_2[CO_3]_3 > La_2[CO_3]_3 >$ bismuth carbonate > copper carbonate $> Ca[CO_3] > Cd[CO_3] \approx Mg[CO_3] >$ zirconium carbonate $> Mn[CO_3] >$ indium carbonate $> Be[CO_3] > Ni[CO_3]$. The relation between the strength of the hardening structure at 298 K and the hydration energy of the metal cation is discussed [26]. The strength of hardened pressed binders prepared from the system H_3BO_3–$Ca[CO_3]$–H_2O is increased by preliminary grinding of H_3BO_3, the addition of $Ca[OH]_2$, $Ba[CO_3]$, glycerol, and mannitol, preliminary decarbonation of $Ca[CO_3]$, and the use of higher curing temperatures. The structure formation is related to H_3BO_3 binding [27]. At 438 K, in the system H_3BO_3–$Ba[CO_3]$ under autoclave conditions, almost all of the H_3BO_3 was bound, and the binder has maximal strength [28].

A review has been compiled on the behavior of the H_3BO_3–MOH–MX–H_2O solutions with M = Li, Na, K, Rb, or NH_4; X = Cl, Br, I, SCN, NO_3, NO_2, $[SO_4]_{1/2}$, $[CrO_4]_{1/2}$, or $[Cr_2O_7]_{1/2}$, using experimental data obtained by the isothermal solubility and other physicochemical methods. The dependence of the nature of the salting-out-salting-in effect of the formed borate component or the formation of solid solutions or compounds with constant compositions on the structural and steric factors and the lyotropic effects have been discussed. The high positive value of ΔE for the Ca^{2+} ion of 1.88 kJ/mol, which is almost twice that for the Na^+ ion, led to the expectation of a negative lyotropic effect on H_3BO_3, even for $Ca[SO_4]$ [29].

The $[B(OH)_4]^- \rightleftharpoons [B_3O_3(OH)_4]^- \rightleftharpoons [B_4O_5(OH)_4]^{2-} \rightleftharpoons [B_5O_6(OH)_4]^-$ equilibrium has been studied in $M[B(OH)_4]$-L-H_2O solutions, where M = Li, Na, or K and L = formic acid, formamide, dimethylformamide, acetamide, urea, glycine, or Trilon B. The equilibrium shifts towards pentaborate on addition of formic acid, towards tetraborate on adding glycine or Trilon B, towards lithium monoborate or sodium tetraborate with dimethylformamide, and towards tetraborate with formamide according to $4\,M[B(OH)_4] + 2\,HCONR_2 \rightleftharpoons M_2[B_4O_5(OH)_4] + 2\,HCOOM + 2\,R_2NH + 5\,H_2O$. Hydrophobicity increases on going from dimethylformamide to acetamide, therefore $Li[B(OH)_4]$ or $Na_2[B_4O_5(OH)_4]$ separate directly and no new complexes are formed [30].

Solubility diagrams have been investigated for the following systems (at 25 °C unless otherwise noted): H_3BO_3–$[NH_4][NO_3]$–H_2O, H_3BO_3–$[NH_4]_2[SO_4]$–H_2O, and H_3BO_3–$[NH_4][SCN]$–H_2O [31]; H_3BO_3–LiOH–LiCl–$Li_2[SO_4]$–H_2O [32]; H_3BO_3–LiOH–$Li[IO_3]$–H_2O [33]; H_3BO_3–LiOH–MgO–H_2O at 40 °C [34]; H_3BO_3–M_2Cl_2–H_2O (M = Na, K; M_2 = Mg, Ca) [35]; H_3BO_3–KBr–H_2O [36]; H_3BO_3–KOH–KX–H_2O (X = Cl, Br, I) [37]; H_3BO_3–KOH–$BaCl_2$–H_2O [38]; H_3BO_3–MgO–

$MgCl_2$–H_2O [39, 40]; H_3BO_3–MgO–H_2O and H_3BO_3–Mg[SO_4]–MgO–H_2O [41]; H_3BO_3–Ca-[B_4O_7]–H_2O and H_3BO_3–Ca[B_6O_{10}]–H_2O (both at 50°C) [42]; H_3BO_3–CaO–propionic acid–H_2O (at 60°C) [43]; H_3BO_3–MOH–formic acid–H_2O (M = Na, K; at 15°C) [44]; H_3BO_3–MOH–C_6H_5OH–H_2O (M = Li, Na, K) [45]; H_3BO_3–MOH–aniline–H_2O, H_3BO_3–MOH–N,N-dimethylaniline–H_2O, H_3BO_3–MOH–N,N-diethylaniline–H_2O [46 to 49]; H_3BO_3–ethylenediamine [50]; H_3BO_3–ethylenediamine–H_2O [51]; H_3BO_3–hexamethylenediamine–H_2O [52]; H_3BO_3–hexamethylenetetramine–H_2O [53 to 56]; H_3BO_3–amine–H_2O [57]; H_3BO_3–MOH–amine–H_2O [58]; H_3BO_3–MOH–amine–H_2O, H_3BO_3–MOH–amide–H_2O [59]; H_3BO_3–formamide–H_2O, H_3BO_3–urea–H_2O [60]; H_3BO_3–NaOH–formamide–H_2O, H_3BO_3–NaOH–dimethylformamide–H_2O [61]; H_3BO_3–NaOH–formamide–H_2O, H_3BO_3–NaOH–acetamide–H_2O, H_3BO_3–NaOH–dimethylformamide–H_2O [62], H_3BO_3–MOH–formamide–H_2O (M = Li, Na, K, NH_4) [63]; H_3BO_3–KOH–formamide–H_2O, H_3BO_3–KOH–dimethylformamide–H_2O [64]; H_3BO_3–acetamide–H_2O, H_3BO_3–NaOH–urea–H_2O, H_3BO_3–NaOH–thiourea–H_2O, H_3BO_3–NaOH–urea–H_2O, H_3BO_3–NaOH–thiourea–H_2O [65]; H_3BO_3–KOH–acetamide–H_2O, H_3BO_3–KOH–urea–H_2O, H_3BO_3–KOH–thiourea–H_2O [66]; H_3BO_3–acetylurea–H_2O [67]; H_3BO_3–MOH–acetylurea–H_2O (M = Li, Na, K, NH_4) [68 to 70]; H_3BO_3–(3-aminopropanol)–H_2O [71]; H_3BO_3–NaOH–diethanolamine–H_2O [72]; H_3BO_3–monoethanolamine–H_2O, H_3BO_3–diethanolamine–H_2O, H_3BO_3–triethanolamine–H_2O [73]; H_3BO_3–MOH–monoethanolamine–H_2O, H_3BO_3–MOH–diethanolamine–H_2O, H_3BO_3–MOH–triethanolamine–H_2O (M = Li, Na, K) [74]; H_3BO_3–glycine–H_2O [67]; H_3BO_3–MOH–glycine–H_2O (M = Li, Na, K) [75 to 77]; H_3BO_3–Trilon B–H_2O [67]; H_3BO_3–MOH–Trilon B–H_2O (M = Li, Na, K, NH_4) [78 to 81].

Steady-state polarization curves and cyclic voltammograms were obtained on Pt microelectrodes in various solutions of nickel ions. The presence of H_3BO_3 lowers the overvoltage for nickel deposition and functions as a catalyst [82].

In H_3BO_3/borax buffer solutions with pH values between 7.4 and 9.2, the following anodic passive oxide films of barrier-type were investigated by physicochemical and electrochemical measurements on Al [83, 84] (also in [NH_4][B_5O_8] solutions [85, 86]); Pb [87, 88]; Sn [89]; Sn/Co [90], Ti and Zr [91, 92]; Cr [93 to 95]; Cr/Zn [96]; Fe [97 to 105]; steel [106]; Fe/Si [107]; Co [108]; Co/Zn [109]; Ni [94, 95, 102, 107, 110, 111]; Ni/Zn [112]; Ni/Cd [113]; Cu [114 to 116].

The sorption of H_3BO_3 from 0.105 M H_3BO_3 solutions at ca. 8 kbar has been studied on the ion-exchanger AN-22x8, a bifunctional sorbent which contains weakly basic primary and secondary amine groups. At 3 kbar, the ion-exchange mechanism of sorption was predominant and at atmospheric pressure, the acidic mechanism [117]. The sorption of H_3BO_3 from 0.2 M NaCl solutions, studied by static methods on glutaraldehyde-crosslinked poly(vinylalcohol) gel sorbents, decreases with an increasing degree of cross-linking [118].

The sorption of borate ions from aqueous and acetone-containing aqueous H_3BO_3 solutions on an AV-17 anion exchanger and by iron hydroxide precipitation was studied under static conditions. It increases with an increasing concentration of acetone in alkaline solutions. The sorption capacity for borate ions from weakly acidic and neutral aqueous-acetone solutions of AV-17 is equal to, and that of iron hydroxide precipitations is lower than that of aqueous solutions [119]. Coprecipitation of borate ions with Mg[$OH]_2$ from aqueous and aqueous-organic solutions was studied as a function of the pH value. Addition of the organic solvent (e.g., acetone, ethanol, dioxane, dimethyl sulfoxide, or dimethylformamide) promotes incorporation of borate into the Mg[$OH]_2$ phase. The degree of the organic solvent effect depends on its nature, i.e., the ability to hydrogen bond with H_2O. The X-ray phase analysis showed coprecipitation of magnesium hexaborate from aqueous acetone [120]. The superequivalent adsorption of carboxylic acids such as formic or acetic acid on the anion exchangers AV-17 (H_3BO_3 form), AM-3, or AM-7 involves associations consisting of several polyborates and possessing more strongly acidic properties [121].

References on pp. 197/201

The capacity of an ABN-11 anion exchanger for $[OH]^-$ exchange for H_3BO_3 in the treatment of thermal waters is ca. 0.6 milliequivalent per gram [122]. The boron sorption properties of anion exchangers containing N-methylglucamine or tris(hydroxymethyl)aminomethane groups are studied at an ionic strength of 1.0 M (NaCl). The dependence of the H_3BO_3 sorption capacity on the acidity of the solution was determined and a sorption maximum between pH 6 and 8 was found. Sorption of H_3BO_3 occurs due to the complexing of the borate anion by the functional group of the anion exchanger [123]. A mechanism involving complex formation by the functional amino and hydroxyl groups is proposed for the sorption of H_3BO_3 on anion exchangers containing N-methylglucamine or tris(hydroxymethyl)aminomethane groups. The maximal sorption capacity is explained in terms of the predominant formation of hydrogen bonds between nonionized exchanger amino groups and H_3BO_3 at a pH between 6 and 8. The nitrogen atom in N-methylglucamine or tris(hydroxymethyl)aminomethane is needed to control the pH value of the exchanger-solution system [124]. A comparative study of the boron sorption kinetics between 25 and 50°C on anion exchangers with different functional groups showed that exchange occurs in the region where both chemical reaction and internal diffusion contribute to the rate. The calculated rate constants and activation energies confirm the mechanisms proposed above involving H_3BO_3 complexation with N-methylglucamine or tris-(hydroxymethyl)aminomethane as functional groups [125].

The boron-selective resin Amberlite IRA-743 was used to remove H_3BO_3 and related compounds from solutions of carbohydrates [126]. In the removal of boron by Amberlite XE-243 from H_3BO_3 and borax production wastewater, the optimum pH was 8.5; the treatment results were used to develop equations for the prediction of boron removal from the wastewater [127].

An anion-exchange membrane can support a transport reaction when the counterion of its fixed charge is capable of complexing a neutral ligand, e.g., the transport of H_3BO_3 through an anion-exchange membrane with $[OH]^-$ or borate counterion as carrier. Borate and polyborate complex formation allows extraction and, in addition, carrier-mediated transport of H_3BO_3 in which the free ligand is recovered [128]. The fixation of H_3BO_3 by two anion exchange membranes (RPA Rhone Poulenc and Permion), initially in the OH-form, was studied. The amount of sorbed boron and the simultaneous conductivity change are in agreement with the nonformation of triborates and the formation of mono- and divalent tetraborates and pentaborates. With regard to Permion membranes, an additional correction for the variation of polyborate/borate must be made in order to take into account the lower apparent stability of polyborates [129].

A study was performed in order to evaluate the degree of the adsorption of 1 g H_3BO_3 by 7.5 g (5.7%), 15 g (17.6%), and 30 g (38.6%) of activated charcoal [130].

Solid solutions of $Mg[OH]_2$-$Ni[OH]_2$ have been used as absorbents of tetraborate ions in the form of $(Ni, Mg)[OH]_{2-2x}[B_4O_5(OH)_4]_x$ with $0.02<x<0.06$ [131]. The ion $[B_4O_5(OH)_4]^{2-}$ is sorbed by an ion exchange mechanism on $Mg_{1-x}Ni_x[OH]_2$ with $0.35<x<0.5$ (exchanging with $[OH]^-$). There is no change in symmetry of the tetraborate ion upon sorption; the selectivity is related to hydrogen bonding between the sorbent Ni–OH and B–OH groups with an intervening H_2O molecule [132, 133]. The best sorption of H_3BO_3 from aqueous solutions by the $Mg_{1-x}Ni_x[OH]_2$ sorbent is at pH 7.8 to 8.2 [134].

A new inorganic ion exchanger, zirconium phosphoborate, was prepared by mixing a 0.1M $ZrOCl_2$ solution with a hot mixture of 0.1M H_3BO_3 and 0.1M H_3PO_4 solutions in the mole ratio 2:1:1. It is used in the quantitative separation of lanthanides [135].

The adsorption of H_3BO_3 from aqueous solutions on zirconium oxide hydrate powders between 25 and 250°C has a maximum at pH$\approx$9 with a maximal capacity of 2.5×10^{-5} mol/m², corresponding to a monomolecular covering. The small reaction enthalpy of 2 kJ/mol corresponds to the small energy of hydration of undissociated boric acid [136].

References for 3.5.7.3:

[1] Hershey, J. P.; Fernandez, M.; Milne, P. J.; Millero, F. J. (Geochim. Cosmochim. Acta **50** [1986] 143/8).

[2] Brandani, V.; Del Re, G.; Di Giacomo, G. (J. Solution Chem. **17** [1988] 429/34).

[3] Rowe, Le Ann M.; Tran, Loc Binh; Atkinson, G. (J. Solution Chem. **18** [1989] 675/89).

[4] Buxton, G. V.; Sellers, R. M. (Radiat. Phys. Chem. **29** [1987] 137/40 from C.A. **106** [1987] No. 129089).

[5] Chiodini, G.; Comodi, P.; Giaquinto, S. (Geothermics **17** [1988] 711/8 from C.A. **110** [1989] No. 234874).

[6] Mallo, P.; Waton, G.; Mellen, R. (Nouv. J. Chim. **9** [1985] 413/7 from C.A. **103** [1985] No. 129875).

[7] Waton, G.; Mallo, P. (Acustica **64** [1987] 219/20 from C.A. **109** [1988] No. 61698).

[8] Mairanovskii, S. G.; Bishimbaeva, G. K.; Odnoral, L. V. (Elektrokhimiya **21** [1985] 216/22; Sov. Elektrochem. **21** [1985] 197/203 from C.A. **103** [1985] No. 13433); Mairanovskii, S. G.; Seilova, K. S.; Globovskaya, T. V.; Loshkarev, Yu. M. (Elektrokhimiya **23** [1987] 981/4 from C.A. **107** [1987] No. 143543).

[9] Oehman, L. O.; Sjoeberg, S. (Mar. Chem. **17** [1985] 91/7 from C.A. **103** [1985] No. 128650).

[10] Rush, J. D.; Kirschenbaum, L. J. (Polyhedron **4** [1985] 1573/8).

[11] Simonson, J. M.; Roy, R. N.; Roy, L. N.; Johnson, D. A. (J. Solution Chem. **16** [1987] 791/803).

[12] Simonson, J. M.; Roy, R. N.; Mrad, D.; Lord, P.; Roy, L. N.; Johnson, D. A. (J. Solution Chem. **17** [1988] 435/46).

[13] Jovancicevic, V.; Bockris, J. O'M. (J. Electrochem. Soc. **133** [1986] 1797/807).

[14] Jovancicevic, V.; Zelenay, P.; Scharifker, B. R. (Electrochim. Acta **32** [1987] 1553/5).

[15] Scharifker, B. R.; Habib, M. A.; Carbajal, J. L.; Bockris, J. O'M. (Surf. Sci. **173** [1986] 97/105 from C.A. **105** [1986] No. 121479).

[16] Bousher, A.; Brimblecombe, P.; Midgley, D. (J. Chem. Soc. Dalton Trans. **1987** 943/6).

[17] Weres, O.; Tsao, L. (J. Phys. Chem. **90** [1986] 3014/8).

[18] Nakai, S.; Aoi, H.; Hayashi, K.; Katoh, T.; Watanabe, T. (J. Nucl. Sci. Technol. **25** [1988] 65/71 from C.A. **108** [1988] No. 120705).

[19] Kamars, A.; Shvarts, E. M. (Latv. PSR Zinat. Akad. Vestis Kim. Ser. **1985** 629/31).

[20] Kamars, A.; Shvarts, E. M. (Latv. PSR Zinat. Akad. Vestis Kim. Ser. **1985** 703/8).

[21] Balz, R.; Brändle, U.; Kammerer, E.; Köhnlein, D.; Lutz, O.; Nolle, A.; Schafitel, R.; Veil, E. (Z. Naturforsch. **41a** [1986] 737/42).

[22] Lutz, O.; Brändle, U.; Helber, H.; Kammerer, E.; Scheiter, A. (Z. Naturforsch. **41a** [1986] 1041/4).

[23] Lajunen, K.; Hakkinen, P.; Purokoski, S. (Finn. Chem. Lett. **13** [1986] 21/5).

[24] Mikhov, B. M. (Electrophoresis [Weinheim, FRG] **7** [1986] 150/1).

[25] Usdowski, E.; Hoefs, J. (Earth Planet. Sci. Lett. **80** [1986] 130/4 from C.A. **106** [1987] No. 9640).

[26] Gode, H.; Naidenova, L. P.; Chemodanov, D. I.; Chernyak, M. Sh. (Latv. PSR Zinat. Akad. Vestis Kim. Ser. **1986** 407/10 from C.A. **105** [1986] 196285).

[27] Gode, H.; Naidenova, L. P.; Chemodanov, D. I. (Latv. PSR Zinat. Akad. Vestis Kim. Ser. **1989** 267/72).

[28] Gode, H.; Naidenova, L. P.; Chemodanov, D. I. (Latv. PSR Zinat. Akad. Vestis Kim. Ser. **1989** 151/5).

[29] Skvortsov, V. G. (Zh. Neorg. Khim. **30** [1985] 2341/52; Russ. J. Inorg. Chem. **30** [1985/86] 1331/7).

[30] Skvortsov, V. G.; Tsekhanskii, R. S.; Molodkin, A. K.; Druzhinin, I. G. (Dokl. Akad. Nauk SSSR **302** [1988] 349/50; Dokl. Chem. Proc. Acad. Sci. USSR **302** [1988] 270/2).

[31] Skvortsov, V. G.; Molodkin, A. K.; Tsekhanskii, R. S.; Sadetdinov, Sh. V.; Nikonov, F. V. (Zh. Neorg. Khim. **30** [1985] 826/9; Russ. J. Inorg. Chem. **30** [1985] 464/5).
[32] Song, Pengsheng; Du, Xianhui (Kexue Tongbao [Foreign Lang. Ed.] **31** [1986] 1338/43 from C.A. **106** [1987] No. 39280).
[33] Arkhipov, S. M.; Kashina, N. I.; Kidyarov, B. I.; Kuzina, V. A. (Zh. Neorg. Khim. **31** [1986] 217/9; Russ. J. Inorg. Chem. **31** [1986] 123/4).
[34] Guo, Z.-Z.; Chen, P.-H.; Chen, Y.-S. (Wuji Huaxue **2** [1986] 113/7 from C.A. **106** [1987] No. 202683).
[35] Karazhanov, N. A.; Omarova, R. S.; Isenzhulova, U. T. (Izv. Akad. Nauk Kaz. SSR Ser. Khim. **1987** No. 1, pp. 21/3 from C.A. **106** [1987] No. 144943).
[36] Dalinina, R. M. (Sintez Svoistva Neorgan. Soedin. Baku **1984** 12/5 from C.A. **103** [1985] No. 110792).
[37] Skvortsov, V. G.; Tsekhanskii, R. S.; Petrova, O. V.; Molodkin, A. K.; Belova, V. F. (Zh. Neorg. Khim. **34** [1989] 500/2; Russ. J. Inorg. Chem. **34** [1989] 279/81).
[38] Gode, H.; Svirkst, J. (Latv. PSR Zinat. Akad. Vestis Kim. Ser. **1985** 548/52).
[39] Gao, Sh.-Y.; Chen, Z.-G.; Feng, J. N. (Wuji Huaxue **2** [1986] 40/52 from C.A. **107** [1987] No. 45794).
[40] Gao, Sh.-Y.; Xu, K.-F.; Li, G.; Feng, J. N. (Huaxue Xuebao **44** [1986] 1229/33 from C.A. **106** [1987] No. 143542).

[41] Song, Pengsheng; Du, Xianhui; Sun, Bai (Kexue Tongbao [Foreign Lang. Ed.] **33** [1988] 1971/3 from C.A. **111** [1989] No. 13176).
[42] Khazikhanova, B. Kh.; Beremzhanov, B. A.; Savich, R. F.; Kalacheva, V. G. (Zh. Neorg. Khim. **30** [1985] 548/50; Russ. J. Inorg. Chem. **30** [1985] 308/10).
[43] Khazikhanova, B. Kh.; Beremzhanov, B. A.; Savich, R. F.; Kalacheva, V. G. (Zh. Neorg. Khim. **30** [1985] 1899/901; Russ. J. Inorg. Chem. **30** [1985/86] 1079/81).
[44] Berniyazova, D. G.; Kalacheva, V. G.; Shvarts, E. M. (Zh. Neorg. Khim. **31** [1986] 1045/8; Russ. J. Inorg. Chem. **31** [1986] 596/8).
[45] Sadetdinov, Sh. V.; Skvortsov, V. G.; Molodkin, A. K.; Shikhranov, D. A.; Polenov, A. D. (Zh. Neorg. Khim. **33** [1988] 2151/3; Russ. J. Inorg. Chem. **33** [1988/89] 1228/9).
[46] Sadetdinov, Sh. V. (Zh. Neorg. Khim. **30** [1985] 795/7; Russ. J. Inorg. Chem. **30** [1985] 445/6).
[47] Sadetdinov, Sh. V.; Polenov, A. D.; Shikhranov, D. A.; Skvortsov, V. G.; Molodkin, A. K.; Tsekhanskii, R. S. (Zh. Neorg. Khim. **30** [1985] 1344/6; Russ. J. Inorg. Chem. **30** [1985/86] 766/8).
[48] Sadetdinov, Sh. V.; Shikhranov, D. A.; Polenov, A. D.; Pavlov, G. P. (Zh. Neorg. Khim. **30** [1985] 1617/9; Russ. J. Inorg. Chem. **30** [1985/86] 923/4).
[49] Sadetdinov, Sh. V.; Skvortsov, V. G.; Shikhranov, D. A.; Polenov, A. D.; Syubaeva, V. V. (Zh. Neorg. Khim. **31** [1986] 1593/6; Russ. J. Inorg. Chem. **31** [1986/87] 913/4).
[50] Rodionov, N. S.; Skvortsov, V. G.; Tsekhanskii, R. S. (Fiz. Khim. Issled. Ravnovesii Rastvorakh, Yaroslavl **1986** 10/4 from C.A. **108** [1988] No. 174435).

[51] Tsekhanskii, R. S.; Skvortsov, V. G.; Molodkin, A. K.; Sukova, L. M.; Rodionov, N. S. (Zh. Neorg. Khim. **30** [1985] 1609/12; Russ. J. Inorg. Chem. **30** [1985] 917/9).
[52] Skvortsov, V. G.; Tsekhanskii, R. S.; Molodkin, A. K.; Rodionov, N. S.; Pavlov, G. P. (Zh. Neorg. Khim. **33** [1988] 1872/5; Russ. J. Inorg. Chem. **33** [1988/89] 1066/7).
[53] Skvortsov, V. G.; Tsekhanskii, R. S.; Molodkin, A. K.; Petrova, O. V.; Akimov, V. M. (Zh. Neorg. Khim. **30** [1985] 277/9; Russ. J. Inorg. Chem. **30** [1985] 158/9).

[54] Batsanov, A. S.; Petrova, O. V.; Struchkov, Yu. T.; Akimov, V. M.; Molodkin, A. K.; Skvortsov, V. G. (Zh. Neorg. Khim. **31** [1986] 1120/6; Russ. J. Inorg. Chem. **31** [1986] 637/40).
[55] Skvortsov, V. G.; Sadetdinov, Sh. V.; Tsekhanskii, R. S.; Petrova, O. V. (Zh. Neorg. Khim. **32** [1987] 2071/3; Russ. J. Inorg. Chem. **32** [1987/88] 1218/9).
[56] Petrova, O. V.; Skvortsov, V. G.; Tsekhanskii, R. S. (Fiz. Khim. Issled. Ravnovesii Rastvorakh, Yaroslavl **1986** 14/7 from C.A. **108** [1988] No. 174434).
[57] Tsekhanskii, R. S.; Skvortsov, V. G.; Molodkin, A. K. (Zh. Neorg. Khim. **31** [1986] 1961/4; Russ. J. Inorg. Chem. **31** [1986/87] 1129/31).
[58] Skvortsov, V. G.; Tsekhanskii, R. S.; Molodkin, A. K.; Druzhinin, I. G. (Dokl. Akad. Nauk USSR **294** [1987] 621/2; Dokl. Chem. Proc. Acad. Sci. USSR **292/297** [1987] 651/2).
[59] Skvortsov, V. G. (Zh. Neorg. Khim. **31** [1986] 3163/72; Russ. J. Inorg. Chem. **31** [1986/87] 1816/22).
[60] Tsekhanskii, R. S.; Skvortsov, V. G.; Molodkin, A. K.; Sukova, L. M. (Zh. Neorg. Khim. **32** [1987] 1488/91; Russ. J. Inorg. Chem. **32** [1987] 893/5).

[61] Tsekhanskii, R. S.; Skvortsov, V. G.; Sukova, L. M.; Molodkin, A. K. (Zh. Neorg. Khim. **32** [1987] 1331/4; Russ. J. Inorg. Chem. **32** [1987] 804/6).
[62] Tsekhanskii, R. S.; Skvortsov, V. G.; Molodkin, A. K.; Sadetdinov, Sh. V. (Zh. Neorg. Khim. **31** [1986] 2987/9; Russ. J. Inorg. Chem. **31** [1986/87] 1720/2).
[63] Skvortsov, V. G.; Tsekhanskii, R. S.; Molodkin, A. K.; Pavlov, G. P.; Rodionov, N. S.; Sharzhanov, A. A. (Zh. Neorg. Khim. **33** [1988] 1033/7; Russ. J. Inorg. Chem. **33** [1988] 585/7).
[64] Skvortsov, V. G.; Tsekhanskii, R. S.; Molodkin, A. K.; Petrova, O. V.; Belova, V. F. (Zh. Neorg. Khim. **34** [1989] 1080/1; Russ. J. Inorg. Chem. **34** [1989] 608/9).
[65] Sadetdinov, Sh. V. (Zh. Neorg. Khim. **30** [1985] 1911/3; Russ. J. Inorg. Chem. **30** [1985/86] 1087/8).
[66] Skvortsov, V. G.; Tsekhanskii, R. S.; Petrova, O. V.; Belova, V. F. (Zh. Neorg. Khim. **34** [1989] 997/9; Russ. J. Inorg. Chem. **34** [1989] 559/60).
[67] Skvortsov, V. G.; Rodionov, N. S.; Molodkin, A. K.; Fedorov, Yu. A.; Tsekhanskii, R. S. (Zh. Neorg. Khim. **30** [1985] 1913/5; Russ. J. Inorg. Chem. **30** [1985/86] 1088/9).
[68] Skvortsov, V. G.; Fedorov, Yu. A.; Molodkin, A. K.; Tsekhanskii, R. S. (Zh. Neorg. Khim. **31** [1986] 1589/90; Russ. J. Inorg. Chem. **31** [1986] 910/1).
[69] Skvortsov, V. G.; Fedorov, Yu. A.; Molodkin, A. K. (Zh. Neorg. Khim. **31** [1986] 2160/3; Russ. J. Inorg. Chem. **31** [1986/87] 1246/7).
[70] Skvortsov, V. G.; Tsekhanskii, R. S.; Fedorov, Yu. A.; Molodkin, A. K.; Sukova, L. M. (Zh. Neorg. Khim. **32** [1987] 2843/8; Russ. J. Inorg. Chem. **31** [1987/88] 1653/6).

[71] Skvortsov, V. G.; Tsekhanskii, R. S.; Sukova, L. M.; Molodkin, A. K.; Rodionov, N. S. (Zh. Neorg. Khim. **31** [1986] 2140/2; Russ. J. Inorg. Chem. **31** [1986/87] 1233/4).
[72] Polenov, A. D.; Skvortsov, V. G.; Molodkin, A. K.; Sadetdinov, Sh. V.; Petrova, O. V. (Zh. Neorg. Khim. **32** [1987] 2069/71; Russ. J. Inorg. Chem. **32** [1987/88] 1217/8).
[73] Rodionov, N. S. (Deposited Doc. VINITI-5-037-84 [1984] 87/8 from C.A. **103** [1985] No. 152515).
[74] Skvortsov, V. G.; Tsekhanskii, R. S.; Molodkin, A. K.; Petrova, O. V. (Zh. Neorg. Khim. **32** [1987] 2317/9; Russ. J. Inorg. Chem. **32** [1987/88] 1355/6).
[75] Skvortsov, V. G.; Molodkin, A. K.; Rodionov, N. S.; Tsekhanskii, R. S. (Zh. Neorg. Khim. **31** [1986] 1591/3; Russ. J. Inorg. Chem. **31** [1986] 911/3).
[76] Skvortsov, V. G.; Molodkin, A. K.; Sadetdinov, Sh. V.; Tsekhanskii, R. S. (Zh. Neorg. Khim. **31** [1986] 1627/9; Russ. J. Inorg. Chem. **31** [1986] 934/5).
[77] Skvortsov, V. G.; Tsekhanskii, R. S.; Molodkin, A. K.; Petrova, O. V.; Akimov, V. M. (Zh. Neorg. Khim. **33** [1988] 1621/3; Russ. J. Inorg. Chem. **33** [1988] 921/3).

[78] Skvortsov, V. G.; Tsekhanskii, R. S.; Molodkin, A. K.; Fedorov, Yu. A.; Sadetinov, Sh. V. (Zh. Neorg. Khim. **30** [1985] 1092/4; Russ. J. Inorg. Chem. **30** [1985] 618/9).
[79] Rodionov, N. S.; Skvortsov, V. G.; Molodkin, A. K.; Tsekhanskii, R. S. (Zh. Neorg. Khim. **30** [1985] 1334/6; Russ. J. Inorg. Chem. **30** [1985] 760/1).
[80] Skvortsov, V. G.; Petrova, O. V.; Molodkin, A. K.; Tsekhanskii, R. S. (Zh. Neorg. Khim. **32** [1987] 1743/5; Russ. J. Inorg. Chem. **32** [1987/88] 1033/4).

[81] Skvortsov, V. G.; Tsekhanskii, R. S.; Molodkin, A. K.; Petrova, O. V. (Zh. Neorg. Khim. **33** [1988] 233/5; Russ. J. Inorg. Chem. **33** [1988] 130/2).
[82] Hoare, J. P. (J. Electrochem. Soc. **133** [1986] 2491/4).
[83] Cabot, P. L.; Centellas, F. A.; Garrido, J. A.; Perez, E. (J. Appl. Electrochem. **17** [1987] 104/12).
[84] John, S.; Balasubramanian, V.; Shenvi, B. A. (Surf. Technol. **26** [1985] 207/16 from C.A. **104** [1986] No. 58370).
[85] Shimizu, K.; Kobayashi, K.; Thompson, G. E.; Wood, G. C. (J. Appl. Electrochem. **15** [1985] 781/3).
[86] Crevecoeur, C.; DeWit, H. J. (J. Electrochem. Soc. **134** [1987] 808/16).
[87] Buchanan, J. S.; Freestone, N. P.; Peter, L. M. (J. Electroanal. Chem. Interfacial Electrochem. **182** [1985] 383/98 from C.A. **102** [1985] No. 102371).
[88] Birss, V. I.; Waudo, W. (Can. J. Chem. **67** [1989] 1098/104).
[89] Varsanyi, M. L.; Jaen, J.; Vertes, A.; Kiss, L. (Electrochim. Acta **30** [1985] 529/33).
[90] Varsanyi, M. L.; Kovacs, E.; Sziraki, L.; Kiss, L. (Magy. Kem. Foly. **91** [1985] 548/54 from C.A. **104** [1986] No. 114197).

[91] Leach, J. S. L.; Panagopoulos, C. N. (Electrochim. Acta **31** [1986] 1577/8).
[92] Panagopoulos, C. N.; Badekas, H. (J. Less-Common Met. **133** [1987] 245/53).
[93] Hoare, J. P. (Proc. AESF Annu. Tech. Conf. **72** [1985] 18 pp. from C.A. **105** [1986] No. 87468).
[94] Hoare, J. P. (J. Electrochem. Soc. **134** [1987] 3102/3).
[95] Hoare, J. P. (Proc. Electrochem. Soc. **87-17** [1987] 269/84 from C.A. **108** [1988] No. 64610).
[96] Nikolova, M.; Vilutiene, V.; Krustev, I.; Sarmaitis, R. (Izv. Khim. **21** [1988] 90/100 from C.A. **109** [1988] No. 45072).
[97] Azumi, K.; Ohtsuka, T.; Sato, N. (J. Electrochem. Soc. **134** [1987] 1352/7).
[98] Höpfner, W.; Plieth, W. J. (Werkst. Korros. **36** [1985] 373/80 from C.A. **103** [1985] No. 168681).
[99] Oltra, R.; Indrianjafy, G.; Lambertin, M.; Colson, J. C.; Keddam, M. (Compt. Rend. [2] **303** II [1986] 673/6).
[100] Vela, M. E.; Vilche, J. R.; Arvia, A. J. (J. Appl. Electrochem. **16** [1986] 490/504).

[101] Bird, H. E. H.; Pearson, B. R.; Brook, P. A. (Corros. Sci. **28** [1988] 81/6 from C.A. **108** [1988] No. 120981).
[102] MacDougall, B.; Bardwell, J. A.; Graham, M. J. (J. Electrochem. Soc. **135** [1988] 340/2).
[103] Bardwell, J. A.; MacDougall, B. (J. Electrochem. Soc. **135** [1988] 2157/61).
[104] MacDougall, B.; Bardwell, J. A. (J. Electrochem. Soc. **135** [1988] 2437/41).
[105] Bardwell, J. A.; MacDougall, B.; Sproule, G. I. (J. Electrochem. Soc. **136** [1989] 1331/6).
[106] Ramasubramanian, N.; Preocanin, N.; Davidson, R. D. (J. Electrochem. Soc. **132** [1985] 793/8).
[107] Janik-Czachor, M. (J. Electrochem. Soc. **132** [1985] 306/9).
[108] Das, D. S.; Subbaiah, T. (J. Appl. Electrochem. **17** [1987] 675/83).
[109] Karwas, C.; Hepel, T. (J. Electrochem. Soc. **136** [1989] 1672/8).
[110] Nishimura, R. (Corrosion [Houston] **43** [1987] 486/92).

[111] Hung, Aina; Chen, Ker Ming (J. Electrochem. Soc. **136** [1989] 72/5).
[112] Karwas, C.; Hepel, T. (J. Electrochem. Soc. **135** [1988] 839/44).
[113] Dueber, R. E.; Fritts, D. H. (J. Electrochem. Soc. **133** [1986] 1292/6).
[114] Figuera, M. G.; Salvarezza, R. C.; Arvia, A. J. (Electrochim. Acta **31** [1986] 665/9).
[115] De Chialvo, M. R. G.; Salvarezza, R. C.; Vasquez Moll, D.; Arvia, A. J. (Electrochim. Acta **30** [1985] 1501/11).
[116] De Chialvo, M. R. G.; De Mele, M. F. L.; Salvarezza, R. C.; Arvia, A. J. (Corros. Sci. **28** [1988] 121/34 from C.A. **108** [1988] No. 158020).
[117] Bogatyrev, V. L. (Zh. Fiz. Khim. **60** [1986] 1981/4; Russ. J. Phys. Chem. 1186/8).
[118] Kisel'gof, G. V.; Arkhangel'skii, L. K.; Shosheva, N. A.; Sorokin, A. Ya.; Kuznetsova, V. A. (Zh. Prikl. Khim. **60** [1987] 2418/21; J. Appl. Chem. [USSR] **60** [1987] 2239/42 from C.A. **108** [1988] No. 82572).
[119] Zhaimina, R. E.; Filippova, Z. O.; Kosenko, G. P. (Izv. Akad. Nauk Kaz. SSR Ser. Khim. **1985** No. 1, pp. 83/7 from C.A. **102** [1985] No. 155398).
[120] Zhaimina, R. E.; Balapanova, B. S. (Izv. Akad. Nauk Kaz. SSR Ser. Khim. **1985** No. 2, pp. 9/13 from C.A. **102** [1985] No. 210161).

[121] Znamenskii, Yu. P.; Davydova, G. N. (Zh. Fiz. Khim. **62** [1988] 1673/5; Russ. J. Phys. Chem. **62** [1988] 853/5).
[122] Meichik, N. R.; Leikin, Yu. A.; Antipov, M. A.; Goryacheva, N. V.; Klimenko, I. A.; Medvedev, S. A.; Galitskaya, N. B. (Zh. Prikl. Khim. **60** [1987] 1970/4; J. Appl. Chem. [USSR] **60** [1987] 1820/4 from C.A. **109** [1988] No. 98461).
[123] Meichik, N. R.; Leikin, Yu. A.; Galitskaya, N. B.; Kosaeva, A. E. (Zh. Fiz. Khim. **63** [1989] 540/3; Russ. J. Phys. Chem. **63** [1989] 297/9).
[124] Meichik, N. R.; Leikin, Yu. A. (Zh. Fiz. Khim. **63** [1989] 543/6; Russ. J. Phys. Chem. **63** [1989] 299/301).
[125] Meichik, N. R.; Leikin, Yu. A.; Galitskaya, N. B.; Kosaeva, A. E. (Zh. Fiz. Khim. **63** [1989] 1871/4; Russ. J. Phys. Chem. **63** [1989] 1024/5).
[126] Hicks, K. B.; Simpson, G. L.; Bradbury, A. G. W. (Carbohydr. Res. **147** [1986] 39/48).
[127] Sahin, S.; Topacik, D. (Bull. Tech. Univ. Istanbul **38** [1985] 1/10 from C.A. **107** [1987] No. 204515).
[128] Langevin, D.; Metayer, M.; Labbe, M.; Pollet, B.; Hankaoui, M.; Selegny, E. (Desalination **68** [1988] 131/48 from C.A. **108** [1988] No. 227417).
[129] Metayer, M.; Langevin, D.; Hankaoui, M.; Pollet, B.; Labbe, M.; Roudesli, S. (React. Polym. Ion Exch. Sorbents **7** [1988] 111/22 from C.A. **108** [1988] No. 193278).
[130] Oderda, G. M.; Klein-Schwartz, W.; Insley, B. M. (J. Toxicol. Clin. Toxicol. **25** [1987] 13/9 from C.A. **107** [1987] No. 53795).

[131] Vol'khin, V. V.; Tomchuk, T. K.; Leont'eva, G. V. (Izv. Akad. Nauk Turkm. SSR Ser. Fiz. Tekh. Khim. Geol. Nauk **1986** No. 3, pp. 44/8 from C.A. **105** [1986] No. 179020).
[132] Vol'khin, V. V.; Leont'eva, G. V.; Tomchuk, T. K. (Izv. Akad. Nauk Turkm. SSR Ser. Fiz. Tekh. Khim. Geol. Nauk **1986** No. 4, pp. 65/8 from C.A. **106** [1987] No. 126484).
[133] Leont'eva, G. V.; Tomchuk, T. K.; Annanurova, M. A.; Vol'khin, V. V. (Izv. Akad. Nauk Turkm. SSR Ser. Fiz. Tekh. Khim. Geol. Nauk **1986** No. 6, pp. 73/7 from C.A. **107** [1987] No. 184304).
[134] Tomchuk, T. K.; Vol'khin, V. V.; Leont'eva, G. V. (Izv. Akad. Nauk Turkm. SSR Ser. Fiz. Tekh. Khim. Geol. Nauk **1986** No. 5, pp. 61/6 from C.A. **106** [1987] No. 126489).
[135] Thind, P. S.; Mittal, S. K.; Gujval, S. (Synth. React. Inorg. Met.-Org. Chem. **18** [1988] 593/607 from C.A. **110** [1989] No. 32843).
[136] Reimann, R.; Bosholm, J. (Kernenergie **32** [1989] 281/3 from C.A. **111** [1989] No. 85722).

3.5.7.4 Reactions with Inorganic Species

Boron nitride can be prepared by thermal decomposition of organic nitrogen compounds in the presence of H_3BO_3; see "Boron Compounds" 4th Suppl. Vol. 3a, 1991, pp. 2/3. The reactions between H_3BO_3 and agents such as urea, biuret, triuret, dicyandiamide, cyanuric acid [1], ammeline, ammelide, melamine, melem, and melone, e.g., at 150°C in an evaporator [2] gave precursors for the boron nitride production. The subsequent pyrolytic processes were investigated by thermogravimetric, differential thermoanalytic, X-ray diffraction analytic, and IR spectroscopic methods. The thermal treatment of the condensation product of H_3BO_3 and glycerol under N_2 gave a ceramic with the partial structure $(C_3H_5BO_3)_n$, containing B–O–C bonds. BN and B_4C were detected above 1300°C under N_2 [3]. The same products were obtained by the pyrolysis of a mixture of H_3BO_3 and diethanolamine or triethanolamine [4]. The conditions for the formation of cubic boron nitride films by plasma excitement of reactive gases such as H_2 and NH_3 using radiofrequency induction and tungsten-filament heating of H_3BO_3 and other boron compounds were studied [5].

The degradation rates of the radiolysis of hydrazine in the presence or absence of atmospheric oxygen in an H_3BO_3 solution are given as a function of the applied radiation dose, pH, and temperature; the reaction mechanism was discussed [6]. Second-order rate constants are reported for the reaction of aqueous solutions of H_3BO_3 with O_3 depending on the pH value [7].

The oxygen reduction on platinum in borate-buffered (pH 8.2) 0.5 M NaCl or $Na[ClO_4]$ solutions is inhibited by Cl^- ions, as was shown by 30 to 40 mV adsorption shifts in the polarization curves to more cathodic potentials in rotating ring-disk electrode studies [8].

H_3BO_3 reacts in an alkaline solution with H_2O_2, as ^{11}B NMR studies show, with the formation of the ions $[B(OH)_4]^-$, $[B(OOH)_4]^-$, $[B_2(O_2)_2(OH)_4]^{2-}$, and $[B_2(O_2)_2(OOH)_4]^{2-}$. An acid solution of H_3BO_3 in H_2O_2 at various concentrations did not form any peroxoboron compounds [9]. As ^{11}B NMR spectra show, H_3BO_3 reacts with H_2O_2 in aqueous solution (containing ca. 25% D_2O as a frequency lock) at 25°C and at an ionic strength of 0.1 M ($K[NO_3]$) at pH ≈ 9 to form the ions $[B(OH)_3(OOH)]^-$, $[B(OH)_2(OOH)_2]^-$, and, in contrast to [9], the acid $B(OH)_2(OOH)$, which is a stronger acid than H_3BO_3 (see also Section 3.7.2, p. 289) [10].

In the presence of H_3BO_3, the H_2O_2 bath stability decreases with an increase of the NaOH concentration [11]. The thermal decomposition of H_2O_2 was studied kinetically at various pH values using H_3BO_3-buffered baths. The bleaching efficiency of H_2O_2 in a borate-silicate buffer was optimized between pH 8.0 and 8.5 and its bleaching time could be reduced by 33% [12].

A solution of H_3BO_3 was adjusted with MOH (M = Na, K, NH_4) solution up to pH 9, then treated with HF solution and cooled to 0°C. After addition of 30% H_2O_2 and adjustment to pH 9 with additional MOH, the peroxofluoroborates $Na_2[B(O_2)F_3]\cdot 4H_2O$, $K_2[B(O_2)F_3]\cdot 4H_2O$, and $[NH_4]_2[B_2(O_2)_3F_2]$ were obtained. The yields were 92%, 74%, and 72%, respectively. IR bands at 1050 cm^{-1} (ν(BF)) and 860 cm^{-1} (ν(O–O)) were assigned and used to discuss structural aspects [13]. Detailed ^{11}B and ^{19}F NMR studies in the system H_3BO_3–MOH–MF–H_2O_2–H_2O with M = Li, Na, K, or NH_4 indicated the formation of the anions $[BF_4]^-$, $[BF_3(OOH)]^-$, $[BF_2(OOH)_2]^-$, $[BF(OOH)_3]^-$, $[B(OOH)_4]^-$, $[BF_3(OH)]^-$, $[B_2F_4(O_2)_2]^{2-}$, and $[B_2F_2(O_2)_2(OOH)_2]^{2-}$. An increase in the H_2O_2 concentration increases the amounts of $[BF(OOH)_3]^-$ and $[B(OOH)_4]^-$. The change in composition of the products was also studied with the increase of F^- concentration and an increase of time [14].

The existence of I_2-H_3BO_3 complexes in saturated steam at 285°C is assumed at sufficiently high levels of H_3BO_3. The ionized species do not behave like normal gaseous molecules; rather, their behavior is likely to be determined by the transformations they undergo in temperature and pressure gradients [15]. Solutions of iodide ion and H_3BO_3 were evaporated to dryness in a gas flow, and the quantities of volatilized (HI) and residual iodine were both pH and flow-rate dependent [16].

H_3BO_3 reacts with $H_2S_2O_7$ (as oleum of the appropriate concentration) to give $H[B(OSO_2OH)_4]$ as a solution in 100% H_2SO_4 or in oleum, which is used to accelerate sulfonation processes [17].

The formation of borosilicate glass from H_3BO_3 and $Si(OR)_4$ and the subsequent hydrolysis by the sol-gel process and heat treatment has been studied by IR and NMR spectroscopy. At the time of gelation, the majority of the boron is in the form of H_3BO_3, which condenses to form borosiloxane bonds only upon heat treatment of the dried gel. For compounds having higher boron contents, a small number of B–O–B bonds are formed which do not convert to B–O–Si bonds upon heating to 500°C [18].

A compound $36\,SiO_2 \cdot 2.5\,H_3BO_3 \cdot 2(H_2NCH_2CH_2NH_2) \cdot 4\,H_2O$, an Al-free ferrierite, with boron as a nonframework constituent, was crystallized in a one-step process from aqueous solutions of SiO_2 in the presence of boric acid and ethylenediamine as templates; a single crystal has the orthorhombic space group Immm–D_{2h}^{25} (No. 71) with a=1855.7(6), b=1388.9(3), and c=724.9(9) pm; R_w=10.6% [19].

The system H_3BO_3–HIO_3–H_2O has been studied at 25°C; the composition of the eutonic solution was determined (HIO_3=72.2 wt%, H_3BO_3=0.83 wt%) [20]. In the system H_3BO_3–LiOH–$Li[IO_3]$–H_2O at 25°C, crystallization branches for $LiBO_2 \cdot 8\,H_2O$, $Li_2[B_4O_7] \cdot 3\,H_2O$, and α-$Li[IO_3]$ were found [21]. By reacting H_3BO_3 and $Li_2[CO_3]$ in a nearly boiling aqueous solution and then drying at ca. 110°C, white hydrated $Li_2[B_4O_7]$ was prepared. Dehydration of this material at 800°C for 30 min gives anhydrous $Li_2[B_4O_7]$, which was used for X-ray fluorescence analysis [22]. By adding H_3BO_3 and $LiOH \cdot H_2O$ to water, reaction at 45°C leads to crystalline $Li_2[B_4O_7] \cdot 3\,H_2O$, formulated as β-$Li_2[B_4O_5(OH)_4] \cdot H_2O$ [23]. For hydrated tetraborates, see Section 3.6.5, p. 273.

H_3BO_3 reacts with $Na_2[CO_3]$ to form tincalconite, $Na_2[B_4O_5(OH)_4] \cdot 3\,H_2O$, by a mechanochemical method [24].

$K_2[B_4O_5(OH)_4] \cdot 2\,H_2O$ was prepared from KOH and H_3BO_3, and its IR spectra and Debye diffractograms were obtained [25]. Crystalline $K_2O \cdot MgO \cdot 3\,B_2O_3 \cdot 9\,H_2O$ was prepared from solidifying mixtures of noncrystalline $MgO \cdot 0.8\,B_2O_3 \cdot 6.76\,H_2O$, $K_2O \cdot 2\,B_2O_3 \cdot 4\,H_2O$, and H_3BO_3 in the presence of small amounts of water; its formation was confirmed by X-ray diffraction analysis [26].

Mixtures of H_3BO_3 with $Cs_2[CO_3]$, with $Cs_2[CO_3]$ and $[NH_4][H_2PO_4]$, with $Na_2[CO_3]$ and $[NH_4][H_2PO_4]$, as well as with $K_2[B_2O_4] \cdot 2.5\,H_2O$ and $[NH_4][H_2PO_4]$ were ground for several months and then exposed to air between 20 and 25°C; reactions were observed in which water of crystallization plays a determining role [27].

In the 100°C isotherm of the H_3BO_3–TlOH system, the compounds $Tl_2[B_4O_6(OH)_2] \cdot 2\,H_2O$ and $Tl_4[B_8O_{12}(OH)_4] \cdot H_2O$ were found [28]. $Tl_2O \cdot 2\,B_2O_3 \cdot 3\,H_2O$ was crystallized from H_3BO_3 and TlOH. From X-ray diffraction analysis and thermal analytical methods, the species was formulated as $Tl_2[B_4O_6(OH)_2] \cdot 2\,H_2O$ [29].

The distribution of borate anions in a solution (e.g., in systems such as H_3BO_3–MgO–H_2O–$Na_2[SO_4]$ from preobrazhenskite and aqueous alkali metal sulfates) strongly depends on the H_2O activity; therefore the "solubility" of borates in salt solutions (e.g., $Na_2[SO_4]$ solution) is increased. Salts such as $Mg_2[B_2O_5] \cdot 2\,H_2O$ crystallized [30].

In the H_3BO_3–$Ca[OH]_2$–propionic acid–H_2O system at 60°C, crystallization branches exist for H_3BO_3, $Ca[B_6O_{10}] \cdot 6\,H_2O$, $Ca[B_6O_{10}] \cdot 5\,H_2O$, $Ca[B_4O_7] \cdot 6\,H_2O$, and calcium propionate [31]; labeled borates were also prepared [32].

References on pp. 204/6

From aqueous solutions of H_3BO_3/KOH (ratio 0.25:1 to 1:1) in excess and $BaCl_2 \cdot 2H_2O$, $Ba[B(OH)_4]_2$ has been prepared; also, IR spectra of $Ca[B_3O_3(OH)_5] \cdot H_2O$ and $Ba[B_3O_3(OH)_5] \cdot H_2O$ were given [33].

The interaction of H_3BO_3 with $[NH_4]Cl$-kaolinite intercalate (5:1 mole ratio) between 563 and 593 K gives an apparent activation energy of 101.65 kJ/mol for 50% replacement of $[NH_4]Cl$ [34]. According to ^1H, ^{11}B, ^{27}Al, and ^{29}Si NMR spectra, the structural bonds of the H_3BO_3-kaolinite intercalates are disordered. The structure contains polymers of $B(O,OH)_3$ triangles and $B(O,OH)_4$ tetrahedra, which are linked by common oxygen atoms [35]. A new microporous aluminoborate was prepared hydrothermally from $CaO–Al_2O_3–B_2O_3$–tetraethylammonium hydroxide–water systems. Hydrothermal crystallization of a reaction mixture with a molecular composition $0.6\,CaO \cdot Al_2O_3 \cdot 6\,B_2O_3 \cdot 0.5$ tetraethylammonium hydroxide $\cdot\, 150\,H_2O$ at 200°C for 6 days yielded a compound $CaO \cdot Al_2O_3 \cdot 9.6\,B_2O_3 \cdot 4.3\,H_2O$, which was characterized by X-ray powder diffraction analysis, IR data, composition and thermal analyses, and gas adsorption [36].

From aqueous mixtures of H_3BO_3 and $M[OC(O)CH_3]_2$ (M = Mn, Co, Ni, Cu, Zn), the crystalline compounds $M[OC(O)CH_3]_2 \cdot 2\,H_3BO_3 \cdot n\,H_2O$ have been isolated and were characterized by X-ray diffraction analysis, differential thermal analysis (DTA), IR, and near UV/VIS spectrometry [37]. The solubility in the system $H_3BO_3–NiCl_2–ZnCl_2–H_2O$ (for a limited concentration range of $NiCl_2$) at 15, 20, and 25°C decreases slightly with increasing contents of $ZnCl_2$ and H_3BO_3 [38].

The thermal decomposition of an $H_3BO_3/Nd(OC_3H_7)_3$ mixture (1:1) at 800°C leads to the formation of $Nd[BO_3]$, which is stable and is not decomposed by water or alkaline solutions [42].

Fe_3BO_6 is formed in addition to $Fe[BO_3]$ in the system $H_3BO_3–Fe_3O_4$ at 600°C [39]. The reaction of $Ca[VO_3]_2 \cdot 4\,H_2O$, which is presumed to pass into solution as $Ca_3[V_{10}O_{28}]$, with H_3BO_3 was studied at various acidities; the rate of hydrolysis depends on the pH value; between pH 5 and 6, essentially no interaction occurs [40]. $Mn[VO_3]_2 \cdot 4\,H_2O$ does not react with H_3BO_3 at 60°C and pH 4 to 5, possibly due to the formation of sparingly soluble manganese borates [41].

For the formation of heteropoly acids of tungsten and vanadium or their salts, respectively, like $H_5[BW_{12}O_{40}] \cdot 30\,H_2O$ or $Cs_2[HBV_6O_{18} \cdot 2.15\,H_2O] \cdot 1.5\,H_2O$, typically starting from, e.g., H_3BO_3 and $Na_2[WO_4]$, see Section 3.6.8, p. 281.

References for 3.5.7.4:

[1] Gontarz, Z.; Podsiadlo, S. (Pol. J. Chem. **58** No. 1/3 [1984] 3/11, 13/21 from C.A. **102** [1985] No. 124446 and No. 124447).
[2] Wada, H.; Ito, Sh.; Kuroda, K.; Kato, Ch. (Chem. Lett. **1985** 691/2).
[3] Wada, H.; Kuroda, K.; Kato, Ch. (Yogyo Kyokaishi **94** [1986] 61/5 from C.A. **104** [1986] No. 93927; Mater. Sci. Res. **20** [1986] 179/85 from C.A. **106** [1987] No. 124507).
[4] Wada, H.; Nojima, K.; Kuroda, K.; Kato, Ch. (Yogyo Kyokaishi **95** [1987] 130/4 from C.A. **106** [1987] No. 89061).
[5] Saitoh, H.; Hirose, T.; Matsui, H.; Hirotsu, Y.; Ichinose, Y. (Surf. Coat. Technol. **39/40** [1989] 265/73 from C.A. **112** [1990] No. 169351).
[6] Langguth, H.; Krauss, K. H. (4th Work. Meet. Radiat. Interact., Leipzig 1987 [1988], pp. 599/604 from C.A. **111** [1989] No. 183966).
[7] Hoigne, J.; Bader, H.; Haag, W. R.; Staehelin, J. (Water Res. **19** [1985] 993/1004 from C.A. **103** [1985] No. 128705).
[8] Kaska, S. M.; Sarangapani, S.; Giner, J. (J. Electrochem. Soc. **136** [1989] 75/83).

[9] Chernyshov, B. N.; Shchetinina, G. P.; Brovkina, O. V.; Ippolitov, E. G. (Koord. Khim. **11** [1985] 31/5).
[10] Pizer, R.; Tihal, C. (Inorg. Chem. **26** [1987] 3639/42).

[11] Grigoriu, A.; Bertes, Anis.; Bertes, Andr.; Doniga, E. (Ind. Usoara Text. Tricotaje Confectii Text. **37** [1986] 79/83 from C.A. **105** [1986] No. 174304).
[12] Sikdar, B.; Adhikari, D.; Das, N. N. (Indian J. Text. Res. **12** [1987] 93/6 from C.A. **107** [1987] No. 156211).
[13] Chaudhuri, M. K.; Das, B. (Inorg. Chem. **24** [1985] 2580/2).
[14] Ippolitov, E. G.; Chernyshov, B. N.; Shchetinina, G. P.; Brovkina, O. V.; Martynyuk, Yu. L.; Gorin, Yu. V. (Ukr. Khim. Zh. [Russ. Ed.] **52** [1986] 818/23 from C.A. **106** [1987] No. 42927).
[15] Turner, D. J. (AECL-9923 [1989] 135/50 from C.A. **111** [1989] No. 182555).
[16] Handy, B. J. (Water Chem. Nucl. React. Syst. **4** [1986] 169/76 from C.A. **109** [1988] No. 100352).
[17] Khelevin, R. N. (Zh. Org. Khim. **22** [1986] 2315/9; J. Org. Chem. [USSR] **22** [1986] 2079/82).
[18] Irvin, A. D.; Holmgren, J. S.; Zerda, T. W.; Jonas, J. (J. Non-Cryst. Solids **89** [1987] 191/205).
[19] Gies, H.; Gunawardane, R. P. (Zeolites **7** [1987] 442/5 from C.A. **108** [1988] No. 15138).
[20] Sultanova, L. Z.; Vinogradov, E. E.; Tarasova, G. N. (Zh. Neorg. Khim. **31** [1986] 223/5; Russ. J. Inorg. Chem. **31** [1986] 126/7).

[21] Arkhipov, S. M.; Kashina, N. I.; Kidyarov, B. I.; Kuzina, V. A. (Zh. Neorg. Khim. **31** [1986] 217/9; Russ. J. Inorg. Chem. 31 [1986] 123/4).
[22] Feng, Deyou (Huaxue Shiji **10** [1988] 371/2, 324 from C.A. **111** [1990] No. 246888).
[23] Karazhanov, N. A.; Omarova, R. S.; Isenzhulova, Yu. T. (Izv. Akad. Nauk Kaz. SSR Ser. Khim. **1988** No. 2, pp. 84/6 from C.A. **108** [1988] No. 211389).
[24] Kolosov, A. S.; Kormilitsina, Z. A.; Avvakumov, E. G. (Izv. Sib. Otd. Akad. Nauk SSSR Ser. Khim. Nauk **1989** No. 6, pp. 123/30 from C.A. **112** [1990] No. 150493).
[25] Gode, H.; Spricis, A. (Latv. PSR Zinat. Akad. Vestis Kim. Ser. **1987** 275/9).
[26] Gode, H.; Spricis, A.; Majore, I. (Latv. PSR Zinat. Akad. Vestis Kim. Ser. **1989** 23/5).
[27] Baluev, A. V.; Mityakhina, V. S.; Rogozev, B. I.; Silin, M. Yu. (Dokl. Akad. Nauk SSSR **291** [1986] 1138/41; Dokl. Phys. Chem. Proc. Acad. Sci. USSR **286/291** [1986] 1117/9).
[28] Touboul, M.; Amoussou, D. (J. Therm. Anal. **31** [1986] 125/30).
[29] Gode, H.; Svirkst, J.; Antropa, A. (Latv. PSR Zinat. Akad. Vestis Kim. Ser. **1985** 545/7).
[30] Perebeinos, A. A.; Kononova, G. N.; Rastegaeva, G. Yu. (Zh. Neorg. Khim. **31** [1986] 3145/9; Russ. J. Inorg. Chem. **31** [1986/87] 1806/9).

[31] Khazikhanova, B. Kh.; Beremzhanov, B. A.; Savich, R. F.; Kalacheva, V. G. (Zh. Neorg. Khim. **30** [1985] 1899/901; Russ. J. Inorg. Chem. **30** [1985/86] 1079/81).
[32] Gode, H.; Kaugare, S. (Latv. PSR Zinat. Akad. Vestis Kim. Ser. **1987** 239/42).
[33] Gode, H.; Svirkst, J. (Latv. PSR Zinat. Akad. Vestis Kim. Ser. **1985** 14/7, 548/52).
[34] Ovramenko, N. A.; Zakharchenko, O. F.; Kalenchuk, V. G.; Shutova, V. I.; Ovcharenko, F. D. (Dokl. Akad. Nauk SSSR **298** [1987] 621/3; Dokl. Chem. Proc. Acad. Sci. USSR **298** [1988] 40/2).
[35] Ovramenko, N. A.; Zakharchenko, O. F.; Litovchenko, A. S.; Trachevskii, V. V.; Shutova, V. I.; Ovcharenko, F. D. (Dokl. Akad. Nauk SSSR **309** [1989] 1141/5; Dokl. Chem. Proc. Acad. Sci. USSR **309** [1989] 364/7).
[36] Wang, Jianhua; Feng, Shouhua; Xu, Ruren (J. Chem. Soc. Chem. Commun. **1989** 265/6).
[37] Wysocka-Lisek, J.; Furtak, W. (Biul. Lubel. Tow. Nauk. Mat. Fiz. Chem. **26** [1984/86] 3/7 from C.A. **108** [1988] No. 15130).

[38] Balej, J. (J. Chem. Soc. Faraday Trans. I **85** [1989] 3327/34).
[39] Kalinskaya, T. V.; Lobanova, L. B.; Kaekina, M. L. (Zh. Neorg. Khim. **30** [1985] 1513/8; Russ. J. Inorg. Chem. **30** [1985] 863/6).
[40] Fedorov, P. I.; Andreev, V. K.; Filippova, N. V.; Ivanova, G. A. (Zh. Neorg. Khim. **30** [1985] 356/9; Russ. J. Inorg. Chem. **30** [1985] 199/201).

[41] Filippova, N. V.; Ivanova, G. A.; Fedorov, P. I.; Andreev, V. K. (Zh. Neorg. Khim. **32** [1987] 631/4; Russ. J. Inorg. Chem. **32** [1987] 354/5).
[42] Badaev, Yu. B.; Kostromina, N. A. (Ukr. Khim. Zh. [Russ. Ed.] **52** [1986] 11/3 from C.A. **104** [1987] No. 198853).

3.5.7.5 Reactions with Organic Species

The H_3BO_3–*n*-propanol–H_2O system has a simple form, but increasing alcohol chain length (C_3 to C_9) leads to the appearance of liquid-liquid, liquid-liquid-solid, and liquid-solid phase equilibrium regions. Distribution isotherms for H_3BO_3 or water of hydration into the organic phase show that, in addition to esterification, H_3BO_3 can undergo hydration and solvation [7]. Solubilities and phase diagrams were determined for the system H_3BO_3–alcohol–solvent (solvent = acetone, benzene, ethyl acetate, butyl acetate) at 25 °C. The phase diagram for propanol has one simple crystallization branch for H_3BO_3 [8]. Solubility diagrams in the systems H_3BO_3–*N,N*-dibutylacetamide, H_3BO_3–*N,N*-diethylhexanoic amide, H_3BO_3–*N,N*-dipentylacetamide, and H_3BO_3–*N,N*-dipentylpropionamide show that extraction capacities decrease with increasing size of the hydrocarbon chain [9]. Distribution coefficients of the H_3BO_3–electrolyte–ROH–diluent–H_2O system at 25 °C show that the influence of the electrolyte on the extraction of H_3BO_3 is controlled by the cation charge and size, the influence of the diluent by its nature (enhanced by increasing alcohol chain length), and by its concentration [10].

The extraction of H_3BO_3 by 3-methylbutane-1,3-diol has been studied as a function of the nature of the organic diluent, the starting and equilibrium concentrations of the diol, the nature and concentration of salting-out agents, median pH value, and H_3BO_3 concentration in solution. For solutions containing NaCl or $MgCl_2$ in large concentrations, the distribution coefficients of H_3BO_3 increase to 14 or 17. $Mg[SO_4]$ displaces the diol in a separate phase, thus making it possible to extract H_3BO_3 with the diol without any organic diluent. H_3BO_3 and 3-methylbutane-1,3-diol esters are isolated from the extract [1]. In the H_3BO_3–(3-methylbutane-1,3-diol)–water system at 25 °C, the equilibrium solid phase is H_3BO_3 which exhibits a significant solubility in the diol (24.52 wt%). Addition of the diol significantly increases the H_3BO_3 solubility in water [2]. The complexes formed in this system were studied in aqueous dimethyl sulfoxide, acetone, or chloroform by ^1H NMR spectroscopy. Complexes with the diol : H_3BO_3 compositions of 1:1, 2:1 (two isomers), and 3:2 were identified. The organic solvents do not interfere with the quantitative determination of the concentrations of the diol-borate complexes and free diol [3].

The quantitative determination of 3-methylbutane-1,3-diol in the $CDCl_3$ phase of the H_2O–$CDCl_3$ extraction system has been studied by ^1H NMR spectroscopy. The distribution coefficients for H_3BO_3 and the 1:1 and 2:1 diol : H_3BO_3 complexes were determined [4]. The distribution of H_3BO_3 between the aqueous and organic phase, in the stratifying $Mg[SO_4]$–(3-methylbutane-1,3-diol)–H_2O system, was studied at 25 °C by using the method of extraction rays plotted on the organic component phase diagram H_3BO_3–(3-methylbutane-1,3-diol)–H_2O. The stratification region of the $Mg[SO_4]$–diol–H_2O system can be used for the extraction of H_3BO_3 without use of an organic diluent [5]. The H_3BO_3 extraction properties were studied for a fraction of low boiling products from the production of synthetic rubber with 9.6 to 13.5% 3-methylbutane-1,3-diol. Under optimal conditions, a chloroform solution of this light fraction was used to extract 79 to 89% of the H_3BO_3 per cycle [6].

Different aliphatic alcohols were tested for the extraction of H_3BO_3 from solutions modeling radioactive waste. In addition to distribution measurements, the number of molecules of extractant per H_3BO_3 molecule in the organic phase was determined assuming extraction by solvation. A method of activity coefficient evaluation in aliphatic alcohol-nonpolar solvent systems is proposed and hydration and solvation effects in these systems are studied. The activity coefficients in systems consisting of alcohols (*n*-octanol, 2-ethylhexanol (EHX), *n*-decanol, or *n*-dodecanol) and nonpolar solvents (toluene) that are the main components of the organic phase in H_3BO_3 extraction from aqueous solutions have been determined [11].

The distribution of H_3BO_3 in the following systems (EHX = 2-ethylhexanol) has been determined: EHX–H_2SO_4–H_2O and EHX–kerosene–H_2SO_4–H_2O [12]; EHX–$Na_2[SO_4]$–H_2O [13]; EHX–*i*-pentanol–$Na_2[SO_4]$–H_2O and EHX–*i*-pentanol–$K_2[SO_4]$–H_2O [14]; EHX–kerosene–H_2SO_4–$Na_2[SO_4]$–H_2O [15]; EHX–kerosene–H_2SO_4–$Mg[SO_4]$–H_2O [16]; EHX–$M_2[SO_4]$–H_2O (M = Li, Na, K; M_2 = Mg, Fe^{2+}), EHX–$M_2[SO_4]_3$ (M = Al, Fe^{3+}), EHX–$Fe_2[SO_4]_3$–H_2O, and EHX–$Al_2[SO_4]_3$–H_2O [17]; EHX–kerosene–H_2SO_4–$Fe_2[SO_4]_2$–H_2O [18]; EHX–kerosene–HCl–$MgCl_2$–H_2O [19]; EHX–$CaCl_2$–H_2O [20]; and EHX–kerosene–$CaCl_2$–H_2O [21]. Investigations of the extraction of H_3BO_3 from aqueous solutions with 2-ethylhexane-1,3-diol in chloroform show a formation constant of log K = 3.27 (±0.18) for the 1:1 complex [22].

The distribution of H_3BO_3 in organic and aqueous phases in the extraction by di-*i*-butylpropane-1,3-diol, 2-octylpropane-1,3-diol, 2-ethyl-2-butylpropane-1,3-diol, and 1-(3-bromophenyl)-2,2-dimethylpropane-1,3-diol was studied. The effects of diluents (chloroform, toluene, acetophenone, *i*-amylalcohol, $Cl_2C{=}CCl_2$, or decanol) and salting out agents (NaCl, KCl, $Mg[SO_4]$, or $MgCl_2$) in the H_3BO_3 extraction were determined. The extraction effectiveness of the substituted propanediols was similar [23]. At pH < 12, H_3BO_3 is practically not extracted by *n*-decane-3,5-diols and *n*-dodecane-3,5-diols with and without organic solvents. The presence of salts increases the extraction. The *n*-alkane-3,5-diols seem to be more effective extraction agents in comparison to isomeric *n*-alkane-1,3-diols [24]. The presence of double bonds in 2,2,5,9-tetramethyl-4,8-decadien-1,3-diol does not decrease its extraction ability for H_3BO_3, which decreases in the order: $CHCl_3 > Cl_2C{=}CCl_2 >$ toluene > acetophenone > decanol > *i*-amylalcohol. The salting-out action varied in the series KCl < NaCl < $Mg[SO_4]$ < $MgCl_2$ [25].

The extraction of H_3BO_3 by 2,2,6-trimethyl-6-methoxydodecane-3,5-diol follows the same pattern as for other diols. The extraction is very effective in highly acidic media. The best diol : H_3BO_3 ratio for the extraction is 3 : 2 [26]. The extraction of H_3BO_3 by tetrachloroethene and toluene solutions of 1-phenyl-4-methyl-4-methoxyalkane-1,3-diols is also effective in acidic solutions; at pH values between 1 and 2, the acid is extracted practically quantitatively. The extraction from weakly acid solutions is also good in the presence of $MgCl_2$ [27].

Monomethylethers of triols such as $CH_3OC(CH_3)_2CH(OH)CH_2CH(OH)C_6H_5$, prepared from ynones and amines with subsequent hydrolysis and reduction of the carbonyl group, are also effective extracting agents for H_3BO_3 [28]. Extraction capacities for H_3BO_3 in 6 N HCl or H_2SO_4 were determined in $Cl_2C{=}CCl_2$ for monoalkyl ethers of triols RR′(R″O)CCH(OH)CH$_2$CH(OH)R‴; they are highly effective [29].

The distribution coefficients of H_3BO_3 were determined between the upper phase rich in polyethylene glycol (PEG) and the lower aqueous $Na_2[CO_3]$ phase; this system is not suitable for H_3BO_3 extraction [30]. Distribution of H_3BO_3 in the stratification region of the system PEG 1000 or PEG 1500–$Mg[SO_4]$–H_2O at 25 °C is such that boric acid is almost completely transferred into the PEG-rich phase [31]. The reaction between PEG and H_3BO_3 in aqueous-organic

References on pp. 221/7

solutions yields predominantly mono- and diborates of PEG. Proton NMR spectra of the reaction products were recorded in acetone/H_2O or dimethyl sulfoxide/H_2O solutions, and the lines of various B–OH ester groups were assigned [32].

The NMR spectra of mono- and dialkylborates, formed by reaction of H_3BO_3 with some aliphatic alcohol mixtures in aqueous acetone solutions were analyzed. Taft inductive constants were obtained, and the possibility of quantitative evaluation of all the forms was demonstrated [33].

The extraction of aqueous H_3BO_3 by 0.4 M 2-hydroxy-5-octyl-benzyl-ethanolamine in organic diluents increases in the order 2-butanone < octanol < C_6H_6 < CCl_4 < hexane [34].

In the H_3BO_3–(3-amino-1-propanol)–H_2O system, a borate with a composition close to the component ratio $HO(CH_2)_3NH_2:B_2O_3:H_2O = 1:1.5:2.5$ has been isolated; the IR spectrum is assigned to the $[B_3O_3(OH)_4]^-$ ion [35]. A review with 80 references on the reactions of H_3BO_3, metaborates, triborates, tetraborates, or pentaborates with amines and amides in aqueous solution was given. The effects of structural and spatial factors on the reactions were considered, and the complex $Na_2[BO(OH)_2 \cdot HCONH_2]$ with four-coordinate boron was obtained [36].

H_3BO_3 reacts with $(C_2H_5)_3N$ in water at 25 °C to form a borate with the composition $(C_2H_5)_3N \cdot 2.5B_2O_3 \cdot 2.5H_2O$, which was characterized as $[(C_2H_5)_3NH][B_5O_6(OH)_4]$; see Section 3.6.6.1, p. 277 [37]. H_3BO_3 reacts with ethylenediamine in H_2O to give $H_2NCH_2CH_2NH_2 \cdot 2.5B_2O_3 \cdot 4H_2O$, which contains the $[B_5O_6(OH)_4]^-$ ion [38]. H_3BO_3 reacts with hexamethylenetetramine ($C_6H_{12}N_4$) in mole ratios between 1:3 and 3:1 in aqueous solution to give $C_6H_{12}N_4 \cdot 2.5B_2O_3 \cdot 3H_2O$ with the structure $[C_6H_{12}N_4H][B_5O_6(OH)_4] \cdot 0.5H_2O$, as IR data suggest [39]. It is monoclinic, space group $P2_1/c-C^5_{2h}$ (No. 14) with a = 957.15(20), b = 1437.2(1), c = 1155.62(6) pm; β = 109.934(4)°; Z = 4; D_m = 1.623 g/cm³ and D_c = 1.636 g/cm³; R = 4.1% and R_w = 6.3% [40].

In the H_3BO_3–(1-hydroxyethyl-1,1-diphosphonic acid)–H_2O system, the solubility isotherm at 25 °C has a crystallization branch for the antimicrobial compound $4C_2H_8O_7P_2 \cdot B_2O_3 \cdot 2H_2O$ [41]. A review on the lyotropic effect of Trilon B, glycine, and acetylurea on H_3BO_3, alkali metal or ammonium borates, and the formation of solid solutions or the conversion into several borate ions is available [42].

Addition of ethane-1,2-diol, propane-1,2-diol, propane-1,3-diol, or glycerol to a sodium borate solution of pH 8 to 10.5, containing $[B_3O_3(OH)_4]^-$ or $[B_3O_3(OH)_5]^{2-}$ ions, results in the formation of a variety of mono- and bis-chelated complexes including spirocyclic anions with five- and six-membered rings sharing a common boron atom. Complexes in which boron atoms of triborate anions are coordinated by diol ligands are also identified. Boron-11 chemical shifts characteristic of chelated and hydroxylic boron centers in mono- and triborate species are reported, taking into account the trends accompanying a change from trigonal to tetrahedral coordination. Proton and ^{13}C NMR spectra of solutions containing the complexes $[LB(OH)_2]^-$ and $[L_2B]^-$ with H_2L = propane-1,2-diol reveal all the possible stereoisomers of this system [43].

The condensation products of H_3BO_3 with glycerol or propane-1,2,3-triol, prepared by heating an equimolar mixture at 150 °C in an evaporator, are transparent glassy solids containing B–O–C bonds [44]. Similar condensation products with coordinated B–N bonds are prepared from H_3BO_3/diethanolamine or H_3BO_3/triethanolamine mixtures [45]. All of these condensation products yield BN, small amounts of B_4C, and glassy solids on pyrolysis above 1300 °C in an N_2 flow [44, 45]. Partial esterification of H_3BO_3 with various alcohols such as propane-2-ol in the presence of a mixture of $(CH_3)_3SiOSi(CH_3)_3$ or $(CH_3)_3SiCl$ gives mono- and dialkoxy derivatives of H_3BO_3 in addition to the trialkoxy derivative and tris(trimethylsilanyl)-borate [46, 47]. Cyclodehydration of H_3BO_3 with ROH (R = *n*-C_3H_7, *i*-C_3H_7, *n*-C_4H_9, *n*-C_6H_{11}, *cyclo*-C_6H_{11}, or $C_6H_5CH_2$) gave 95 to 99.9% tris(alkoxy)boroxins, $[-O-B(OR)-]_3$ [48].

An ^{11}B NMR study of the reaction of $[B(OH)_4]^-$ with various polyols has been performed considering the association constants. The complexation is not solely restricted to hydroxy groups on adjacent carbon atoms, but also to alternate ones, e.g., 2-methylpentane-2,4-diol. For equimolar formulations, the following values of K_1 and K_2 (in dm^3/mol) have been calculated: ethanediol, 0.74 and 0.29; propane-1,2-diol, 2.2 and 0.55; propane-1,3-diol, 1.1 and 0.08; glycerol, 6.7 and 1.3; mannitol, 137 and 21; and sorbitol, 278 and 24 [49]. The ^{11}B and ^{13}C NMR studies of the reaction of $[B(OH)_4]^-$ with poly(vinyl alcohol) in aqueous solution show that the reaction involves alternate OH groups in the polymer. The reaction is most facile when the OH groups are in the *meso* configuration, and the results provide confirmation of the assignments of the ^{13}C resonances inferred from Bernoullian statistics by previous workers [50].

The interaction of poly(vinyl alcohol) with H_3BO_3 in different solvents has been studied [51]. The association between poly(vinyl alcohol) and H_3BO_3 proceeds via formation of $[B(OH)_4]^-$ at 0.02 M H_3BO_3. Only a small portion of the OH groups of poly(vinyl alcohol) participate in the association, but this feature is not due to the steric effect. The association is minimally dependent on the molecular weight of poly(vinyl alcohol) [52]. The sorption capacity of OH-containing sorbents based on poly(vinyl alcohol) and poly(vinyleneglycol) with respect to H_3BO_3 was studied for the recovery of boron from natural waters and brines in alkaline and acidic media. The maximal sorption capacity G of poly(vinyleneglycol) is higher than that of poly(vinyl alcohol). The sharp decline in G of poly(vinyl alcohol) with decreasing pH indicates that the sorbent can be easily regenerated in acids [53].

Sorption of boron has been studied as a function of pH with various OH-containing solvents and is controlled primarily by the H_3BO_3 concentration and the pH; molecular weight of the poly(vinyl alcohol) and poly(vinyleneglycol) sorbents had little effect [54]. The reaction of borate ions with isotactic poly(vinyl alcohol) (I) was compared with that of atactic poly(vinyl alcohol) (II); the extent of reaction was estimated by measuring the H^+ concentration of the reaction solution at three temperatures. The concentration for I was higher than that for II. The free energy, enthalpy, and entropy functions for both systems were negative, and those for I were lower than those for II. These results suggest an easy reaction of borate ions with isotactically arranged adjacent OH groups [55].

Complexation of poly(vinyl alcohol) (PVA) with borate ions was studied by pH measurements in aqueous NaCl solutions at 10, 25, and 40°C. The formation constants are of the order of 10 for the mono-diol (K_1) and 100 for the bis-diol (K_2) type complexes, respectively. The value of K_2 increases with increasing NaCl concentration, but K_1 remains nearly constant. The entropy and enthalpy changes in complexation are negative for both complexes, the latter being more exothermic for the bis-diol than for the mono-diol. The effects of the NaCl concentration on the formation constants and thermodynamic quantities for these complexes were interpreted in terms of the polyelectrolyte nature of the poly(vinyl alcohol) chains complexed with negatively charged borate ions [56].

A method for the immobilization of activated sludge uses poly(vinyl alcohol)/H_3BO_3. Synthetic wastewater was treated at high loading rates of total organic carbon (TOC) of 0.5 to 2.35 $kg \cdot m^{-3} \cdot d^{-1}$, using the immobilized activated-sludge process. TOC and total nitrogen were removed at efficiencies of 93 and 30 to 40%, respectively. The kinetic constants for the process were 0.594 g mixed liquor suspended solids/g TOC and 0.0219/day. The cost of chemicals required for the immobilization of activated sludge by the poly(vinyl alcohol)-H_3BO_3 method was extremely low [57]. The cross-linked structure in poly(vinyl alcohol)-H_3BO_3 was deduced by comparison of the ^{11}B NMR spectrum of a poly(vinyl alcohol)–(pentane-2,4-diol) model system with that of the polymer system. The complexation and cross-linking of poly(vinyl alcohol) with H_3BO_3 was exothermic with a reaction enthalpy of −8.3 kcal/mol. There were two ^{11}B NMR signal components which had different spin relaxation rates. The relation-

References on pp. 221/7

ships between NMR parameters and viscosity in the poly(vinyl alcohol)–H_3BO_3 system were discussed [58].

Certain combinations of solutions of poly(vinyl alcohol) and polysaccharides with sodium borate result in fluids that exhibit a maximal viscosity followed by shear-thinning as the shear rate is increased. The ^{11}B NMR spectra are useful in elucidating the nature of the borate/hydroxyl-dyad complexes, including their stereoselectivity. The ^{11}B resonance peaks allow quantitative determination of the number of complexes. Dynamic mechanical properties are included, and a physical picture of network structure building and breaking during flow of associating polymers is shown [59].

The gelation curve of the reactions of H_3BO_3 (or borax) with poly(vinyl alcohol), strength of self-setting sand mixtures of boron compounds (hardening agents) and poly(vinyl alcohol), effects of alcohols on the hardening rate, hardening with air, CO_2, and N_2, and low-temperature baking were studied; low-temperature air hardening gave high-strength sand mixtures with poly(vinyl alcohol) and boron compounds with sorbitol as retarder [60]. The strength of a poly(vinyl alcohol) film increases by treating it with aqueous H_3BO_3 solution. This effect vanishes quickly when the film is immersed in water; hence, it appears that the cross-linking of poly(vinyl alcohol) by H_3BO_3 is very susceptible to breaking by water. The temperature of thermal treatment has a large effect on the film strength, the most effective temperature being 160°C [61].

Electrochemical deposition of poly(vinyl alcohol) coatings under galvanostatic conditions in the presence of H_3BO_3 proceeds via discharging of the H^+ ion and increasing the concentration of $[B(OH)_4]^-$ in the electrode layer. The latter ions form hydrogen bonds with poly(vinyl alcohol) to give a hydrogel coating without cross-linking [62]. The cross-linking mechanism of an aqueous alkaline solution of poly(vinyl alcohol) in the presence of H_3BO_3 was investigated by ^{11}B NMR spectroscopy. Isopropanol and pentane-2,4-diol were chosen as monomeric and dimeric model compounds of poly(vinyl alcohol), respectively. Monohydroxy alcohols did not react with H_3BO_3, but di- and polyhydroxy alcohols (having the structure of 1,3-diols) did. The temperature dependence of the chemical shift of the signals showed that the product was a so-called mono-diol type but not a bis-diol type, inconsistent with the conventional concept of the poly(vinyl alcohol)-H_3BO_3 reaction [63]. The gel-sol transition of poly(vinyl alcohol)-borate aqueous solutions was investigated as a function of polymer concentration, molecular weight, boric acid concentration, and pH. The enthalpy of cross-linking formation was estimated from a modified Eldridge-Ferry theory to be ca. 7 kcal/mol of junctions, which is consistent with literature data. The critical gelation concentration above which the solution is capable of gelation was also estimated as a function of the degree of polymerization [64].

The formation of a complex consisting of poly(vinyl alcohol), I_2, and H_3BO_3 in the presence of poly(D-glutamic acid) and poly(acrylic acid) (PAA) and its light-polarizing properties were investigated by the UV and polarizing microscope. The formation of the complex was greatly accelerated by poly(D-glutamic acid) (PGA) and poly(acrylic acid) (PAA), due to the enhancement of the aggregation of poly(vinyl alcohol) (PVA) chains. The increase in absorbance of the complex formed 24 h after preparation of the PVA–PAA system was larger than that of the complex formed by simultaneous mixing. The complex formed in the presence of poly(acrylic acid) was not dependent on the decomposition point, but on the saponification of the poly(vinyl alcohol) [65]. H_3BO_3 induced a blue color (λ_{max} at 670 nm) in the PVA–I_2–I^- system contrary to hydroxy acids like tartaric acid or H_3PO_4 (λ_{max} at 590 nm). Unlike starch-iodine blue, PVA-iodine blue did not exhibit dichroism in the visible range [66]. A colorimetric method based on the formation of the blue PVA-I_2-H_3BO_3 complex (having absorptivity to dissolved Cd^{2+} ions) was developed for a spectrophotometrical determination of poly(vinyl alcohol) in $Cd[OH]_2$ pastes in aqueous solutions with concentrations of 2 to 40 mg/L [67].

Monocomplexes (Z; with a $B(OH)_2$ structural element) and spirocomplexes (Z′; containing two cyclitol moieties) between H_3BO_3 (at pH 12 in the form of $[B(OH)_4]^-$, symbolized by Y) and some cyclitols (X), e.g., cyclohexane-*cis*,*cis*-1,3,5-triol or *epi*-inositol, were investigated by ^{11}B NMR spectroscopy; X + Y → Z and 2X + Y = Z′. $K'_s = [Z'] \cdot [Y]^{-1} \cdot [X]^{-2}$; $K'_m = [Z] \cdot [Y]^{-1} \cdot [X]^{-1}$; and $K_i \cdot K'_c = [XB(OH)^-] \cdot [X]^{-1} \cdot [Y]^{-1}$. For cyclohexane-*cis*-1,2-diol, $K'_m = 1.8$ and for cyclohexane-*cis*,*cis*-1,3,5-triol, $K_i \cdot K'_c = 15.6$. Modified neglect of diatomic overlap (MNDO) calculations were performed in order to determine the enthalpies of inversion. Three different complexes of *myo*-inositol were identified with $K'_m = 0.31$, $K'_s = 0.027$, and $K_i \cdot K'_c = 7.758$. For *myo*-inositol, an iterative method of calculation supports the proposed borate complex formulation [68].

The following stability constants for the complexation of borate ions with pentitols have been calculated by an iterative method and complemented by MNDO semiempirical LCAO-MO calculations; for xylitol: $K_1 = 192.2$, $K_3 = 17.1$, $K_1 \cdot K_2 = 7995.5$, and $K_1 \cdot K_4 = 422.8$; for ribitol: $K_1 = 42.3$, $K_3 = 42.5$, $K_1 \cdot K_2 = 1885.8$, and $K_1 \cdot K_4 = 25.4$; for arabitol: $K_1 = 89.8$, $K_3 = 17.1$, $K_1 \cdot K_2 = 194.6$, and $K_1 \cdot K_4 = 269.4$. $[X\{B(OH)_4\}_2]^{2-}$ is assumed to have two five-membered rings each containing boron (X = cyclitol). For another 1:1 complex (with mono six-membered ring), the ^{11}B resonance of which was at higher field than that of $[B(OH)_4]^-$, see [49]. Ribitol forms the zigzag 2,4-borate or the sickle 3,4-borate, arabitol the zigzag 2,3-borate, and xylitol the zigzag 2,4-borate, the sickle 1,2-borate, and the five-membered 3,4-borate [69]. Stability constants of other pentitols and pentoses with H_3BO_3 were determined at 25 °C and ionic strength 0.1 M (NaCl) by potentiometric titration. The available structural factors of polyols affecting the stability of chelates were examined; liquid structures of the complexes are discussed [70].

H_3BO_3 reacts with butane-1,3-diol (H_2L) in aqueous acetone to give a cyclic borate (–L–)BOH, 1,3,2-dioxaborepane. Heating of this species or dehydrating it over P_4O_{10} gave a dimer. The intermediates in these reactions were discussed, and ^1H NMR chemical shifts for several cyclic boric acid esters were determined [71]. From H_3BO_3, pentaerythritol (H_4L), and ethanolamine (Z) in aqueous solution under various conditions, polymeric complexes $(HZ[BL] \cdot m H_2O)_n$ with m = 1, 1.25, 2, or 3 were prepared. The complexes were characterized by X-ray phase analysis, IR, ^1H and ^{13}C NMR spectra, conductometric titrations, and thermal decomposition studies [72].

Stability constants were determined for the H_3BO_3-mannitol complexes; the results of a graphical and a pH-metric method were compared [73]. The acid strengths of the complexes of H_3BO_3 and some polyols are compared by potentiometric titration. The H_3BO_3-alditol complexes are the most acidic, because acyclic alditols except D-ribose and D-fructose easily form *cis*-esters with $[B(OH)_4]^-$ ions; the electromotive force (EMF) difference for the titration of H_3BO_3-alditol complexes is very large due to the weak acidity of the alditols [74].

The heats of reactions of the borate ion with polyols and carbohydrates (see Table 3/21, pp. 212/3) were determined by thermometric titrimetry at 25 °C and ionic strengths of 1.0 M or 0.1 M ($Na[NO_3]$); from these enthalpies and equilibrium constants, taken from the literature, entropies were calculated. Examination of the results indicates a leading role of solute-solvent external factors, e.g., desolvations of the borate ion, in determining the stabilities of the complexes. Among the various factors for the ligands, some appear to be of minor importance, while the contribution of others, such as the strength of the borate-hexose C–O–B bonds, is fairly constant throughout one series. The present data also seem to confirm the presence of the furanose cyclic form of the carbohydrate in the complex [75].

Table 3/21

Values of Overall Stability Constants and Thermodynamics of Borate Complexes in Aqueous Solution at 25°C and I = 0.1 M.

For β_i (= β_1, β_2), see the explanations for Table 3/20, p. 193.

hydroxy compound	borate complex	log K_i [75]	log β_i [76]	log β_i [78]	$-\Delta G_i^\circ$ (kJ/mol) [75]	$-\Delta H_i^\circ$ (kJ/mol) [75]	$-\Delta S_i^\circ$ ($J \cdot mol^{-1} \cdot K^{-1}$) [75]
ethylene glycol	$[BL]^-$	0.23(1)			1.30(8)	5.8(5)	15(2)
propane-1,2-diol	$[BL]^-$	0.46(1)			2.64(8)	9.3(6)	22(2)
	$[BL_2]^-$	−0.35(4)			−2.0(2)	38.9(17)	138(4)
glycerol	$[BL]^-$	1.30(2)		1.39	7.41(12)	10.2(5)	9.6(16)
	$[BL_2]^-$	0.18(5)			1.0(3)	28.9(17)	92(4)
D-mannitol	$[BL]^-$	3.04(8)			17.3(5)	19.7(2)	7.9(16)
	$[BL_2]^-$	1.50(9)			8.5(5)	23.4(5)	50(2)
D-galactose	$[BL]^-$	2.21(3)	1.99	1.97	12.6(2)	24.7(17)	42(4)
	$[BL_2]^-$	0.18(17)	2.56	2.52	1.0(8)	−48.5(40)	−167(12)
D-glucose	$[BL]^-$	2.27(5)	1.80	2.07	12.9(3)	17(1)	12.5(4)
	$[BL_2]^-$	0.49(5)	3.05	2.80	2.8(3)	−15(3)	−58(8)
D-fructose	$[BL]^-$	3.58(5)	2.82	—	20.4(3)	3(1)	−59(4)
	$[BL_2]^-$	1.36(5)	4.97	—	7.7(3)	33(1)	84(4)
D-mannose	$[BL]^-$	4.42(5)	2.01	—	25.2(3)	6.81(2)	−61.5(8)
	$[BL_2]^-$		2.76	2.74		25.15(8)	−26.8(8)
L-sorbose	$[BL]^-$	5.80(5)	<3.05	—	33.1(3)		
	$[BL_2]^-$		5.75				
D-arabinose	$[BL]^-$		2.14	—			
	$[BL_2]^-$		2.99	—			
erythritol	$[BL]^-$		1.85	1.99			
	$[BL_2]^-$		2.91	—			
lactose	$[BL]^-$		1.43	1.51			
	$[BL_2]^-$		2.17	—			
maltose	$[BL]^-$		1.41	1.36			
	$[BL_2]^-$		1.89	—			
lactulose	$[BL]^-$		2.91	—			
	$[BL_2]^-$		5.14	—			
melibiose	$[BL]^-$		—	1.82			
	$[BL_2]^-$		—	2.44			

Table 3/21 (continued)

hydroxy compound	borate complex	log K_i [75]	log β_i [76]	log β_i [78]	$-\Delta G_i^\circ$ (kJ/mol) [75]	$-\Delta H_i^\circ$ (kJ/mol) [75]	$-\Delta S_i^\circ$ ($J \cdot mol^{-1} \cdot K^{-1}$) [75]
turanose	$[BL]^-$		—	1.91			
	$[BL_2]^-$		—	2.47			
raffinose	$[BL]^-$		1.35	—			
	$[BL_2]^-$		1.67	—			
melezitose	$[BL]^-$		1.05	—			
	$[BL_2]^-$		1.14	—			
saccharose	$[BL]^-$		0.86	0.75			
	$[BL_2]^-$		0.70	—			
gentibiose	$[BL]^-$			1.14			
cellobiose	$[BL]^-$			1.25			
trehalose	$[BL]^-$			1.04			
meso-inositol	$[BL]^-$			1.57			
gluconic acid	$[BL]^-$			2.83			
	$[BL_2]^-$			4.46			
glucoronic acid	$[BL]^-$			1.71			
glucaric acid	$[BL]^-$			2.16			
	$[BL_2]^-$			3.58			

The stability constants β_1 and β_2 of borate complexes of several mono-, di-, and trisaccharides in aqueous solution at 25 °C and of ionic strength 0.1M (KCl) have been determined by the potentiometric method and calculated by using an extended version of the Antikainen equation (see Table 3/21). Raffinose, melezitose, saccharose, and other fructose-containing oligosaccharides gave complexes of low stability, whereas lactulose and fructose gave very stable complexes. Fructose moieties in oligosaccharides complex the borate anion with their 2- and 3-OH groups [76]. These results are in agreement with an ^{11}B NMR study on 1:1 and 1:2 borate complexes of pentoses and hexoses. In both species, all sugars are complexed in the furanose form. Those of arabino-galacto-fructo series and of xylo-gluco-sorbo series gave the same type of complex involving the anomeric OH group and the nearest ring CHOH. The very stable 1:2 complex of D-ribose with log $\beta_1 = 2.26$ (log $\beta_2 = 4.00$) is a mixture of two species, borate bound either at C-1, C-2 or at C-2, C-3; cf. D-xylose with log $\beta_1 = 1.95$ (log $\beta_2 = 3.74$) as opposed to D-lyxose with log $\beta_1 = 2.15$ (log $\beta_2 = 3.39$), which possesses two *cis*-CHOH groups and belongs to the lyxo-manno-tagato series. For other values, see Table 3/21 [77].

The anionic complex formation of H_3BO_3 with disaccharides, sugar acids, or polyols (see Table 3/22, pp. 214/5) [78] or with lactitol and maltitol [79] was studied potentiometrically at 25 °C in aqueous KCl medium of ionic strength 0.1M; the equilibrium constants for both 1:1 and 2:1 complexes were calculated by the computer program MINIQUAD 75 [78, 79].

References on pp. 221/7

Table 3/22
Boron-11 Chemical Shifts, Line Width, and Association Constants of Borate Esters (D_2O, 25°C; pD=11) [89].

alcohol	ester type	chemical shift[a)] (ppm)		line width (Hz)		association constant (L/mol)	
		LB^-	L_2B^-	LB^-	L_2B^-	LB^-	L_2B^-
glycol[b)]	1,2	−13.9	−10.1	13	16	1.0	0.1
propane-1,2-diol[b)]	1,2	−14.0	−10.2	12	26	1.8	1.5
pinacol[b)]	2,3	−14.8	−11.9	14	45	3.5	8.5
propane-1,3-diol[b)]	1,3	−18.3	−18.9	9		1.2	0.05
glycerol	1,2	−13.6	−9.6	18	30	25	3.0
	1,3	−18.5	−19.0	9		3.7	0.05
mannitol	*erythro*-2,3	−14.7					
	threo-3,4	−13.7	−9.6				
glucitol	*threo*	−13.6	−9.5				
	erythro-4,5	−14.6					
cis-cyclopentane-1,2-diol	1,2	−13.6	−9.4				
cis-1,2-cyclohexane-1,2-diol	1,2	−14.2	−10.7				
all-*cis*-cyclohexane-1,3,5-triol	1,3,5	−18.1		13		14	
epi-inositol	1,3,5	−19.4				10000	
glycerate	2,3	−13.1	−8.7	23		6.2	0.62
arabinonate	*threo*-2,3	−13.2	−9.3	26	78	44	13
	erythro-3,4	−14.2		38		10	
	4,5	−13.5		24		18	
	2,4/3,5	−18.1		16		0.31	
ribonate	*erythro*-2,3	−13.8	−9.9	29	95	13	4.3
	erythro-3,4	−14.4		49		6.6	

	4,5	−13.3	−9.3	41	47	10	0.98
	2,4	−18.0		19		14	
lyxonate	*erythro*-2,3	−14.9				6.9	
	threo-3,4	−13.6	−9.7	20	58	230	27
	2,4/3,5	−18.2		11		5.1	
mannonate	*threo*-3,4	−13.4	−9.2	42	111	1200	48
	erythro-4,5	−14.5		35		140	
	2,4/3,5/4,6	−18.3		34		54	
gluconate	*threo*	−13.4	−9.4	35	85	240	31
	erythro-4,5	−14.2		36		72	
	2,4	−18.2		21		19	
gulonate	*erythro*-2,3	−14.6		30		65	
	threo	−13.5	−9.5	39	81	540	79
	3,5	−18.2		21		17	
meso-tartrate	*erythro*-2,3	−13.0	−8.7	55	120	2.2	0.36
(S,S)-tartrate	*threo*-2,3	−12.6	−7.7	58	160	11	17
meso-3,4-dihydroxyadipate	*erythro*-3,4	−13.9	−10.4	31	74	3.3	0.42
rac. 3,4-dihydroxyadipate	*threo*-3,4	−13.8	−10.1	27	87	15	3.2
idarate	*threo*	−13.2	−9.1	57	120	120	20
	2,4	−18.0	−18.4	10	18		—
galactarate	*threo*-2,3	−13.3	−9.1	32	93	260	7.7
	erythro-3,4	−14.4		26		42	
glucarate	*threo*	−13.2	−9.0	59	168	180	31
	erythro-4,5	−14.0				20	
	2,4	−17.9				11	

[a] Relative to 0.1M boric acid as external reference. – [b] Corrected for $\delta_{B^-} = -17.1$ ppm.

References on pp. 221/7

Proton and ^{13}C NMR studies on D-fructose-H_3BO_3 complexes in aqueous solution indicate that the 2:1 complexes are favored and that the anomeric OH groups are always involved and the 3-OH groups certainly [80]. The formation of borate complexes on the addition of solutions of H_3BO_3 and NaOH to solutions of guar galactomannan in D_2O was studied by ^{13}C NMR spectroscopy [81]. Borate mono- and dicomplexes of a series of glycosides (particularly CH_3-β-D-mannopyranoside and CH_3-α-D-galactopyranoside) and of D-galacto-D-mannan from guar in aqueous alkaline solution have been identified by ^{11}B NMR spectroscopy; the associated formation constants have been determined [82]. Also, complex formation between guar 4,6-D-galacto-D-mannan (and similar compounds) with the $[B(OH)_4]^-$ ion in pH 12 buffer solutions has been studied using ^{11}B NMR spectroscopy. The association constant K_1 is 7.4 at 295 K; the thermodynamic data are $\Delta H^\circ = -15.1$ kJ/mol and $\Delta S^\circ = -34.6$ $J \cdot mol^{-1} \cdot K^{-1}$ [83].

Reversible gel formation in the galactomannan-borate system (guaran or hydroxypropylguaran and borate) has been studied by ^{11}B NMR spectroscopy; the existence of 5- and 6-membered ring mono-diol- and bis-diol-complexes was demonstrated [84]. When complexation led to the formation of intramolecular cross-links, the usual law of mass action, adequate for small molecules, must be revised. The ^{11}B NMR spectroscopy of a borate-poly(glyceryl methacrylate) system was used to determine the concentration of free borate 1:1 and 1:2 complexes in a saline solution. The concentration of 1:1 complexes was proportional to the free ligand concentration, but the formation of 1:2 complexes from 1:1 complexes was independent of the global ligand concentration [85]. The formation of a weak gel complex between guaran and borate was also studied by dialysis experiments; the association constant was found to be $K_1 = 11.4$ L/mol in 1M NaCl [86].

The phase diagrams, observed in the galactomannan–H_3BO_3–NaOH system, are described by a model that is based on the parameter relevant for both reversible gelation and demixing [87]. The chemistry of binding between guar galactomannans and borates was also investigated by dynamic light scattering, and high resolution ^{13}C NMR spectroscopy detected complexation between both the sugars (CH_3-β-D-galactopyranoside and CH_3-α-D-mannopyranoside) with borate [88].

Boron-11 NMR spectroscopy enables the distinction and direct identification of a variety of boric acid esters and, simultaneously, the determination of the corresponding association constants [89].

The pH dependence of the stability of boric acid esters of a series of polyols and polyhydroxycarboxylates has been studied. Addition of protonic acids favors the boric acid ester. The association constants for the borate (B^-) mono- and diesters are defined as $K_1 = [LB^-] \cdot [B^-]^{-1} \cdot [L]^{-1}$ and $K_2 = [L_2B^-] \cdot [LB^-]^{-1} \cdot [L]^{-1}$. The association constants have been investigated in aqueous alkaline solution by ^{11}B NMR spectroscopy. The prevailing borate mono- and diesters for each compound were identified. Boron-11 NMR constants have been determined. Rules are given for predicting the ^{11}B chemical shift, the relative stability, and the structure of the boric acid esters in aqueous solution [89]. For data, see Table 3/22, pp. 214/5 [89].

A combination of ^{11}B and ^{13}C NMR spectrometry has been used to elucidate the nature of the ester formation between borates and carbohydrates in aqueous solution between pH 6 and 12 at 25°C. At high carbohydrate/borate ratios, D-mannitol shows the selective formation of bis(D-mannitol)-3,4:3′,4′-borate, whereas D-glucitol forms a mixture of six different bis(D-glucitol)borates consisting of pairs of diastereomers of the 2,3-β- and 1,2-α-furanose species, respectively. At low carbohydrate/borate ratios, mono-, di-, and, sometimes, triborates are formed, e.g., mixtures of α-D-glucofuranose-1,2:3,5-diborate and α-D-glucofuranose-1,2:5,6-diborate for D-glucose, and a mixture of β-D-fructopyranose-1,2:4,5-diborate and β-D-fructopyranose-2,3:4,5-diborate for D-fructose. The overall stability constants for the boric acid diesters of D-mannitol, D-glucitol, and D-fructose are about two orders of magnitude higher than that of D-glucose [90]. Proton and ^{13}C NMR spectroscopy were used to establish substituent effects upon boric acid ester formation of polyhydroxycarboxylates in water; the data show that the *threo*-3,4-diol function of gluconate, glucarate, and idarate is preferred [91]. The increased abilities of polyhydroxycarboxylates to coordinate Ca^{2+} ions in aqueous alkaline solution upon addition of borate were studied by ^{11}B NMR spectroscopy. The capacity for sequestering Ca^{2+} was determined along with ion-selective electrode measurements of Ca^{2+}. The synergistic effect was caused by the formation of mono- and di-calcium complexes of the boric acid diesters of the polyhydroxycarboxylates [92].

A combination of ^{1}H, ^{11}B, and ^{13}C NMR spectroscopy was used to determine the structures of compounds present in the aqueous glucarate–borate–Ca^{2+} system. Boric acid mono- and diester formation was found to occur preferably at the *threo*-3,4-position of glucarate. Two diastereoisomeric boric acid diesters of glucarate are the major Ca^{2+} coordinating species [93]. As was shown by ^{11}B NMR spectroscopy, Na^+, K^+, and Ag^+ ions do not exhibit preferential coordination with borate–D-glucarate systems in aqueous alkaline media, while Mg^{2+}, Ca^{2+}, Sr^{2+}, Ba^{2+}, Ni^{2+}, Co^{2+}, and Cd^{2+} are coordinated, and Cu^{2+}, Zn^{2+}, Pr^{3+}, and Pb^{2+} compete with borate for diol functions, but Al^{3+}, Sb^{3+}, Ge^{4+}, and Sn^{4+} are more strongly coordinated by free D-glucarate than by its boric acid diester [94]. An NMR study of H_3BO_3–1,3-dione systems in dimethyl sulfoxide shows that the formation of boric acid diesters (of the enol form) is preferred over that of monoesters. The 1,3-dione moiety of the boric acid diester closely resembles that of the free enol tautomer [95]. The conformations and pseudorotations of 1,3-dioxolanes have been studied and used as model compounds for boric acid monoesters of vicinal diols. The ^{13}C chemical shift of the C(2) atom in 1,3-dioxolanes correlate well with the ^{11}B chemical shifts in the corresponding boric acid monoesters [96].

H_3BO_3 reacts with diols (H_2L; e.g., propane-1,2-diol) in chloroform or tetrahydrofuran to give B_2L_3, and with triols (H_3L'; e.g., glycerol) to give BL′ as monomers or polymers. Pyrocatechol (1,2-$(OH)_2C_6H_4$) supposedly does not react with H_3BO_3 [98]. By heating H_3BO_3 with phenol and pyrocatechol (H_2L), the pyrocatechol ester LB–O–C_6H_5 is prepared [99]. Five-, six-, and seven-membered bis-pyrocatechol spirocomplexes $[C_6H_9N_2][BL_2]$ have been obtained by the reaction of substituted pyrocatechols 1,2-$(OH)_2C_6H_3R$ (H_2L; R = H, 3-OH, 4-CH_3, 3-CHO) with (2-amino-4-methylpyridine)-borane ($C_6H_8N_2 \cdot BH_3$) in $HC(OC_2H_5)_3$; spiroborates of 2,3-dihydroxynaphthalene, salicylic acid, 2,6-dihydroxybenzoic acid, and 2,2′-dihydroxybiphenyl are also prepared [100]. The complexing ability of glutaric acid, 1:1 and 1:2 mixtures of glutaric acid and H_3BO_3, and pure H_3BO_3 for Ca^{2+} ions at pH 4 is low in comparison to sodium tripolyphosphate [97].

The isotachophoretic separation of pyrocatechol (1,2-$(OH)_2C_6H_4$), homocatechol (1,2-$(OH)_2$-3-CH_3-C_6H_3), and 2,3-dihydroxynaphthalene, obtained with a buffer of pH 7.5 as leading electrolyte and H_3BO_3 as terminating electrolyte, has a reproducibility of the zone length of 3.6% for 3 mmol 1,2-$(OH)_2C_6H_4$; 1,3-$(OH)_2C_6H_4$ and 1,4-$(OH)_2C_6H_4$ could not be detected, since they interact only weakly with H_3BO_3 [101].

Heating of H_3BO_3 with equimolar amounts H_2L (1,2-$(OH)_2C_6H_4$ or 2,3-dihydroxynaphthalene) and MHL′ (H_2L' = phthalic acid; M = Li, Na, K, Rb, Cs, H) in toluene with azeotropic removal of water gave complexes of the types $M[BLL'] \cdot nH_2O$. The complexes were characterized by IR spectra and thermal analysis. Dehydration occurs at 50 °C, whereas at 250 °C the borate anion began to decompose, and $M_2B_2O_4$ was formed at 800 °C [102].

H_3BO_3 reacts with 2-$(HOCH_2)C_6H_4OH$ (H_2L) and $SrCl_2$ in the presence of KOH to give $Sr[B(OH)_2L]_2$, which was characterized by IR spectra, X-ray diffraction and thermal analyses [103]. H_3BO_3, 2-hydroxymethyl-phenol (H_2L), and Q, which may be NH_3, $NH_2CH_2CH_2NH_2$,

References on pp. 221/7

$NH_2CH_2CH_2OH$, decylamine, dodecylamine, diisopropylamine, or dinonylamine, react to form complexes $(QH)[BL_2]\cdot nH_2O$ (n=0, 0.5, 4) which were characterized by IR and ^1H spectra, solubility and electrical conductivity studies, X-ray diffraction and thermal analyses [104].

H_3BO_3 reacts with pyrocatechol (1,2-$(OH)_2C_6H_4$; H_2L) and $C_6H_5CH_2NH_2$, or R_2NH, or Q (which may be dimethylquinoline or 4-nitrophthalimide) in the presence of an amine to form complexes of the types $C_6H_5CH_2NH_3[BL_2]\cdot 1.5H_2O$, or $R_2NH_2[BL_2]\cdot nH_2O$ ($R=CH_3$, n=0.5; $R=C_2H_5$, n=1), or $QH[BL_2]\cdot nH_2O$ (n=0.5 or 1), respectively; these were characterized by electrical conductivity and solubility measurements, IR spectra, X-ray diffraction and thermal analyses. Water of crystallization is lost between 80 and 150°C, and the anhydrous compounds are stable up to 150 to 180°C [105]. The complexes $(QH)[BL_2]$ with H_2L=1,2-$(OH)_2C_6H_4$ and Q=$(CH_3)_2NH$, $(C_2H_5)_2NH$, $C_6H_5CH_2NH_2$, or 2,4-dimethylquinoline were studied by IR spectroscopy, X-ray phase analysis, and thermal analysis; the compounds decompose between 300 and 400°C; the secondary ammonium salts sublimed prior to decomposition [106]. Complexes of the type $Q[BL_2]$ with H_2L=pyrocatechol (1,2-$(OH)_2C_6H_4$), homocatechol (1,2-$(OH)_2$-3-CH_3-C_6H_3), naphthol, and others and with Q=H, K, $C_6H_5NH_3$ or 2-amino-4-methylpyridinium have been investigated by ^1H or ^{13}C NMR and mass spectral techniques [100, 107].

Although H_3BO_3 is often used as a component for buffer mixtures, it complexes with *o*-quinones. Electrochemical and spectrophotometric experiments have shown that borate buffers at pH 9.16 exhibit two types of interference with these systems, one with the *o*-quinones and one with the corresponding hydroquinones. The following complexation constants have been determined: with phenanthrenequinone-3-sulfonate, $K<2.5\times10^{16}$; with 1,2-naphthoquinone-3-sulfonate, $K<3.9\times10^{17}$; and with 1,2-naphthoquinone-4-sulfonate, $K\approx3\times10^{12}$ [108].

The complex formation between the orthoborate ion and lactic acid was investigated by analyzing the pH resulting from the combination of the two species; no neutral boric acid complex was found [110].

The ^{11}B NMR spectra were obtained for dilute aqueous solutions of H_3BO_3 and borate, respectively, with some typical hydroxycarboxylic acids such as lactic, malic, tartaric, and D(−)- or L(+)-succinic acid as a function of the pH value; typical shifts were found for the esters occurring at the different pH ranges. The origin of the two ^{11}B NMR signals observable in wine could be detected [109].

H_3BO_3 reacts with α-hydroxy acids, oxalic acid, 3-methylpyrocatechol (1,2-$(OH)_2$-3-CH_3-C_6H_3), phenylhydroxamic acid, citric acid, and tartaric acid in dimethylformamide to give spiro compounds. The spirane structure has been established by ^1H, ^{11}B, and ^{13}C NMR spectroscopy. The pK_a values of these compounds, as determined by potentiometric titration with triethylamine, correspond to strong acids ($1.5<pK_a<2$ in dimethylformamide, $-2.5<pK_a<0.3$ in dimethyl sulfoxide). In boron-spiranes, an important decrease in the acidity is observed when the α-hydroxy acid ligand is replaced by the ethanediol moiety [111].

Isothermal solubility methods were used to determine the density, reflection indices, and pH values of saturated solutions of the H_3BO_3–butyric acid–H_2O system at 10 and 20°C. A crystallization branch for H_3BO_3 was determined; butyric acid has a salting-out effect on H_3BO_3 [112].

$[NH_4][BL(OH)_2]\cdot 2H_2O$ (H_2L=mandelic acid) [113], **$M[BL_2]_2\cdot nH_2O$** (M=Ni, n=4; M=Mn, n=0; H_2L=mandelic acid) [114], **$QH[BL_2]\cdot nH_2O$** (Q=NH_3, $C_6H_5NH_2$, n=2; Q=$(C_2H_5)_2NH$, $(n\text{-}C_3H_7)_2NH$, n=0; H_2L=citric acid) [115], **$M[BL_2]\cdot nH_2O$** (M=Li, Na, K, NH_4, Rb; n=0 to 3; H_2L=citric acid), **$M[BL_2]_2\cdot 8H_2O$** (M=Mn, Fe, Co, Ni, Cu, Zn, Cd; H_2L=citric acid) [116 to 119], **$M_4[L(LH)(HO)B_3O_3\text{–}O\text{–}B_3O_3(OH)(LH)L]\cdot nH_2O$** (M=Na, n=8; M=K, n=2; H_2L=salicylic acid) [120], **$Zn(H_2O)_6[BL_2]_2\cdot 4H_2O$** ($H_2L$=salicylic acid) [121], **$M[BL_2]\cdot nH_2O$** (M=Na, n=0, 2; M=K, n=1; M=Rb, n=2; M=Cs, n=2.5; H_2L= 5-hydroxysalicylic or gentisic acid) [122],

$M[BL_2]_2 \cdot nH_2O$ (M = Mg, Sr, Hg, n = 4; M = Ba, n = 2; H_2L = 4-aminosalicylic acid) [123], and **$Ag[BL_2] \cdot nH_2O \cdot xC_2H_5OH$** (HL = 4-hydroxy- or 4-methylsalicylic acid) were prepared and characterized by IR, UV, and ^1H NMR data, density, electrical conductivity and cryoscopic studies, by X-ray diffractometry, and, especially, by thermal analyses such as thermogravimetry (TG), differential thermal analysis (DTA), and differential thermogravimetry (DTG) [124]. Some complex borates with salicylic acid and its derivatives and other polyols have been investigated by structural analyses: **$[Zn(H_2O)_6][BL_2]_2 \cdot 4H_2O$** ($H_2L$ = salicylic acid) [121] and **$[Cu(H_2O)_6][BL]_2 \cdot 4H_2O$** ($H_4L$ = pentaerythritol) [125].

Electrical conductivities of aqueous, methanolic, and ethanolic solutions of $M[BL_2]$ (M = Na, K, Rb, Cs; H_2L = 5-hydroxysalicylic acid) were determined at 25 °C. The complex anion decomposes in water, but simple dissociation into cation and anion occurs in ethanol, and, to some extent, in methanol [126].

The influence of isotopic substitution of ^{11}B for ^{10}B on the IR spectra of $M[BL_2]$ with M = Na and H_2L = 5-hydroxysalicylic acid, of Cs and K disaligeninborates, Ba monosaligeninborates, and of analogous compounds has been determined. The observation of absorption bands between 800 and 1100 cm^{-1} can be used as the special criterion for the presence of tetrahedrally coordinated boron [127].

H_3BO_3 decreases the diffusion of salicylic acid through a lipid membrane due to the formation of a 1:1 complex. The stability constant of the salicylic acid-borate complex was determined by a UV spectrometric method; the spectrum was almost identical with that of salicylic acid [128]. In the presence of H_3BO_3, a well-defined diffusion-controlled voltammetric reduction wave on a glassy carbon electrode is noticed between pH 6 and 7 due to the reduction of the salicylic acid borate complex to salicylaldehyde [131]. The equilibrium constants of the reaction $H_3BO_3 + [HL]^- \rightleftharpoons [LB(OH)_2]^- + H_2O$ (H_2L = salicylic or 4-nitrosalicylic acid) were determined at pH 5.2 and 25 °C (I = 0.1 M) spectrophotometrically. The constants increase with decreasing acidity of H_2L [129].

The reaction of H_3BO_3 with 3-formylsalicylic acid, glycolic acid, or benzilic acid in the presence of various large cations leads to bis- and mono-3-formylsalicylato-, glycolato-, and benzilato borates. In the presence of pyridine (C_5H_5N), the neutral complex $[BL(OH)C_5H_5N]$ with L = 3-formylsalicylato has been formed and characterized [130].

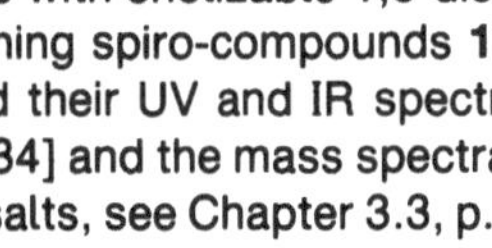

17

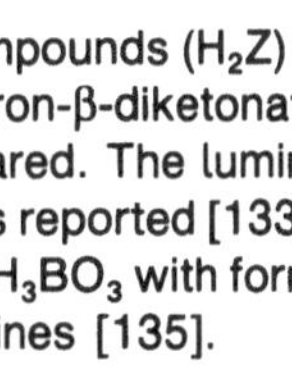

18

H_3BO_3 reacts with enolizable 1,3-dicarbonyl and diol compounds (H_2Z) to give a new type of boron-containing spiro-compounds **17** [132]. Various boron-β-diketonates (**18**) have been synthesized and their UV and IR spectra have been compared. The luminescent properties were studied [134] and the mass spectral fragmentation was reported [133]. For bis(β-diketonato)boronium salts, see Chapter 3.3, p. 68. The reaction of H_3BO_3 with formazanes in CH_3OH yields 1,1-diacetoxy-1,5-diaryl-4-furyl-1-boro-2,3,5,6-tetrazines [135].

Treatment of H_3BO_3 with acetic anhydride gives boron acetates, which react with dibasic quadridentate Schiff bases to form complexes with four-coordinate boron [136]. For complexes of H_3BO_3 with hydroxy-substituted triarylformazans, see [137].

The reaction between H_3BO_3 and urea was studied in aqueous solution by solubility measurements and IR spectrometry. The product at pH 4.24 is a 1:1 complex with a stability constant of 0.057 L/mol at 25 °C. The considerable increase in solubility of H_3BO_3 in neutral medium was explained by interactions between urea and different polyborate species [138]. Between 95 and 110 °C, H_3BO_3 reacts with urea by condensation with water elimination to give a linear polymer $[-B(OH)NHCONH-]_n$, which was verified by employing D_3BO_3 [139]. The structure of the compound which is formed in the H_3BO_3–urea–water system between 52 and 75 °C has the formula $[(H_2NCONH_3)_x(NH_4)_{1-x}][B_5O_6(OH)_4]\cdot 2H_2O\cdot 8H_2NCONH_2$ ($x<1$), as is indicated by IR spectral data. In the system H_3BO_3–formamide–H_2O, the compound $B_2O_3 \cdot 2H_2O\cdot 2HCONH_2$ is formed at 25 °C in which $HCONH_2$ is oxygen-bonded to boron, and the latter is both four- and three-coordinate [140]. For the reaction of H_3BO_3 with hydrobenzoin, see [141]. A solution of H_3BO_3 in acetic anhydride reacts with ketoamines (HL) in 1:1 mole ratio to give complexes of the types $[B(OC(O)CH_3)_2L]$, in which the central boron atom is always four-coordinate [142]. H_3BO_3 reacts with benzohydroxamic acid derivatives, RCONHOH, in NaOH to form salts of the general formula $Na[(OH)_2B\{-OC(R)NO-\}]$ [143].

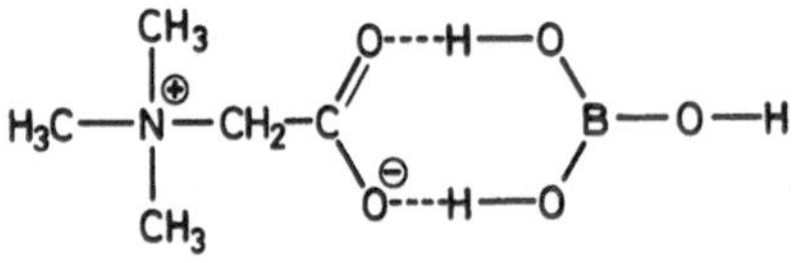

19

The betaine borate **19** has been prepared; it undergoes a second order phase transition at 142.5 K. Crystals below this temperature belong to the ferroelastic Aizu species mmmF2/m. The paraelastic phase is orthorhombic, space group Pmcn–D_{2h}^{16} (No. 62) with a = 776.9(1), b = 987.3(2), c = 1197.4(2) pm; Z = 4; D_m = 1.292 g/cm³ and D_c = 1.295 g/cm³; refined to R = R_w = 4.1% for 519 reflections. The B–O distances are 134.5, 134.5, and 134.8 pm. The ferroelastic phase (T = 130 K) is monoclinic, space group $P2_1/c$–C_{2h}^5 (No. 14) with a = 761.5(5), b = 987.2(3), c = 1194.7(5) pm; β = 92.98(8)°; Z = 4; D_c = 1.325 g/cm³; refined to R = 8.3% and to R_w = 8.7% for 507 reflections. The B–O distances are 135.9, 136.7, and 137.0 pm. In both structures, the betaine moieties are connected to $B(OH)_3$ groups via hydrogen bonds to form chains running parallel to (001). These chains are linked to each other by van der Waals forces [144].

A discussion (with 24 references) on the photochemical behavior of luminescent dyes in sol-gel and H_3BO_3 glasses is given [145].

Nonlinear optical interactions in a low-melting H_3BO_3 glass doped with the organic dye fluorescein have been investigated. This material has a very large third-order nonlinear optical susceptibility of ca. 1 esu which is shown to be electronic and not thermal in origin. Measurements of the tensor nature of the nonlinear susceptibility are presented [146]. A passive one-way aberration correction in order to reconstruct the wave front uses degenerate four-wave mixing in fluorescein-doped H_3BO_3 glass [147]. Fluorescein-doped H_3BO_3 glass has a small saturation intensity of ca. 15 mW/cm²; the saturated absorption depends on the state of polarization of the saturating beam, even though the unsaturated absorption is polarization intensive. Two-beam coupling due to the nonlinearity of saturable absorption was demonstrated. The magnitude of the coupling is maximized by including a frequency shift between the two beams of ca. 0.1 Hz [148]. Phase-conjugate reflectivity of ca. 0.5% was achieved using this fluorescein-doped H_3BO_3 glass. The phase-conjugate reflectivity could be improved through the use of a dye/host system that does not display excited state absorption. To improve the vector character of the phase-conjugation process, one should use a dye/host system with smaller site-to-site variation [149].

The transfer of excitation energy from sodium fluorescein to Rhodamin 6G in H_3BO_3 glass at room temperature was studied by fluorescence decay data analysis. The results show that at low acceptor concentration energy transfer occurs according to the theoretical calculation, which accounts for the effects of donor-donor transport on trapping by acceptors. At high acceptor concentration, direct transfer to an acceptor takes place [150]. Optical phase conjugation was demonstrated in Rhodamin 6G-doped H_3BO_3 glass using a continuous-wave Ar ion laser. The dependence of the phase-conjugated signal on the intensity and wavelength of the pump beam is studied. The role of amplitude and phase grating contribution to the phase-conjugated signal is discussed [151].

The first observation of two-color, photon-gated spectral hole-burning in an organic system is reported: carbazole in H_3BO_3 glass. This new hole-burning mechanism involves stepwise biphotonic photoionization of carbazole at 1.4 K. The quantum yield for photoionization increases exponentially with the energy of the second proton [152].

Two-step laser burning of stable holes in the absorption spectrum of perylene in H_3BO_3 at 5 K was obtained using a pulsed dye laser tuned to the perylene O-O transition. The hole-burning includes two-proton ionization, producing perylene cation radicals. Photoexcitation proceeded only along the singlet state route. The photoreaction is irreversible, and efficient even at low power excitations (ca. 0.5 mW/cm^2) [153]. A detailed spectroscopic analysis of the electronic absorption spectrum between 190 and 500 nm of the perylene radical cation produced in an H_3BO_3 glass film by photo-oxidation is reported. In order to interpret the observed spectrum, the electronic energy levels and oscillator strengths were calculated using the Pariser-Parr free electron molecular orbital method with limited configuration interaction (CI). The effect of the variation of solute concentration and irradiation time of the films on the relative intensities of the absorption bands was also studied. The experimentally observed transition energies of four strong bands show a linear relationship with the theoretically calculated ones. The agreement between experimental and theoretical results is good [154].

The electronic absorption spectra between 100 and 900 nm of anthraquinone and its cation or of dibenzanthracene in H_3BO_3 glasses were recorded. Using the Pariser-Parr molecular orbital method with limited configuration interaction, the electronic energy levels were calculated in order to interpret the experimental spectra of the molecules. Agreement between the experimentally observed bands and the theoretical calculations was found [155].

References for 3.5.7.5:

[1] Ignass, R. T.; Shvarts, E. M. (Latv. PSR Zinat. Akad. Vestis Kim. Ser. **1985** 449/55).

[2] Marchenko, M. M.; Shvarts, E. M. (Latv. PSR Zinat. Akad. Vestis Kim. Ser. **1987** 357/8).

[3] Kamars, A.; Shvarts, E. M. (Latv. PSR Zinat. Akad. Vestis Kim. Ser. **1989** 387/93).

[4] Kamars, A.; Shvarts, E. M. (Latv. PSR Zinat. Akad. Vestis Kim. Ser. **1989** 394/401).

[5] Kursina, M. M.; Shvarts, E. M. (Latv. PSR Zinat. Akad. Vestis Kim. Ser. **1988** 291/6).

[6] Ignass, R. T.; Shvarts, E. M.; Idlis, G.; Batalin, O. E. (Latv. PSR Zinat. Akad. Vestis Kim. Ser. **1989** 730/5).

[7] Tanasheva, M. R.; Beremzhanov, B. A.; Tsygankova, I. I.; Kazymbetova, M. S. (Zh. Obshch. Khim. **57** [1987] 992/5; J. Gen. Chem. [USSR] **57** [1987] 882/5 from C.A. **107** [1987] No. 84747).

[8] Tanasheva, M. R.; Beremzhanov, B. A.; Tsygankova, I. I. (Zh. Neorg. Khim. **33** [1988] 486/8; Russ. J. Inorg. Chem. **33** [1988] 272/4).

[9] Kotov, G. N.; Beremzhanov, B. A.; Tanasheva, M. R. (Zh. Neorg. Khim. **32** [1987] 2319/22; Russ. J. Inorg. Chem. **32** [1987/88] 1356/8).

[10] Tanasheva, M. R.; Beremzhanov, B. A.; Tsygankova, I. I. (Zh. Obshch. Khim. **58** [1988] 500/2; J. Gen. Chem. [USSR] **58** [1988] 436/8 from C.A. **108** [1988] No. 211283).

[11] Hejda, J.; Jedináková, V. (Collect. Czech. Chem. Commun. **52** [1987] 2142/8; J. Radioanal. Nucl. Chem. **121** [1988] 441/6).
[12] Vinogradov, E. E.; Kuliev, A. A.; Shamiryan, P. S. (Zh. Neorg. Khim. **31** [1986] 1004/9; Russ. J. Inorg. Chem. **31** [1986] 571/5).
[13] Shamiryan, P. S.; Tarasova, G. N.; Vinogradov, E. E. (Zh. Neorg. Khim. **31** [1986] 515/7; Russ. J. Inorg. Chem. **31** [1986] 293/5).
[14] Vinogradov, E. E.; Shamiryan, P. S. (Zh. Neorg. Khim. **31** [1986] 1255/6; Russ. J. Inorg. Chem. **31** [1986] 713/5).
[15] Shamiryan, P. S.; Vinogradov, E. E. (Zh. Neorg. Khim. **31** [1986] 1599/601; Russ. J. Inorg. Chem. **31** [1986] 916/7).
[16] Vinogradov, E. E.; Shamiryan, P. S. (Zh. Neorg. Khim. **31** [1986] 1596/8; Russ. J. Inorg. Chem. **31** [1986] 914/6).
[17] Shamiryan, P. S.; Lepeshkov, I. N.; Vinogradov, E. E. (Arm. Khim. Zh. **39** [1986] 486/9, 570/4, 588/90 from C.A. **106** [1987] Nos. 144887 to 144889).
[18] Vinogradov, E. E.; Shamiryan, P. S. (Zh. Neorg. Khim. **32** [1987] 2522/6; Russ. J. Inorg. Chem. **32** [1987/88] 1469/71).
[19] Vinogradov, E. E.; Dikhanov, E. T.; Ivanov, A. A.; Tarasova, G. N. (Zh. Neorg. Khim. **31** [1986] 1257/61; Russ. J. Inorg. Chem. **31** [1986] 715/7).
[20] Tarasova, G. N.; Vinogradov, E. E.; Kristanova, L. Ts.; Balarev, Kh. Kh.; Lepeshkov, I. N. (Zh. Neorg. Khim. **33** [1988] 2669/71; Russ. J. Inorg. Chem. **33** [1988/89] 1531/3).

[21] Tarasova, G. N.; Vinogradov, E. E.; Lepeshkov, I. N. (Zh. Neorg. Khim. **33** [1988] 2612/6; Russ. J. Inorg. Chem. **33** [1988/89] 1498/501).
[22] Petrovska, N. T.; Grizo, A. N. (J. Serb. Chem. Soc. **54** [1989] 319/24 from C.A. **113** [1990] No. 47478).
[23] Bernane, A.; Shvarts, E. M.; Timoteus, H.; Hansens, T. (Latv. PSR Zinat. Akad. Vestis Kim. Ser. **1987** 48/51).
[24] Shvarts, E. M.; Kalve, I.; Putnina, A.; Timoteus, H. (Latv. PSR Zinat. Akad. Vestis Kim. Ser. **1987** 471/6).
[25] Bernane, A.; Shvarts, E. M.; Timoteus, H. (Latv. PSR Zinat. Akad. Vestis Kim. Ser. **1989** 447/50).
[26] Shvarts, E. M.; Zanina, A. S.; Shergina, S. I.; Kotlyarevskii, I. L. (Latv. PSR Zinat. Akad. Vestis Kim. Ser. **1986** 308/11).
[27] Tel'zhenskaya, P. N.; Shvarts, E. M.; Shergina, S. I.; Sokolov, I. E.; Zanina, A. S.; Kotlyarevskii, I. L. (Latv. PSR Zinat. Akad. Vestis Kim. Ser. **1985** 326/9).
[28] Kotlyarevskii, I. L.; Zanina, A. S.; Shergina, S. I.; Sokolov, I. E.; Shvarts, E. M.; Ignass, R. T.; Tel'zhenskaya, P. N. (Izv. Akad. Nauk SSSR Ser. Khim. **1987** 621/3; Bull. Akad. Sci. USSR Dir. Chem. Sci. **36** [1987] 568/70).
[29] Shvarts, E. M.; Ignass, R. T.; Tel'zhenskaya, P. N.; Shergina, S. I.; Sokolov, I. E.; Zanina, A. S.; Kotlyarevskii, I. L. (Zh. Prikl. Khim. **60** [1987] 1567/71; J. Appl. Chem. [USSR] **60** [1987] 1472/6 from C.A. **107** [1987] No. 162783).
[30] Kursina, M.; Shvarts, E. M. (Latv. PSR Zinat. Akad. Vestis Kim. Ser. **1988** 547/51).

[31] Kursina, M.; Shvarts, E. M. (Latv. PSR Zinat. Akad. Vestis Kim. Ser. **1988** 654/8; **1989** 538/42).
[32] Kamars, A.; Shvarts, E. M. (Latv. PSR Zinat. Akad. Vestis Kim. Ser. **1989** 243/4).
[33] Kamars, A.; Shvarts, E. M. (Latv. PSR Zinat. Akad. Vestis Kim. Ser. **1985** 703/8).
[34] Alekperov, E. R. (Azerb. Khim. Zh. **1988** No. 1, pp. 109/11 from C.A. **111** [1989] No. 141689).
[35] Skvortsov, V. G.; Tsekhanskii, R. S.; Sukova, L. M.; Molodkin, A. K.; Rodionov, N. S. (Zh. Neorg. Khim. **31** [1986] 2140/2; Russ. J. Inorg. Chem. **31** [1986/87] 1233/5).

[36] Skvortsov, V. G. (Zh. Neorg. Khim. **31** [1986] 3163/72; Russ. J. Inorg. Chem. **31** [1986/87] 1816/22).
[37] Skvortsov, V. G.; Tsekhanskii, R. S.; Molodkin, A. K.; Sukova, L. M.; Petrova, O. V.; Dolganev, V. P. (Zh. Neorg. Khim. **32** [1987] 491/4; Russ. J. Inorg. Chem. **32** [1987] 272/4).
[38] Tsekhanskii, R. S.; Skvortsov, V. G.; Molodkin, A. K.; Sukova, L. M.; Rodionov, N. S. (Zh. Neorg. Khim. **30** [1985] 1609/12; Russ. J. Inorg. Chem. **30** [1985] 917/9).
[39] Skvortsov, V. G.; Tsekhanskii, R. S.; Molodkin, A. K.; Petrova, O. V.; Akimov, V. M. (Zh. Neorg. Khim. **30** [1985] 277/9; Russ. J. Inorg. Chem. **30** [1985] 158/9).
[40] Batsanov, A. S.; Petrova, O. V.; Struchkov, Yu. T.; Akimov, V. M.; Molodkin, A. K.; Skvortsov, V. G. (Zh. Neorg. Khim. **31** [1986] 1120/5; Russ. J. Inorg. Chem. **31** [1986] 637/40).

[41] Skvortsov, V. G.; Sadetdinov, Sh. V.; Molodkin, A. K.; Mikhailov, V. I.; Pavlov, G. P.; Nikiforov, N. G. (Zh. Neorg. Khim. **34** [1989] 1083/6; Russ. J. Inorg. Chem. **34** [1989] 610/2).
[42] Skvortsov, V. G. (Zh. Neorg. Khim. **34** [1989] 974/8; Russ. J. Inorg. Chem. **34** [1989] 546/9).
[43] Coddington, J. M.; Taylor, M. J. (J. Coord. Chem. **20** [1989] 27/38).
[44] Wada, H.; Ito, Sh.; Kuroda, K.; Kato, Ch. (Chem. Lett. **1985** 691/2).
[45] Wada, H.; Nojima, K.; Kuroda, K.; Kato, Ch. (Yogyo Kyokaishi **95** [1987] 130/4 from C.A. **106** [1987] No. 89061).
[46] Wada, H.; Kim, B. K.; Kuroda, K.; Kato, Ch. (Rikogaku Kenkyusho Hokoku Waseda Daigaku No. 106 [1984] 63/73 from C.A. **103** [1985] No. 215364).
[47] Wada, H.; Araki, S.; Kuroda, K.; Kato, Ch. (Polyhedron **4** [1985] 653/6).
[48] Cifuentes, L. O.; Bahamondes, J. (Bol. Soc. Quim. Peru **51** [1985] 189/92 from C.A. **107** [1987] No. 115764).
[49] Dawber, J. G.; Green, S. I. E. (J. Chem. Soc. Faraday Trans. I **82** [1986] 3407/13).
[50] Bowcher, T. L.; Dawber, J. G. (Polym. Commun. **30** [1989] 215/7 from C.A. **111** [1989] No. 154497).

[51] Dubina, L. G.; Mikul'skii, G. F.; Khomutov, L. I. (Protsessy Strukturoobraz. Polim. Sistemakh, Saratov **1986** 39/52 from C.A. **107** [1987] No. 7759).
[52] Kisel'gof, G. V.; Arkhangel'skii, L. K.; Korol'kova, S. V. (Zh. Prikl. Khim. **58** [1985] 2731/3; J. Appl. Chem. [USSR] **58** [1985] 2524/6 from C.A. **104** [1986] No. 149820).
[53] Kisel'gof, G. V.; Arkhangel'skii, L. K.; Bochkova, N. A. (Zh. Prikl. Khim. **59** [1986] 909/12; J. Appl. Chem. [USSR] **59** [1986] 839/41 from C.A. **105** [1986] No. 8693).
[54] Kisel'gof, G. V.; Arkhangel'skii, L. K.; Bochkova, N. A.; Lavrova, N. K.; Korol'kova, S. V.; Skorokhodov, S. S.; Stepanov, V. V. (Ionnyi Obmen Ionometriya No. 5 [1986] 80/9 from C.A. **106** [1987] No. 73485).
[55] Matsuzawa, S.; Yamaura, K.; Tanigami, T.; Somura, T.; Nakata, M. (Polym. Commun. **28** No. 4 [1987] 105/6 from C.A. **107** [1987] No. 40480).
[56] Ochiai, H.; Kohno, R.; Murakami, I. (Polym. Commun. **27** No. 12 [1986] 366/8 from C.A. **106** [1987] No. 33834).
[57] Hashimoto, S.; Furukawa, K. (Biotechnol. Bioeng. **30** [1987] 52/9 from C.A. **107** [1987] No. 120455).
[58] Sinton, S. W. (Macromolecules **20** [1987] 2430/41 from C.A. **107** [1987] No. 155104).
[59] Maerker, J. M.; Sinton, S. W. (J. Rheol. [N. Y.] **30** [1986] 77/99 from C.A. **104** [1986] No. 226993).
[60] Liu, Ch.-Y. (Zhuzao **1985** No. 5, pp. 26/32 from C.A. **104** [1986] No. 153757).

[61] Ikeda, T. (Hiroshima Joshi Daigaku Kaseigakubu Kiyo No. 24 [1988] 1/8 from C.A. **111** [1989] No. 154 914).
[62] Tikhonova, L. S.; Belozerova, O. A.; Kuznetsova, O. G.; Khaikin, S. Ya.; Zytner, Ya. D.; Makarov, K. A. (Zh. Prikl. Khim. **61** [1988] 575/9; J. Appl. Chem. [USSR] **61** [1988] 516/9 from C.A. **108** [1988] No. 188461).
[63] Shibayama, M.; Sato, M.; Kimura, Y.; Fujiwara, H.; Nomura, S. (Polymer **29** [1988] 336/40 from C.A. **108** [1988] No. 132426).
[64] Shibayama, M.; Yoshizawa, H.; Kurokawa, H.; Fujiwara, H.; Nomura, S. (Polymer **29** [1988] 2066/71 from C.A. **110** [1989] No. 39623).
[65] Cho, Ch. S.; Han, S. Y. (Pollimo **10** [1986] 261/8 from C.A. **105** [1986] No. 98066).
[66] Pal, M. K.; Pal, P. K. (Makromol. Chem. **188** [1987] 1735/42).
[67] Baumgartner, C. E. (Anal. Chem. **59** [1987] 2716/8).
[68] Bell, C. F.; Beauchamp, R. D.; Short, E. L. (Carbohydr. Res. **147** [1986] 191/203).
[69] Bell, C. F.; Beauchamp, R. D.; Short, E. L. (Carbohydr. Res. **185** [1989] 39/50).
[70] Huttunen, E. (Ann. Acad. Sci. Fenn. Ser. A 2 **201** [1984] 1/45 from C.A. **102** [1985] No. 120917).

[71] Kamars, A. A.; Shvarts, E. M. (Latv. PSR Zinat. Akad. Vestis Kim. Ser. **1986** 700/7).
[72] Belousova, R. G.; Shvarts, E. M.; Vitola, I. I.; Kamars, A. A. (Latv. PSR Zinat. Akad. Vestis Kim. Ser. **1989** 529/37).
[73] Verchere, J. F. (Bull. Union Physiciens No. 684 [1986] 871/9 from C.A. **105** [1986] No. 49960).
[74] Tanihara, N.; Takita, K.; Tanaka, H.; Kawamura, M.; Kashima, T. (Kyoritsu Yakka Daigaku Kenkyu Nenpo No. 31 **1987** 9/16 from C.A. **106** [1987] No. 221183).
[75] Aruga, R. (Talanta **32** [1985] 517/9; J. Chem. Soc. Dalton Trans. **1988** 2971/4).
[76] Verchere, J. F.; Hlaibi, M. (Polyhedron **6** [1987] 1415/20).
[77] Chapelle, S.; Verchere, J. F. (Tetrahedron **44** [1988] 4469/82).
[78] Lajunen, K.; Hakkinen, P.; Purokoski, S. (Finn. Chem. Lett. **13** [1986] 21/5).
[79] Hakkinen, P.; Lajunen, K.; Purokoski, S. (Finn. Chem. Lett. **15** [1988] 7/12).
[80] Pelmore, H.; Symons, M. C. R. (Carbohydr. Res. **155** [1986] 206/11).

[81] Noble, O.; Taravel, F. R. (Carbohydr. Res. **166** [1987] 1/11).
[82] Gey, C.; Noble, O.; Perez, S.; Taravel, F. R. (Carbohydr. Res. **173** [1988] 175/84).
[83] Noble, O.; Taravel, F. R. (Carbohydr. Res. **184** [1988] 236/43).
[84] Pezron, E.; Ricard, A.; Lafuma, F.; Audebert, R. (Macromolecules **21** [1988] 1121/5).
[85] Pezron, E.; Leibler, L.; Ricard, A.; Lafuma, F.; Audebert, R. (Macromolecules **22** [1989] 1169/74).
[86] Pezron, E.; Ricard, A.; Lafuma, F.; Audebert, R. (Biol. Synth. Polym. Networks **1988** 113/26 from C.A. **110** [1989] No. 75927).
[87] Pezron, E.; Ricard, A.; Leibler, L.; Lafuma, F.; Audebert, R. (Food Hydrocolloids **1** [1986] 481/3 from C.A. **110** [1989] No. 154737).
[88] Kramer, J.; Prud'homme, R. K.; Wiltzius, P.; Mirau, P.; Knoll, S. (Colloid Polym. Sci. **266** [1988] 145/55 from C.A. **109** [1988] No. 73806).
[89] Van Duin, M.; Peters, J. A.; Kieboom, A. P. G.; Van Bekkum, H. (Tetrahedron **41** [1985] 3411/21).
[90] Makkee, M.; Kieboom, A. P. G.; Van Bekkum, H. (Recl. J. R. Neth. Chem. Soc. **104** [1985] 230/5 from C.A. **104** [1986] No. 207567).

[91] Van Duin, M.; Peters, J. A.; Kieboom, A. P. G.; Van Bekkum, H. (Recl. J. R. Neth. Chem. Soc. **105** [1986] 488/93 from C.A. **107** [1987] No. 59381).

[92] Van Duin, M.; Peters, J. A.; Kieboom, A. P. G.; Van Bekkum, H. (Carbohydr. Res. **162** [1987] 65/78).
[93] Van Duin, M.; Peters, J. A.; Kieboom, A. P. G.; Van Bekkum, H. (J. Chem. Soc. Perkin Trans. II **1987** 473/8).
[94] Van Duin, M.; Peters, J. A.; Kieboom, A. P. G.; Van Bekkum, H. (J. Chem. Soc. Dalton Trans. **1987** 2051/7).
[95] Van Duin, M.; Peters, J. A.; Sinnema, A.; Kieboom, A. P. G.; Van Bekkum, H. (Recl. J. R. Neth. Chem. Soc. **106** [1987] 495/7 from C.A. **109** [1988] No. 93092).
[96] Van Duin, M.; Hoefnagel, M. A.; Baas, J. M. A.; Van de Graf, B. (Recl. J. R. Neth. Chem. Soc. **106** [1987] 607/12 from C.A. **109** [1988] No. 210282).
[97] Dijkgraaf, P. J. M.; Verkuylen, M. E. C. G.; Van der Wiele, K. (Carbohydr. Res. **163** [1987] 127/31).
[98] Gunduz, N.; Kilic, A. (Chim. Acta Turc. **14** [1986] 95/100 from C.A. **107** [1987] No. 189505).
[99] Grachek, V. I.; Naumova, S. F.; Motol'ko, G. R.; Smolyakov, A. V.; Bukanova, N. N.; Kozlov, N. S. (Dokl. Akad. Nauk BSSR **29** [1985] 536/9 from C.A. **104** [1986] No. 148952).
[100] Okamoto, Y.; Kinoshita, T.; Takei, Y.; Matsumoto, Y. (Polyhedron **5** [1986] 2051/7).

[101] Tanaka, S.; Kaneta, T.; Yoshida, H. (Anal. Sci. **5** [1989] 217/8 from C.A. **112** [1990] No. 30059).
[102] Dzhioshvili, B. D.; Pirtskhalava, N. I.; Turiashvili, L. G.; Metreveli, N. G. (Soobshch. Akad. Nauk Gruz. SSR **129** [1988] 549/52 from C.A. **110** [1989] No. 192 891; **133** [1989] 73/6 from C.A. **111** [1989] No. 69747).
[103] Grundshtein, V. V.; Shvarts, E. M.; Vitola, I. (Latv. PSR Zinat. Akad. Vestis Kim. Ser. **1986** 534/6).
[104] Grundshtein, V. V.; Shvarts, E. M.; Lange, I. (Latv. PSR Zinat. Akad. Vestis Kim. Ser. **1989** 33/8).
[105] Sagulenko, V. S.; Shvarts, E. M.; Kalacheva, V. G.; Skorikov, S. N.; Vitola, I. (Latv. PSR Zinat. Akad. Vestis Kim. Ser. **1988** 25/9).
[106] Shvarts, E. M.; Sultanova, L. Z.; Sagulenko, V. S.; Gorenshtein, R. G.; Skorikov, S. N. (Zh. Neorg. Khim. **34** [1989] 606/10; Russ. J. Inorg. Chem. **34** [1989] 337/40).
[107] Okamoto, Y.; Takei, Y.; Tagaki, K. (Polyhedron **6** [1987] 2119/28).
[108] Bailey, S. I.; Ritchie, I. M. (Electrochim. Acta **32** [1987] 1027/33).
[109] Lutz, O.; Veil, E.; Brändle, U.; Helber, H.; Kammerer, E.; Scheiter, A. (Z. Naturforsch. **41a** [1986] 1041/4).
[110] Paal, T.; Mate, M. (Magy. Kem. Foly. **94** [1988] 143/4).

[111] Lamande, L.; Boyer, D.; Muñoz, A. (J. Organomet. Chem. **329** [1987] 1/29).
[112] Khazikhamova, B. Kh.; Azimbaeva, G. E.; Kalacheva, V. G. (Zh. Neorg. Khim. **31** [1986] 1601/3; Russ. J. Inorg. Chem. **31** [1987] 917/8).
[113] Berniyazova, D. G.; Shvarts, E. M.; Kalacheva, V. G. (Zh. Neorg. Khim. **34** [1989] 602/5; Russ. J. Inorg. Chem. **34** [1989] 335/7).
[114] Berniyazova, D. G.; Kalacheva, V. G.; Shvarts, E. M.; Vitola, I. M.; Skorikov, S. N.; Leonov, I. D. (Zh. Neorg. Khim. **32** [1987] 1543/6; Russ. J. Inorg. Chem. **32** [1987] 924/6).
[115] Sergeeva, G. S.; Burnashova, N. N.; Shvarts, E. M. (Latv. PSR Zinat. Akad. Vestis Kim. Ser. **1989** 26/9).
[116] Shvarts, E. M.; Piloyan, G. O.; Vitola, I. M.; Drozdova, O. V. (Thermochim. Acta **92** [1985] 693/6; Therm. Anal. Proc. 8th Int. Conf., Bratislava 1985, Vol.1, pp. 693/9 from C.A. **107** [1987] No. 88542).

[117] Shvarts, E. M.; Vitola, I. M.; Sergeeva, G. S.; Piloyan, G. O.; Drozdova, O. V. (J. Therm. Anal. **31** [1986] 351/9).
[118] Sergeeva, S. G.; Vitola, I. M.; Shvarts, E. M. (Latv. PSR Zinat. Akad. Vestis Kim. Ser. **1986** 537/40).
[119] Zorkii, P. M.; Chernikova, N. Yu.; Zviedre, I. I. (Koord. Khim. **13** [1987] 966/71).
[120] Mardanenko, V. K.; Shvarts, E. M. (Latv. PSR Zinat. Akad. Vestis Kim. Ser. **1985** 556/60).

[121] Zviedre, I. I.; Bel'skii, V. K.; Mardanenko, V. K. (Latv. PSR Zinat. Akad. Vestis Kim. Ser. **1985** 387/93).
[122] Terauda, A. A.; Shvarts, E. M.; Vitola, I. M.; Lange, I. Ya.; Vladyko, G. V.; Boreko, E. I.; Sadovnikova, N. N. (Latv. PSR Zinat. Akad. Vestis Kim. Ser. **1985** 695/702).
[123] Mardanenko, V. K.; Shvarts, E. M.; Tarasenko, Yu. A. (Zh. Neorg. Khim. **31** [1986] 2534/8; Russ. J. Inorg. Chem. **31** [1986/87] 1462/5).
[124] Lokenbana, M.; Vitola, I. M.; Lange, I. Ya.; Shvarts, E. M. (Latv. PSR Zinat. Akad. Vestis Kim. Ser. **1986** 667/9).
[125] Zviedre, I. I.; Bel'skii, V. K.; Shvarts, E. M. (Latv. PSR Zinat. Akad. Vestis Kim. Ser. **1985** 672/7).
[126] Terauda, A. A.; Shvarts, E. M. (Latv. PSR Zinat. Akad. Vestis Kim. Ser. **1988** 663/9).
[127] Lange, I. Ya.; Terauda, A. A.; Grundshtein, V. V. (Latv. PSR Zinat. Akad. Vestis Kim. Ser. **1987** 467/70).
[128] Paál, T.; Mate, M. (Magy. Kem. Foly. **91** [1985] 41/3; 569/71 from C.A. **103** [1985] No. 66151; **105** [1986] No. 6542).
[129] Lukkari, O.; Tamminen, J. (Finn. Chem. Lett. **15** [1988] 13/7).
[130] Dey, K.; Gangopadhyay, A.; Biswas, A. K. (J. Bangladesh. Acad. Sci. **11** [1987] 55/66 from C.A. **109** [1988] No. 6562).

[131] Krishnamoorthy, S.; Noel, M. (J. Electrochem. Soc. India **38** [1989] 189/93 from C.A. **112** [1990] No. 107281).
[132] Hartmann, H. (J. Prakt. Chem. **328** [1986] 755/62).
[133] Schade, W.; Ilge, H.-D.; Hartmann, A. (J. Prakt. Chem. **328** [1986] 941/4).
[134] Karasev, V. E.; Korotkikh, O. A. (Zh. Neorg. Khim. **30** [1985] 2269/72; Russ. J. Inorg. Chem. **30** [1985/86] 1290/2).
[135] Nguyen Dinh Trieu; Ha Thi Diep; Ngyugen Hong Tien (Tap Chi Hoa Hoc **25** [1987] 12/3 from C.A. **109** [1988] No. 129090).
[136] Ghose, B. N. (Synth. React. Inorg. Met.-Org. Chem. **16** [1986] 1383/93 from C.A. **107** [1987] No. 154361).
[137] Stepanov, B. I.; Avramenko, G. V.; Salekh Khamud; Mustafaeva, Sh. I. (Zh. Obshch. Khim. **56** [1986] 390/2; J. Gen. Chem. [USSR] **56** [1986] 339/41 from C.A. **105** [1986] No. 107263).
[138] Orszagh, I.; Beck, M.; Holly, S. (Magy. Kem. Foly. **91** [1985] 1383/93 from C.A. **104** [1986] No. 136818).
[139] Germanskii, A. M.; Ershov, V. A.; Bogdanov, S. P.; Pikalov, S. N.; Ushveridze, L. A. (Zh. Prikl. Khim. **61** [1988] 686/8; J. Appl. Chem. [USSR] **61** [1988] 619/21 from C.A. **108** [1988] No. 205141).
[140] Tsekhanskii, R. S.; Skvortsov, V. G.; Molodkin, A. K.; Sukova, L. M. (Zh. Neorg. Khim. **32** [1987] 1488/91; Russ. J. Inorg. Chem. **32** [1987] 893/5).

[141] Schaumburg, G. D.; Battarai, K. M.; Gautam, D. P. (J. Nepal Chem. Soc. **3** [1983] 11/30 from C.A. **103** [1985] No. 142055).
[142] Singh, H. B.; Tandon, J. P. (Indian J. Chem. A **26** [1987] 1054/5 from C.A. **110** [1989] No. 154347).

[143] Das, M. K.; Chakraborty, S. (Synth. React. Inorg. Met.-Org. Chem. **18** [1988] 225/30 from C.A. **110** [1989] No. 192888).
[144] Zobetz, E.; Preisinger, A. (Monatsh. Chem. **120** [1989] 291/8).
[145] Reisfeld, R.; Eyal, M.; Gvishi, R.; Jørgensen, C. K. (Photochem. Photophys. Coord. Compd. Proc. 7th Int. Symp., Schloss Elmau, FRG, 1987, pp. 313/6 from C.A. **109** [1988] No. 137567).
[146] Tompkin, W. R.; Boyd, R. W. (Proc. SPIE-Int. Soc. Opt. Eng. No. 824 [1988] 2/6 from C.A. **108** [1988] No. 158677).
[147] MacDonald, K. R.; Tompkin, W. R.; Boyd, R. W. (Opt. Lett. **13** [1988] 485/7).
[148] Kramer, M. A.; Tompkin, W. R.; Boyd, R. W. (Phys. Rev. [3] A **34** [1986] 2026/31).
[149] Boyd, R. W.; Gauthier, D. J.; Kauranen, M.; Malcuit, M. S.; Tompkin, W. R. (Proc. SPIE-Int. Soc. Opt. Eng. No. 1060 [1989] 58/65 from C.A. **111** [1989] No. 204820).
[150] Pandey, K. K.; Joshi, H. C.; Pant, T. C. (Chem. Phys. Lett. **148** [1988] 472/8).

[151] Kumar, G. R.; Singh, B. P.; Sharma, K. K. (Opt. Commun. **73** [1989] 81/4 from C.A. **111** [1989] No. 204881).
[152] Lee, H. W. H.; Gehrtz, M.; Marinero, E. E.; Mörner, W. E. (Chem. Phys. Lett. **118** [1985] 611/6).
[153] Al'shits, E. I.; Kharlamov, B. M.; Personov, R. I. (Opt. Spectrosk. **65** [1988] 548/51; Opt. Spectrosc. [USSR] **65** [1988] 326/8 from C.A. **109** [1988] No. 240485).
[154] Jain, V. K.; Zaidi, Z. H. (Spectrochim. Acta A **43** [1987] 1275/9; A **44** [1988] 1159/63).
[155] Khan, S. A.; Babu, R.; Zaidi, Z. H. (Acta Cienc. Indica Phys. **10** [1984] 167/9 from C.A. **103** [1985] No. 95368; **11** [1985] 40/4 from C.A. **107** [1987] No. 48566).

3.5.7.6 Applications

3.5.7.6.1 Toxicity

Toxicology and carcinogenesis studies were conducted by feeding H_3BO_3 to mice [1]. Respiratory symptoms and eye irritation were found among workers exposed to boric acid dusts [2].

3.5.7.6.2 Fireproofing, Flammability, Thermal Degradation

For fireproofing and flame-retardant applications of H_3BO_3, see also Section 3.5.7.6.3, p. 228. For the preparation of heat stabilizers from B_2O_3 and phenols, see Section 3.2.5.3, p. 45.

Boric acid promotes the dehydration of cellulose above 250 °C, but does not catalyze intramolecular dehydration reactions. The stabilizing effect during the active thermal degradation phase between ca. 280 and 330 °C was explained by the formation of degradation products containing quasi-aromatic polyborates incorporated into polyconjugated systems [3]. The main product of thermal transformation of guaiacylpropan-1-ol in the presence of H_3BO_3 is diisoeugenol (70% yield). Although free radicals are assumed, the product is apparently produced as a result of acid-base interactions with the formation of reactive centers at the α-C atoms, followed by reaction on the C(6) atom of the aromatic nucleus [4]. Thermal analysis and IR spectroscopy of hydrolytic lignin and HCl-lignin, heated to between 200 and 600 °C, shows that the presence of H_3BO_3 suppresses the thermal degradation reactions of the lignin samples at lower temperatures due to formation of complexes with lignin, but catalyzes these reactions between 300 and 350 °C [5].

Thermal analyses of cellulose treated with H_3BO_3 are conducted in N_2 and in He with 2.5 to 21% oxygen. H_3BO_3 accelerates the beginning of the weight loss, increases the char yield, and inhibits the char oxidation. The main weight loss and the char yield may be explained by a chain reaction, ending with grafting termination [6].

3.5.7.6.3 Impregnation with Boric Acid

Vacuum impregnation of plywood from red lauan by H_3BO_3/borate gives fairly good results for fire retardancy [7]. In order to meet the standard treated plywood fire performance requirements of 1200 s for the weight loss from 30 to 70%, the interpolated value for the concentration of borax/boric acid (3:2) solutions is 9.0% [8].

$2ZnO \cdot 3B_2O_3 \cdot 3H_2O$, prepared from H_3BO_3, promotes the formation of a strong char and suppresses smoke production in halogen-free polymers. The release of its water of hydration at high temperatures might contribute to the intumescence of the char and also inhibits the flaming combustion. The material can be used in combination with alumina trihydrate as an effective flame retardant and smoke suppressant in polyethylene, ethylene-vinyl acetate copolymer, epoxy resins, and acrylic polymers [9]. Flame-retardant boric acid-isophthalic acid-maleic anhydride-glycol polyesters were prepared by a three-step synthesis [10].

Treatments with H_3BO_3 are very effective in eliminating established decay in wood windows, more in sapwood than heartwood, but distributions were not significantly affected by painting [11]. Treatment of green bamboo with a 1:1:1 mixture of H_3BO_3, borax, and pentachlorophenol reduces losses in pump yield from 12.3 to 4.5% over a storage period of 12 months. In flowered and dried bamboo, losses were less than for green bamboo [12]. A coating composition for protecting tree wounds contains 2% H_3BO_3 [13]. For tropical hardwood lumbers, better penetration and retention were obtained with TimBor (H_3BO_3 and disodium octaborate) as compared to AmBor-S (ammonium pentaborate/$Na_2[SO_4]$ solution) [15]. Treatment of unseasoned virola lumber with a solution of TimBor (containing 1.6% sodium pentachlorophenolate) gave adequate protection against lyctid beetles and stain fungi [14].

Adsorption isotherms of nitrogen are determined at 77 K on samples of activated charcoal cloth prepared from viscose rayon cloth after impregnation with aqueous solutions containing NaCl/H_3BO_3 or borax/H_3BO_3. Application of the α_s-method of isotherm analysis reveals that the absorbents possess an appreciable surface area (up to 300 m^2/g) located outside of the micropore structure. These results indicate that the presence of Na^+ ions is essential for the generation of mesoporosity in borate-containing cloths. However, the pH value of the impregnating solutions is also important. At low pH values, as they exist in H_3BO_3–NaCl system, the rate of activation is limited, whereas rapid and uncontrolled development of mesoporosity is promoted at quite low Na^+/B ratios. In H_3BO_3–borax systems and at Na^+/B ratios over 0.4, the borate species readily and relatively easily penetrate the fiber structure [16]. The influence of aqueous H_3BO_3 solutions as impregnants of rayon cloths on the subsequent activation of the derived carbonized materials in CO_2 gas at 850 °C was also studied. In terms of the activation rate generated, borax was more effective as an additive to H_3BO_3 solutions than NaCl. The results were interpreted in terms of the catalytic abilities of various alkali metal ions in the system studied, and the complexity of the borate species generated at different pH values in the aqueous impregnating solutions employed [17].

A preservative combination containing 0.5% H_3BO_3 and 0.5% sodium benzoate was safely used to replace NaCl in the preservation of goatskins and sheepskins [18].

3.5.7.6.4 Coatings and Energy Transfer

Self-aligned shallow junctions on silicon having high dopant concentrations may be accomplished by depositing oxidized forms of dopants such as H_3BO_3 on patterned wafers, reducing the deposited material, e.g., annealing between 600 and 650°C in a reducing atmosphere, and carrying out thermal activation and drive-in steps [19].

Steel containing 2.25% Cr and 1% Mo was coated with H_3BO_3 by vapor deposition, preheated to form B_2O_3, and oxidized in dry flowing oxygen at 600°C for 100 h; the B_2O_3 has an inhibiting effect on the oxidation process; the mechanism depends on the B_2O_3 layer fracture [21].

H_3BO_3 is an emulsifier for an aluminium paste used in the manufacture of autoclaved cellular concrete [20]. H_3BO_3 is a low linear energy transfer radiation-enhancement agent in vivo [23]. The ablation properties of insulating materials such as H_3BO_3 have been studied experimentally and theoretically for arc lengths up to 150 mm and currents up to 16 MA [22].

For a coating composition containing 2% H_3BO_3 to protect tree wounds, see [13].

3.5.7.6.5 Boric Acid in Pressured Water Reactors

A significant fraction of the operating pressured water reactor (PWR) steam generators are using H_3BO_3 as an inhibitor to control stress corrosion cracking, intergranular attack, or denting. H_3BO_3 is applied via crevice flushing, low power soaks, on-line, or by a combination of these methods. The data on its physical properties, its effect on corrosion, and how it should be correctly applied in nuclear steam generators are given [24].

In the nuclear industry, H_3BO_3 dissolved in the reactor coolant is used as a soluble reactivity and neutron-absorbing control agent. In pressured water reactor plants, the dissolved H_3BO_3 is referred to as a soluble poison or chemical shim, due to the high capacity for thermal neutron capture exhibited by the ^{10}B isotope contained in the boric acid molecule. This slow reactivity change or chemical shim control would otherwise have to be performed using control rods, a much more expensive proposition. Reactivity changes are controlled by the ^{10}B isotope by virtue of its very high cross section (3837 barns) for thermal neutron absorption. However, natural boron contains only 20 at% of the ^{10}B isotope. The remaining 80% of ^{11}B isotope with a cross section of 0.005 barns is of essentially no use as a neutron absorber. Since ^{11}B makes up the bulk of the total boron present, one should eliminate this isotope; operating pressured water reactor plants need only a fraction of the H_3BO_3 concentration to accomplish identical nuclear operations. However, to achieve the elimination of ^{11}B from natural H_3BO_3, an isotope separation must be performed [25].

An important cause of tube degradation in recirculating pressured water reactor steam generators with crevices is the intergranular attack on Alloy 600 tubing in the crevice region. The crevice environment has high concentrations of a caustic species, which can be neutralized by boric acid via off-line flushing [26]. H_3BO_3, which is used to combat other forms of corrosion, was identified as a potentially useful additive to the feed water or soak water for neutralizating caustics and for inhibiting intergranular attack on Alloy 600 tubes [27]. A cold-worked hump in the middle of slow strain rate tensile specimens of Alloy 600 greatly increased the intergranular stress corrosion cracking sensitivity in simulated pressured water reactors using H_3BO_3-LiOH-H_2 high temperature water solutions for steam generator tubing [28]. Corrosion of Alloy 600 tubes is caused by the buildup of impurities in the tube sheet crevices. Procedures were developed to flush impurities from the crevices. The same procedures were

References on pp. 230/2

demonstrated in the laboratory to concentrate boric acid in the crevices when the acid is added to the steam generator bulk water during crevice flushing. Boric acid in the crevices decreases or stops intergranular stress corrosion cracking in laboratory tests [29].

The effects of borated water leakage on carbon and low-alloy steel components in pressured water reactors applications have been reviewed. Boric acid corrosion field experience and laboratory test result data are also given [30].

An emergency boration system is described for the rapid injection of H_3BO_3, capable of reducing temperature and pressure transients associated with control rod failure and loss of coolant flow in a pressured water reactor [31]. The H_3BO_3 concentration variation during fuel burnup (Russian WWER type reactor), caused by a small perturbation of the multiplication factor, was estimated with ca. 1% error in mean zonal flows [32].

The $^{239}Pu/^{235}U$ neutron fission rate ratio profile was measured in the region of a plane interface between a tank containing ordinary water and a tank containing an aqueous solution of H_3BO_3, as part of reactor neutron moderator studies. The results indicate a ca. 8 mm shift of the central point of the fission rate ratio profile [33]. A method was developed by which radioactive wastes containing borate generated in pressured water reactor power plants can be treated chemically in such a manner that insoluble borates can be formed. These products are then suitable for storage, transportation, and disposal [34].

3.5.7.6.6 Catalysis

Catalysts for the epoxidation of olefins are made from H_3BO_3, pyrocatechol (1,2-$(OH)_2$-C_6H_4), and molybdenyl acetylacetonate [35]. Heterogeneous H_3BO_3-$[MoO_4]^{2-}$ mixed resins based upon styrene and methacrylamide are new catalysts for liquid-phase oxygen-transfer reactions to olefins [36]. Solid-state polyamidation of dodecamethylenediammonium adipate was studied in the presence of acid catalysts; H_3BO_3 was the most effective catalyst, followed by H_2SO_4 [37].

In the rigid foam extrusion of PVC, H_3BO_3 enhances the uniformity of the cells. The efficiency of the blowing agents $Na[HCO_3]$ or azodicarbamide is increased slightly [38]. H_3BO_3 promotes the decomposition of the blowing agent [39].

H_3BO_3 is a rapid reversible competitive inhibitor of urease at pH values between 6.25 and 9.3 [40]. The oxidation of a methionine residue in subtilisin-type proteinases by the alkaline aqueous H_3BO_3 system is an active site-directed reaction [41].

The photohydration of styrenes and of phenylacetylenes in 0.5 M H_3BO_3/borax buffer solution has been investigated [42].

References for 3.5.7.6:

[1] National Toxicology Program (Natl. Toxicol. Program Tech. Rept. Ser. No. 324 [1987] 1/126 from C.A. **108** [1988] No. 107932).

[2] Smith, T. J. (JOM J. Occup. Med. **26** [1984] 584/6 from C.A. **101** [1984] No. 215714).

[3] Rossinskaya, G. A.; Skripchenko, T. N.; Dobele, G.; Domburgs, G.; Yurk'yan, V. V. (Khim. Drev. **1988** No. 5, pp. 85/91 from C.A. **109** [1988] No. 212564).

[4] Dobele, G.; Skripchenko, T. N.; Domburgs, G.; Liepins, M.; Osadshaya, T. N. (Khim. Drev. **1988** No. 6, pp. 83/6 from C.A. **110** [1989] No. 116927).

[5] Sharapova, T. E.; Domburgs, G.; Brods, L.; Kronberga, V. (Khim. Drev. **1989** No. 3, pp. 59/66 from C.A. **111** [1989] No. 176489).

[6] Hirata, T.; Werner, K. E. (J. Appl. Polym. Sci. **33** [1987] 1533/56 from C.A. **106** [1987] No. 178292).
[7] Satonaka, S.; Nagino, H. (Enshurin Kenkyu Hokoku Hokkaido Daigaku Nogakubu **43** [1986] 57/72 from C.A. **105** [1986] No. 80985).
[8] Dev, I.; Lal, R. (J. Timber Develop. Assoc. India **34** No. 3 [1988] 1/4 from C.A. **110** [1989] No. 156305).
[9] Shen, K. K. (Proc. Int. Conf. Fire Saf. **12** [1987] 340/65 from C.A. **110** [1989] No. 39809; Plast. Compound. **11** No. 7 [1988] 26/8, 30/2, 34 from C.A. **110** [1989] No. 58581).
[10] Agrawal, J. P.; Bansod, V. P.; Satpute, R. S.; Khare, Y. (J. Polym. Mater. **5** [1988] 279/84 from C.A. **110** [1989] No. 154967).

[11] Dietz, M. G.; Schmidt, E. L. (For. Prod. J. **38** No. 5 [1988] 9/14 from C.A. **109** [1988] No. 39585).
[12] Kumar, S.; Kalra, K. K.; Dobriyal, P. B. (J. Timber Develop. Assoc. India **31** No. 4 [1985] 5/12 from C.A. **104** [1986] No. 111565).
[13] Hesko, J. (Czech. 220 938 [1981/86] 1/2 from C.A. **105** [1986] No. 56340).
[14] Williams, L. H.; Mauldin, J. K. (For. Prod. J. **36** No. 11/12 [1986] 24/8 from C.A. **106** [1987] No. 103988).
[15] Barnes, H. M.; Williams, L. H. (For. Prod. J. **38** No. 9 [1988] 13/9 from C.A. **109** [1988] No. 212547).
[16] Freeman, J. J.; Gimblett, F. R. G.; Roberts, R. A.; Sing, K. S. W. (Carbon **25** [1987] 559/63 from C.A. **107** [1987] No. 162476; Extend. Abstr. Program Bienn. Conf. Carbon **17** [1985], 245/6 from C.A. **103** [1985] No. 216798).
[17] Freeman, J. J.; Gimblett, F. R. G. (Carbon **25** [1987] 565/8 from C.A. **107** [1987] No. 200215).
[18] Rizvi, N.; Qureshi, A. W. (Pakistan J. Sci. Ind. Res. **32** [1989] 358/60 from C.A. **112** [1990] No. 8959).
[19] Anonymous (Res. Discl. No. 290 [1988] 424 from C.A. **109** [1988] No. 102705).
[20] Paul, F.; Craciunescu, L.; Dinu, M. H.; Gheorghe, M.; Florescu, A.; Ghindescu, C. (Mater. Constr. [Bucharest] **17** [1987] 99/101 from C.A. **108** [1988] No. 26301).

[21] Simms, N. J.; Little, J. A. (J. Mater. Sci. Lett. **6** [1987] 171/4 from C.A. **106** [1987] No. 141815; Corrosion [Houston] **45** [1989] 498/515 from C.A. **111** [1989] No. 101034).
[22] Stokes, A. D.; Sibilsky, H.; Kovitya, P. (J. Phys. D **22** [1989] 1702/7).
[23] Slatkin, D. N.; Stoner, R. D.; Rosander, K. M.; Kalef-Ezra, J. A.; Laissue, J. A. (Proc. Acad. Sci. USA **85** [1988] 4020/4 from C.A. **109** [1988] No. 34564).
[24] Hermer, R. E. (EPRI-NP-5558 [1987] 1/157 from C.A. **110** [1989] No. 103497).
[25] Battaglia, J. A.; Waters, R. M.; von Hollen, J. M.; Lamatia, L. A.; Bergmann, C. A.; Ditommaso, S. M. (EPRI-NP-6458 [1989] 1/103 from C.A. **112** [1990] No. 225132).
[26] Rubright, M. M. (EPRI-NP-4635-Vol. 3 [1986] 1/33 from C.A. **106** [1987] No. 54349).
[27] Agrawal, A. K.; Means, J. L.; Adair, J. H.; Miller, J. F.; Grotta, H. M.; Berry, W. E. (EPRI-NP-4635-Vol. 2 [1986] 1/133 from C.A. **106** [1987] No. 54359).
[28] Totsuka, N.; Lunarska, E.; Cragnolino, G. A.; Szklarska-Smialowska, Z. (Scr. Metall. **20** [1986] 1035/40 from C.A. **105** [1986] No. 64777).
[29] Partridge, M. J.; Paine, J. P. N.; Williams, C. L. (Proc. 3rd Int. Symp. Environ. Degrad. Mater. Nucl. Power System-Water React., Traverse City, Mich., 1987 [1988], pp. 473/9 from C.A. **110** [1989] No. 80692).
[30] O'Neill, A. S.; Hall, J. F. (EPRI-NP-5985 [1989] 1/42 from C.A. **110** [1989] No. 180989).

[31] Anonymous (Res. Discl. No. 261 [1986] 5 from C.A. **104** [1986] No. 57957).
[32] Airapetyan, A. A. (At. Energ. **64** [1988] 448 from C.A. **109** [1988] No. 100380).

[33] Haidar, N. H. S. (J. Nucl. Sci. Technol. **22** [1985] 484/90 from C.A. **103** [1985] No. 28878).
[34] Iji, T.; Kodama, H.; Pierlas, R.; Jaouen, C.; Kuribayashi, H.; Kurumada, N. (Proc. Symp. Waste Manage. **1985** 211/6 from C.A. **105** [1986] No. 234165).
[35] Tempesti, E.; Giuffre, L.; Modica, G.; Montoneri, E. (Eur. Appl. 159619 [1984/85] 1/31 from C.A. **105** [1986] No. 116968).
[36] Tempesti, E.; Giuffre, L.; Mazzocchia, C.; Di Renzo, F.; Gronchi, P. (Stud. Surf. Sci. Catal. **41** [1988] 403/8 from C.A. **110** [1989] No. 10046).
[37] Papaspyrides, C. D.; Kampouris, E. M. (Polymer **27** [1986] 1433/6 from C.A. **105** [1986] No. 209440).
[38] Kim, K. U.; Park, T. S.; Kim, B. Ch. (J. Polym. Eng. **7** [1986] 1/12 from C.A. **106** [1987] No. 139227).
[39] Kim, B. Ch.; Kim, K. U.; Hong, S. I. (Pollimo **10** [1986] 215/21 from C.A. **105** [1986] No. 80079).
[40] Breitenbach, J. M.; Hausinger, R. P. (Biochem. J. **250** [1988] 917/20).

[41] Hausdorf, S.; Krüger, K.; Küttner, G.; Holzhütter, H. G.; Frömmel, C.; Höhne, W. (Biochim. Biophys. Acta (P) **952** [1988] 20/6).
[42] McEwen, J.; Yates, K. (J. Am. Chem. Soc. **109** [1987] 5800/8).

3.5.7.7 Analytical Methods for H_3BO_3 or Using H_3BO_3

3.5.7.7.1 Determinations of H_3BO_3

The gravimetric determination of H_3BO_3 by Gautier and Pignard [1], involving precipitation with a solution containing $BaCl_2$, tartaric acid, $[NH_4]Cl$, and ammonia, was improved by aging the precipitation for 16 to 20 h; the standard deviation was 0.40 to 0.47 mg for determining 40 to 75 mg H_3BO_3 [2]. Optimum conditions for the quantitative analysis of mixtures of H_3BO_3, HBF_4, and H_2SO_4 have been studied. The determination of H_3BO_3 and HBF_4 is most effective, when twice the stoichiometric amount of $CaCl_2$ is added to the mixture, and titration with NaOH is carried out at the boiling point [4].

2-Hydroxynaphthyl-2′-methyl-1-phenyl(azobenzene)-4′-sulfonic acid (ORL dye) was successfully employed as an adsorption indicator in the titrimetric precipitation determination of borate with Pb^{2+} ions in the pH range 8.2 to 9.0. A sudden color change from yellowish orange (suspension) to pink (precipitation) marks the end point of the titration [3].

The H_3BO_3 content of boron powder for certification was first determined by an acid-base titration method; second, the moisture content of boron samples was measured with a moisture evolution analyzer; and third, boron powder samples were acid digested in a microwave oven [5]. H_3BO_3 from cosmetic products is selectively extracted by complexation with mannitol and titrated potentiometrically with NaOH solution [6].

A precise method for the determination of H_3BO_3 in metal borides by potentiometric titration of the mannitol-H_3BO_3 complex with NaOH or KOH involves evaluation by a multiparametric curve-fitting procedure on the basis of model functions for this type of titration. The initial H_3BO_3 concentration, or parameters such as the conditional acidity constant or the association number of the mannitol-H_3BO_3 complex can be determined. For a sample of titanium boride, the standard deviation was 0.15% for a boron content of 29.18% [7].

H_3BO_3 in nuclear reactor coolant was determined coulometrically by constant-current titration with electrochemically generated $[OH]^-$ ions, using Pt electrodes and electrolytes containing mannitol and $Na_2[SO_4]$. Corrections were derived for the presence of K^+ and NH_3. The mean square error was ca. 0.02 g/L for the determination of 6 to 8 g/L H_3BO_3 [8].

Potentiometric acid-base titrations, employing a glass electrode and coulometrically generated $[OH]^-$ ions, are described for the determination of trace amounts of weak acids. The evaluation of the concentration and of the acidity constant of protolytic impurities is based on the deviation from constancy of the mononuclear protolysis quotient of H_3BO_3. Concentration levels of ca. 8×10^{-6} M could be determined with errors of $\pm2\times10^{-6}$ M or better. The main contaminant is CO_2 [9]. The Gran potentiometric method for the determination of H_3BO_3 [12], which has the advantage of not requiring mannitol addition for the final titration, was re-examined with a 0.25% relative standard deviation for the titration of 0.005 and 0.05 M H_3BO_3 solutions of ionic strength 0.2 M NaCl [13].

H_3BO_3 is determined in nickel electroplating baths by potentiometric titration without removing Ni. Mannitol was added in order to form a stronger acid, and the end point of the titration with NaOH or KOH was detected by monitoring the pH change using a glass electrode. The optimal concentration of H_3BO_3 is 10 to 40 g/L [10]. H_3BO_3 was determined in Watts Ni and Zn plating baths by potentiometry with a $[BF_4]^-$ ion-selective electrode. Complete conversion of H_3BO_3 into $[BF_4]^-$ was achieved by using hydrofluoric acid. The average standard deviation was 0.5 g/L for determining 37 to 54 g/L H_3BO_3. The results are comparable to those of the mannitol titration [11].

Urushi, a natural oriental lacquer, is used as a membrane matrix for a $[BF_4]^-$ ion-selective electrode. The electrode was applied to the measurement of the formation rate of $[BF_4]^-$ from boric acids [14].

As an indicator for controlling the H_3BO_3 concentration in baths for nickel electroplating, the best results (the most sharp color change) were found using potassium ferrocyanide rather than Trilon B or sodium oxalate [15]. In an electroplating bath, in a second manifold after Ni^{2+} and Fe^{2+}, H_3BO_3 is measured indirectly by the reaction of 1-hydroxy-8-amino-naphthalene-3,6-disulfonic acid with salicylaldehyde to form 1-(salicylideneamino)-8-hydroxynaphthalene-3,6-disulfonic acid (Azomethine H) in the presence of H_3BO_3. A linear calibration graph is obtained between 0.05 and 0.8 M H_3BO_3 versus absorbance at 420 nm with a twofold on-line dilution. Fe^{2+} does not interfere, while Ni^{2+} suppresses the peak height due to the H_3BO_3, and the characteristic Ni^{2+} absorbance at 395 nm extends to 420 nm, resulting in a background absorbance [16].

A procedure for the determination of H_3BO_3 in irrigation waters involves the prior distillation as $B(OCH_3)_3$ using an $H_2SO_4:CH_3OH$ ratio of 1:3 for esterification. After distillation (15 min) in a flow of dry air at a temperature of 80 °C, the determination is accomplished by the hydrolysis of $B(OCH_3)_3$ and subsequent spectrophotometric determination with Azomethine H. The selectivity is better than that of the direct analytical method at 410 nm. High concentrations of K^+ ($B:K\leq1:50$) were found to interfere [17].

The conditions and standard results for the calorimetric determination of borate in water by the standard Azomethine H method were defined for natural waters from results with river and lake water samples, and with synthetic humic acids solutions with an added H_3BO_3 standard. Light has little influence on color development. Between 12 and 24 °C, the absorbance maximum for the Azomethine H complex was reached in 1 h and was constant for 1 h. At boron concentrations between 20 and 200 μg/L, the standard deviation is 3 μg/L and the determination limit is 16 μg/L [18].

A comparison is made of two techniques for eliminating interferences due to the color of the raw surface water samples in the determination of H_3BO_3 using Azomethine H. The second-order derivatives spectrophotometry is advantageous when compared to the conventional

References on pp. 237/8

color-correction measurement with respect to the expenditure of time and the samples and reagents required. The detection limit and the standard deviation of the two techniques are practically the same [19].

Micro amounts of boron in steels were determined by using Azomethine H (1-(salicylidene-amino)-8-hydroxynaphthalene-3,6-disulfonic acid). After esterification, boron is distilled as $B(OCH_3)_3$ into a 4% NaOH solution as absorbent which was then dried. Azomethine H, dissolved in $[NH_4][OC(O)CH_3]/CH_3COOH$ buffer (pH 5.0), was added. After 90 min, the absorbance was measured at 415 nm; 0.0001 to 0.004% boron in steels could be determined with good reproducibility with relative standard deviations between 1.5 and 6.1% [20].

The kinetics of the formation of the boric acid-Resorcin H complex (Resorcin H is (2,4-dihydroxybenzene-1-azo-1′)-8′-hydroxynaphthalene-3,6′-disulfonic acid; H_4Z), i.e., $[B(OH)Z]^{2-}$, and its association with diphenylguanidine (DPG) to form $B(OH)Z(DPG)_2$ was studied spectrophotometrically at pH 5 to 6; both reactions are of first order. The overall reaction is $H_3BO_3 + [H_2Z]^{2-} + 2[DPG]^+ \rightleftharpoons B(OH)Z(DPG)_2 + 2H_2O$ [21].

An automated flow-injection pseudo-titrimetric determination for H_3BO_3, using a flow-injection photometric analyzer controlled by a microcomputer, is described. The method is based on the injection of 200 μL of a sample in a flowing stream of mannitol-bromothymol blue titrant and measuring the peak width in time units. Equivalent times between 10 and 70 s are measured with relative standard deviations of 0.1 to 0.4%, and the analytical range lies between 1.5 and 309 mg/100 mL of H_3BO_3. The method was evaluated by performing recovery studies in mixtures and assays of preparations which were compared with the official classical titrimetric method [22].

Sodium tetraborate was determined in bulk and in pharmaceuticals by a spectrophotometric method based on the reaction with *N*-p-tolylsulfonyl-2-(2,4-diketo-3-amyl)naphthoquinonimine, and absorbance measurement at 634 nm. Beer's law was obeyed in the drug concentration range between 0.76 and 1.07 mg/100 mL, and the detection limit of the method was 0.68 μg/mL [23]. The enormous fluorescence enhancement of the potent anti-mutagen ellagic acid by borax offers a simple method for the visualization in the course of a quantitative fluorimetric method for the determination of boron [24].

The esterification reaction between H_3BO_3 and CH_3OH in a concentrated H_2SO_4 medium and the vaporization of the resultant $B(OCH_3)_3$ are used in the determination of boron, especially from nuclear reactor waters, by measuring the $[BO_2]^\bullet$ radical emission at 548 nm by flame atomic emission spectrometry with a detection limit of 0.15 μg [26], or by AAS (atomic absorption spectroscopy) with an NO/HC≡CH flame at 249.7 nm with a sensitivity of 1.45 mg/L, a detection limit of 1.50 μg, and a relative standard deviation of ca. 5%; matrix effects are produced only by fluorides [27].

A method for the concentration and separation of boron in the form of H_3BO_3 from steel includes extraction with 2-*i*-propyl-5-methylhexane-1,3-diol/chloroform solution, subsequent reextraction with NaOH or KOH, and photometrical determination with 1,1′-dianthrimide [25].

The time for the determination of boron in steel can be shortened by simultaneous distillation of $B(OCH_3)_3$ and evaporation of the distillate, which is collected in sodium hydroxide solution, for analysis by inductively coupled plasma atomic emission spectrometry (ICPAES). The limit of detection is 35 ng boron per 0.5 g steel [28]. In another publication, boron is converted to $B(OCH_3)_3$, distilled and condensed, and selectively volatilized at 50°C into the plasma for ICPAES without the interference of methanol which quenches the plasma. The 3σ detection limit is 40 ng/mL boron, the calibration graph is linear up to 10 μg/mL, and the relative standard deviation for 10 replicate measurements of 2.0 μg/mL boron is 3.0% [29].

Inductively coupled plasma atomic emission spectrometry (ICPAES) was also used for the determination of H_3BO_3 in calcium borogluconate veterinary medicines (besides Ca, Mg, and P). The samples are diluted, acidified, and sprayed directly into the plasma. The reproducibility relative confidence intervals for a single sample assay are 1.8% B, 1.4% Ca, 1.4% Mg, and 2.6% P [30]. A continuous-flow system for the determination of boron by ICPAES, after on-line generation of $B(OCH_3)_3$ followed by on-line separation from the matrix, was developed. The esterification reaction was carried out with methanol in 14 M H_2SO_4 without applying external heat. The gaseous $B(OCH_3)_3$ and CH_3OH were separated from the liquid in a filter funnel by aeration with argon through a sintered glass frit from which the reaction mixture flowed out. This argon stream was then fed into the inductively coupled plasma at the torch entrance normally used for the nebulizer gas. Contrary to a batch system, no interference from fluorides or iron was found [31].

Single-column ion chromatography allows the separation and direct conductometric detection of mixtures containing borate, silicate, germanate, F^-, and Cl^- ions by using a resin-based anion-exchange column and a sodium hydroxide/sodium benzoate eluent. Judicious selection of sample preparation and of eluent composition is important. Glasses are analyzed for H_3BO_3 and F^- contents resulting in good agreement with certified values [32].

H_3BO_3 was detected by using a conductometric detector after ion-exclusion chromatographic separation on a cation-exchange resin. Use of eluents of polyol compounds permitted the determination of the weak acid by a conductometric detector. With 0.1M fructose as an eluent, it was possible to detect 0.5 ppb of H_3BO_3 (as boron) [33].

Borate in waters, hydrometallurgical process solutions, and effluents is determined by ion-exclusion chromatography. The procedure is faster and more accurate than classical methods of determination, and is also free from interferences. The precision of the method is good; the relative standard deviation is found to be 0.029. The preparation of the sample requires only a dilution step [34].

Ion-exclusion and ion-exchange columns are used to separate H_3BO_3 from the anions Cl^-, $[NO_3]^-$, and I^-. The fully automated system is controlled by a microprocessor housed in one of the system's pumps. Flow paths between the columns and detectors are directed by two six-port valves. Detection was by refractive index for H_3BO_3 and by conductometry for the anions. The instrumentation was evaluated for performance using a set of solutions over the pH range 2 to 13 and gave a wide linear dynamic range and reproducibility of 1% relative standard deviation [35].

3.5.7.7.2 Determinations in H_3BO_3

Using ICPMS (inductively coupled plasma mass spectrometry), trace elements can be determined from $Li[NO_3]/H_3BO_3$ solutions. Three dissolution techniques compared well with the direct analysis of the solid sample [36].

The influence of temperature, time, and sample to reagent mass ratio on the distillation of H_3BO_3 in the presence of HF and ethanol has been studied by using a headspace autoclave or a flow system. A method has been developed for atomic absorption spectroscopy (AAS) determination of some elemental impurities in H_3BO_3 after matrix removal. The limits of detection are 1×10^{-7} to 2×10^{-6}% [37]. Trace impurities such as Pb, Cr, Mn, Fe, Co, Ni, Cu, Zn, or Cd in pure H_3BO_3 were determined by AAS with air/HC≡CH flames after extraction with diethylammonium diethyldithiocarbamate into n-$C_4H_9OC(O)CH_3$. The sample solution was treated with $HClO_4$, $Fe[ClO_4]_3$, and $Hg[SCN]_2$, and the absorbance was measured at 460 nm. The preconcentration efficiency was enhanced by dissolution of H_3BO_3 in $[NH_4]F$ or NaOH solutions [38].

References on pp. 237/8

3.5.7.7.3 Determination of Ions or Organic Compounds in the Presence or with the Aid of H_3BO_3

The indirect catalytic determination of trace amounts of F^- ions, based on the Zr^{IV}-catalyzed oxidation of iodide with sodium peroxoborate in HCl/H_3BO_3, was modified [40]. The determination of trace amounts of Cl^- ions in electrolytes consisting of ethylene glycol, H_3BO_3, adipic acid, NH_3, and H_3PO_4 was achieved by absorbance measurements [41]. The increased ability of polyhydroxycarboxylates to coordinate Ca^{2+} ions in aqueous solution upon addition of H_3BO_3 was used for ion-selective electrode measurements of Ca^{2+} ions [42].

Solid crystalline H_3BO_3 has been tested as an NH_3 absorbent in the determination of nitrogen in an effort to replace the Kjeldahl method [43]. A method for the rapid determination of free KOH in $[Ag(CN)_2]^-$ electrolytes is based on the thermometric titration with a 0.5 M H_3BO_3 solution [44]. Single-column ion chromatography of inorganic anions using high pH borate-gluconate buffers and conductivity detection produces system peaks as the result of complexation of polyvalent cations such as Ca^{2+} [45]. The matrix effect of H_3BO_3 (with two absorption lines at 249.68 and 249.77 nm, which are near the iron line) was investigated for the determination of iron ions in the primary circuit coolant water of nuclear power plants. There is no absorption at the iron line (248.33 nm) even at 12 g/L H_3BO_3, but H_3BO_3 decreased the intensity of the iron peak and caused a higher background [46]. The determination time of iron impurities in cryolite was decreased due to the dissolution of the sample in a mixture of sulfosalicylic acid and H_3BO_3 [47]. Trace amounts of Mn^{2+} ions can be determined spectrophotometrically by oxidation of indigocarmine by H_2O_2 catalyzed by Mn-diethylenetriamine complexes and activated by $[HOOB(OH)_3]^-$ from H_3BO_3 [48]. The spinel-forming cations Mg, Ca, Sr, Ba, La, and Ce have an effect on the determination of Mo by atomic absorption spectrometry (AAS); addition of H_3BO_3 minimizes this effect [49].

High-performance liquid chromatographic determination of electrically neutral carbohydrates was carried out by conductometric methods employing an H_3BO_3 solution as an eluent, complexing the carbohydrates to sufficiently strong acids. This method can be applied for the simultaneous determination of carbohydrates and organic acids [50]. Monosaccharides, disaccharides, and acidic polyols were investigated as potential additives to an H_3BO_3 eluent in anion-exchange chromatography. The addition of a polyol increases the eluent strength of H_3BO_3. In particular, acidic polyol compounds gave stronger eluents when they were used with H_3BO_3 than when they were used alone. This promotion of the eluent strength is due to the formation of polyol-borate complexes [51]. In nonsuppressed ion chromatography with conductometric detection, polyol-borate eluents fulfil the conditions for a sensitive analysis; the tartrate-borate eluent is especially suitable, because it has a high eluent strength and a low background conductance at low pH values [52]. For the determination of boron in tetraalkylborates by d.c. argon plasma emission spectroscopy at 249.733 nm with H_3BO_3 as a standard, see [39].

The structural features of a gluconate-borate eluent for single-column ion chromatography were studied by NMR spectroscopy and potentiometric titration, and its chromatographic performance on two different columns was compared with that of other eluents for the separation and determination of monovalent inorganic ions. The gluconate-borate eluent consists of a mixture of sodium gluconate and a borax buffer at pH 8.5; the NMR and potentiometric studies suggest the formation of a cyclic anionic complex between gluconate and borate [53]. The chromatographic efficiencies of four different carbohydrate-borate eluents at pH values between 8.0 and 9.5 were compared. The carbohydrates studied were gluconate, mannoic acid, glucose, and mannitol. The mannoic acid-borate eluent was as efficient as the original gluconate-borate eluent, but the glucose-borate and mannitol-borate eluents gave poor results. For each eluent several carbohydrate-borate complexes were responsible for the elution of the anions Cl^-, Br^-, $[NO_2]^-$, $[NO_3]^-$, $[HPO_4]^{2-}$, or $[SO_4]^{2-}$ [54].

Neutral sugars and alcohols can be detected with an indirect conductometric method, using polyol-borate eluents. A moderately involved theoretical explanation of the phenomena was given, in which contributions to the conductivity from protons were not included. A simplified model is given. Most analytes show a decrease in conductivity, but some (like fructose) may increase conductivity [55].

References for 3.5.7.7:

[1] Gautier, J. A.; Pignard, P. (Compt. Rend. **235** [1952] 242/4).
[2] Yurkin, V. G. (Zavodsk. Lab. **52** No. 3 [1986] 19/20 from C.A. **104** [1986] No. 236431).
[3] Kumar, R.; Chaturvedi, S.; Agarwal, D. C.; Chaturvedi, G. K. (Indian J. Chem. A **25** [1986] 793/5 from C.A. **105** [1986] No. 202266).
[4] Kaya, M. (Kim. Sanayi **29** No. 141/144 [1985] 116/27 from C.A. **105** [1986] No. 17411).
[5] Larson, J. D.; Fossey, J. L.; Hinds, S. C.; Lantz, L. L.; Smith, R. E. (BDX-613-3932 [1988] 1/25 from C.A. **111** [1989] No. 166379).
[6] Crisp, S.; Lyall, P. K.; Pindar, A. G.; Tinsley, H. M. (Analyst [London] **110** [1985] 209/10).
[7] Drescher, A.; Kucharkowski, R.; Kluge, W.; Schwarz, T. (Anal. Chim. Acta **219** [1989] 273/9).
[8] Hajduckova, A.; Stefanovska, V.; Burol, R. (Jad. Energ. **33** [1987] 335/9 from C.A. **108** [1988] No. 123590).
[9] Ciavatta, L.; Porto, R.; Vasca, E. (Ann. Chim. [Rome] **79** [1989] 217/30 from C.A. **112** [1990] No. 110985).
[10] Nabivanets, B. I.; Gorina, D. O.; Sobol, T. A. (Vestn. Kievsk. Politekh. Inst. Khim. Mashinostr. Tekhnol. **25** [1988] 25/6 from C.A. **110** [1989] No. 241706).

[11] Johnson, M. L.; Ward, W. G. (Met. Finish. **86** No. 11 [1988] 49/51 from C.A. **111** [1989] No. 49515).
[12] Gran, G. (Analyst **77** [1952] 661/71).
[13] Pressinotti, Queenie S. H. Chui; Lichtig, J. (Fertilizantes **10** [1988] 2/7 from C.A. **110** [1989] No. 38150).
[14] Hiiro, K.; Tanaka, T.; Wakida, Sh. (Anal. Sci. **2** [1986] 145/8 from C.A. **105** [1986] No. 53586).
[15] Aryanina, T. G.; Murav'eva, S. A.; Shchipanova, E. V. (Svetotekhnika **1989** No. 3, pp. 23/4 from C.A. **111** [1989] No. 49579).
[16] Whitman, D. A.; Christian, G. D.; Ruzicka, J. (Analyst [London] **113** [1988] 1821/6).
[17] Monzo, J.; Pomares, F.; De la Guardia, M. (Analyst [London] **113** [1988] 1069/72).
[18] Randow, F. F. E. (Z. Wasser Abwasser Forsch. **18** [1985] 190/3 from C.A. **103** [1985] No. 146856).
[19] Wollin, K. M. (Acta Hydrochim. Hydrobiol. **16** [1988] 39/43 from C.A. **108** [1988] No. 192489).
[20] Hosoya, M.; Takeuchi, M. (Bunseki Kagaku **35** [1986] 854/7 from C.A. **106** [1987] No. 187998).

[21] Flyantikova, G. V.; Chekirda, T. N. (Zh. Neorg. Khim. **30** [1985] 93/7; Russ. J. Inorg. Chem. **30** [1985] 52/4).
[22] Anagnostopolou, P. I.; Koupparis, M. A. (J. Pharm. Sci. **74** [1985] 886/8 from C.A. **103** [1985] No. 188664).
[23] Keitlin, I. M.; Petenko, V. V.; Artemchenko, S. S.; Nichvoloda, V. M. (Farm. Zh. [Kiev] **1989** No. 6, pp. 48/51 from C.A. **112** [1990] No. 125351).
[24] Wolfbeis, O. S.; Hoichmuth, P. (Monatsh. Chem. **117** [1986] 369/74).

[25] Shvarts, E. M.; Bernane, A. A.; Dzene, A. Je. (Latv. PSR Zinat. Akad. Vestis Kim. Ser. **1985** 52/6).
[26] Castillo, J. R.; Mir, J. M.; Martinez, C.; Bendicho, C. (Analyst [London] **110** [1985] 1435/8).
[27] Castillo, J. R.; Mir, J. M.; Bendicho, C.; Martinez, C. (At. Spectrosc. **6** [1985] 152/6).
[28] Hosoya, M.; Tozawa, K.; Takada, K. (Talanta **33** [1986] 691/3).
[29] Kumamaru, T.; Matsuo, H.; Okamoto, Y.; Yamamoto, M.; Yamamoto, Y. (Anal. Chim. Acta **186** [1986] 267/72).
[30] Lyons, D. J.; Spann, K. P. (J. Assoc. Off. Anal. Chem. **68** [1985] 160/2 from C.A. **102** [1985] No. 191221).

[31] Novozamsky, I.; Van Eck, R.; Van der Lee, J. J.; Houba, V. J. G.; Ayaga, G. O. (At. Spectrosc. **9** [1988] 97/9).
[32] McCrory-Joy, C. (Anal. Chim. Acta **181** [1986] 277/82).
[33] Okada, T.; Kuwamoto, T. (Fresenius Z. Anal. Chem. **325** [1986] 683/5).
[34] Cameron, A.; Pohlandt-Watson, C. (MINTEK-M 284 [1986] 1/11 from C.A. **106** [1987] No. 168029 and C.A. **107** [1987] No. 249104).
[35] Jones, W. R.; Heckenberg, A. L.; Jandik, P. (J. Chromatogr. **366** [1986] 225/33).
[36] Stotesbury, S. J.; Pickering, J. M.; Grifferty, M. A. (J. Anal. At. Spectrom. **4** [1989] 457/60 from C.A. **112** [1990] No. 90715).
[37] Krasil'shchik, V. Z.; Zhiteleva, O. A.; Shlyakova, E. Yu.; Chupakhin, M. S. (Zh. Anal. Khim. **43** [1988] 1008/11 from C.A. **110** [1989] No. 68685).
[38] Arpadjan, S.; Novkirischka, M.; Alexandrov, S.; Petkov, P. (Z. Chem. [Leipzig] **26** [1986] 430/3).
[39] Yoo, A.; Moore, C. E.; Jones, R.; Synsteby, A. (Appl. Spectrosc. **40** [1986] 1073/4).
[40] Kellner, J.; Prikryl, F.; Kis, M. (Chem. Listy **82** [1988] 1203/9 from C.A. **110** [1989] No. 107261).

[41] Gavrilyuk, A. I.; Kosheleva, S. I.; Kirpichenko, O. V.; Buchko, O. A.; Kurilo, O. S. (Ukr. Khim. Zh. [Russ. Ed.] **51** [1985] 76/7 from C.A. **102** [1985] No. 178476).
[42] Van Duim, M.; Peters, J. A.; Kieboom, A. P. G.; Van Bekkum, H. (Carbohydr. Res. **162** [1987] 65/78).
[43] Siemer, D. D. (Analyst [London] **111** [1986] 1013/5).
[44] Lorenz, E.; Geissler, M. (Z. Chem. [Leipzig] **26** [1986] 297).
[45] Erkelenz, C.; Billiet, H. A. H.; De Galan, L.; De Leer, E. W. B. (J. Chromatogr. **404** [1987] 67/72).
[46] Smolander, K.; Jarnstrom, R.; Sajalinna, A. (Fresenius Z. Anal. Chem. **335** [1989] 392/4).
[47] Chesnokova, S. M.; Smirnova, G. I. (Steklo Keram. **1989** No. 10, p. 29 from C.A. **113** [1990] No. 33931).
[48] Batyr, D. G.; Isak, V. G.; Kirienko, A. A.; Povar, I. G. (Koord. Khim. **11** [1985] 628/30 from C.A. **103** [1985] No. 152785).
[49] Abdallah, A. M.; El-Defrawy, M. M.; Mostafa, M. A.; Sakla, A. B. (Anal. Chim. Acta **174** [1985] 347/52).
[50] Okada, T.; Kuwamoto, T. (Anal. Chem. **58** [1986] 1375/9).

[51] Okada, T. (J. Chromatogr. **403** [1987] 27/33).
[52] Okada, T.; Kuwamoto, T. (J. Chromatogr. **403** [1987] 35/45).
[53] Schmuckler, G.; Jagö, A. L.; Girard, J. E.; Buell, P. E. (J. Chromatogr. **356** [1986] 413/9).
[54] Girard, J. E.; Rebbani, N.; Buell, P. E.; Al-Khalidi, A. H. E. (J. Chromatogr. **448** [1988] 355/63).
[55] Bertrand, G. L.; Armstrong, D. W. (Anal. Chem. **61** [1989] 631/2).

3.5.8 Boric Acid Esters

3.5.8.1 General Remarks

The automated synthesis of tris(organyloxy)boranes (boric acid esters), $B(OR)_3$, from the corresponding alcohols ROH and BCl_3 has been presented (R = alkyl) [1]. A variety of $B(OR)_3$ was prepared and screened for antimicrobiological activity [2].

An electromigration study at 25 °C of $SbCl_3$ in $B(OCH_3)_3$ and of $SnCl_4$ or $TiCl_4$ in $B(OC_3H_7\text{-}i)_3$ and $B(OC_4H_9\text{-}n)_3$, respectively, indicates the formation of adduct ion pairs. Ion solvation shell compounds depend on the Lewis acid strength. Weak Lewis acids such as $SbCl_3$ form $[\{SbCl_2 \cdot 2E\} \cdot mSbCl_3 \cdot E \cdot nSbCl_3]^+$ and $[SbCl_4 \cdot mSbCl_3 \cdot E \cdot nSbCl_3]^-$ ion pairs, whereas the strong Lewis acids $SnCl_4$ or $TiCl_4$ form $[\{MCl_3 \cdot 3E\} \cdot \{mMCl_4 \cdot 2E\} \cdot xE]^+$ and $[\{MCl_5 \cdot E\} \cdot \{mMCl_4 \cdot 2E\} \cdot xE]^-$ ion pairs ($E = B(OR)_3$). No carbonium ions are formed. In the presence of stronger Lewis acids, a partial conversion of boric acid esters to borinic acid esters occurs in addition to alkyl halide formation [8].

Quantum-chemical calculations for $B(OR)_3$ (R = H, alkyl) catalysts show that their relative catalytic activity in transesterification of an acid ester is determined by the strength of the B–OR bond in the anions $[B(OR)_4]^-$ and by the charge on the oxygen atom [12].

$B(OR)_3$ compounds react with benzylchlorides and CO in the presence of KI and a rhodium(I) catalyst to form different products [14]. Electronically excited aryl nitriles and halides react with $B(OR)_3$ by an electron transfer reaction path to give reduction products [15]. For bis(1,3-diketonato)boronium perchlorates from boric acid esters, see [13, 75].

3.5.8.2 Boric Acid Trimethyl Ester, $B(OCH_3)_3$

For reactions with some Lewis acids such as $SbCl_3$, $SnCl_4$, or $TiCl_4$, see above.

By AM1 calculations, the heat of formation was obtained for **$B(OCH_3)_3$** as −215.8 kcal/mol (observed −214.9 kcal/mol) for C_3 symmetry with a B–O distance of 135.9 (observed 136.7) pm, with an O–C distance of 142.8 (observed 142.4) pm, and an angle ∢C–O–B of 116.7°. The ionization potential was calculated by AM1 to be 11.15 eV [3].

The entropies and enthalpies of evaporation and the boiling points of some $B(OR)_3$ compounds were determined by measuring the temperature dependence of their vapor pressures. The surface tensions and the parachors were also determined. The high entropy values indicate an association in the liquid phase. The surface tension decreases with increasing number of carbon atoms [4]. The temperature dependence of the vapor pressure p (in Pa) was calculated according to the equation $p = A - B/T$ and the following data were obtained for $B(OCH_3)_3$: $A = 10.36$, $B = 1830$, $T = 293$ to 341 K [5].

Liquid-liquid equilibria for the solvent-solute pair $B(OCH_3)_3/CH_3OH$ were measured at 25 °C with *n*-hexane, *n*-heptane, and *n*-nonane [18]. Vapor-liquid equilibria for the binary systems $B(OCH_3)_3$–*n*-heptane [19] and $B(OCH_3)_3$–cyclohexane [20] were measured at 101.325 kPa; the data were tested for thermodynamic consistency and were also correlated by the Wilson equations. The best overall fit for the cyclohexane system was with the Wilson temperature-indepent parameters $G_{12} = 0.608$ and $G_{21} = 1.051$. Vapor-liquid equilibria for the binary system $B(OCH_3)_3$-($HClC{=}CCl_2$) were also measured at 101.325 kPa. The data approached ideality and can be represented reasonably well by a constant relative volability of $\alpha_{12} = 1.782$. For

References on pp. 247/50

$B(OCH_3)_3$, the boiling point was measured to be 341.62 K and the refractive index at 298.15 K to be 1.35441 [21].

IR spectra in the gas phase are reported between 500 and 3100 cm^{-1} for $B(OCH_3)_3$ (frequencies near 3000 cm^{-1}, see Table 3/23), $B(OCD_3)_3$ (Table 3/24), and $B(OCHD_2)_3$ (Table 3/25). The spectrum of the CD_3 species permits a clear assignment of the modes, while the CHD_2 spectrum is less susceptible to analysis. The spectra are compared to those of $ClB(OCH_3)_2$ and $Cl_2BO{-}CH_3$ [22].

Table 3/23
Infrared Frequencies Observed in $B(OCH_3)_3$ near 3000 cm^{-1} [22].

ν in cm^{-1}	intensity	assignment*)
3030 sh		$\nu_{as}(CH_3)$ [a′]
2980		$\nu_{as}(CH_3)$ [a″]
2966 br	s	$\nu_{as}(CH_3)$, $2\delta_s(CH_3)$, $2\delta_{as}(CH_3)$
2885	s	
2815 sh, br	w	

*) [a′] and [a″] refer to the local methyl group symmetry.

Table 3/24
Infrared Frequencies Observed in $B(OCD_3)_3$ [22].

ν in cm^{-1}	intensity	assignment*)	ν in cm^{-1}	intensity	assignment*)
2899 br	vw	2×1449	1670	w	
2805 br	w	2×1404	1578	w	
2736 br	vw		1449	vs	$\nu_{as}(^{10}BO_3)$ (e′)
2574	vw		1404	vvs	$\nu_{as}(^{11}BO_3)$ (e′)
2493	vw		1349 sh		
2405	vw		1283	w	
2360			1170 sh		
2280 sh		$\nu_{as}(CD_3)$ [a′]	1120	s	$\delta_s(CD_3)$
2245.2	ms	2×1122	1066	w	$\delta_{as}(CD_3)$
2235.5	ms	$\nu_{as}(CD_3)$ [a″]	1000.6	ms	ν(CO) (e′)
2224 sh			918.6	mw	$\rho(CD_3)$ (e′)
2205 sh			908.2	mw	$\rho(CD_3)$ (a″)
2144	m	2×1066 (?)	≈860	vvw	
2086	ms	$\nu_s(CD_3)$	775	w	
1994	w		701.9	m	$\delta_\perp(^{10}BO_3)$ (a″)
1945 sh			678.3	ms	$\delta_\perp(^{11}BO_3)$ (a″)
1900 sh			648.3	mw	$\nu_{22}+\nu_{28}$
1805	vw				

*) [a′] and [a″] refer to the local methyl group symmetry, (e′) and (a″) to the overall symmetry, assumed to be C_{3h}.

Table 3/25
Infrared Frequencies Observed in $B(OCHD_2)_3$ [22].

ν in cm^{-1}	intensity	assignment	ν in cm^{-1}	intensity	assignment
3012 sh	w	ν(CH)[1)]	1326	vs	δ(CH)
2967?, sh			1186	m	
2956.0	ms	ν(CH)[2)]	1179	m	
2954.3	ms	ν(CH)[2)]	1171	m	
2950.4	ms	ν(CH)[2)]	1143	m	ν(CO?)
2265 sh, br		$\nu_{as}(CD_2)$	1091	ms	$\delta_s(CD_2)$
2233 br	w	$\nu_{as}(CD_2)$	1044	ms	ν(CO)
2186 br	mw	$\nu_s(CD_2)$, $2\delta_s(CD_2)$	993	mw	
2140 br	mw	$\nu_s(CD_2)$, $2\delta_s(CD_2)$	925	w	
2018	vw		846	w	
1675	vw		703 sh		$\delta_\perp(^{10}BO_3)$
1460	vs	$\nu_{as}(^{10}BO_3)$	681	m	$\delta_\perp(^{11}BO_3)$
1422	vvs	$\nu_{as}(^{11}BO_3)$	657	m	
1386	s	δ(CH)	650	m	$\nu_{22}+\nu_{28}$

[1)] Bond lying in, or close to, the skeletal plane. – [2)] Bonds lying distant from the skeletal plane.

The isotope enrichment and separation of ^{10}B and ^{11}B starts with laser irradiation of $B(OCH_3)_3$ or $B(OC_2H_5)_3$ [6]. The qualitative fluoride ion affinity is calculated from ion cyclotron double-resonance and bracketing experiments and the following order was found. The values for the enthalpy changes, which give the halide ion bond dissociation energies in $L–F^-$, include the following L: $B(OC_2H_5)_3$ (44.3 kcal/mol) ≈ SO_2 > $B(OCH_3)_3$ (42 ± 3 kcal/mol) ≈ OCF_2 > PF_3 [7]. Proton, ^{11}B, and ^{13}C chemical shifts are given for pyridine and 2-ethylpyridine complexes of $B(OCH_3)_3$ [9].

Using $B(OCH_3)_3$ as a reagent, the chemical-ionization mass spectra (CIMS) of some *cis*-cyclic glycols, cyclopentane-1,2-diols, cyclohexane-1,2-diols, mono- and disaccharides have been studied [23]. From vertical neutralization and reionization mass spectrometry (NRMS), structural information is obtained for the radical ion $[B(OCH_3)_3]^{\bullet+}$, and especially for $[CH_3O]^\bullet$ radicals [24]. Collision-induced dissociative ionization (CIDI) mass spectra of $B(OCH_3)_3$ also showed $[CH_3O]^\bullet$ radicals with 31 amu and much weaker peaks for $[CH_3]^\bullet$ radicals [25]. Exposure of $B(OCH_3)_3$ as a dilute solution in $CFCl_3$ at 77 K to radiation-electron spin resonance gives either the radical cation $[CH_2OB(OCH_3)(CH_3OH)]^{\bullet+}$ or the radical $[CH_2OB(OCH_3)]^\bullet$ and CH_3OH [27]. Efficient free-radical addition of $B(OCH_3)_3$ to fluorinated alkenes has been studied in order to show substituent effects on free radicals [28].

The mass spectral fragmentation of alkyl borates appears to be controlled by the ability of oxygen to stabilize the positive charge of the fragment ions. Medium-resolution mass spectrometry suggests structural assignments which are consistent with this assumption. For example, the ions with m/z = 59 and m/z = 43 for $B(OCH_3)_3$ are assigned to $[CH_3O–B{=}OH]^+$ and $[CH_3O{=}BH]^+$, respectively, and not to $[BO_3]^+$ and $[BO_2]^+$, as was reported previously. Several additional experimental procedures were used in order to further verify the structural assignments. Ion-molecular reactions indicate that many of the fragment ions contain a $[–B{=}O]^+$ unit, which is susceptible to nucleophilic attack. Proton affinity bracketing methods were used to obtain thermodynamic values for the doubly bonded boron molecules. For $CH_3OB{=}O$, a proton

References on pp. 247/50

affinity of 183±3 kcal/mol and a heat of formation of −110±26 kcal/mol were determined. For $CH_3B{=}CH_2$, a previously reported proton affinity of 220.5±2 kcal/mol suggests a heat of formation of 31±6 kcal/mol. These values, as well as bond strengths, are compared with those of similar carbon compounds [26].

$B(OCH_3)_3$ in CH_3OH reacts with $Na[HF_2]$ to give $Na[F_3B{-}OCH_3]$, which forms $Na[BF_{4-x}(OCH_3)_x]$ (x=0 to 4); $B(OCH_3)_3$ in CH_3OH reacts at −40°C with $Na[H_2F_3]$ or $K[H_2F_3]$ to give $Na[F_3B{-}OCH_3]$ or $K[F_3B{-}OCH_3]$, respectively. $Na[F_3B{-}OCH_3]$ decomposed on heating at 150 to 240°C forming $B(OCH_3)_3$ and small amounts of $FB(OCH_3)_2$ and $F_2B{-}OCH_3$ [29].

The system $B(OCH_3)_3{-}CH_3OH{-}Br_2$ is a reagent for α-monobromation of ketones at room temperature [30].

The $B(OCH_3)_3$-modified silica surface of aerosols was studied by chemical analysis for boron, temperature-programmed desorption, and ^{11}B or ^{13}C NMR spectroscopy. The nature of the CH_3OH complexes with boron atoms in the surface dimethylborosilanyl groups was discussed [32]. An IR spectral study showed that hydrolysis of these modified silica surface groups results in the breaking of B–O–C, B–C, or Si–O–B bonds, although breaking of B–C bonds is accelerated by the presence of oxygen [33].

The methylation of $B(OCH_3)_3$ with CH_3MgBr gives 72 to 92% methylboronic ester, more than 2% dimethylborinic ester, and more than 1% $B(CH_3)_3$. The methylation of $B(OCH_3)_3$ with CH_3Li at −78°C in ether gave $Li[(CH_3)_xB(OCH_3)_{4-x}]$ (x=0 to 4) [31].

Various tertiary amines have been reduced by thf·BH_3 in the presence of $B(OCH_3)_3$ at 0°C [34]. Reaction of $B(OCH_3)_3$ with borabicyclononane gave 87% methyl derivative in addition to 79% $HB(OCH_3)_2$ [35].

Polydimethylsilanes, $[-Si(CH_3)_2-]_n$, are converted into the corresponding polycarbosilanes, $[-SiH(CH_3)CH_2-]_n$, with different molecular masses and melting points by monomeric $B(OCH_3)_3$ catalysis at 380°C in 9 to 21 h [17].

$B(OCH_3)_3$ reacts with 2,2′,2″-nitrilotriphenol in CH_3CN to form a boron complex with an intermolecular N→B dative bond of 168 pm distance [36]. $B(OCH_3)_3$ reacts in ether at −60°C with 4-lithium-2,5-di-*n*-hexylbenzeneboronic acid [37] and with dioxolanes in the presence of an acid catalyst to yield 2,2-dialkoxypropanes [38].

Heat-resistant vinylchloride polymers are prepared in the presence of $B(OCH_3)_3$ [16].

3.5.8.3 Boric Acid Triethyl Ester, $B(OC_2H_5)_3$

By AM1 calculations, the heat of formation was obtained for $\mathbf{B(OC_2H_5)_3}$ as −233.5 (observed −239.5) kcal/mol for C_3 symmetry with r(BO)=135.9 pm, r(OC)=143.4 pm, and ∢(BOC)=117.9°. The ionization potential was calculated by AM1 to be 11.07 eV [3].

The entropies and enthalpies of evaporation and the boiling points of some $B(OR)_3$ compounds were determined by measuring the temperature dependence of their vapor pressures. The surface tensions and the parachors were also determined. The high entropy values indicate an association in the liquid phase. The surface tension decreases with increasing number of carbon atoms [4]. The temperature dependence of the vapor pressure p (in Pa) was calculated according to the equation p=A−B/T and the following data were obtained for $B(OC_2H_5)_3$: A=10.52, B=2153, T=310 to 390 K [5].

A method was developed for the preparation of high-purity $B(OC_2H_5)_3$ by the reaction of BCl_3 with C_2H_5OH, followed by extraction [39]. The total content of impurities in $B(OC_2H_5)_3$ has been determined by differential scanning microcalorimetry [40]. The liquid-liquid equilibrium in the $B(OC_2H_5)_3$–C_2H_5OH–HCl system has been investigated [41]. $B(OC_2H_5)_3$ does not react with hydrolyzates of $Si(OC_2H_5)_4$ [42]. Adsorption isotherms were determined for $B(OC_2H_5)_3$ and $Si(OC_2H_5)_4$ on a special microporous carbon (BAU) between 403 and 433 K. Heats of adsorption are 1.5 times the heats of condensation of $B(OC_2H_5)_3$ and $Si(OC_2H_5)_4$ indicating physical adsorption. The separation coefficients change significantly with the composition of the gas mixture [43].

The thermal decomposition of $B(OC_2H_5)_3$ was carried out in an argon carrier gas in a reaction chamber consisting of a horizontal cylindrical quartz channel, on the inside surface of which the B_2O_3 product was deposited. A mathematical model of the process is presented which includes as parameters the flow rate of the gas-liquid mixture, the initial reagent concentrations, the channel geometry, the temperature profile on the deposition surface, and a first approximation for the expression of the rate constant of the thermal decomposition of $B(OC_2H_5)_3$ [46]. The kinetics of this pyrolysis was studied under the auspices of determining the mechanism of formation of hydrocarbon impurities during the high-temperature synthesis of B_2O_3 [47].

The catalytic activity of a tungsten-carbene catalyst for the polymerization of alkynes increases in the presence of $B(OC_2H_5)_3$ [48].

$B(OC_2H_5)_3$ reacts with $(C_2H_5)_2O$–$Al(C_2H_5)_3$ to give a 90% yield of $B(C_2H_5)_3$ [44], and with an ultrasonic (43 kHz, 180 W), I_2-activated mixture of C_2H_5Br and aluminium powder, formulated as $(C_2H_5)_3Al_2Br_3$, to give a 100% yield of $B(C_2H_5)_3$ [45].

$B(OC_2H_5)_3$ reacts with diols to give 2-ethoxy-1,3,2-dioxaborolanes, C_2H_5O–B(–OZO–), which, on azeotropic distillation in toluene with aminophenols, HO–Y–NR_2, give mixed boric acid esters of the types (–OZO–)B–O–Y–NR_2, e.g., 2-(2′-aminophenoxy)-4,4,5,5-tetramethyl-1,3,2-dioxaborolane. The structure of this compound is discussed on the basis of IR and Raman spectra, and mass spectral, ^1H and ^{11}B NMR data, and with reference to earlier X-ray crystal investigations, see p. 246 [50]. The presence of a boron-nitrogen coordination is proposed in all cases [51]. The relationships between the proton-acceptor properties and the thermodynamic characteristics of formation of BF_3 and $B(OC_2H_5)_3$ complexes with N-containing ligands in tetrahydrofuran solutions have a linear character which points to the absence of specific interactions. The effect of the electron density of the vertical atom on the stability and properties of the complex was also studied [49].

The isotope enrichment and separation of ^{10}B and ^{11}B starts with laser irradiation of $B(OC_2H_5)_3$ [6]. The qualitative fluoride ion affinity is calculated from ion cyclotron double-resonance and bracketing experiments. The value for the enthalpy change, which gives the halide ion bond dissociation energy in $(C_2H_5O)_3B$–F^-, is 44.3 kcal/mol [7].

3.5.8.4 Boric Acid Triphenyl Ester, $B(OC_6H_5)_3$

The ionization potential for $\mathbf{B(OC_6H_5)_3}$ was calculated by AM1 to be 10.88 eV [3].

The entropies and enthalpies of evaporation and the boiling points of some $B(OR)_3$ compounds including $B(OC_6H_5)_3$ were determined by measuring the temperature dependence of their vapor pressures. The surface tensions and the parachors were also determined. The high entropy values indicate an association in the liquid phase. The surface tension decreases with increasing number of carbon atoms [4].

$B(OC_6H_5)_3$ solution was used in the determination of environmental C_2 to C_4 diols by gas chromatography/mass spectroscopy [52]. The good antioxidant properties of $B(OC_6H_5)_3$ have been shown by calculations of rate constants [53]. In the telomerization of *i*-C_3H_7–SCH_2CH_2-OH with butadiene in $C_6H_5CH_3$ over $Pd(acac)_2/Al(CH_3)_3$ (acac = acetylacetonate), the selectivity of $B(OC_6H_5)_3$ has been tested [54]. The yield and selectivity of the oxidation of alkyl-substituted benzenes to ketones or phenols in the presence of HBr is increased through the use of $B(OC_6H_5)_3$ [55]. $B(OC_6H_5)_3$ is a component of F-Cl-hydrocarbon compositions and serves as an inhibitor in their decomposition at 150°C [56].

Proton, ^{11}B, and ^{13}C chemical shifts are given for pyridine and 2-ethylpyridine complexes of $B(OC_6H_5)_3$ [9].

Heat-resistent vinylchloride polymers are prepared in the presence of $B(OC_6H_5)_3$ [16].

3.5.8.5 Additional Boric Acid Trialkyl Esters

By AM1 calculations, the heat of formation for **$B(OC_3H_7\text{-}n)_3$** was found as −252.7 (observed −259.4) kcal/mol [3]. Polydimethylsilanes, $[-Si(CH_3)_2-]_n$, are converted into the corresponding polycarbosilanes, $[-SiH(CH_3)CH_2-]_n$, with different molecular masses and melting points by monomeric $B(OC_3H_7\text{-}n)$ catalysis at 380°C in 9 to 21 h [17].

The temperature dependence of the vapor pressure p (in Pa) was calculated according to the equation $p = A - B/T$ and the following data were obtained. For $B(OC_3H_7\text{-}n)_3$: A = 9.73, B = 2134, and T = 361 to 448 K; for **$B(OC_3H_7\text{-}i)_3$**: A = 10.04, B = 1970, and T = 315 to 391 K [5].

The entropies and enthalpies of evaporation and the boiling points of some $B(OR)_3$ compounds were determined by measuring the temperature dependence of their vapor pressures. The surface tensions and the parachors were also determined. The high entropy values indicate an association in the liquid phase. The surface tension decreases with increasing number of carbon atoms [4]. $Na[B(C_6H_5)_4]$ has been prepared in 80% yield by the reaction of $B(OC_3H_7\text{-}i)_3$, chlorobenzene, and cycloheptane with a dispersion of sodium in cyclopentane at 118°C over 50 min and keeping at room temperature for 3 h [57]. The reactions of $B(OC_3H_7\text{-}i)_3$ with one to three equivalents of sulfur-containing Schiff bases to form more complex boronic esters were investigated [58]. For reaction of $B(OC_3H_7\text{-}i)_3$ with some Lewis acids such as $SbCl_3$, $SnCl_4$, or $TiCl_4$; see Section 3.5.8.1, p. 239.

The reaction between **$B(OC_5H_{11}\text{-}i)_3$** and KH is catalyzed by $B(OC_3H_7\text{-}i)_3$ [59].

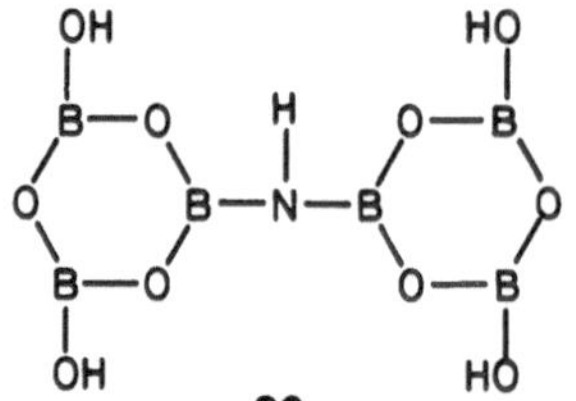

20

By heating of **$B(OC_4H_9\text{-}n)_3$** with B_2O_3, a compound **$BO_{1.1}(OC_4H_9\text{-}n)_{0.8}$** has been obtained. This compound reacts with CH_3CONH_2, $C_6H_5CH_2NH_2$, 1,3-$(CONH_2)_2C_6H_4$, or biuret to give boric acid imides, e.g., **20** [60]. $B(OC_4H_9\text{-}n)_3$ was used to synthesize organoboron derivatives of pyrimidines [61]. Decomposition of cyclohexane hydroperoxides and 1-methylcyclohexane hydroperoxides in *cis*- and *trans*-1,4-dimethylcyclohexane and in the presence of $B(OC_4H_9\text{-}n)_3$ results in a nonstereospecific attack at the tertiary C–H bonds of the 1,4-dimethylcyclohexane

[62]. For reaction of $B(OC_4H_9\text{-}n)_3$ with some Lewis acids such as $SbCl_3$, $SnCl_4$, or $TiCl_4$; see Section 3.5.8.1, p. 239.

$1\text{-}(CH_3O)_2B\text{–}O\text{–}C_6H_9$-*cyclo*, $1\text{-}(CH_3O)_2B\text{–}O\text{–}C_5H_7$-*cyclo*, $(CH_3)_2C{=}CH\text{–}O\text{–}B(OCH_3)_2$, and **$CH_2{=}CH\text{–}O\text{–}B(OCH_3)_2$** have been prepared from the lithium enolates of the corrresponding ketones and $ClB(OCH_3)_2$. They add to aldehydes in a stereoconvergent reaction leading to *syn*-aldehydes [69].

Formation of $M[t\text{-}C_4H_9O\text{–}B(OR)_3]$ associates ($R = n\text{-}C_3H_7$ to $n\text{-}C_{13}H_{27}$; M = Li, Na, K) in solid state and solution was studied by rheological dielectric mass spectrometry and small angle X-ray scattering methods [10]; extremes in viscosity and dielectric constant versus composition curves show solvation and association between 293 and 353 K [11].

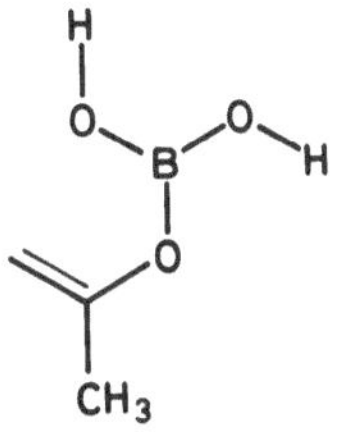

21a

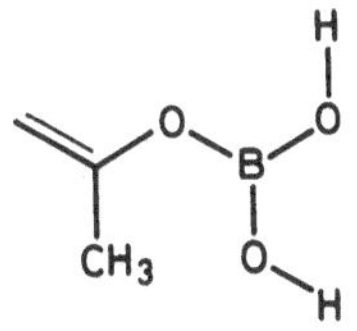

21b

$Pd[P(C_6H_5)_3]_4$ catalyzes the allylic alkylation of carbon nucleophiles with allylic borates or B_2O_3 and allyl alcohols in tetrahydrofuran at 65 °C under neutral conditions [63]. Ab initio STO-3G calculations on modified neglect of diatomic overlap (MNDO) optimized geometries of (E)-enol borates indicated both U (**21a**) and extended (**21b**) conformations. The U conformation is 1 to 2 kcal/mol more stable than the extended one. On the other hand, (Z)-enol borates exist exclusively in the extended conformation, which is maintained in a chair transition state in the aldol addition reaction [64].

3.5.8.6 Cyclic Esters of the Type ROB(–OZO–)

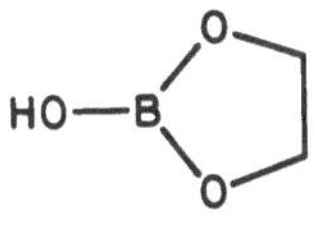

22

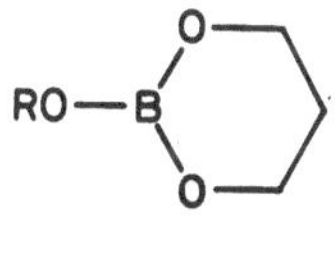

23

The ^{13}C NMR chemical shifts of 2-hydroxy-1,3,2-dioxaborolane, **22** (half-chair), and 2-hydroxy-1,3,2-borinane, **23** (R = H; chair), and their methyl derivatives are reported. Substituent effects on the chemical shifts can be regarded as an accurate and successful application to structure elucidation and both configurational and conformational analysis [65]. For further 1,3,2-dioxaborolanes and 1,3,2-borinanes, see **11**, p. 172, and **13**, p. 175, respectively.

2-Methoxy-1,3,2-dioxaborinane (**23**, $R = CH_3O$) has been prepared from a 1:1 mixture of **23** (R = H) and trimethoxyborane by stirring, addition of anhydrous ether, cooling of the solution in a dry ice-acetone bath, dropwise addition of CH_3MgCl, stirring, allowing to warm to room temperature, adding pentane, stirring, protonating with anhydrous HCl in ether at 0 °C, stirring, decanting, washing with pentane, extracting with NaOH in water, discarding of the pentane, removal of the water at 14 Torr and 60 °C until a colorless solid was obtained. The complex was destroyed by addition of concentrated HCl/methanol. The methyl ester of the boronic acid was distilled from the solution, and then converted into the methylboronic acid [31].

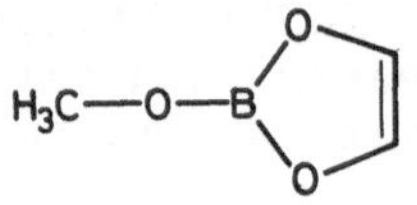

24

2-Methoxy-1,3,2-dioxaborole (**24**) has been prepared by shaking a mixture of 10 mmol of 9,10-dihydro-9,10-ethanoanthracene-11,12-diol, 7 g molecular sieves 4 Å, and 50 mL dry benzene for 1 h. After the addition of 20 mmol $B(OCH_3)_3$, the shaking was continued for an additional 2 h. The molecular sieves were removed by filtration and washed with dry benzene. The solvent was removed under reduced pressure, and the residue of 9,10-dihydroanthraceno[9,10-d]-2-methoxy-1,3,2-dioxaborolane was purified by sublimation. The yield was more than 96%, melting point 177°C. This compound was converted by sublimation into vapor (220°C/10 Pa) in a quartz thermolysis apparatus and thermolyzed by passage through a quartz tube at 550°/10 Pa. The anthracene formed during the thermolysis condensed in the portion of the tube, while **24** was collected in a cold trap. The yield was more than 95%; NMR data ($CDCl_3$; in ppm): $\delta^1H = 3.74$ (s, 3H, OCH_3) and 6.76 (s, broad, 2H, =CHO–); $\delta^{13}C = 53.28$ (d, $^1J(C,H) = 144.9$ Hz, CH_3) and 132.57 (d, $^1J(C,H) = 205.8$ Hz, =CHO–); mass spectrum (70 eV): m/z = 69 ($[M]^+ - 15$). By immobilizing this borolane in a polymeric support, sugar aldehydes condense to higher monosaccharides [66].

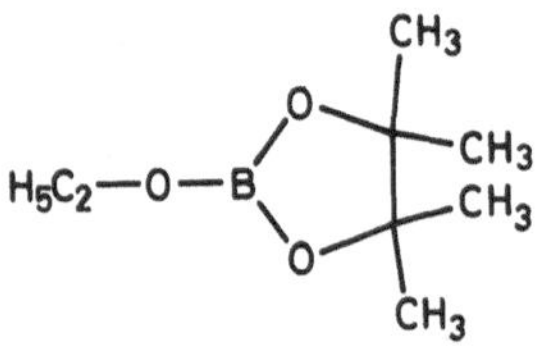

25

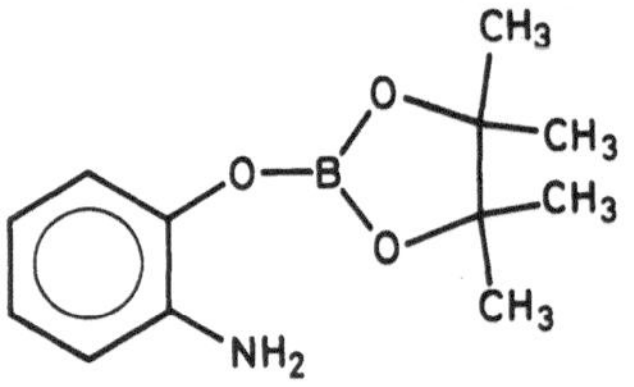

26

2-Ethoxy-4,4,5,5-tetramethyl-1,3,2-dioxaborolane (**25**) reacts with 2-aminophenol to yield 2-(2′-aminophenoxy)-4,4,5,5-tetramethyl-1,3,2-dioxaborolane (**26**), the crystal structure of which was determined as orthorhombic, space group $Pna2_1 - C^9_{2v}$ (No. 33) with a = 727.59, b = 1031.10, and c = 1681.50 pm; $D_c = 1.148$ g/cm³ [50].

2-(4-Hydroxyphenyl)-1,3,2-dioxaborolane, melting point 108°C, has been prepared from 4-hydroxyphenylboronic acid and ethylene glycol in toluene by azeotropic removal of the resultant water and concentration of the solution [67].

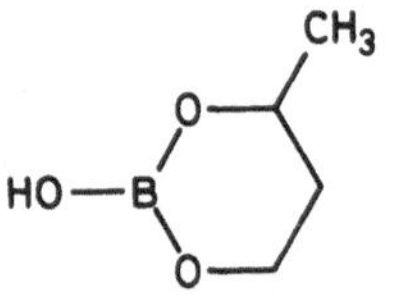

27

H_3BO_3 reacts with β-diols such as butane-1,3-diol in aqueous acetone to give **27**, which, on heating or dehydrating over P_4O_{10} gave the dimer. The intermediates are discussed, and the ^1H NMR chemical shifts for several cyclic boric acid esters were determined [68].

Aldehyde-derived enol-borates polymerize easily; nevertheless, their in situ addition to aldehydes is possible. Thus, 4-alkoxy-1,3,2-dioxaborinanes are formed, which are internally protected aldols [69].

For 2,5-diaryl-1,3,2-dioxaborinanes, a new series of liquid crystals, see p. 175 [70].

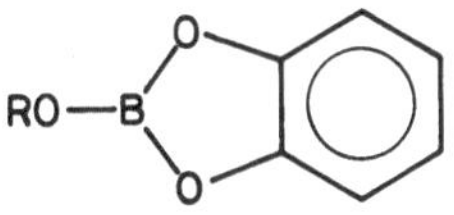

28

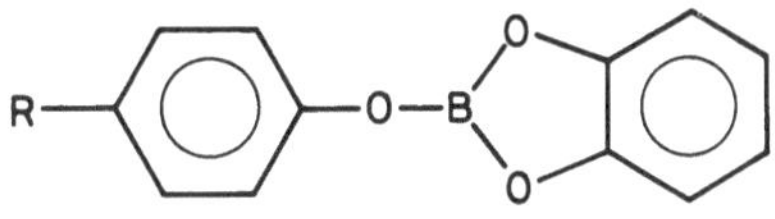

29

Proton, ^{11}B, and ^{13}C NMR chemical shifts are given for pyridine and 2-ethylpyridine complexes of 2-organyloxy-1,3,2-benzodioxaboroles [9]. Reactions of such species with bidentate Schiff bases, with azenes, or with 5-benzylthiocarbazates yield tetracoordinated stable 1:1 complexes containing B-N bonds [71].

Compound **28** (R = n-C_4H_9) and $B(OC_4H_9\text{-}n)_3$ were studied in relation to their molecular structures by Raman spectroscopy and further compounds of the type **28** (R = CH_2=$CHCH_2$, $CH_3OCH_2CH_2$, CH≡CCH_2, and cyclooctyl) are described [72]. Carbon-13 NMR spectra of compounds **29** in comparison to spectra of 1,2-$(OH)_2C_6H_4$ and (4-$HOC_6H_4)_2C(CH_3)_2$ showed that the boron atom shifts the signals of the carbon atom towards low field; this shift is larger the closer the boron atom is to the carbon atom [73]. Physicochemical investigations and mass spectral data on some similar compounds have been reported [74]. For compound **28** with H instead of OR, see Section 3.5.6.3.2, p. 167; for 2-alkyl- and 2-aryl-substituted compounds **28** (R instead of OR), see Section 3.5.6.4.2.3, p. 173.

References for 3.5.8:

[1] Efremov, A. A.; Grinberg, E. E.; Tartakovskii, V. L.; Bessarabov, A. M.; Makarov, V. V.; Tarasova, E. S. (Vysokochist. Veshchestva **1988** No. 2, pp. 55/9 from C.A. **110** [1989] No. 24021).

[2] Watanabe, S.; Fujita, T.; Sakamoto, M. (J. Am. Oil Chem. Soc. **65** [1988] 1479/82 from C.A. **110** [1989] No. 4495).

[3] Dewar, M. J. S.; Jie, Caoxian; Zoebisch, E. G. (Organometallics **7** [1988] 513/21).

[4] Grinberg, E. E.; Chernaya, N. G.; Efremov, A. A. (Vysokochist. Veshchestva **1988** No. 3, pp. 180/4 from C.A. **110** [1989] No. 141886).

[5] Omiadze, A. P.; Grinberg, E. E.; Krasavin, V. P.; Efremov, A. A.; Rodina, G. L.; Gabisiani, G. G. (Zh. Fiz. Khim. **60** [1986] 1575/8; Russ. J. Phys. Chem. **60** [1986] 947/50).

[6] Lahoda, E. J.; Burgman, H. A.; Snyder, T. S. (Ger. Offen. 3445858 [1983/85] 1/11 from C.A. **103** [1985] No. 131034).

[7] Larson, J. W.; McMahon, T. B. (J. Am. Chem. Soc. **107** [1985] 766/73).

[8] Lysenko, Yu. A.; Nevechera, I. V. (Zh. Obshch. Khim. **57** [1987] 2415/9; J. Gen. Chem. [USSR] **57** [1987] 2153/6).

[9] Farfan, N.; Contreras, R. (J. Chem. Soc. Perkin Trans. **1987** 771/3).

[10] Bol'shakov, G. F.; Dmitrieva, Z. T. (Morphol. Polym. Proc. 17th Europhys. Conf. Macromol. Phys., Prague 1985 [1986], pp. 739/46 from C.A. **106** [1987] No. 9686).

[11] Dmitrieva, Z. T.; Bol'shakov, G. F.; Levus, Yu. I. (Dokl. Akad. Nauk SSSR **281** [1985] 883/7).

[12] Korshak, V. V.; Stankevich, I. V.; Chistyakov, A. L.; Vasnev, V. A.; Vinogradova, S. V.; Markova, G. D.; Nametov, V. M. (Dokl. Akad. Nauk SSSR **301** [1988] 1135/8).

[13] Ilge, H. D.; Hartmann, H. (Z. Chem. [Leipzig] **26** [1986] 399/400).

[14] Alper, H.; Hamel, N.; Smith, D. J. H.; David, J. H.; Woell, J. B. (Tetrahedron Lett. **26** [1985] 2273/4).
[15] Schuster, G. B.; Kropp, M. (AD-A196946 [1988] 1/6 from C.A. **111** [1989] No. 114599).
[16] Adachi, T.; Saeki, H.; Kakei, H.; Mito, Y. (Jpn. Kokai Tokkyo Koho 86-261339 [1985/86] 1/4 from C.A. **107** [1987] No. 41946).
[17] Duboudin, F.; Birot, M.; Babot, O.; Dunogues, J.; Calas, R. (J. Organomet. Chem. **341** [1988] 125/32).
[18] Schmidt, M. B.; Plank, C. A.; Laukhuf, W. L. S. (J. Chem. Eng. Data **30** [1985] 251/3).
[19] Niswanger, D. S.; Plank, C. A.; Laukhuf, W. L. S. (J. Chem. Eng. Data **30** [1985] 209/11).
[20] Kraher, M. V.; Plank, C. A.; Laukhuf, W. L. S. (J. Chem. Eng. Data **31** [1986] 387/8).

[21] Owensby, G. S.; Plank, C. A.; Laukhuf, W. L. S. (J. Chem. Eng. Data **34** [1989] 213/4).
[22] McKean, D. C.; Coats, A. M. (Spectrochim. Acta A **45** [1989] 409/19).
[23] Suming, H.; Yaozu, C.; Longfei, J.; Shuman, X. (Org. Mass Spectrom. **20** [1985] 719/23).
[24] Wesdemiotis, C.; Feng, R.; Williams, E. R.; McLafferty, F. W. (Org. Mass Spectrom. **21** [1986] 689/95).
[25] Holmes, J. L.; Terlouw, J. K. (Org. Mass Spectrom. **21** [1986] 776/8).
[26] Hettich, R. L.; Cole, T.; Freiser, B. S. (Int. J. Mass Spectrom. Ion Processes **81** [1987] 203/15 from C.A. [1987] No. 154345).
[27] Ganghi, N. S.; Rao, D. N. R.; Symons, M. C. R. (J. Chem. Soc. Faraday Trans. I **82** [1986] 2367/76).
[28] Chambers, R. D.; Grievson, B.; Kelly, N. M. (J. Chem. Soc. Perkin Trans. I **1985** 2209/13).
[29] Plakhotnik, V. N.; Parkhomenko, N. G.; Kharchenko, L. V.; Kopanev, V. D. (Koord. Khim. **12** [1986] 1341/4 from C.A. **105** [1986] No. 237401).
[30] Zav'yalov, S. I.; Dorofeeva, O. V.; Rumyantseva, E. E.; Sitkareva, I. V.; Zavozin, A. G. (Izv. Akad. Nauk SSSR Ser. Khim. **1987** 2395/6; Bull. Acad. Sci. USSR Div. Chem. Sci. **36** [1987] 2224 from C.A. **109** [1988] No. 22617).

[31] Brown, H. C.; Cole, T. E. (Organometallics **4** [1985] 816/21).
[32] Kasperskii, V. A.; Brei, V. V.; Gorlov, Yu. I.; Chuiko, A. A. (Teor. Eksperim. Khim. **24** [1988] 118/21 from C.A. **108** [1988] No. 174065).
[33] Kasperskii, V. A.; Brei, V. V.; Chuiko, A. A. (Zh. Prikl. Spektrosk. **49** [1988] 460/4 from C.A. **110** [1989] No. 29364).
[34] Oh, I. H.; Yoon, N. M. (Bull. Korean Chem. Soc. **10** No. 1 [1989] 12/8 from C.A. **111** [1989] No. 77148).
[35] Soderquist, J. A.; Negron, A. (J. Org. Chem. **52** [1987] 3441/2).
[36] Müller, E.; Bürgi, H.-B. (Helv. Chim. Acta **70** [1987] 499/510).
[37] Rehahn, M.; Schlüter, A.-D.; Wegner, G.; Feast, W. J. (Polymer **30** [1989] 1060/2 from C.A. **111** [1989] No. 115872).
[38] Messina, G.; Moretti, M. D.; Sanna, G.; Soma, G.; Sanna, S. R. (Eur. Appl. 251092 [1986/88] 1/12 from C.A. **109** [1988] No. 148879).
[39] Chernaya, N. G.; Tartakovskii, V. L.; Grinberg, E. E.; Bessarabov, A. M.; Efremov, A. A. (Vysokochist. Veshchestva **1988** No. 2, pp. 51/4 from C.A. **109** [1988] No. 42388).
[40] Novozhilov, V. A.; Allakhverdov, R. G.; Efremov, A. A.; Grinberg, E. E. (Tr. IREA No. 47 [1985] 36/9 from C.A. **106** [1987] No. 84687).

[41] Tartakovskii, V. L.; Grinberg, E. E.; Bessarabov, A. M.; Efremov, A. A.; Chernaya, N. G. (Nauch. Tr. VNII Khim. Reaktivov Osobo Chist. Khim. Veshchestv No. 51 [1989] 44/50 from C.A. **113** [1990] No. 121528).
[42] Rastorguev, Yu.; Kuznetsova, R. V.; Kuznetsov, A. I.; Shalumov, B. Z.; Efremov, A. A.; Zhukova, L. A.; Smetina, G. F. (Zh. Prikl. Khim. **58** [1985] 630/4 from C.A. **103** [1985] No. 54490).

[43] Trubnikov, I. B.; Fedorov, V. A.; Zhuravlev, O. E. (Vysokochist. Veshchestva **1989** No. 2, pp. 107/9 from C.A. **111** [1989] No. 141173).
[44] Lin, Y. T. (J. Organomet. Chem. **317** [1986] 277/83).
[45] Liou, K. F.; Yang, P. H.; Lin, Y. T. (J. Organomet. Chem. **294** [1985] 145/9).
[46] Minkina, V. G.; Aloyan, S. G.; Kupryazhkina, T. N.; Ivanov, M. Ya.; Ovsyannikov, N. A. (Teor. Osn. Khim. Tekhnol. **22** [1988] 705/6 from C.A. **109** [1988] No. 214974).
[47] Ivanov, M. Ya.; Ivanova, N. Yu.; Kompaniets, V. Z.; Konoplev, A. A.; Kupryazhkina, T. N.; Ovsyannikov, N. A.; Potapova, E. V.; Tutundzhyan, A. K. (Vysokochist. Vestchestva **1988** No. 2, pp. 88/92 from C.A. **109** [1988] No. 211130).
[48] Soum, A.; Fontanille, M.; Rudler, H.; Gouarderes, R. (Macromol. Chem. Rapid Commun. **7** [1986] 525/9).
[49] Plakhotnik, V. N.; Dulova, V. I.; Gulivets, I. L. (Koord. Khim. **12** [1986] 318/20 from C.A. **104** [1986] No. 156858).
[50] Seeger, K.; Heller, G. (Z. Kristallogr. **172** [1985] 105/9).

[51] Heller, G.; Seeger, K. (Z. Naturforsch. **43b** [1988] 547/56).
[52] Hayakawa, S.; Okuma, K.; Araki, K.; Niinomi, J.; Kanamaru, T. (Mie-ken Kankyo Kagaku Senta Kenkyu Hokoku [1987] No. 7, pp. 51/6 from C.A. **107** [1987] No. 102305).
[53] König, T.; Männel, D.; Habicher, W. D.; Schwertlick, K. (Polym. Degrad. Stab. **22** [1988] 137/45).
[54] Dzhemilev, U. M.; Kunakova, R. V.; Baibulatova, N. Z. (Zh. Org. Khim. **22** [1986] 1591/7; J. Org. Chem. [USSR] **22** [1986/87] 1431/7).
[55] Drake, C. A. (U.S. 4532360 [1984/85] 1/6 from C.A. **103** [1985] No. 141621).
[56] Enjo, H. E.; Harada, T. (Eur. Appl. 136 683 [1983/85] 1/28 from C.A. **103** [1985] No. 72876; Jpn. Kokai Tokkyo Koho 86-83280 [1984/86] 1/5 from C.A. **105** [1986] No. 136081).
[57] Beatty, R. P. (Can. 1248141 [1984/89] 1/10 from C.A. **111** [1989] No. 58015).
[58] Chaturvedi, K. K.; Singh, R. V.; Tandon, J. P. (J. Prakt. Chem. **326** [1984] 817/22; Synth. React. Inorg. Met.-Org. Chem. **14** [1984] 1085/97 from C.A. **102** [1984] No. 113557).
[59] Brown, C. A.; Krishnamurthy, S. (J. Org. Chem. **51** [1986] 238/40).
[60] Tokarev, B. V.; Orlov, A. V. (Zh. Vses. Khim. Obshch. im. D. I. Mendeleeva **33** [1988] 353/5 from C.A. **111** [1989] No. 7609).

[61] Schinazi, R. F.; Prusoff, W. H. (J. Org. Chem. **50** [1985] 841/7).
[62] Kunzelmann, A.; Lauterbach, G.; Potekhin, V. M.; Pritzkow, W.; Schmidt-Renner, W.; Vasina, L. F. (J. Prakt. Chem. **328** [1986] 772/6).
[63] Lu, Xiyan; Jiang, Xiaohui; Tao, Xiaochan (J. Organomet. Chem. **344** [1988] 109/18).
[64] Hoffmann, R. W.; Ditrich, K.; Fröch, S.; Cremer, D. (Tetrahedron **41** [1985] 5517/24).
[65] Rossi, K.; Pihlaja, K. (Acta Chem. Scand. B **39** [1985] 671/83).
[66] Wulff, G.; Hansen, A. (Carbohydr. Res. **164** [1987] 123/40).
[67] Wulff, G.; Poll, H. G. (Macromol. Chem. **188** [1987] 741/8).
[68] Kamars, A.; Shvarts, E. M. (Latv. PSR Zinat. Akad. Vestis Kim. Ser. **1986** 700/7).
[69] Hoffmann, R. W.; Fröch, S. (Tetrahedron Lett. **26** [1985] 1643/6); Hoffmann, R. W.; Ditrich, K.; Fröch, S. (Liebigs Ann. Chem. **1987** 977/85).
[70] Matsubara, H.; Seto, K.; Tahara, T.; Takahashi, S. (Bull. Chem. Soc. Jpn. **62** [1989] 3896/901).

[71] Bhal, L.; Tandon, J. P. (Indian J. Chem. A **24** [1985] 562/4; Indian J. Chem. A **25** [1986] 54/6; J. Prakt. Chem. **328** [1986] 911/20).
[72] Zhu, Z.-Y.; Fang, Y.-X.; He, D.-J.; Zhou, W.-K.; Zhang, G.-Y.; Ding, H.-X. (Huaxue Xuebao **43** [1985] 640/4 from C.A. **105** [1986] No. 24299).

[73] Abramov, A. F.; Grachek, V. I.; Motol'ko, G. R.; Naumova, S. F. (Zh. Obshch. Khim. **58** [1988] 1571/4; J. Gen. Chem. [USSR] **58** [1988] 1399/401 from C.A. **110** [1988] No. 212886).
[74] Singh, Y. P.; Rupani, P.; Singh, A.; Rai, A. K.; Mehrotra, R. C.; Rogers, R. D.; Atwood, J. L. (Inorg. Chem. **25** [1986] 3076/81).
[75] Hartmann, H.; Schumann, T.; Dusi, R.; Bartsch, U.; Ilge, H.-D. (Z. Chem. **26** [1986] 330/1).

3.5.9 Diorganylborinic Acid Anhydrides (Diboryloxanes), $R_2B–O–BR_2$, and Related Species

Ab initio Hartree-Fock-Roothaan calculations with large polarized basis sets gave ^{17}O nuclear quadrupole coupling constants, e^2qQ/h, for planar borinic acid anhydride, **$H_2B–O–BH_2$**, which is a molecular model for bridging oxygen atoms in B_2O_3 (which vary strongly with the angle). The calculated e^2qQ/h values of 4.8 and 6.1 MHz for the B–O–B angle of 120° and 132°, respectively, are compared with the experimental data of 4.7 and 5.8 MHz for the two nonequivalent oxygens in vitreous B_2O_3, for which angles B–O–B of 120° (boroxin rings) and 128° to 132° (BO_3 triangles, linked to boroxin rings) have been obtained by X-ray and neutron diffraction experiments. The calculated ^{17}O NMR shielding constant is 87 ppm for $H_2B–O–BH_2$ (angle B–O–B of 132°) and increases with the angle at a rate of about 1 ppm/degree. Less accurate, smaller-basis set calculations in lower symmetries indicate that the ^{17}O electric-field-gradient values are not changed significantly by ring closure, but are effected by nonzero dihedral angles between the BH_2 planes [1].

$(CH_3)_2B–O–B(CH_3)_2$ is dimeric at −160 °C, as the study of crystals generated by the Bridgman technique showed. The bond lengths are: r(BO) = 158.0(1) pm in the ring and 137.7(1) pm exocyclic; r(BC) = 156.3 to 159.5(1) pm. The bonding angles are: in the ring, ∢(OBO) = 86.1(1)° and ∢(BOB) = 93.9(1)°; outside of the ring, ∢(BOB) = 133.2(1)° and ∢(OBC) = 119.8(1)°. The compound seems to be the first B–O example with an aggregation to a BOBO ring. The ^1H NMR spectrum shows the monomeric species, which below −68 °C must be interpreted as the spectrum of the dimer [2].

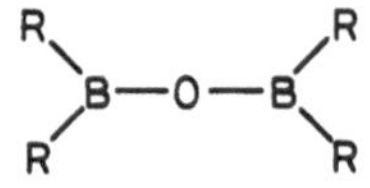

30

30	R
a	C_2H_5
b	C_6H_5
c	OC_6H_5
d	$OC(O)CH_3$

Compound **30a** ($R = C_2H_5$) was formed as a by-product in the reaction of $[–(C_2H_5)BO–]_3$ with octakis-*O*-(diethylboryl)-sucrose [4]. $C_2H_5B(OH)_2$ was treated with $B(C_2H_5)_3$ to give a mixture of **30a** and $[–(C_2H_5)BO–]_3$, which is suitable for the introduction of *O*-diethylboryl- and *O*-ethylboranediyl groups into hydroxy compounds [5]. **30a** and the compound (–Z–)B–O–B(–Z–), Z = 1,5-cyclooctanediyl, react with $AlCl_3$ or $AlBr_3$ to give halogeno(diorgano)boranes and dimeric diorganylboryloxyaluminiumdihalogenide [3] as well as with nitrogen bases such as pyridine, 4-methylpyridine, or quinuclidine to form stable liquid or solid adducts, as was shown by isolation and/or ^{11}B and ^{13}C NMR spectroscopy [6]; they also react with pyrazole, 3-methylpyrazole, and indazole to form stable 1:1 adducts. Proton, ^{11}B, and ^{13}C NMR spectra and X-ray analyses show their principal structure contains a B_2ON_2 heterocycle [8]. The structures of the adducts of (–Z–)B–O–B(–Z–), Z = 1,5-cyclooctanediyl, with pyridine and quinuclidine show phase transformations [7].

30a (R = C_2H_5) and **30b** (R = C_6H_5) react at elevated temperatures with excess pyrazole to give an essentially quantitative yield of the corresponding pyrazabole, $R_2B(\mu\text{-}N_2C_3H_3)_2BR_2$ ($C_3H_4N_2$ = pyrazole); see "Boron Compounds" 4th Suppl. Vol. 3b, 1992, pp. 33ff. [9]. **30b** reacts with dimethylamine or azetidine in the presence of formaldehyde to form substituted 2,2-diphenyl-1,3-dioxa-5-azonia-2-boratocyclohexanes [10]. The three-component reaction of **30b** with α-dicarbonyl compounds and guanidine or benzamidine leads to diphenylboron chelates [11]. Cyclocondensation of **30b** with N-hydroxy-N-(1-hydroxycyclohexyl)-methylsalicylamide results in the formation of a seven-membered chelate ring [12].

A copolymer between **30c** (R = OC_6H_5) and HCHO was prepared and has good heat and water resistance properties [13]. Oxybis(diacetoxyborane), **30d** (R = $OC(O)CH_3$), reacts with pyrocatechol (1,2-$(OH)_2C_6H_4$) or diols to yield mixed ligand complexes of boron [14]. Heating of **30d** in mixture with diphenylsilanediol and some dibasic tridentate Schiff bases in equimolar ratios in dry toluene gave tricentered diborosiloxanes [15].

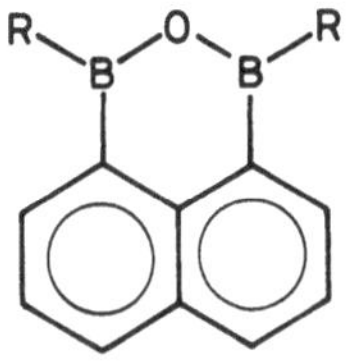

31

31	R
a	OH
b	OCH_3
c	Cl
d	CH_3

The reaction of 1,8-naphthalene diboronic anhydride, **31a**, R = OH (prepared from 1,8-diiodonaphthalene and $B(OCH_3)_3$), with methanol in benzene leads to the "dimethyl ester" **31b** (R = OCH_3), melting point 80 to 82 °C; IR: ν = 1330 cm^{-1} (B–O–B). NMR data (C_6D_6; δ in ppm): δ^{1}H = 3.64 (s, 6H), 7.35 (d, 2β-H), 7.71 (d, 2γ-H), and 8.38 (d, 2α-H); δ^{11}B (in CD_2Cl_2 versus $(CH_3)_2O\text{–}BF_3$) = 29.6 ppm; mass spectrum: m/z = 226. The compound **31d** resists attempts to break the B–O–B linkage with $(CH_3)_2O\text{–}BF_3$, but forms complexes with nitrogen bases such as pyridine. **31b** reacts with PCl_5 and CCl_4 to yield **31c** (R = Cl), melting point 180 °C (decomposition), and with excess CH_3MgBr in tetrahydrofuran to form **31d** (R = CH_3), melting point 97 to 99 °C; IR: ν = 1307 cm^{-1} (B–O–B). NMR data (in ppm; CD_2Cl_2): δ^{1}H = 1.06 (s, 6H), 7.64 (d, 2β-H), 8.09 (d, 2γ-H), and 8.31 (d, 2α-H); δ^{11}B (in CD_2Cl_2 versus $(CH_3)_2O\text{–}BF_3$) = 50.7; mass spectrum: m/z = 194 [16].

Trialkylboroxins $[\text{–B(R)–O–}]_3$ (R = CH_3 or C_2H_5; see pp. 253/4) react with enolizable 1,3-diketones $(R'CO)_2CH_2$ (R' = CH_3 or C_6H_5) with fast degradation of the boroxin rings to give monochelated diboroxanes **32** with H-bridged bicyclic structures which in solution are in equilibrium with partially ring-opened compounds [17].

References on pp. 257/8

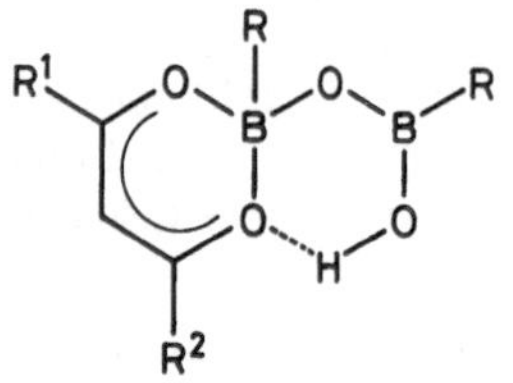

32

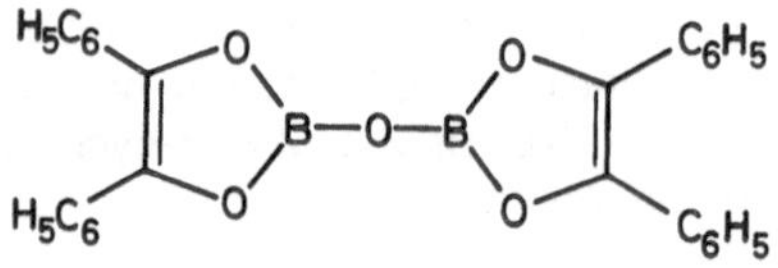

33

Compound **33** was prepared by treating hydrobenzoin or benzoin with boric acid, followed by monobromination and dehydrobromination [18]. For the preparation of bis(9-bora-10-oxa-phenanthrenyl)oxides (formally containing an RB–O–BR structure) from metallated 2-hydroxy-biphenyls or 2′-hydroxy-*m*-terphenyls with BBr_3 or BCl_3, see [19].

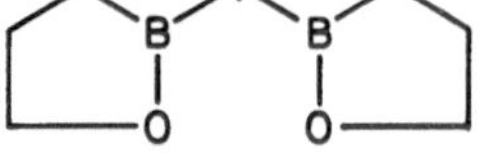

34

Raman spectra of bis(oxaborolanes), **34** with Y = O, $OCH_2–C{\equiv}C–CH_2O$, or $O–(CH_2)_n–O$ (n = 2, 3, or 4) were studied in relation to their molecular structures. **34** with Y = $O–CH_2–CH_2–O$ can exist in two tautomeric (cyclic and network-like) forms [20].

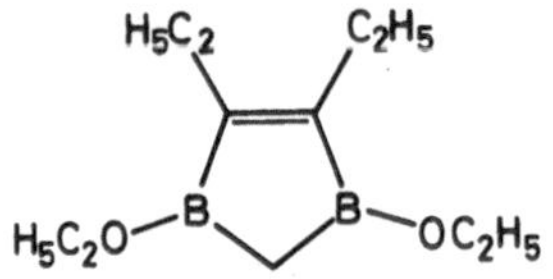

35

Redox reactions between $CH_2(BI_2)_2$ and 3-hexyne lead to 4,5-diethyl-1,3-diiodo-2,3-dihy-dro-1,3-diborole **35** (I instead of OC_2H_5). The iodine of the B–I function can be substituted by OC_2H_5 to give **35** (and also by other nucleophiles). An analogous reaction of $CH_2(BI_2)_2$ occurs with 2-butyne, but a series of other alkynes failed to react [22].

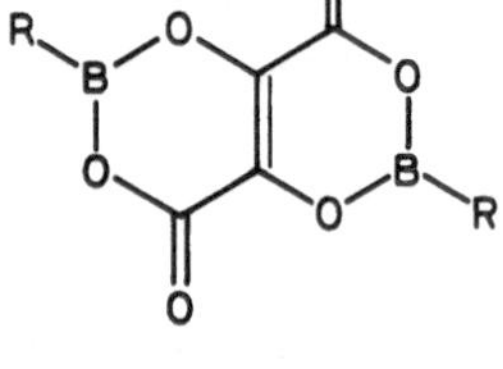

36

The course of the reversible aggregation of 2,6-di-*n*-propyl-1,3,5,7-tetraoxa-2,6-dibora-4,8-octalindione, **36** (R = *n*-C_3H_7), at low temperatures has been investigated; compounds **36** (R = C_2H_5, C_6H_5, or *t*-C_4H_9) are described [23]. Phase changes resulting from solid-state thermochromic reactions of **36** (R = CH_3, C_2H_5, *n*-C_3H_7, *i*-C_3H_7, *i*-C_4H_9, *cyclo*-C_6H_{11}, *n*-C_8H_{17}, *n*-$C_{16}H_{33}$, or $C_6H_5CH_2$) were used to study the mechanisms underlying phase transformations and hysteresis [24]. The molecular structures of the colorless **bis(4-methylphenyl)boryl acetylacetonate** and the orange colored **9-borabicyclo[3.3.1]non-9-yl acetylacetonate** have been determined by X-ray analysis. Bis(4-methylphenyl)boryl acetylacetonate crystallizes in the monoclinic space group $P2_1/c-C^5_{2h}$ (No. 14) with a = 951.5(2), b = 1316.4(5), and c = 1428.1(4) pm; β = 102.16(2)°; D_c = 1.11 g/cm³. 9-Borabicyclo[3.3.1]non-9-yl acetylacetonate crystallizes in the triclinic space group $P\bar{1}-C^1_i$ (No. 2) with a = 732.7(2), b = 883.3(2), and c = 1052.4(3) pm; α = 93.09(2)°, β = 108.61(2)°, and γ = 97.89(2)°; Z = 2; D_c = 1.14 g/cm³ [25]. A 1,3-dibora-2,4-dioxetane has been prepared by hydrolysis and dehydration of (2,4,6-tri-*t*-butylphenyl)dimethoxyborane, prepared from 2,4,6-tri-*t*-butylphenyllithium and $B(OCH_3)_3$ in hexane. The compound reacts with water and oxygen; when heated to 85 °C, it decomposes with the loss of two molecules of 2-methylpropene. Photolysis in the presence of trapping agents, or irradiation at −196 °C led to products which may arise from the intermediate oxoborane (see p. 148) [26].

3.5.10 Triorganylboroxins, $[-B(R)-O-]_3$, Tris(organyloxy)boroxins, $[-B(OR)-O-]_3$, and Trihydroxyboroxin, $[-B(OH)-O-]_3$

The heat of formation for $[-B(H)-O-]_3$, **37a** (R = H), was calculated by AM1 calculations as −286.3 kcal/mol (observed −288.8 kcal/mol) for D_{3h} symmetry with r(BO) = 135.7 pm and r(BH) = 118.7 pm; the ionization potential was calculated to be 12.34 eV [21].

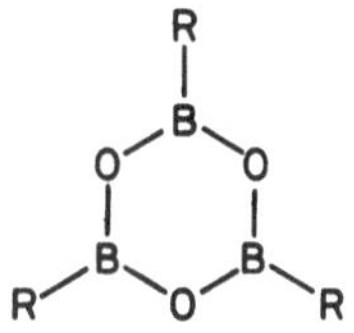

37

37	R
a	H
b	CH_3
c	C_2H_5

Trimethylboroxin, **37b** (R = CH_3), was prepared by various routes. It is obtained in 81% yield by the atmospheric carbonylation of $H_3B-S(CH_3)_2$ in the presence of a small amount of a borohydride species in tetrahydrofuran and has a boiling point of 79 to 80 °C/748 Torr and a refraction index of n^{20}_D = 1.3638. NMR data: δ¹H (neat) = 0.15 to 0.35 ppm (broad, 9 H); δ¹¹B (in CD_2Cl_2 versus $(CH_3)_2O-BF_3$) = 32.5 ppm (s). The reaction of CH_3Li with selected trialkoxyboranes or of methyl-Grignard reagent with 2-methoxy-1,3,2-dioxaborinane yields the corresponding esters, $CH_3-B(OR)_2$, which are readily hydrolyzed to methylboronic acid, $CH_3-B(OH)_2$ (see Section 3.5.6.2.2.1, p. 158), in 79 or 73% yield, respectively. This compound is dehydrated with CaH_2 (reflux in ether, 58% yield) or by mixing with $SOCl_2$ (exothermic reaction, 63% yield) to give trimethylboroxin. Each of these routes was studied in detail [27].

37b (R = CH_3) reacts with 1-alkynyl-lithium to an "ate" complex which can be iodinated [28]. **37b** (R = CH_3) and **37c** (R = C_2H_5) react with enolizable 1,3-diketones, $(R'CO)_2CH_2$ (R' = CH_3 or C_6H_5), with fast degradation of the boroxin rings to give monochelated or bischelated diboroxanes with H-bridged bicyclic structures [17].

$[-B(C_2H_5)-O-]_3$ undergoes a reversible first order phase transition step at −82 °C (heating up) or −89 °C (cooling down), as shown by X-ray diffraction analysis. Both phases belong to the hexagonal space group $P6_3/m-C_{6h}^2$ (No. 176). The low-temperature phase at −90°C has the cell dimensions a = 902.9(3) and c = 692.3(3) pm; Z = 2; D_c = 1.13 g/cm³; refined to R = 4.4% and to R_w = 4.9% for 586 reflections. The high-temperature phase at −85°C has the cell dimensions a = 1557.6(4) and c = 703.8(4) pm; Z = 6; D_c = 1.13 g/cm³; refined to R = 4.8% and R_w = 5.0% for 826 reflections [30]. The compound is used to prove the presence of *cis*-diol groups [31].

$C_2H_5B(OH)_2$ was treated with $(C_2H_5)_3B$ to give a mixture of $[(C_2H_5)_2B]_2O/[-B(C_2H_5)-O-]_3$, which is suitable for the introduction of *O*-diethylboryl and *O*-ethylboranediyl groups into hydroxy compounds [32]. $[-B(C_2H_5)-O-]_3$ reacts with octakis-*O*-(diethylboryl)-sucrose to yield differently composed mixtures, containing, e.g., substituted *O*-(ethylborane-diyl)-sucroses and $(C_2H_5)_2B-O-B(C_2H_5)_2$ [4].

Reactions of four molar equivalents of $[-B(R)-O-]_3$ (R = C_2H_5, *n*-C_4H_9, or C_6H_5) with three equivalents of $AlCl_3$ or $AlBr_3$ yield new organoboron-oxohaloaluminium compounds [33].

$[-B(R)-O-]_3$ (R = C_2H_5 or C_6H_5) reacts with pyrazole to yield pyrazaboles in which the two boron atoms are also bridged by an −O−B(R)−O− link [34]. At room temperature, $[-B(C_2H_5)-O-]_3$ reacts with pyrazole beyond mere complexation; additional pyrazole and even minor increases in temperature promote the condensation process. **$[-B(C_6H_5)-O-]_3$** reacts with pyrazole ($C_3H_4N_2$) at room temperature to form an adduct, and at elevated temperatures to yield the triply bridged pyrazabole $C_6H_5B[\mu-O-B(C_6H_5)-O-](\mu-N_2C_3H_3)_2BC_6H_5$ [9, 35]. Triply bridged pyrazaboles $RB[\mu-O-B(R)-O-](\mu-N_2C_3H_3)_2BR$ were reacted with $SOCl_2$ to yield RClB-$(\mu-N_2C_3H_3)_2$BClR, which reacts with alkali metal alkoxides, R′OM (M = alkali metal), or with alcohols, R′OH, in the presence of triethylamine to give $R(R'O)B(\mu-N_2C_3H_3)_2B(OR')R$, the first examples of B-alkoxy pyrazaboles (R = CH_3 or C_2H_5) [36]. For details on pyrazaboles, see "Boron Compounds" 4th Suppl. Vol. 3b, 1992, pp. 33/40, 58/9, and 248/50; and "Boron Compounds" 4th Suppl. Vol. 4, 1991, pp. 68/9 and 100/2.

The triply bridged pyrazoboles hydrolyze with traces of water at elevated temperatures to form boroxins [36]. Both $[-B(C_2H_5)-O-]_3$ and $[-B(C_6H_5)-O-]_3$ also form 1:1 molar adducts of the types $[-B(R)-O-]_3 \cdot L$ with L = hydrazine, N,N′-dimethylhydrazine, or N,N-dimethylhydrazine. The complexes exhibit only one ^{11}B NMR signal at room temperature, suggesting that the species show fluxional attachment of L with the nitrogen atom coordinating to all three boron atoms of the boroxin ring. N_2H_4 also forms 2:1 molar complexes, while $(CH_3)_2NNH_2$ forms a 2:1 complex only with $[-B(C_6H_5)-O-]_3$ [37]. The structures of the cited 1:1 adducts of $[-B(R)-O-]_3$ with pyrazole [9, 35] have been reinvestigated. Detailed NMR analyses reveal that, contrary to recent reports, they form normal 1:1 adducts, in which only one nitrogen atom and one boron atom are involved in the bonding [38].

Reactions of phenylboronic acid anhydride with hexamethylcyclotrisilazane in the presence of 1% KOH between 180 and 280°C in 4 h gave 40% $[-B(C_6H_5)-O-]_3$, in addition to cyclotrisiloxane, cyclosilazasiloxane, C_6H_6, and CH_4 [39]. The structure of the compound has been solved; it crystallizes in the monoclinic space group $P2_1/c-C_{2h}^5$ (No. 14) with a = 1071.5(2), b = 1365.2(3), and c = 1170.3(2) pm; β = 100.38(1)°; Z = 4; D_c = 1.230 g/cm³; for 130 variables and 758 unique reflections the structure is refined to R = 4.1%, and for all 3851 reflections to R_w = 10.0%. The central B_3O_3 ring is found to be nearly planar but has a small envelope distortion; one of the boron atoms is displaced by 11.9(7) pm out of the plane of the other five atoms. The B−O bond lengths are 137.8, 138.4, 138.6, 138.6, 138.9, and 139.0 pm. The determined molecular dimensions are similar to those found for the related compounds 1,3,5-triphenylbenzene and triphenyl-*s*-triazine, but the conformations of the three molecules differ. The phenyl rings in the hydrocarbon adopt a propeller-like arrangement, while in the boroxin and triazine the phenyl groups are nearly coplanar with the central B_3O_3 or C_3N_3 ring. Possible relationships between thermal-motion descriptions and the intramolecular potential-energy surfaces for these molecules are discussed [40].

Boron nuclear quadrupole double resonance spectra of $[-B(C_6H_5)-O-]_3$ show that the extent of the O–B π-bond is approximately half as strong as the N–B π-bond in borazine: $^{10}B(7\leftrightarrow 1,6\leftrightarrow 1)$ at 3135(15); $^{10}B(7\leftrightarrow 3,6\leftrightarrow 3)$ at 2310(10); ^{11}B at 1654(5); ^{10}B at 1448(10). For tri-coordinate ^{11}B, $e^2qQ/h=3148(20)$ and $\eta=0.56(2)$; all frequencies in kHz [41].

$[-B(C_6H_5)-O-]_3$ cyclizes hydromethylphosphines to yield cyclic boryloxyalkylphosphines with electron-acceptor substituents [42], dioxaboraphosphinanes [43], or triethylammonium 1,4-diphenyl-3-(2′-hydroxyphenyl)-2,8,9-trioxa-1-borata-4-phospha-6,7-benzocyclo[3.3.1]-nonane [44].

Tris(2,4,6-triethylphenyl)boroxin, melting point 41 to 43 °C, has been prepared by heating dihydroxy-(2,4,6-triethylphenyl)borane for 15 h to 100 °C at 5×10^{-3} Torr; ^1H NMR spectrum ($CDCl_3$; δ in ppm): δ=1.19 (t, 9H), 2.60 (q, 2H), 2.75 (q, 4H), 6.88 (s, 2H); mass spectrum: $[M]^+$ [45]. **2,4,6-Tris[2-(methoxymethyl)phenyl]boroxin**, melting point 70 °C, has been prepared from $NaOCH_3$ and 2,4,6-tris[(2-bromomethyl)phenyl]boroxin. **2,4,6-Tris{2-[(ethylthio)methyl]phenyl}boroxin**, melting point 54 °C, was obtained from ethanediol, sodium methylate, and 2,4,6-tris[(2-bromomethyl)phenyl]boroxin [46].

Cyclodehydration of H_3BO_3 with ROH (R=*n*-C_3H_7, *i*-C_3H_7, *n*-C_4H_9, *n*-C_6H_{11}, or $C_6H_5CH_2$) gave 95 to 99.6% yields of *B*-alkoxyboroxins, **$[-B(OR)-O-]_3$ (38**, R=alkyl) [47].

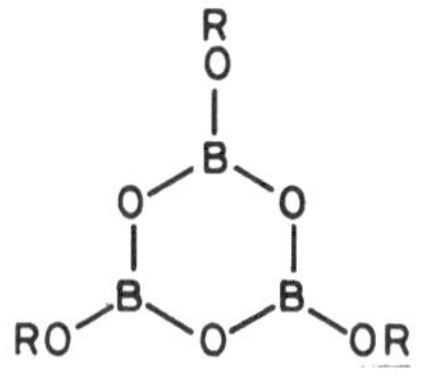

38

$[-O-B(OH)-]_3$ (38 with R=H), trimeric metaboric acid, $H_3B_3O_6$, seems to be the main reaction product during the oxidation of hot-pressed B_4C by water vapor at 900 °C between 1.5 and 20 kPa, as proved by its weight loss; the oxidation rate can be expressed by surface chemical reaction-controlled kinetics with an apparent activation energy of 200 kJ/mol [48].

Thermodynamic analysis of the composition of monomeric gaseous phases, used for heterodiffusion syntheses of polyboroxanes, shows the presence of $[-B(OH)-O-]_3$ between 1200 and 1350 K [49]. $[-B(OH)-O-]_3$ has been tabulated with distances r(BO)=136 pm, r(OH)=100 pm, and the bond angles ∢(B–O–B)=∢(B–O–H)=120° [50].

Pyrolysis of samples containing B_2H_6 and oxygen at ≤280 °C leads to a strong IR absorption at ca. 1380 cm^{-1} which is assigned (by comparison with literature spectra) to the E′ mode ν_9 of boroxin. Oxygen-18 data were obtained for the major isotopomers of $[-B(OH)-O-]_3$; their band positions at 1353 to 1420 cm^{-1} in an Ar matrix support the D_{3h} structure of boroxin [51].

Ab initio Hartree-Fock calculations gave a value of 2.56 for the nuclear quadrupole coupling constant of $[-B(OH)-O-]_3$; the distances are r(B–OH)=2.595, r(BO)=2.561, and r(OH)=1.814 a.u.; the angles are ∢(OBO)=120° and ∢(BOH)=111.36° [52]. Ab initio Hartree-Fock calculations with less accurate smaller-basis sets on $[-B(OH)-O-]_3$ rings have also been performed [1]. By AM1 calculations, the heat of formation of $[-B(OH)-O-]_3$ was found as −529.0 kcal/mol (observed −543.6 kcal/mol) for C_{3h} symmetry with r(B–OH)=133.3 pm and r(BO)=136.0 pm and with ∢(OBO)=119.1°; the ionization potential was calculated to be 13.02 eV [21].

In the cells Pt/H_2/HBO_2, $NaBO_2$, KBr/AgBr/Ag and Pt/H_2/HBO_2, $NaBO_2$, CsI/AgI/Ag (both H_2O/propylene glycol), the standard potentials of the Ag/AgBr and Ag/AgI electrodes, respectively, have been determined; $[-B(OH)-O-]_3$ was not formed [53].

References on pp. 257/8

$[-B(OH)-O-]_3$ is a catalyst in the oxidation of cycloalkanes by molecular oxygen to give cycloalkanoles and cycloalkanones in an autoclave [54]. For reactions of HBO_2, see Section 3.5.6.1., p. 156.

The reaction of 2-C_6H_5-6-R-C_6H_3OM (M = Li or Na) with BBr_3 or BCl_3 in toluene gives **$[-B(O-C_6H_3-(C_6H_5\text{-}2)(R\text{-}6)-O-]_3$** as by-products (R = H or C_6H_5) [19].

3.5.11 Higher Boric Acids and $(HO)_2B-B(OH)_2$

For diborates, see Section 3.6.3, p. 269, for triborates, see Section 3.6.4, p. 270, and for tetraborates, see Section 3.6.5, p. 273.

Molecular mimicry of the geometry and charge distribution of **$H_3B_3O_6$** (C_{3h} symmetry), **$H_4B_2O_5$** (C_2 symmetry), and **$H_8B_2O_7$** (C_v symmetry) have been determined [57]. Differential thermal analysis (DTA) and thermogravimetry (TG) curves of the classical dehydration of H_3BO_3 at ca. 1000°C were compared with laser-heated decomposition. While thermal heating is interpreted by the reaction via a polyboric acid: $4\,H_3BO_3 \rightarrow H_2B_4O_7 + 5\,H_2O$, and $2\,H_2B_4O_7 \rightarrow 4\,B_2O_3 + 2\,H_2O$, the laser irradiation goes directly by $2\,H_3BO_3 \rightarrow B_2O_3 + 3\,H_2O$ [58].

AM1 calculations for the heat of formation for **$(HO)_2B-B(OH)_2$** gave −313.7 kcal/mol (observed −315.0 kcal/mol) for D_{2h} symmetry with r(BB) = 163.1 pm and r(BO) = 135.1 pm, and with the bond angles ∢(OBB) = 125.7° and ∢(HBO) = 106.6°; the ionization potential was calculated to be 11.42 eV [21].

3.5.12 Compounds of the Type $B_3H_6-OC(=O)R$

B_6H_{14} reacts with RCOOH (R = H, CH_3, C_2H_5, or C_6H_5) to give compounds of type **39** [55]. **$B_3H_6-OC(=O)C_6H_5$ (39**, R = C_6H_5), produced from benzoic acid and in situ generated B_6H_{14} or B_4H_{10}, was the first monosubstituted derivative of B_3H_7; the X-ray structure determination is reported. It crystallizes in the monoclinic space group P2/c − C_{2h}^4 (No. 13) with a = 992.3(12), b = 583.7(11), and c = 1742.3(13) pm; β = 117.55(8)°; Z = 4; D_c = 1.18 g/cm³ [56].

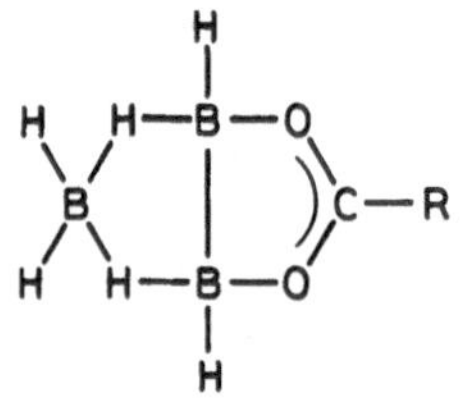

39

3.5.13 Tetrakis[organoboranediylbis(oxy)]cyclobutanes

MNDO and STO-3G calculations explain the formation of nonpropellanic tetrakis[organoboranediyl-bis(oxy)]cyclobutanes from BCl_3 and tetraoxycyclobutane without a trace of the isomeric tetrakisdioxabora[3.3.2]propellanes [29].

References for 3.5.9 to 3.5.13:

[1] Tossell, J. A.; Lazzeretti, P. (J. Non-Cryst. Solids **99** [1988] 267/75).
[2] Borrmann, H.; Simon, A.; Vahrenkamp, H. (Angew. Chem. **101** [1989] 182/3).
[3] Köster, R.; Tsay, Y. H.; Krüger, C.; Serwatowski, J. (Chem. Ber. **119** [1986] 1174/88).
[4] Taba, K. M.; Köster, R.; Dahlhoff, W. V. (Liebigs Ann. Chem. **1987** 463/6).
[5] Köster, R.; Idelman, P. (Inorg. Synth. **24** [1986] 83/7).
[6] Yalpani, M.; Serwatowski, J.; Köster, R. (Chem. Ber. **122** [1989] 3/7).
[7] Yalpani, M.; Köster, R. (Chem. Ber. **122** [1989] 9/17).
[8] Yalpani, M.; Köster, R.; Boese, R. (Chem. Ber. **122** [1989] 19/24).
[9] Bielawski, J.; Niedenzu, K. (Inorg. Chem. **25** [1986] 1771/4).
[10] Kliegel, W.; Rettig, S. J.; Trotter, J. (Can. J. Chem. **66** [1988] 377/84).

[11] Kliegel, W.; Motzkus, H. W. (Chem. Ber. **121** [1988] 1865/7).
[12] Kliegel, W.; Tajerbashi, M.; Rettig, S. J.; Trotter, J. (Can. J. Chem. **66** [1988] 2621/30).
[13] Huang, Ch.-X.; Jin, Z.-Sh. (Guti Runhua **5** [1985] 17/21 from C.A. **104** [1986] No. 20140).
[14] Singh, H. B.; Tandon, J. P. (Synth. React. Inorg. Met.-Org. Chem. **15** [1985] 391/400).
[15] Chaturvedi, V.; Tandon, J. P. (Indian J. Chem. A **24** [1985] 641/4).
[16] Katz, H. E. (J. Org. Chem. **50** [1985] 2575/6).
[17] Köster, R.; Angermund, K.; Sporzynski, A.; Serwatowski, J. (Chem. Ber. **119** [1986] 1931/52).
[18] Schaumburg, G. D.; Battarai, K. M.; Gautam, D. P. (J. Nepal. Chem. Soc. **3** [1983] 11/20 from C.A. **103** [1985] No. 142055).
[19] Maringgele, W.; Meller, A.; Noltemeyer, M.; Sheldrick, G. M. (Z. Anorg. Allg. Chem. **536** [1986] 24/34).
[20] Zhu, Z.-Y.; Fang, Y.-X.; He, D.-J.; Zhou, W.-K.; Zhang, G.-Y.; Ding, H.-X. (Huaxue Xuebao **43** [1985] 640/4 from C.A. **105** [1986] No. 24299); Zhu, Z.-Y.; Zhou, W.-K.; Fang, Y.-X.; He, D.-J.; Zhang, G.-Y.; Ding, H.-X. (Huaxue Xuebao **43** [1985] 1212/4 from C.A. **105** [1986] No. 133941).

[21] Dewar, M. J. S.; Jie, Caoxian; Zoebisch, E. G. (Organometallics **7** [1988] 513/21).
[22] Siebert, W.; Ender, U.; Herter, W. (Z. Naturforsch. **40b** [1985] 326/30).
[23] Yalpani, M.; Klotzbuecher, W. E. (Z. Naturforsch. **40b** [1985] 222/8).
[24] Yalpani, M.; Scheidt, W. R.; Seevogel, K. (J. Am. Chem. Soc. **107** [1985] 1684/90).
[25] Boese, R.; Köster, R.; Yalpani, M. (Chem. Ber. **118** [1985] 670/5).
[26] Pachaly, B.; West, R. (J. Am. Chem. Soc. **107** [1985] 2987/8).
[27] Brown, H. C.; Cole, T. E. (Organometallics **4** [1985] 816/21).
[28] Brown, H. C.; Basavaiah, D.; Bhat, N. G. (J. Org. Chem. **51** [1986] 4518/21).
[29] Stanger, A.; Apeloig, Y.; Ginsburg, D. (Helv. Chim. Acta **68** [1985] 1179/85).
[30] Boese, R.; Polk, M.; Bläser, D. (Angew. Chem. **99** [1987] 239/41).

[31] Yalpani, M.; Wilke, G. (Chem. Ber. **118** [1985] 661/9).
[32] Köster, R.; Idelman, P. (Inorg. Synth. **24** [1986] 83/7).
[33] Köster, R.; Angermund, K.; Serwatowski, J.; Sporzynski, A. (Chem. Ber. **119** [1986] 1301/14).
[34] Bielawski, J.; Niedenzu, K. (Inorg. Chem. **25** [1986] 85/7).
[35] Bielawski, J.; Das, M. K.; Hanecker, E.; Niedenzu, K.; Nöth, H. (Inorg. Chem. **25** [1986] 4623/8).
[36] Hsu, L. Y.; Mariategui, J. F.; Niedenzu, K.; Shore, S. G. (Inorg. Chem. **26** [1987] 143/7; AD-A 173 131/4 GAR [1986] 1/21 from C.A. **108** [1988] No. 186797).

[37] Das, M. K.; Mariategui, J. F.; Niedenzu, K. (Inorg. Chem. **26** [1987] 3114/6; AD-A 182688 [1987] 1/11 from C.A. **110** [1989] No. 95304).
[38] Yalpani, M.; Köster, R. (Chem. Ber. **122** [1988] 1553/65).
[39] Gruzinova, E. A.; Svatikov, M. Yu.; Kotrelev, G. V.; Zhdanov, A. A. (Izv. Akad. Nauk SSSR Ser. Khim. **1987** 1145/6; Bull. Acad. Sci. USSR Ser. Chem. **36** [1987/88] 1059/60).
[40] Pratt Brock, C.; Minton, R. P.; Niedenzu, K. (Acta Crystallogr. C **43** [1987] 1775/9; AD-A 177470/2/GAR [1987] 1/17 from C.A. **109** [1988] No. 23000).

[41] Lötz, A.; Voitländer, J.; Stephenson, D.; Smith, J. A. S. (Z. Naturforsch. **41a** [1986] 200/2).
[42] Ignat'eva, S. N.; Nikonov, G. N.; Erastov, O. A.; Arbuzov, B. A. (Izv. Akad. Nauk SSSR Ser. Khim. **1985** 1102/6; Bull. Acad. Sci. USSR Ser. Chem. **34** [1985/86] 1005/9).
[43] Arbuzov, B. A.; Nikonov, G. N.; Erastov, O. A. (Izv. Akad. Nauk SSSR Ser. Khim. **1986** 643/8; Bull. Acad. Sci. USSR Div. Chem. Sci. **35** [1986] 588/92).
[44] Nikonov, G. N.; Karasik, A. A.; Erastov, O. A.; Arbuzov, B. A. (Izv. Akad. Nauk SSSR Ser. Khim. **1987** 2118/20; Bull. Acad. Sci. USSR Div. Chem. Sci. **36** [1987/88] 1969/71).
[45] Schacht, W.; Kaufmann, D. (Chem. Ber. **120** [1987] 1331/8).
[46] Lauer, M.; Böhnke, H.; Grotstollen, R.; Salehnia, M.; Wulff, G. (Chem. Ber. **118** [1985] 246/60).
[47] Cifuentes, L. O.; Bahamondes, J. (Bol. Soc. Quim. Peru **51** [1985] 189/92 from C.A. **107** [1987] No. 115764).
[48] Sato, T.; Haryu, K.; Endo, T.; Shimada, M. (Zairyo **37** No. 412 [1988] 77/82 from C.A. **108** [1988] No. 208946).
[49] Zav'yalov, V. V.; Tarasevich, B. P.; Sirotkin, O. S.; Khitrov, M. Yu. (Zh. Obshch. Khim. **55** [1985] 2157/61; J. Gen. Chem. [USSR] **55** [1985] 1915/9 from C.A. **104** [1986] No. 61063).
[50] Tarasevich, B. P.; Sirotkina, N. Z.; Il'in, A. S.; Kusnetsov, E. V. (Vysokomol. Soedin. A **27** [1985] 451/63; Polym. Sci. [USSR] **27** [1985] 503/17).

[51] Ault, B. S. (J. Mol. Struct. **159** [1987] 297/302).
[52] Gajhede, M. (Chem. Phys. Lett. **120** [1985] 266/71).
[53] Sastry, V. V.; Kalidas, C. (J. Chem. Eng. Data **30** [1985] 91/4).
[54] Tamaru, A.; Habu, H.; Matsuoka, T.; Harada, S.; Watanabe, K. (Jpn. Kokai Tokkyo Koho 86-227541 [1985/86] 1/10 from C.A. **106** [1987] No. 137984).
[55] Brellochs, B.; Binder, H. (Angew. Chem. **100** [1988] 270/1).
[56] Binder, H.; Brellochs, B.; Frei, B.; Simon, A.; Hettich, B. (Chem. Ber. **122** [1989] 1049/56).
[57] Zhang, Z. G.; Boisen, M. B.; Finger, L. W.; Gibbs, G. V. (Am. Mineral. **70** [1985] 1238/47).
[58] Jovancicevic, V.; Zelenay, P.; Scharifker, B. R. (Electrochim. Acta **32** [1987] 1553/5).

3.6 Hydrated Anionic Species and Species Derived Thereof

For a previous rewiew, see "Boron Compounds" 3rd Suppl. Vol. 2, 1987, pp. 150/82.

For the formulations of the borates with or without parentheses or brackets, cf. Chapter 3.4, p. 70.

3.6.1 General Studies and Remarks

A systematic survey of the structures of water-containing borates as well as of the structures of anhydrous borates is available (550 references) [1]. A review with 265 references on the properties of small- and macro-anionic metal oxoborates in the crystalline state and in solution was compiled [2]. Also, a review on sodium borates as hazardous materials [3] and a booklet about hydrated alkaline earth metal borates were published [4]. The crystal and molecular structures of synthetic tetraborates and hexaborates as determined by X-ray techniques were reviewed [17]. A review with 34 references on the features of ^{10}B NMR and ^{11}B NMR spectra of hydrated borates discusses its usefulness for structural studies [21]. The optimal conditions for obtaining well resolved ^{11}B NMR spectra are reported for a wide range of borates and borosilicates. Rapid sample spinning, together with power proton decoupling, were found to be needed for rapid acquisition of these spectra [22].

A review (54 references) is available in which many of the structural features of sol-gel derived inorganic polymers such as borates are explained on the basis of the stability of the M–O–M condensation products in their synthesis environments (M = metal). Structures which emerge in solution reflect a successive series of hydrolysis, condensation, and, depending on the acid or base condensation, restructuring reactions. M–O–M bonds which are unstable with respect to hydrolysis or alcoholysis are generally absent. During consolidation, the condensation process continues within a very stiff matrix in a relatively inert environment. Under these conditions, metastable species may be formed which are stabilized temporarily by the high matrix viscosity and the absence of water and alcohol [23].

Based on calculated values of the index of reliability, 21 out of 25 borate minerals show adherence to the Gladstone-Dale rule [5]. Calcium borate hydrates such as inyoite are subject to hydrolysis, and, therefore, their solubilities cannot be determined directly but only by dissolving small amounts of the crystalline borate up to saturation. In the absence of other ions, the solubilities of calcium borates can be calculated from the solubility isotherms [6]. The point of intersection of the crystallization lines can be used for the solution of some problems of alkaline earth metal borate synthesis in alkali metal or ammonium borate solutions [7].

The nature of the interaction of magnesium borate minerals such as sulfoborite, kaliborite, and preobrazhenskite with water was studied by thermogravimetry, X-ray phase analysis, and chemical analyses. Initially the borate was completely dissolved in water, then the borate anion and the magnesium cation were formed by hydrolysis, and finally new magnesium borates precipitated [8]. The kinetics of dehydration of ascharite, kaliborite, and preobrazhenskite were studied; the activation energies depend on the degree of dehydration [9]. Thermal analysis of alkaline earth metal polyborates, including inyoite, kurnakovite, inderborite, $Mg[B_6O_9(OH)_2]\cdot 6.5\,H_2O$, $Ca[B_6O_9(OH)_2]\cdot 5\,H_2O$, $Sr[B_6O_9(OH)_2]\cdot 4\,H_2O$, $Ba[B_6O_9(OH)_2]\cdot 4\,H_2O$, kaliborite, colemanite, and hydroboracite, shows that the dehydration occurs in two steps. Borates with a $[B_3O_3(OH)_5]^{2-}$ anion and ring polymerization have low temperature endothermal effects (200 °C) and low hydration activation energy values, indicating that both water molecules and OH groups are weakly bonded. Borates with a $[B_3O_4(OH)_3]^{2-}$ structural anion and chain polymerization have high dehydration temperatures (270 to 640 °C) and high activation energy values, indicating that both H_2O molecules and OH groups are strongly bonded in the crystal lattice. The alkaline earth metal hexaborates occupy an intermediate position in terms of dehydration temperatures and activation energy values. Among these, the dehydration of the strontium and barium hexaborates is different from those of magnesium and calcium; kinetic parameters of their first dehydration steps are given [10].

$3\,B_2O_3\cdot MgO\cdot 7\,H_2O$ crystallizes during the first period from a concentrated solution of $MgCl_2$ containing 5.67% $2\,B_2O_3\cdot MgO$, corresponding to its kinetic curve, and subsequently $2\,B_2O_3\cdot 2\,MgO\cdot MgCl_2\cdot 14\,H_2O$ precipitates. The kinetic equation of the two borates was

References on pp. 261/2

derived, and the mechanisms of crystallization reactions are discussed [11]. Thermodynamic analyses of the system B_2O_3–CaO–SiO_2–CO_2–H_2O show that optimal conditions for endogenic datolite and danburite mineralization are: high content of boron in solution, low CO_2 content, and relatively high temperatures [12].

The $^{11}B/^{10}B$ ratio is given for various boron minerals from different genetic types of ore deposits [13]. Boron isotopic analyses for nine marine borate mineral evaporites show a composition of $^{11}B/^{10}B = 4.15 \pm 0.02$, while for 25 nonmarine evaporite borates it is $^{11}B/^{10}B = 4.02 \pm 0.04$; thus, it should be possible to use the boron isotopic composition as a tracer for boron of marine evaporite origin [14].

An empirical method is described to calculate the ΔH_f values of B_2O_3 in various minerals, including calciborite, kurchatovite, and danburite [15].

Based on the hypothesis that crystal structures may be ordered according to the polymerization of the coordination polyhedra with higher bond-valences, a hierarchical classification is set up for mineral structures that are based on triangular and octahedral cation coordination groups. The minerals are arranged according to their basic heteropolyhedral cluster, or fundamental building block (FBB), and the dimensional character of the cluster polymerization. The FBB, repeated by glide planes and screw axes, forms the structure module, a complex anionic polyhedral array in which the excess charge is balanced by (extra-modular) large low-valence cations. The mode of polymerization of the FBB is related to the Lewis basicity of the simple oxoanions that constitute this FBB. The Lewis basicity of the structure module may be related, via the valence-matching principle, to the Lewis acidity of the extra-modular cations. With anion coordination numbers selected in a manner that they result in most-nearly-equal bond-valences for the extra-modular cations, the coordination number and Lewis acidity of the extra-modular cations can be predicted. The following borate minerals fall into the class of infinite frameworks: fluoborite, painite, jeremejevite, warwickite, the ludwigite type, pinakiolite, orthopinakiolite, takeuchite, wightmanite, and the szaibelyite group [16].

Hydrated thallium(I) borates, for example $Tl[B_3O_4(OH)_2] \cdot 0.5H_2O$ (see Section 3.6.4.2, p. 271) and $Tl_2[B_4O_7] \cdot 1.5H_2O$ (see Section 3.6.8, p. 280), show the existence of macrochains and can be considered as condensed inorganic heterocycles [18].

The geometries of a variety of borate polyanions, extracted from borate crystals and protonated in order to achieve quasi-neutral or neutral molecules, were optimized using quantum mechanical molecular orbital (MO) methods. The calculated bond lengths and angles for neutral molecules are in good agreement with those found in borate minerals. However, calculations on molecules with negative charges as large as 2 esu yield bridging angles that depart by as much as 15°, and B–O bond lengths that depart by as much as 3 pm from observed values. Furthermore, when these molecules can be neutralized by further protonation, the calculated angles agree with observed values. A regression analysis of bond strengths and calculated B–O bond lengths yields bond strength-bond length parameters that are identical with those obtained for crystals. Deformation electron-density maps calculated for various borate molecules mimic experimental maps of comparable units in crystals; however, they fail to provide evidence for the bonding of discrete OBO units into endless chains [19].

A reaction involving a monomer, a dimer, and a six-membered ring of triangles is examined. An analysis of the energetics of the reaction indicates that an important component of the destabilization energy of the ring is bond angle strain. A variant of a hybrid orbital model is developed and used to rank bond lengths in distorted borate triangles and tetrahedra in crystals; an inescapable conclusion drawn is that the bond lengths and angles of polyanions in borate crystals behave as if they are primarily determined by short-range forces. The following geometries have been calculated on an STO-3G basis:

a) The ion $[H_6B_2O_7]^{2-}$, consisting of two borate tetrahedra each bonded to three protons, assuming C_{2v} symmetry;
b) the ion $[H_4B_3O_7]^-$, consisting of a ring of two triangles and one tetrahedron, assuming C_2 symmetry; and
c) the ion $[H_5B_3O_8]^{2-}$, consisting of a triangle and two tetrahedra, assuming C_v symmetry [19].

IR spectra show that in the structure of *N*-substituted ammonium borates in the corresponding amine–H_3BO_3–H_2O system, irrespective of the acidity of the medium (from pH 11.5 to 6.6), pentaborate rings predominate over the other polyborate ions in the products in each case [20].

The ^{11}B NMR chemical shifts and line widths were measured in very dilute aqueous solutions of boric acid (0.1 molal: $\delta = 0$ ppm) and sodium borate, $NaBO_2$ (1 molal: $\delta = -17.6$ ppm). The chemical shift for $Na_2[B_4O_7]$ solutions with concentrations below 0.1M changes from $\delta =$ -9.9 to -8.4 ppm, and for $[NH_4][B_5O_8]$ solutions with concentrations below 0.1M from $\delta = -1.5$ to -2.9 ppm. An additional weak signal is observed at higher concentrations at $\delta = -18.4$ ppm. These results can be explained by taking pH-dependent weighted averages over the species $B(OH)_3$ and $[B(OH)_4]^-$ [24].

The ^{11}B-^{10}B primary isotope effect on the magnetic shielding is smaller than 3×10^{-8}. The H_2O-D_2O solvent isotope effect on the longitudinal relaxation times (T_1) was established for ^{11}B in the species mentioned, and from the ratios of T_1, the quadrupolar origin of the relaxation mechanism was inferred [24]. Addition of some typical hydroxycarboxylic acids yields chemical shifts between -9.4 and -10.0 ppm [25]. Boron-11 chemical shifts, line widths, and association constants were determined in D_2O, pD = 11, for boric acid esters of 25 polyols or polyhydroxycarboxylates [26].

The nonpolar stationary-anion chromatography with a coated PRP-1 column and eluents containing methanol, potassium hydrogen phthalate, and tris(hydroxymethylamino)methane was used for borates at $pH > 7$ [27].

The carrier-mediated transport of H_3BO_3 through an anion-exchange membrane with $[OH]^-$ or borate counter ions as carrier recovers the free ligand and allows extraction [28]. The amounts of sorbed boron and the simultaneous conductivity change are in agreement with the nonformation of triborates and the formation of mono- and divalent tetra- and pentaborates [29].

References for 3.6.1:

[1] Heller, G. (Top. Curr. Chem. **131** [1986] 39/98).
[2] Tarasevich, B. P.; Kuznetsov, E. V. (Usp. Khim. **56** [1987] 353/92).
[3] Anonymous (Dangerous Prop. Ind. Mater. Rept. **8** No. 1 [1988] 68/72 from C.A. **108** [1988] No. 172560).
[4] Gode, H. (Book about Hydrated Alkaline Earth Metal Borates, Zinatne, Riga 1986, 167 pp. from C.A. **106** [1987] No. 42979).
[5] Gode, H.; Sprice, D. (Latv. PSR Zinat. Akad. Vestis Kim. Ser. **1987** 62/6).
[6] Gode, H.; Jaunzeme, A. (Latv. PSR Zinat. Akad. Vestis Kim. Ser. **1987** 271/4).
[7] Gode, H. (Latv. PSR Zinat. Akad. Vestis Kim. Ser. **1987** 643/9).
[8] Perebeinos, A. A.; Kononova, G. N.; Rastegaeva, G. Yu. (Zh. Neorg. Khim. **31** [1986] 2996/3005; Russ. J. Inorg. Chem. **31** [1986/87] 1726/31).
[9] Saiko, I. G.; Kononova, G. N.; Zakgeim, A. Yu. (Zh. Prikl. Khim. **58** [1985] 654/6; J. Appl. Chem. [USSR] **58** [1985] 558/60 from C.A. **102** [1985] No. 210052).
[10] Tkachev, K. V.; Leont'eva, I. A. (Latv. PSR Zinat. Akad. Vestis Kim. Ser. **1989** 30/2).

[11] Gao, Shiyang; Li, Qixin; Xia, Shuping (Wuji Huaxue **4** No. 2 [1988] 64/73 from C.A. **110** [1989] No. 63236).
[12] Malinko, S. V. (Mineral. Zh. **7** [1985] 36/45 from C.A. **103** [1985] No. 9147); Semenov, Yu. V.; Malinko, S. V.; Kiseleva, I. A.; Khodakovskii, I. L. (Geokhimiya **1987** 1182/90 from C.A. **107** [1987] No. 137835).
[13] Malinko, S. V.; Lisitsyn, A. E.; Sumin, L. V. (Sov. Geol. **1987** No. 3, pp. 89/97 from C.A. **107** [1987] No. 158511).
[14] Swihart, G. H.; Moore, P. B.; Callis, E. L. (Geochim. Cosmochim. Acta **50** [1986] 1297/301).
[15] Zuev, V. V. (Geokhimiya **1986** 1160/9 from C.A. **106** [1987] No. 87787).
[16] Hawthrone, F. C. (Can. Mineral. **24** [1986] 625/34).
[17] Silina, E.; Ozolins, G.; Tetere, I.; Zincenko, R. V. (Latv. PSR Zinat. Akad. Vestis Kim. Ser. **1987** 228/38).
[18] Touboul, M. (Phosphorus Sulfur **28** [1986] 145/9).
[19] Zhang, Z. G.; Boisen, M. B.; Finger, L. W.; Gibbs, G. V. (Am. Mineral. **70** [1985] 1238/47).
[20] Tsekhanskii, R. S.; Skvortsov, V. G.; Molodkin, A. K. (Zh. Neorg. Khim. **31** [1986] 1961/4; Russ. J. Inorg. Chem. **31** [1986/87] 1129/31).

[21] Bray, P. J. (AIP Conf. Proc. No. 140 [1986] 142/67 from C.A. **105** [1986] No. 34119).
[22] Turner, G. L.; Smith, K. A.; Kirkpatrick, R. J.; Oldfield, E. (J. Magn. Reson. **67** [1986] 544/50).
[23] Brinker, C. J.; Bunker, B. C.; Tallant, D. R.; Ward, K. J.; Kirkpatrick, R. J. (ACS Symp. Ser. No. 360 [1988] 314/32 from C.A. **108** [1988] No. 157052).
[24] Balz, R.; Brändle, U.; Kammerer, E.; Köhnlein, D.; Lutz, O.; Nolle, A.; Schafitel, R.; Veil, E. (Z. Naturforsch. **41a** [1986] 737/42).
[25] Lutz, O.; Veil, E.; Brändle, U.; Helber, H.; Kammerer, E.; Scheiter, A. (Z. Naturforsch. **41a** [1986] 1041/4).
[26] Van Duin, M.; Peters, J. A.; Kieboom, A. P. G.; Van Bekkum, H. (Tetrahedron **41** [1985] 3411/21).
[27] Papp, E.; Fehervari, A. (J. Chromatogr. **447** [1988] 315/22).
[28] Langevin, D.; Metayer, M.; Labbe, M.; Pollet, B.; Hankaoui, M.; Selegny, E. (Desalination **68** No. 2/3 [1988] 131/48 from C.A. **108** [1988] No. 227417).
[29] Metayer, M.; Langevin, D.; Hankaoui, M.; Pollet, B.; Labbe, M.; Roudesli, S. (React. Polym. Ion Exch. Sorbents **7** [1988] 111/22 from C.A. **108** [1988] No. 193278).

3.6.2 Monoborates, Derivatives Thereof, and the Radical $[B(OH)_4]^{\bullet}$

3.6.2.1 The Ion $[B(OH)_4]^-$ and Derivatives Thereof

Molecular mimicry of the geometry and charge density distribution of the $\mathbf{[B(OH)_4]^-}$ ion in borate minerals was reviewed (47 references) [4].

For aqueous $[B(OH)_4]^-$, tabulated as aqueous $[BO_2]^-$, cf. [61, 62]. Correlation algorithms permit the prediction of species-dependent parameters in revised equations of state (r_e = 138 pm, $\omega^{obs} \times 10^{-5} = 1.2208$ cal/mol, $\omega^{cal} \times 10^{-5} = 1.7595$ cal/mol, and the fit coefficients σ = −0.6 cm³/mol and $\xi = -7.4 \times 10^{-2}$ $cm^3 \cdot K \cdot mol^{-1}$), which can be used together with values at 25 °C and 1 bar of the standard partial molal thermodynamic properties (entropy $\overline{S}^\circ = -8.9$ cal $\cdot mol^{-1} \cdot K^{-1}$, volume $\overline{V}^\circ = -14.5$ cm³/mol, and heat capacity $\overline{C}_p^\circ = -41.0$ $cal \cdot mol^{-1} \cdot K^{-1}$) to calculate its standard partial molal thermodynamic properties at temperatures up to 1000°C and pressures up to 5 kbar [1].

Work by Pindas and Shamir [2] suggests that the fundamental stretching vibrations $\nu_1(A_1)$ and $\nu_3(E_2)$ of the tetrahedral ion $[B(OH)_4]^-$ in aqueous solutions of $H_2{}^{16}O$ behave in an anomalous fashion on ^{18}O substitution, since these frequencies were observed some 40 cm^{-1} higher than expected for the ^{18}O form. However, a new determination of the Raman and IR spectra of $[B(^{16}OH)_4]^-$ with $\nu_1=746$ and $\nu_3=955$ cm^{-1} and of $[B(^{18}OH)_4]^-$ (produced by alkaline hydrolysis of BCl_3 in $H_2{}^{18}O$) with $\nu_1=706$ and $\nu_3=914$ cm^{-1} are in good agreement with published values [3]. Torsional barriers and force fields have been calculated for $[B(OH)_4]^-$ by using self-consistent field (SCF) theory at the 6-31G* basis level [8].

The surface-enhanced Raman spectrum of the ion $[B(OH)_4]^-$ was obtained by adding metal ions to a silver sol containing this anion, and the enhancement factor was estimated to be larger than 10^4 to 10^5. This is caused by metal ions, which form a complex with $[B(OH)_4]^-$ on the surface of the Ag particles, probably leading to the formation of an Ag–O bond [13].

The chemical equilibrium model of Harvie et al. [5], based on the semiempirical equations of Pitzer [6], is extended to include borate species and is valid up to high ionic strength (ca. 14 M) and high borate concentration. Excellent agreement of the existing electromotive force (EMF) and isopiestic and solubility data in the system $[B(OH)_4]^-$–$[SO_4]^{2-}$–$[CO_3]^{2-}$–Cl^-–Na^+–K^+–Mg^{2+}–Ca^{2+}–H^+–H_2O is obtained, although only the presence of $[B(OH)_4]^-$, $[B_3O_3(OH)_4]^-$, and $[B_4O_5(OH)_4]^{2-}$, but not that of $[B_5O_6(OH)_4]^-$, was assumed. Calculated mineral solubilities are generally within 10% of their experimental values, even at high inonic strengths [7].

$[B(OH)_4]^-$ ions accelerate the thermal decay of HOI [9].

By thermolysis and low-temperature X-ray structural analysis, the positions of all hydrogen atoms were found in the compounds $Na[B(OH)_4]\cdot5H_2O$, $Na[B(OH)_4]\cdot2H_2O$, and $Na[B(OH)_4]$; the H-bonding system was described quantitatively. A second modification of $Na[B(OH)_4]$ was found which shows a strong deviation from the T_d symmetry. The structural data of $Na[B(OH)_4]\cdot2H_2O$ [10] must be corrected in some points [11].

The stability constants for the formation of ion-pairs with $[B(OH)_4]^-$ were determined by potentiometry and nonapproximative linearization of titration data to be $K^*\{Li[B(OH)_4]\}=0.89\pm0.02$, $K^*\{Na[B(OH)_4]\}=0.44\pm0.01$, $K^*\{[MgB(OH)_4]^+\}=13.6\pm0.7$, $K^*\{[CaB(OH)_4]^+\}=11.4\pm0.15$, and $K^*\{[SrB(OH)_4]^+\}=3.47\pm0.006$ [12].

$Ca[^{10}B(OH)_4]_2$ and **$Ca[^{11}B(OH)_4]_2$** were prepared in high yield from aqueous CaO and freshly prepared labeled $Na[B(OH)_4]$ solutions [63].

$Ba[B(OH)_4]_2\cdot H_2O$, or $BaO\cdot B_2O_3\cdot5H_2O$, was obtained by the interaction of potassium borate and $BaCl_2$ in aqueous solution [14]. It crystallizes in the monoclinic space group $P2_1/b-C^5_{2h}$ (No. 14) with a = 596.1(3), b = 1516(1), and c = 854.3(6) pm; γ = 102.32(5)°; Z = 4; D_c = 2.76 g/cm³; refined to R = 2.2%. In the structure the boron atoms are tetrahedrally coordinated; the barium atoms are coordinated to eight OH groups and two H_2O molecules, and hydrogen bonds are linking [15]. In an older determination of this compound by X-ray powder diffractometry, it was indexed to the space group $P2_1/c-C^5_{2h}$ (No. 14) with a = 594.1(5), b = 855.3(6), and c = 1507.1(26) pm; β = 102.406(104)°; Z = 4; D_m = 2.77 g/cm³ and D_c = 2.78 g/cm³ [16].

$Sr[B(OH)_4]_2$, or $SrO\cdot B_2O_3\cdot4H_2O$, is monoclinic with a = 817.0(7), b = 1599.0(5), and c = 1087.3(7) pm; β = 118.84(7)°; Z = 8; D_m = 2.61 g/cm³ and D_c = 2.62 g/cm³. The structure was solved by direct methods and refined by least-squares [17].

The thermal decomposition of sulfoborite, **$Mg_3[(OH)_2|SO_4|\{B(OH)_4\}_2]$**, was investigated by thermal analysis, infrared spectra, and X-ray phase analysis. Dehydration occurs in two stages

References on pp. 267/9

with the loss of 0.5 and 4 moles water with decomposition of the structure; at 580 °C, $Mg[SO_4] \cdot 0.5H_2O$ crystallizes in addition to $Mg[B_4O_7]$ [18].

Henmilite, $\mathbf{Ca_2Cu[(OH)_4|\{B(OH)_4\}_2]}$, or $Ca_2Cu[BO_3]_2 \cdot 6H_2O$, is a new bluish violet mineral from Fuka, Japan, and has been investigated by X-ray powder diffractometry. It is triclinic, space group $P\bar{1}-C_i^1$ (No. 2) with a=576.17(5), b=797.74(6), and c=564.88(4) pm; α=109.611(6)°, β=91.473(7)°, and γ=83.686(7)°; Z=1; D_c=2.523 g/cm³; R=2.6% and R_w=3.4% using 2430 observed reflections. The structure is built from isolated $[B(OH)_4]^-$ groups and square-planar coordinated $[Cu(OH)_4]^{2-}$ ions; Ca^{2+} is surrounded by eight OH groups. Two of the resulting polyhedra share an edge to form dimers, which are connected to each other through $[B(OH)_4]^-$ tetrahedra to form chains parallel to the c axis. The chains are linked through square-planar $Cu(OH)_4$ groups into a three-dimensional structure [19].

The new mineral moydite, $\mathbf{(Y,RE)[CO_3|B(OH)_4]}$, where RE means a rare earth element, is orthorhombic, space group $Pbca-D_{2h}^{15}$ (No. 61) with a=908.9(1), b=1224.4(1), and c=892.6(1) pm; Z=8; R=5.1% and R_w=3.7% for 1278 reflections. Moydite has a layered structure with sheets of corner-sharing YO_9 polyhedra, reinforced by trigonal-planar carbonate groups. Successive sheets are linked by isolated $B(OH)_4$ tetrahedra which attach themselves by opposite edges [20].

Hydrothermal crystallization in the B_2O_3–Na_2O–Al_2O_3–SiO_2–H_2O system ($xB_2O_3 \cdot 17Na_2O \cdot Al_2O_3 \cdot 2SiO_2 \cdot yH_2O$, where x=0.1 to 10; y=260 to 290) reveals the predominant formation of tetrahydroxoborate sodalite, $\mathbf{Na_8[\{B(OH)_4\}_2|(AlSiO_4)_6]}$, between 473 and 773 K at 100 to 150 MPa. The high-temperature form (T≥310 K) of tetrahydroxoborate sodalite shows cubic symmetry in space group $P\bar{4}3n-T_d^4$ (No. 218) with a=901.0(1) pm, whereas the superstructure diffraction patterns of the low-temperature form (T≤270 K) was indexed by transformation of the cubic subcell to orthorhombic symmetry with a=2551.0(1), b=1275.0(1), and c=902.0(1) pm. Silicon-29 magic angle spinning nuclear magnetic resonance spectroscopy (MAS-NMR) shows complete ordering of the Si and Al atoms in the tetrahedral framework of the tetrahydroxoborate sodalite; the observed single peak of the ^{11}B MAS-NMR spectrum correlates with the location of $[B(OH)_4]^-$ at the center of the sodalite cages. Substitution of boron for aluminium in the tetrahedral sites of the sodalite framework and threefold coordinated boron within the sodalite cages or the framework are excluded [21].

An ^{11}B NMR study of the formation of complex compounds between $[B(OH)_4]^-$ and various polyols was carried out; a 1:1 borate-polyol complex produces a peak at δ=6.6 ppm, a 1:2 complex at δ=10.3 ppm, while the peak due to free $[B(OH)_4]^-$ is at δ=2.7 ppm (versus external aqueous H_3BO_3). The association constants are for ethanediol, K_1=0.74, K_2=0.29 dm³/mol; for propane-1,2-diol, K_1=2.2, K_2=0.55 dm³/mol; for propane-1,3-diol, K_1=1.1, K_2=0.08 dm³/mol; for glycerol, K_1=6.7, K_2=1.3 dm³/mol; for mannitol, K_1=137, K_2=21 dm³/mol; and for sorbitol, K_1=278 and K_2=24 dm³/mol [22]. The reactions of two chiral diols with $[B(OH)_4]^-$ were studied by polarimetry, circular dichroism, and ^{11}B NMR spectroscopy [23]. Another ^{11}B NMR study identified several complexes between $[B(OH)_4]^-$ and some cyclitols such as cyclohexane-*cis*-1,2-diol, cyclohexane-*cis*,*cis*-1,3,5-triol, *myo*-inositol, and *epi*-inositol [24]. Aqueous H_2O_2 transforms organoboranes such as tri(*n*-hexyl)borane in diglyme or tetrahydrofuran in the presence of dilute NaOH or $Na_2[B_4O_7]$ quantitatively into the corresponding alcohols and $Na[B(OH)_4]$ [25].

Formally, $\mathbf{H[B(O{-}SO_2OH)_4]}$ can be understood as a derivative of the ion $[B(OH)_4]^-$. A solution of $H[B(O{-}SO_2OH)_4]$ in 100% H_2SO_4 or in oleum was prepared from H_3BO_3 and $H_2S_2O_7$ (as oleum of the appropriate concentration) and was used to accelerate sulfonation processes [64].

3.6.2.2 The Radical $[B(OH)_4]^•$

Flash photolysis of an aqueous solution of $Na[B(OH)_4]$ and $K_2[S_2O_8]$ at pH 11.5 and 23 ± 1 °C gives rise to a new transient species with absorption maxima at 590 nm (ε = 80 m²/mol) and 650 nm (ε = 90 m²/mol). The pK_a and reduction potential of the transient species, the radical **$[B(OH)_4]^•$**, are 10.75 ± 0.02 and (estimated) 1.4 eV, respectively, as was shown from the rate constants of oxidation reactions of bromobenzene, anisole, 1-aminopropane, 4-nitroaniline, 4-aminobenzoic acid, aniline, 4-methylaniline, 4-methoxyaniline, hydroquinone, or 4-aminophenol with the one-electron borate radical. Salt-effect studies are consistent with the reactive species being $[B(OH)_3O]^{•-}$ which is the result of the equilibrium $[B(OH)_4]^• \rightleftharpoons [B(OH)_3O]^{•-} + H^+$ [26].

3.6.2.3 $[BSiO_4(OH)]^{2-}$ Structure

The mineral datolite, **$Ca[BSiO_4(OH)]$**, or $Ca[BO_2(OH)] \cdot SiO_2$, shows the following parameters: isotropic chemical shift δ_{iso} (1H) = 4.3 ppm, r(O–H···O) = 293 pm (from diffraction data 321 pm). From ^1H magic angle spinning nuclear magnetic resonance (MAS-NMR) measurements at 200 MHz, $\Delta_{v1/2}$ = 1.270 Hz, and at 500 MHz, $\Delta_{v1/2}$ = 1.350 Hz ($\Delta_{v1/2}$ are the half-height linewidths). The average hydrogen density in the crystal lattice is C_H = 11.3 atoms/nm³ [27].

3.6.2.4 Ions of the Type $[HB(OR)_3]^-$ and $[HB(OC(O)CH_3)_3]^-$

$Li[HB(OC_2H_5)_3]$ is a reducing agent for α,β-unsaturated compounds [28]. The stereoselectivity of **$Li[HB(OC_3H_7\text{-}i)_3]$** in the reduction of representative cyclic ketones was examined and compared with that of the potassium derivative. The reaction of **$K[HB(OR)_3]$** (R = i-C_3H_7, s-C_4H_9, t-C_4H_9, or 2-methylcyclohexyl) with LiCl in tetrahydrofuran was examined to establish the generality of this preparation of $Li[HB(OR)_3]$. While the metal ion exchange reaction between **$K[HB(OC_3H_7\text{-}i)_3]$** and LiCl proceeds instantly at room temperature and the resultant $Li[HB(OC_3H_7\text{-}i)_3]$ is very stable towards redistribution, the lithium compounds with R = s-C_4H_9, t-C_4H_9, or 2-methylcyclohexyl are unstable due to increasing steric requirements and they rearrange to $[H_3B(OR)]^-$ and $[B(OR)_4]^-$ [29].

$K[HB(OC_3H_7\text{-}i)_3]$ reduces ruthenium formyl complexes in CH_2Cl_2 at −30 °C [30].

Tris(acetoxy)hydroborate ions, e.g., **$[HB(OC(O)CH_3)_3]^-$**, are mild and stereoselective reducing agents for acyclic β-hydroxyketones; the sodium [31] and the tetramethylammonium salts [32] were employed.

$[(CH_3)_4N][HB(OC(O)CH_3)_3]$, melting point 96.5 to 98 °C, $\delta^{11}B$ = 0.71 ppm, was prepared from $[(CH_3)_4N][BH_4]$ and acetic acid anhydride [32]. It reduces acyclic β-hydroxyketones to their corresponding *anti*-diols with high diastereoselectivity [33]. Lithium tris(β-naphthoxy)-hydroborate, prepared from β-naphthol and $Li[BH_4]$, is stable and has reducing power [34].

References on pp. 267/9

3.6.2.5 Ions of the Type $[HBR(-OZO-)]^-$ and $[HB(-Z-)-OR]^-$

Chiral cyclic boronic acid esters RB(–OZO–), possessing chirality on alkyl or diol moieties, were prepared via potassium alkanediyldioxy(monoalkyl)hydroborates, **K[HBR(–OZO–)]**. These salts as well as the anlogous potassium alkanediyl(alkoxy)hydroborates, **K[HB(–Z–)–OR]**, were obtained by the reaction of excess KH with the appropriate alkylboronic acid esters or cyclic boronic acid esters, respectively [35].

3.6.2.6 $[H_3B{-}OH]^-$ and Ions of the Type $[H_3B{-}OC(O)R]^-$, $[R_3B{-}OR']^-$, and $[R_2BR'{-}OR'']^-$

The ion $[BH_4]^-$ does not react at a significant rate with $[Fe(CN)_6]^{3-}$ above a pH value of ca. 13 in the temperature range of 20 to 60 °C. With a decrease in the pH of the medium, hydrolysis products of $[BH_4]^-$ such as **$[H_3B{-}OH]^-$** are formed, which are responsible for reducing $[Fe(CN)_6]^{3-}$, especially in alkaline solutions, as kinetic studies by gas volumetry, spectrophotometry, and chemical analytical reactions show [36]. A steroid-peptide ester linkage can also be reduced by $Na[H_3B{-}OH]$ [37].

The reaction between $Na[BH_4]$ and isobutyric acid in diglyme was studied by ^{11}B NMR spectroscopy and showed considerable rearrangement into the monoacyloxy compound **$Na[H_3B{-}OC(O)C_3H_7\text{-}i]$** [38]. Reduction of α,β-unsaturated carbonyl compounds with **$Na[H_3B{-}OC(O)CH_3]$**, prepared in situ from $Na[BH_4]$ and CH_3COOH, gave the corresponding allylic alcohols [39].

Triorganylboranes, BR_3, react with $[R'O]^-$ ions to give tetrahedral adducts **$[R_3B(OR')]^-$**, which decompose with the formation of $[R_2BO]^-$ [40]. Sodium (E)-1-alkenyl(dialkyl)methoxyborates react with RZnCl or $ZnCl_2$ to give (E,E)-1,3-dienes [41].

3.6.2.7 Ions of the Type $[B(OR)_4]^-$ and $[R'O{-}B(OR)_3]^-$

$[H_3NCH_2CH_2NH_3][B(OCH_3)_4]_2$ has a heat of melting of 344 J/g; the energy storage capacity is 464 J/g between 50 and 90 °C [42].

Tetrakis(3-alkylphenyl)borates were synthesized, and their properties have been studied [43].

Infrared, NMR, and mass spectral data have been given for the compounds **$Li[t\text{-}C_4H_9O{-}B(OR)_3]$** with $R = n\text{-}C_3H_7$, $n\text{-}C_4H_9$, $n\text{-}C_5H_{11}$, $n\text{-}C_6H_{13}$, $n\text{-}C_7H_{15}$, $n\text{-}C_8H_{17}$, $n\text{-}C_{10}H_{21}$, $n\text{-}C_{11}H_{23}$, $n\text{-}C_{13}H_{27}$, or $C_6H_5CH_2$ [44]; based on the analysis of the mass spectra, a model of a three-dimensional supramolecular structure with cylindrical voids of these complexes was proposed [45]. Depending on the dielectric constants of the media (heptane, isooctane, decane, hexadecane, CCl_4, and C_2Cl_6) between 293 and 353 K and 8 to 25 mmol lithium complex, the molecules are associated forming such a structure in solvents of low dielectric constants. Long-chain associations form bridging RO or $t\text{-}C_4H_9O$ groups, or dissociate into ions [46]. A small-angle X-ray-scattering (SAXS) study has been performed. The structuring by solvents confirmed the association of these complexes via their alkyl substituents; the degree of association depends on solvent dielectric constants and the length of the alkyl chains. The association radius of gyration and maximal dimension were determined in structured solutions. The shapes of complex associations are nearly cylindrical; the average volume of these complex associations is 1 to 1.5×10^4 Å^3 [47].

The IR spectrum and viscosity of $Li[t\text{-}C_4H_9O{-}B(OR)_3]$ (0.02 to 0.08 M; $R = n\text{-}C_4H_9$, $n\text{-}C_6H_{13}$, or $n\text{-}C_7H_{15}$) were determined in heptane, toluene, or CCl_4. Structured solutions or gels exhibit no bands between 3000 and 3700 cm^{-1}. Addition of alcohols causes the appearance of bands characteristic for R–O(H)···H bonding. The existence of several bands indicates the presence of hydrogen bonds of varying strength. The O_2 adsorption-desorption properties of these gels ($R = n\text{-}C_4H_9$, $n\text{-}C_7H_{15}$, or $n\text{-}C_{10}H_{21}$) were studied between 298 and 548 K. The n O_2/Li[t-C_4H_9O–$B(OR)_3$] sorption complex mole ratio is ca. 2, regardless of the chain length of R. The desorption occurs in two clearly defined steps if $R = n\text{-}C_4H_9$, but in only one step if $R = n\text{-}C_{10}H_{21}$ [48]. Two new weak bands at 3588 (narrow) and 3065 to 3095 cm^{-1} (broad) appear in the solvate spectra. The broad band's intensity increases as the chain length of R increases. The H-bond stability depends on the Li–O interaction and the solution viscosity. A model of the transition state during the alcohol solvation is presented [49]. The kinetic parameters of the O_2 thermal desorption were determined [50], and the properties of the complexes of Li[t-C_4H_9O–$B(OR)_3$] were studied by differential thermal analysis (DTA) [51]. The viscoelastic properties of structured solutions of the species and their solvates [52] as well as their rheokinetic properties were studied [53] and the change in volume of the cross-linking solutions was noted [54]; an X-ray diffractometrical study of these solutions has been reported [55].

[$Cu(C_{12}H_8N_2)_2$][$(HO)_3B$–OC(O)H] can be obtained from [$Cu(C_{12}H_8N_2)P(C_6H_5)_3$][$BH_4$] and either aqueous HCOOH in the presence of 1,10-phenanthroline ($=C_{12}H_8N_2$) or CO_2 in moist tetrahydrofuran in the presence of 1,10-phenanthroline [56].

3.6.2.8 Ions of the Type $[B(-OZO-)_2]^-$

Some complex borates with salicylic acid or its derivatives and other polyols were investigated by structural analysis. Hexaquozinc di(salicylato)borate tetrahydrate, **[$Zn(H_2O)_6$]-[$B(-OC_6H_4C(O)O-)_2$]$_2$·4H_2O**, is monoclinic in the space group $P2_1/b-C_{2h}^5$ (No. 14) with a = 985.5(2), b = 2888.2(5), and c = 1239.7(2) pm; $\gamma = 93.54(1)^\circ$; Z = 4; $D_m = 1.50$ g/cm³ and $D_c = 1.53$ g/cm³. The structure was solved by direct methods and refined to R = 9.9% by full-matrix least-squares. The Zn atoms are in octahedral coordination with the oxygen atoms of six H_2O molecules; the hydrogen bonding is described [57]. Hexaquocopper pentaerythritolatoborate tetrahydrate, **[$Cu(H_2O)_6$][$(C_5H_8O_4)_2B$]$_2$·4H_2O**, is triclinic, space group $P1-C_i^1$ (No. 2) with a = 563.6(1), b = 1115.5(2), and c = 884.7(2) pm; $\alpha = 109.06(1)^\circ$, $\beta = 104.24(1)^\circ$, and $\gamma = 92.6(1)^\circ$; Z = 1; $D_m = 1.86$ g/cm³ and $D_c = 1.74$ g/cm³; R = 2.9%. The structure consists of $[Cu(H_2O)_6]^{2+}$ cations with Cu^{2+} in tetragonal bipyramidal coordination, polymeric $[(C_5H_8O_4)_2B]_n^{n-}$ anions, and H_2O molecules [58].

For cations with OBO structures, see Chapter 3.3, pp. 68 ff. [59, 60].

References for 3.6.2:

[1] Shock, E. L.; Helgeson, H. C. (Geochim. Cosmochim. Acta **52** [1988] 2009/36).
[2] Pindas, S.; Shamir, J. (J. Chem. Phys. **56** [1972] 2017/9).
[3] Campbell, N. J.; Flanagan, J.; Griffith, W. P. (J. Chem. Phys. **83** [1985] 3712/3).
[4] Zhang, Z. G.; Boisen, M. B.; Finger, L. W.; Gibbs, G. V. (Am. Mineral. **70** [1985] 1238/47).
[5] Harvie, C. R.; Møller, N.; Weare, J. H. (Geochim. Cosmochim. Acta **48** [1984] 723/51).
[6] Pitzer, K. (J. Phys. Chem. **77** [1973] 268/77).
[7] Felmy, A. R.; Weare, J. H. (Geochim. Cosmochim. Acta **50** [1986] 2771/83).
[8] Hess, A. C.; McMillan, P. F.; O'Keeffe, M. (J. Phys. Chem. **92** [1988] 1785/91).
[9] Buxton, G. V.; Kilner, C.; Sellers, R. M. (AERE-R-11974 [1986] 151/66 from C.A. **106** [1987] No. 109630).
[10] Block, S.; Perloff, A. (Acta Crystallogr. **16** [1963] 1233/6).

[11] Felsche, J.; Ketterer, B.; Schmid, R. (Z. Kristallogr. **174** [1986] 50/1).
[12] Rogers, H. R.; Van den Berg, C. M. G. (Talanta **35** [1988] 271/5).
[13] Zhang, Chunping; Yu, Fengqi; Zhang, Guangyin (J. Raman Spectrosc. **20** [1989] 431/4).
[14] Gode, H. K.; Shvirkst, Ya. Ya. (Latv. PSR Zinat. Akad. Vestis Kim. Ser. **1985** 548/52).
[15] Simonov, M. A.; Karpov, O. G.; Shvirkst, Ya. Ya.; Gode, H. K. (Kristallografiya **34** [1989] 1292/4; Soviet Phys. Cryst. **34** [1989] 778/80).
[16] Shvirkst, Ya. Ya.; Kondrat'eva, V. V.; Gode, H. K. (Latv. PSR Zinat. Akad. Vestis Kim. Ser. **1987** 218/20).
[17] Apinitis, S.; Gode, H. K. (Latv. PSR Zinat. Akad. Vestis Kim. Ser. **1989** 702/4).
[18] Perebeinos, A. A.; Kononova, G. N.; Saiko, I. G. (Zh. Neorg. Khim. **30** [1985] 2534/7; Russ. J. Inorg. Chem. **30** [1985/86] 1444/6).
[19] Nakai, I.; Okada, H.; Masutomi, K.; Koyama, E.; Nagashima, K. (Am. Mineral. **71** [1986] 1234/6); Nakai, I. (Am. Mineral. **71** [1986] 1236/9).
[20] Grice, J. D.; Ercit, T. C. (Can. Mineral. **24** [1986] 675/8).

[21] Buhl, J. C.; Engelhardt, G.; Felsche, J. (Zeolites **9** [1989] 40/4).
[22] Dawber, J. G.; Green, S. I. E. (J. Chem. Soc. Faraday Trans. I **82** [1986] 3407/13).
[23] Dawber, J. G. (J. Chem. Soc. Faraday Trans. I **83** [1987] 771/7).
[24] Bell, C. F.; Beauchamp, R. D.; Short, E. L. (Carbohydr. Res. **147** [1986] 191/203).
[25] Brown, H. C.; Snyder, C.; Subba Rao, B. C.; Zweifel, G. (Tetrahedron **42** [1986] 5505/10).
[26] Padmajam, S.; Ramakrishnan, V.; Rajaram, J.; Kuraicose, J. C. (J. Chem. Soc. Faraday Trans. I **85** [1989] 2249/54).
[27] Yesinowski, J. P.; Eckert, H.; Rossman, G. R. (J. Am. Chem. Soc. **110** [1988] 1367/75).
[28] Mieczkowski, J.; Wroebel, J. T. (Bull. Pol. Acad. Sci. Chem. **35** No. 3/4 [1987] 113/9).
[29] Cha, J. S.; Kim, J. E.; Lee, J. Ch.; Yoon, M. S. (Bull. Korean Chem. Soc. **7** [1986] 66/9).
[30] Baratt, D. S.; Cole-Hamilton, J. D. (J. Chem. Soc. Chem. Commun. **1985** 458/9).

[31] Evans, D. A.; DiMare, M. (J. Am. Chem. Soc. **108** [1986] 2476/8).
[32] Evans, D. A.; Chapman, K. T. (Tetrahedron Lett. **27** [1986] 5939/42).
[33] Evans, D. A.; Chapman, K. T.; Carreira, E. M. (J. Am. Chem. Soc. **110** [1988] 3560/78).
[34] Shim, S. Ch.; Choi, J. H. (Bull. Korean Chem. Soc. **8** [1987] 4/5).
[35] Brown, H. C.; Park, W. S.; Cho, B. T. (J. Org. Chem. **51** [1986] 3278/82); Brown, H. C.; Cho, B. T.; Park, W. S. (J. Org. Chem. **52** [1987] 4020/4).
[36] Khain, V. S.; Volkov, A. A. (Zh. Neorg. Khim. **32** [1987] 717/21; Russ. J. Inorg. Chem. **32** [1987] 402/4).
[37] Bounds, P. L.; Pollack, R. M. (Biochemistry **26** [1987] 2263/9).
[38] Malvik, A. C.; Obenius, U.; Henriksson, U. (J. Org. Chem. **53** [1988] 221/3).
[39] Nutaitis, C. F.; Bernardo, J. E. (J. Org. Chem. **54** [1989] 5629/30).
[40] Hayes, R. N.; Sheldon, J. C.; Bowie, J. H. (Organometallics **5** [1986] 162/7).

[41] Molander, G. A.; Zinke, P. W. (Organometallics **5** [1986] 2161/2).
[42] Stringer, K.; Wilson, J. W.; Morgan, R. (Int. J. Energ. Res. **13** [1989] 741/2 from C.A. **112** [1990] No. 122214).
[43] Yoo, A. L. (Diss. Loyola Univ., Chicago, Ill., USA 1986, pp. 1/150 from C.A. **105** [1986] No. 153110).
[44] Dmitrieva, Z. T.; Rodionova, N. I.; Turov, Yu. P.; Kadychagov, P. B. (Koord. Khim. **13** [1987] 297/303).
[45] Dmitrieva, Z. T.; Kadychagov, P. B.; Turov, Yu. P.; Mezhibor, N. G. (Koord. Khim. **12** [1986] 878/81; Sov. J. Coord. Chem. **12** [1986] 503/6).
[46] Dmitrieva, Z. T.; Levus, U. I.; Tikhonova, L. D. (Kolloidn. Zh. **48** [1986] 549/52 from C.A. **105** [1986] No. 198021).

[47] Dmitrieva, Z. T.; Tailashev, A. S. (Izv. Akad. Nauk SSSR Ser. Khim. **1987** 2756/60; Bull. Acad. Sci. USSR Div. Chem. Sci. **1987** 2553/7).
[48] Dmitrieva, Z. T. (Mol. Cryst. Liq. Cryst. B **161** [1987/88] 529/42; C.A. **110** [1989] No. 122139).
[49] Dmitrieva, Z. T.; Ryzhikova, I. G. (Zh. Strukt. Khim. **30** [1989] 153/5; J. Struct. Chem. [USSR] **30** [1989] 493/5).
[50] Bol'shakov, G. F.; Dmitrieva, Z. T.; Ryzhikova, I. G. (J. Chem. Soc. Faraday Trans. I **85** [1989] 3119/23).

[51] Ryzhikova, I. G.; Martynova, V. A.; Dmitrieva, Z. T. (Fiz. Khim. Svoistva Dispers. Sistem Ikh Primenenie Tomsk **1988** 25/34 from C.A. **111** [1989] No. 208173).
[52] Korobeinikova, N. V.; Dmitrieva, Z. T. (Struktura Rastvorov Dispersii: Svoistva Kolloid. Sistem Neft. Rastvorov Polimerov Novosibirsk **1988** 65/70 from C.A. **111** [1989] No. 141764).
[53] Dmitrieva, Z. T.; Tikhonova, L. D.; Levus, Yu. I. (Struktura Rastvorov Dispersii: Svoistva Kolloid Sistem Neft. Rastvorov Polimerov Novosibirsk **1988** 48/55 from C.A. **111** [1989] No. 121857).
[54] Dmitrieva, Z. T.; Mezhibor, N. G.; Vavilkin, A. S. (Struktura Rastvorov Dispersii: Svoistva Kolloid. Sistem Neft. Rastvorov Polimerov Novosibirsk **1988** 56/60 from C.A. **111** [1989] No. 202673).
[55] Dmitrieva, Z. T.; Chernyavskii, V. N. (Spektroskopiya Stekloobrazuyushch. Sistem Riga **1988** 76/81 from C.A. **111** [1989] No. 161586).
[56] La Monica, G.; Angaroni, M. A.; Cariati, F.; Genini, S. (Inorg. Chim. Acta **143** [1988] 239/45).
[57] Zviedre, I. I.; Bel'skii, V. K.; Mardanenko, V. K. (Latv. PSR Zinat. Akad. Vestis Kim. Ser. **1985** 387/93).
[58] Zviedre, I. I.; Bel'skii, V. K.; Shvarts, E. M. (Latv. PSR Zinat. Akad. Vestis Kim. Ser. **1985** 672/7).
[59] Mukhopadhyay, D.; Sur, B.; Bhattacharyya, R. G. (Indian J. Chem. A **24** [1985] 425/6).
[60] Karasev, V. E.; Korotkikh, O. A. (Zh. Neorg. Khim. **30** [1985] 2269/72; Russ. J. Inorg. Chem. **30** [1985/86] 1290/2).

[61] Edwards, J. O.; Morrison, G. C.; Ross, V. F.; Schultz, J. W. (J. Am. Chem. Soc. **77** [1955] 266/8).
[62] Edwards, J. O. (J. Am. Chem. Soc. **75** [1953] 6151/4).
[63] Gode, H.; Kaugare, S. (Latv. PSR Zinat. Akad. Vestis Kim. Ser. **1987** 239/42).
[64] Khelevin, R. N. (Zh. Org. Khim. **22** [1986] 2315/9; J. Org. Chem. [USSR] **22** [1986] 2079/82).

3.6.3 Diborate Ions

The diborate ions in this section are not discrete ions, but have to be considered as structural units only; see also Section 3.6.1, pp. 259/61.

Homilite, $\mathbf{Ca_{2.00}Fe^{II}_{0.90}Mn^{II}_{0.03}[B_{2.00}Si_{2.00}O_{9.86}(OH)_{0.14}]}$, a mineral in the monoclinic space group $P2_1/a-C^5_{2h}$ (No. 14), has the cell dimensions a = 978.6(2), b = 762.1(2), and c = 977.6(1) pm; β = 90.61(2)°; Z = 2; D_c = 3.451 g/cm³; R = 3.3% and R_w = 2.7% for 1596 reflections. The structure consists of sheets built up from alternating SiO_4 and BO_4 tetrahedra ("O_2BOSiO_2 units") parallel to the (001) plane. The B–O distances are 155.3, 153.0, 150.7, and 142.7 pm for the non-

 References on p. 270

bridging bonds. The calcium and iron atoms are located between the sheets and form a CaO_6 tetragonal antiprism and an FeO_6 octahedron, respectively [5].

The mineral sussexite, **$Mn^{II}_2(OH)[B_2O_4(OH)]$**, or $B_2O_3 \cdot 2MnO \cdot H_2O$, also crystallizes in the monoclinic space group $P2_1/a-C^5_{2h}$ (No. 14) with a=1277, b=1070, and c=325 pm; β=95° [1].

The mineral szaibelyite, **$Mg_2(OH)[B_2O_4(OH)]$**, also crystallizes in the space group $P2_1/a-C^5_{2h}$ (No. 14) with a=1257.7(2), b=1039.3(2), and c=313.9(1) pm; β=95.88(2)° [1].

Szaibelyite from the Taiga skarn-magnetite deposit in southern Yukatan, **$Mg_{1.88}Fe_{0.042}$-$Mn_{0.01}Ca_{0.05}(OH)[B_{2.01}O_4(OH)] \cdot 0.2H_2O$**, has the cell dimensions a=1256.2, b=1041.8, and c=312.8 pm; β=95.1°; the IR spectrum and thermal analysis curves are given [2]. Szaibelyite, containing ca. 1.5 wt% MnO and ca. 3.5 wt% FeO, forms an isomorphous series with sussexite, $Mn^{II}_2(OH)[B_2O_4(OH)]$, and possibly an unknown mineral, $B_2O_3 \cdot 2FeO \cdot H_2O$ [3]. The pseudosymmetry and polymorphism of szaibelyite has been described earlier [4].

The crystal structure of the mineral danburite, **$Ca[B_2Si_2O_8] \cdot nH_2O$**, with two corner-sharing tetrahedra in the $[B_2O_7]^{8-}$ group, was studied at temperatures below 900°C [6].

The mineral lüneburgite, **$Mg_3[B_2O(OH)_4|(PO_4)_2] \cdot 6H_2O$**, discovered in a mirabilite deposit in Shansi, China, has been studied by microscopic laser spectral analysis, optical microscopy, scanning electron microscopy (SEM), differential thermal, chemical, and IR spectral analyses. It is monoclinic with a=971.0, b=755.3, and c=992.2 pm; β=97.10°. The strongest lines in the powder diffraction pattern are 5.00(10), 4.87(9), 3.25(8), 3.03(8), 2.83(9), and 1.943(7); the main IR absorption bands are at 3399, 3232, 1663, 1122, 1074, and 1019 cm^{-1} [7].

References for 3.6.3:

[1] Hawthorne, F. C. (Can. Mineral. **24** [1986] 625/42).
[2] Lisitsyn, A. E.; Rudnev, V. V.; Gaft, A. L.; Dobrovol'skaya, N. V.; Dara, O. M.; Tkacheva, T. V. (Zap. Vses. Mineral. Obshch. **114** [1985] 62/73 from C.A. **103** [1985] No. 9130).
[3] Kwak, T. A. P.; Nicholson, M. (Mineral. Mag. **52** No. 368 [1988] 713/6 from C.A. **110** [1989] No. 11179).
[4] Takéuchi, Y.; Kudoh, Y. (Am. Mineral. **60** [1975] 273/9).
[5] Miyawaki, R.; Nakai, I.; Nagashima, K. (Acta Crystallogr. C **41** [1985] 13/5).
[6] Sugiyama, K.; Takéuchi, Y. (Z. Kristallogr. **173** [1985] 293/304).
[7] Wei, Dang Yan (Yanshi Kuangwu Ji Ceshi **4** [1985] 313/8, 1 plate from C.A. **104** [1986] No. 189866).

3.6.4 Triborate Ions

For triborates, see also Section 3.6.1, pp. 259/60.

In the original literature, different representations for triborates are used. In this section, alternative formulations, e.g., $K[B_3O_3(OH)_4] \cdot H_2O$ and $K[B_3O_5] \cdot 3H_2O$, are both given.

3.6.4.1 The Ion $[B_3O_3(OH)_4]^-$

The mobilities of the ion $[B_3O_3(OH)_4]^-$ (**40**) at different temperatures and different ionic strengths are calculated using previously published equations for calculating the mobilities of composed ions [16].

$K[B_3O_3(OH)_4]\cdot H_2O$, or $K[B_3O_5]\cdot 3H_2O$, was prepared in quantitative yield in a solid state reaction in an atmosphere of more than 60% (preferably 100%) relative humidity at 25 to 150°F (−4 to 65°C) by mixing $K[B_5O_8]\cdot 4H_2O$ and, e.g., $K_2[B_4O_7]\cdot 4H_2O$ (both ground to 0.05 nm particle size; 1:1 mole ratio) in a reaction chamber for 72 h. At 50 to 60% humidity and room temperature, no reaction of the same mixture was observed after two months [1]. The structure of this compound has been determined at −110°C; it crystallizes in the monoclinic space group C2/c−C_{2h}^6 (No. 15) with a=1554.0(5), b=682.1(2), and c=1427.3(4) pm; β=104.44(2)°; Z=8; D_m=1.78(5) g/cm³ and D_c=1.86 g/cm³; from 1165 reflections R=2.9% and R_w=4.1%. The structure contains an isolated anion $[B_3O_3(OH)_4]^-$, **40**, formed from a nearly planar B_3O_3 ring consisting of one tetrahedral (B–O distances 145.3, 146.2, 147.5, and 148.9 pm) and two trigonal (B–O distances 134.7, 138.3, 139.3 pm; 135.2, 135.6, 139.2 pm) boron atoms. There are two independent K^+ ions, one is eight-coordinate to two free water molecules and six oxygen atoms from the anion and one is approximately octahedrally coordinated to six oxygen atoms from the anion. Six intermolecular hydrogen bonds connect the crystalline network [2].

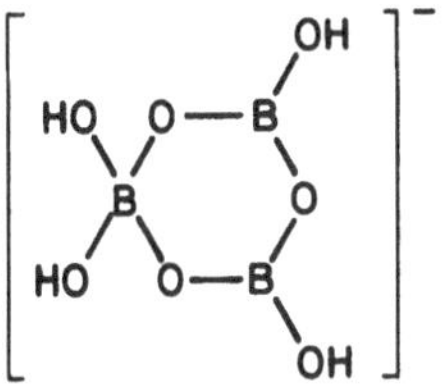

40

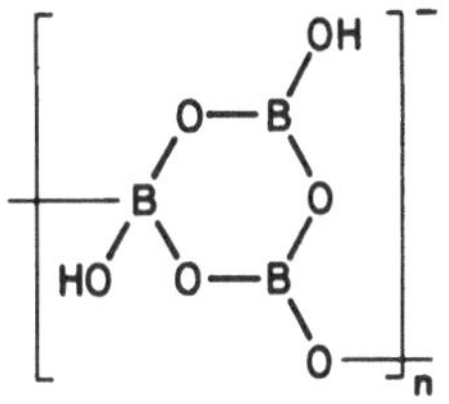

41

3.6.4.2 The Ion $[B_3O_4(OH)_2]^-$

In **$Tl[B_3O_4(OH)_2]\cdot 0.5H_2O$**, the structural unit is the well-known B_3O_3 ring which is linked to form an infinite chain as shown in **41** [3].

3.6.4.3 The Ions $[B_3O_3(OH)_5]^{2-}$ and $[B_3O_4(OH)_3]^{2-}$

There exist two minerals of the formula **$Mg[B_3O_3(OH)_5]\cdot 5H_2O$**: the monoclinic inderite and the triclinic kurnakovite.

Synthetic $Mg[B_3O_3(OH)_5]\cdot 5H_2O$ is crystallized from a 0.3M solution of $K_2[B_4O_7]$ and $MgCl_2$, while 0.1M solutions gave supersaturated solutions. It is suggested that this deviation is caused by a partial conversion of the tetraborate ion into the triborate ion under the action of Mg^{2+} ions [4]. The structure of synthetic $Mg[B_3O_3(OH)_5]\cdot 5H_2O$ was determined by X-ray diffraction methods. It is monoclinic with a=1193(3), b=1290(10), and c=674(5) pm; β=104.5° and contains the ion **42**. The structure of synthetic $Mg[B_3O_3(OH)_5]\cdot 5H_2O$ is identical with that of the mineral inderite [5].

The specific heat of the triclinic mineral kurnakovite, $Mg[B_3O_3(OH)_5]\cdot 5H_2O$ or $2MgO\cdot 3B_2O_3\cdot 15H_2O$, was measured between 65 and 310 K and the enthalpy, Gibbs energy, and entropy data at intervals of 5 K were calculated. At 299.15 K, the following values were obtained: enthalpy, ΔH=118.8±0.2 kJ/mol; Gibbs energy, ΔG=327±5.4 kJ/mol; entropy, ΔS=723.4±4.8 J·mol⁻¹·K⁻¹ [17].

References on p. 273

$Ba[B_3O_3(OH)_5]\cdot H_2O$, or $3B_2O_3\cdot 2BaO\cdot 7H_2O$, measured density $D_m = 2.90$ g/cm³, was prepared from $Ba(OH)_2$, NH_3, and H_3BO_3 (2:1:6 mole ratio) in aqueous solution. It has an ion analogous to the mineral meyerhofferite, **$3B_2O_3\cdot 2CaO\cdot 7H_2O$**, as was shown by differential thermal analysis (DTA), thermogravimetry, infrared spectra, and X-ray diffraction analysis [6].

Synthetic strontium-colemanite, $Sr_2B_6O_{11}\cdot 5H_2O$, or **$Sr[B_3O_4(OH)_3]\cdot H_2O$**, was prepared from aqueous solutions of $Sr(OH)_2$ and H_3BO_3 under thermal conditions and in concentration ranges in which the parent ring $[B_3O_3(OH)_5]^{2-}$, **42**, exists. The polyborate skeleton of the colemanite structure may be described as a succession of $[B_3O_4(OH)_3]^{2-}$ rings (linked together by a common oxygen atom) so as to form infinite polyborate chains. The formation of such chains may be considered as resulting from polycondensation (dehydration) reactions of isolated ring ions $[B_3O_3(OH)_5]^{2-}$. $Sr_2[B_6O_{11}]\cdot 5H_2O$ is monoclinic, space group $P2_1/a-C^5_{2h}$ (No. 14) with a = 879.4(1), b = 1147.1(1), c = 626.30(8) pm, and β = 108.53(2)°; Z = 2. It is isotypic with the mineral colemanite, $Ca_2B_6O_{11}\cdot 5H_2O$, or $Ca[B_3O_4(OH)_3]\cdot H_2O$. The IR spectrum is too complex to allow detailed assignments. The thermal decomposition of the two compounds was compared [7].

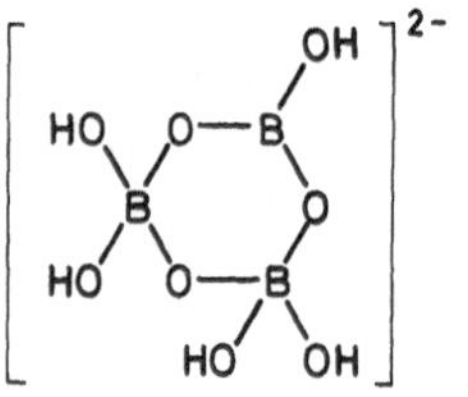

42

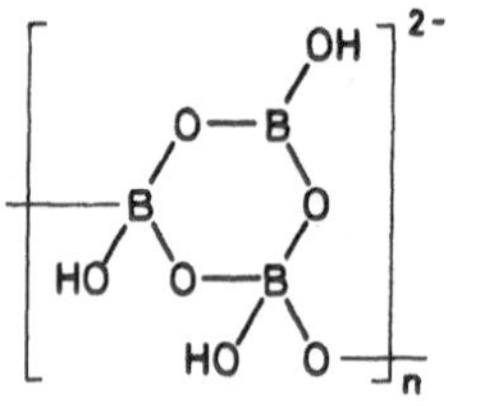

43

Boron-11 NMR studies on powdered ferroelectric colemanite, **$Ca[B_3O_4(OH)_3]\cdot H_2O$**, containing the anion **43**, show the effects of the radiofrequence pulse length on the NMR line intensity measurement for the spin I = 3/2. The pulse length is shown to be a critical parameter for the quantitative determination in a quadrupolar spin system [8]. The dissolution kinetics of original and calcined samples of colemanite in CO_2-saturated water depend on particle size, calcination temperature, and reaction time; the activation energy of the sample calcined at 400°C was 57.7 kJ/mol [9].

The compounds **$CaMg[B_3O_4(OH)_3]_2\cdot 3H_2O$**, or $CaMg[B_6O_{11}]\cdot 6H_2O$, and **$CaMg[B_3O_3(OH)_5]_2\cdot 6H_2O$**, or $CaMg[B_6O_{11}]\cdot 11H_2O$, were studied by differential thermal analysis (DTA), thermogravimetry (TG), differential thermogravimetry (DTG), and X-ray phase analysis in order to explain the effect of structural features and the nature of the water on the stability of their structures during heating in air: $CaMg[B_3O_3(OH)_5]_2\cdot 6H_2O$ is stable only up to 100±10°C, whereas $CaMg[B_3O_4(OH)_3]_2\cdot 3H_2O$ is stable to 300±20°C. These differences in their thermal stabilities are explained by structural differences [10]. The thermal deformation and crystal structures of $Ca[B_3O_4(OH)_3]\cdot H_2O$ and $CaMg[B_3O_4(OH)_3]_2\cdot 3H_2O$ were compared. The thermal expansion depends on the crystal structure characteristics. Conclusions concerning the regularities of the chain borate **43** and the thermal deformation are given [11]. A thermal expansion study of $CaMg[B_3O_4(OH)_3]_2\cdot 3H_2O$ between 18 and 300°C shows that dissociation occurs at 270°C [12].

The thermal decomposition of colemanite monocrystals has also been investigated by optical microscopy, thermal, X-ray, and IR methods between room temperature and 405°C. The first endothermic peak at 369°C is due to the formation of H_2O from the OH groups. A peak at 386°C is accompanied by a 24.1 wt% loss due to the release of water by breaking of water bonds and borate chain bonds and subsequent removal of both kinds of the formed water from the anhydrous phase of the preserved borate structure [13].

The mineral hydroboracite (?), **$CaMg[B_3O_4(OH)_3]_2$**, is monoclinic and has the lattice parameters a=1176, b=666, c=823 pm, and β=102.81° [14]. The low-temperature heat capacity of hydroboracite (?), **$CaMg[B_3O_3(OH)_5]_2 \cdot H_2O$**, was determined experimentally to be C_p°=452.03 ±0.73 $J \cdot mol^{-1} \cdot K^{-1}$; ΔS°=382.68±0.44 $J \cdot mol^{-1} \cdot K^{-1}$, and ΔH°=65.993±0.037 kJ/mol [15]. For hydroboracite, **$CaMg[B_3O_4(OH)_3]_2 \cdot 3H_2O$**, see Table 3/26, p. 294.

References for 3.6.4:

[1] Salentine, C. G. (U.S. 4640827 [1985/87] 1/3 from C.A. **106** [1987] No. 140544).
[2] Salentine, C. G. (Inorg. Chem. **26** [1987] 128/32).
[3] Touboul, M. (Phosphorus Sulfur **28** [1986] 145/9).
[4] Gode, H.; Bernane, A. (Latv. PSR Zinat. Akad. Vestis Kim. Ser. **1989** 273/4).
[5] Gode, H.; Majore, I.; Sokolova, E. V.; Yamnova, N. A. (Latv. PSR Zinat. Akad. Vestis Kim. Ser. **1986** 532/3).
[6] Gode, H.; Shvirkst, Ya. Ya. (Latv. PSR Zinat. Akad. Vestis Kim. Ser. **1985** 14/7).
[7] Liegeois-Duyckaerts, M.; Rulmont, A. (Eur. J. Solid State Inorg. Chem. **25** [1988] 289/96 from C.A. **110** [1989] No. 184715).
[8] Man, P. P. (J. Magn. Reson. **77** [1988] 148/54 from C.A. **108** [1988] No. 230749).
[9] Alkan, M.; Kocakerim, M. M.; Colak, S. (J. Chem. Technol. Biotechnol. Chem. Technol. **35A** [1985] 382/6 from C.A. **104** [1986] No. 54092).
[10] Kondrat'eva, V. V. (Trudy Leningr. Obshch. Estestvoispyt. **79** No. 2 [1986] 93/9 from C.A. **107** [1987] No. 227909).

[11] Filatov, S. K.; Kondrat'eva, V. V. (Latv. PSR Zinat. Akad. Vestis Kim. Ser. **1987** 280/3).
[12] Kondrat'eva, V. V.; Filatov, S. K. (Izv. Akad. Nauk SSSR Neorg. Mater. **22** [1986] 273/6 from C.A. **104** [1986] No. 120313).
[13] Waclawska, I.; Stoch, L.; Paulik, J.; Paulik, F. (Thermochim. Acta **126** [1988] 307/18).
[14] Fischer, W.; Thomas, R.; Loos, G. (Chem. Erde **48** [1988] 333/8 from C.A. **110** [1989] No. 26763).
[15] Gurevich, V. M.; Gorbunov, V. E.; Aksemova, T. D.; Gavrichev, K. S.; Khodanovskii, I. L. (Zh. Fiz. Khim. **62** [1988] 3110/3; Russ. J. Phys. Chem. **62** [1988/89] 1625/6).
[16] Michov, B. (Electrophoresis **9** [1988] 105/6).
[17] Huang, Shufeng; Zhang, Qiang; Li, Yarong; Chen, Peiheng (Huaxue Xuebao **46** [1988] 967/71 from C.A. **110** [1989] No. 42175).

3.6.5 Tetraborate Ions

For the formation of tetraborates in H_3BO_3-containing systems, see also Section 3.5.7.4, p. 203.

In the original literature, different writings of tetraborates can be found. In this section, alternative formulations, e.g., $Li_2[B_4O_5(OH)_4] \cdot H_2O$ and $Li_2[B_4O_7] \cdot 3H_2O$, or for borax, Na_2-$[B_4O_7] \cdot 10H_2O$ and $Na_2[B_4O_5(OH)_4] \cdot 8H_2O$, are both used.

3.6.5.1 The Ion $[B_4O_5(OH)_4]^{2-}$

Ab initio Hartree-Fock calculations by SCF/DZ, CT/DZ, SCF/DZP, and by CI/DZP (SCF = self-consistent field, CI = configuration interaction, CT = charge transfer, DZ = double-zeta level, DZP = double-zeta plus polarization level with an extended basis set) gave a value of 2.65 MHz for the nuclear quadrupole coupling constant of $[B_4O_5(OH)_4]^{2-}$ (**44**) with C_{2v} symmetry; the calculated B–O distances are 2.719, 2.761, and 2.804 a.u.; 2.553 and 2.595 a.u.; the O–B–O angles are 104.2°, 109.4°, 110.8°, and 111.9°; 116.1° and 125.8° [1].

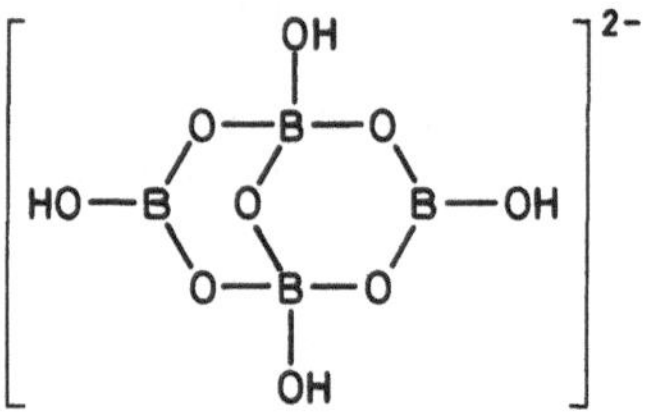

44

Crystalline **β-$Li_2[B_4O_5(OH)_4]\cdot H_2O$**, or β-$Li_2[B_4O_7]\cdot 3H_2O$, was prepared by introducing solid H_3BO_3 and $LiOH\cdot H_2O$ into water (pH 6.28, c = 1×10^{-6} to 2×10^{-6} M) and reacting the mixture at a constant temperature of 45°C. The species is an isomer of the stable α-$Li_2[B_4O_7]\cdot 3H_2O$; the transformation of the β-isomer into the α-form at 80°C was studied, and the physical and chemical properties of both compounds are discussed [2].

The NMR-NQR spectra of **$Na_2[B_4O_5(OH)_4]\cdot 8H_2O$**, or $Na_2[B_4O_7]\cdot 10H_2O$, and **$K_2[B_4O_5(OH)_4]\cdot 2H_2O$**, or $K_2[B_4O_7]\cdot 4H_2O$, are reported for 110 K; as quadrupole coupling constants show, the boron atoms are located at both trigonal and tetragonal sites in the two cited compounds [6]. The efficiency of the double NQR method for studying the transition of ^{10}B, ^{11}B, ^{23}Na, and ^{39}K in both compounds was demonstrated. An automated method is described for interpreting the spectra of nuclei with a spin number of three; the method was used to determine the quadrupole couplings and asymmetry parameters of ^{10}B at two nonequivalent couplings and asymmetry parameters of ^{10}B at two nonequivalent sites in $Na_2[B_4O_7]\cdot 10H_2O$ and four such sites in $K_2[B_4O_7]\cdot 4H_2O$ [7]. On the other hand, the magnetic relaxation times determined were interpreted in terms of the presence of a $[\{B_4O_5(OH)_4\}_n]^{2n-}$ polyanion [8].

Examination of the intracrystalline site preference of hydrogen isotopes in borax show that the hydrogen atoms in **44** are much more enriched in deuterium than those of the $[Na_2(OH_2)_8]^{2+}$ ion; the mechanism of the dehydration of borax is discussed [3].

Twice recrystallized $Na_2[B_4O_5(OH)_4]\cdot 8H_2O$, or $Na_2[B_4O_7]\cdot 10H_2O$, dried and kept over a saturated solution of saccharose and NaCl, was characterized as a pH standard of 9.278 ± 0.0042, 9.178 ± 0.0027, and 9.043 ± 0.0054 at 15, 25, and 45°C, respectively, for the primary standardization of pH meters. At 25°C the buffer capacity was determined to be 0.020 mol/pH, and the dilution factor $pH_{1/2}$ = 0.01 [4].

Very pure samples of $K_2[B_4O_5(OH)_4]\cdot 2H_2O$, prepared from KOH and H_3BO_3, were used to redetermine the density as 1.94 g/cm³ and refraction index values as N_g = 1.478 and N_p = 1.465; infrared spectra, Debye crystallograms and derivatograms were also obtained [9].

$K_{1.67}Na_{0.33}[B_4O_5(OH)_4]\cdot 3H_2O$ crystallizes in the hexagonal space group $P\bar{6}2c-D^2_{3h}$ (No. 188) with a = 1127.8(8) and c = 1580.6(10) pm; Z = 6; D_m = 1.83 g/cm³ (flotation) and D_c = 1.821 g/cm³; refined to R = 4.3% and R_w = 6.3%. The structure is strongly related to that of the mineral tincalconite, $Na_2[B_4O_5(OH)_4]\cdot 3H_2O$ [5].

Crystalline **$Fe_2^{III}[B_4O_5(OH)_4]_3 \cdot 10H_2O$**, or $Fe_2O_3 \cdot 6B_2O_3 \cdot 16H_2O$, ($I_{max}$=40.9°), **$K_2Fe_2^{III}[B_4O_5(OH)_4]_4 \cdot 24H_2O$**, or $K_2O \cdot Fe_2O_3 \cdot 8B_2O_3 \cdot 32H_2O$, ($I_{max}$=41.8°), and amorphous $Fe(BO_2)_3 \cdot 4H_2O$, or $Fe_2O_3 \cdot 3B_2O_3 \cdot 8H_2O$, were prepared from $K_2[B_4O_7] \cdot 4H_2O$ and $FeCl_3$ between pH 7.1 and 11.7 and were characterized by infrared and X-ray diffraction data [10].

Ethylenediammonium tetraborate, **$[NH_3CH_2CH_2NH_3][B_4O_5(OH)_4]$**, is monoclinic, space group B2/b-C_{2h}^6 (No. 15) with a=1185.8(2), b=1022.2(2), and c=883.6(1) pm, γ=92.69(2)°; Z=4; D_c=1.57 g/cm³. The structure was solved by direct methods and refined to R=3.3% by full-matrix least-squares; the anion **44** has mm2 symmetry. The tetraborate ions are linked to the ethylenediammonium ions by two hydrogen bonds. The B–O distances are 136.0, 136.6, and 137.2 pm ($HOBO_2$); 143.5, 145.9, 148.8, and 150.1 pm ($HOBO_3$) [11].

Guanidinium tetraborate dihydrate, **$[C(NH_2)_3]_2[B_4O_5(OH)_4] \cdot 2H_2O$**, is triclinic, space group P$\bar{1}$-C_i^1 (No. 2) with a=844.3(10), b=1256.0(15), and c=729.0(10) pm; α=106.76(5)°, β=103.87(6)°, and γ=90.38(7)°; Z=2; D_m=1.60(1) g/cm³ and D_c=1.611 g/cm³; final R=5.2% for 1712 reflections. The discrete anion contains two tetrahedral boron atoms with an average B–O distance of 147.5 pm, and two trigonal boron atoms with an average B–O distance of 136.5 pm. All of the hydrogen atoms (except two hydrogen atoms per one cation) participate in hydrogen bonding [12].

3.6.5.2 The Ions $[B_4O_6(OH)_2]^{2-}$, $[B_8O_{12}(OH)_4]^{4-}$, and $[B_{12}O_{18}(OH)_6]^{6-}$

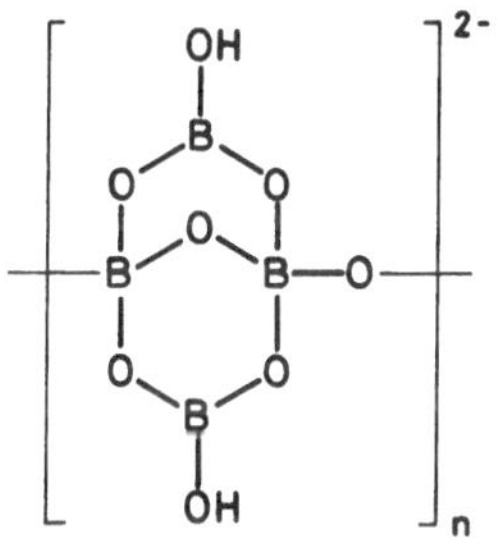

45

Cross-polarization magic angle spinning nuclear magnetic resonance (CP-MAS-NMR) measurements can discriminate between quadrupole nuclei in sites with large versus small electrostatic field gradients. This was demonstrated on the mineral kernite, **$Na_2[B_4O_6(OH)_2] \cdot 3H_2O$**, containing the anion **45** [13].

$Tl_2[B_4O_6(OH)_2] \cdot 2H_2O$, crystallized from aqueous solutions of H_3BO_3 and TlOH, has been characterized by X-ray diffraction analysis.

Thermal analytic methods show loss of H_2O below 265 °C and decomposition of the remaining $Tl_2[B_4O_7]$ between 510 and 520°C with melting to give $Tl_4[B_2O_5]$ and B_2O_3 [14]. In the 100°C isotherm of the Tl_2O–B_2O_3–H_2O system, two distinct compounds were found: $Tl_2[B_4O_6(OH)_2] \cdot 2H_2O$, or $Tl_2[B_4O_7] \cdot 3H_2O$, and **$Tl_4[B_8O_{12}(OH)_4] \cdot H_2O$**, or $Tl_2[B_4O_7] \cdot 1.5H_2O$; the latter polyion may be considered as the dimer of the first (see Section 3.6.8, p. 280). Thermal gravimetry and solution experiments, both under pressure, allow the prediction that another hydrated Tl(I) diborate, $Tl_2[B_4O_7] \cdot H_2O$, exists with the possible structural formula **$Tl_6[B_{12}O_{18}(OH)_6]$**. Actually, only monocrystals of $Tl_4[B_8O_{12}(OH)_4] \cdot H_2O$ were obtained hydrothermally from $Tl_2[B_4O_6(OH)_2] \cdot 2H_2O$ [15].

References on p. 276

References for 3.6.5:

[1] Gajhede, M. (Chem. Phys. Lett. **120** [1985] 266/71).
[2] Ou, Y.-Ch. (Xiangtan Daxue Ziran Kexue Xuebao **1** [1986] 132/40; J. of Xiangtan Univ. **1** [1986] 132/40).
[3] Pradhananga, T. M.; Matsuo, S. (J. Phys. Chem. **89** [1985] 72/6).
[4] Csefalvayova, B.; Thurzo, A.; Tarabkova, E. (Chem. Listy **83** [1989] 550/3).
[5] Smykalla, C.; Behm, H. (Z. Kristallogr. **183** [1988] 51/61).
[6] Anferov, V. P.; Beloglazov, G. S.; Grecheshkin, V. S. (Sovrem. Metody YaMR EPR Khim. Tverd. Tela. Mater. 4th Vses. Koord. Soveshch, Noginsk, USSR, 1985, pp. 135/7 from C.A. **103** [1985] No. 226096).
[7] Anferov, V. P.; Grecheshkin, V. S.; Beloglazov, G. S. (Zh. Fiz. Khim. **60** [1986] 2750/5; Russ. J. Phys. Chem. **60** [1986/87] 1653/6).
[8] Anferov, V. P.; Beloglazov, G. S.; Grecheshkin, V. S. (Izv. Vyssh. Uchebn. Zaved. Fiz. **29** No. 12 [1986] 3/9 from C.A. **106** [1987] No. 187673).
[9] Gode, H.; Sprice. D. (Latv. PSR Zinat. Akad. Vestis Kim. Ser. **1987** 275/9).
[10] Chudnovskaya, O. N.; Baev, A. K.; Barkatina, E. N. (Vestsi Akad. Navuk BSSR Ser. Khim. Navuk **1985** No. 2, pp. 3/7).

[11] Silina, Ye. Ya.; Bel'skii, V. K.; Tetere, I. V.; Ozolins, G. V. (Latv. PSR Zinat. Akad. Vestis Kim. Ser. **1985** 399/404).
[12] Weakley, T. J. R. (Acta Crystallogr. C **41** [1985] 377/9).
[13] Woessner, D. E. (Z. Phys. Chem. [N.F.] **152** [1987] 51/8).
[14] Gode, H.; Shvirkst, Ya. Ya.; Antropa, A. (Latv. PSR Zinat. Akad. Vestis Kim. Ser. **1985** 545/7).
[15] Touboul, M.; Amoussou, D. (J. Therm. Anal. **31** [1986] 125/30).

3.6.6 Pentaborate Ions

In the original literature, different representations for pentaborates are used. In this section, alternative formulations, e.g., $K[B_5O_6(OH)_4]\cdot 2H_2O$ and $K[B_5O_8]\cdot 4H_2O$, are both given. In papers dealing with applications, e.g., in the field of nonlinear optics, typically the latter formulation, which does not take in account structural features of H_2O distribution, is preferred.

3.6.6.1 The Ion $[B_5O_6(OH)_4]^-$

$\mathbf{K[B_5O_6(OH)_4]\cdot 2H_2O}$ and $\mathbf{[NH_4][B_5O_6(OH)_4]\cdot 2H_2O}$ (both with the anion **46**; see p. 278) were prepared from a suspension of H_3BO_3 in H_2O with KOH or aqueous NH_3, respectively, at pH 9 followed by the addition of acetylacetone; their characterization and structural assessment are based on the results of elemental analysis, of conductivity measurements, and of infrared and laser-Raman spectroscopy [3]. Ammonium pentaborate is an effective intumescent char-forming flame retardant additive for thermoplastic polyurethanes [11]. For NMR spectra of $[NH_4][B_5O_8]$, see Section 3.6.1, p. 261.

Large twin-free single crystals of $K[B_5O_6(OH)_4]\cdot 2H_2O$ were grown in an aqueous solution by slow cooling. Crystals of 30 × 27 × 25 mm size have a transmittance of 72 to 95% between 190 and 1400 nm [1]. The layered growth of the (010) face of these single crystals was studied by optical microscopy. The overgrowth of the (010) face, examined by structural analysis, was mainly the growth of twins with the ($0\bar{1}1$) face [2].

Structure-sensitive parameters of a saturated solution of potassium pentaborate were measured at room temperature at various acidity values. The osmotic pressure, the density, and the surface tension all have a sharply monotonic character and can be correlated with the crystal growth and nucleation processes. The best crystals grow from the least associated solutions, i.e., from solutions which have the largest value of osmotic pressure [4].

A review of the nonlinear optical properties and typical applications of $K[B_5O_6(OH)_4]\cdot 2H_2O$, or $K[B_5O_8]\cdot 4H_2O$ and **$K[B_5O_6(OD)_4]\cdot 2D_2O$**, or $K[B_5O_8]\cdot 4D_2O$, is available. Formulas are presented for the calculation of the phase-matching angles of the biaxial crystals upon propagation of the radiation in principal planes [5]. Optical nonlinear transformation of third-harmonic generation laser radiation into UV wavelengths has been tested with $K[B_5O_8]\cdot 4H_2O$ [6]. Nd^{3+}-doped $YAlO_3$ laser picosecond 5th-harmonic generation was investigated by three different methods in $K[H_2PO_4]$ and $K[B_5O_8]\cdot 4H_2O$ crystals. The optimal method was explained which yielded the 5th-harmonic ($\lambda = 216$ nm) energy of 0.5 mJ; the pulse duration was 15 ps [7].

Based on the ionic group theory, the second harmonic generation (SHG) coefficients of the $K[B_5O_8]\cdot 4H_2O$ crystals were calculated to give the values: $d_{31} = 2.61\times10^{-10}$ esu, $d_{32} = 0.07\times10^{-10}$ esu, and $d_{33} = 3.26\times10^{-10}$ esu, which agree fairly well with the experimental data. The calculated data indicate that the hydrogen bond of the B_5O_{10} ionic group has a large effect on the coefficients of the $K[B_5O_8]\cdot 4H_2O$ crystal; thus it is possible to determine the extent to which the hydrogen belongs to the ionic group. The probability that the hydrogen belongs to the ionic group is about 40 to 60%. Furthermore, the reasons for too small SHG coefficients of $K[B_5O_8]\cdot 4H_2O$ were analyzed. In addition to the small microscopic SHG coefficients of the B_5O_{10} group, the space arrangement of this group is in an unfavorable configuration so that the largest microscopic SHG coefficients, χ_{123}, cancel each other. Some structural requirements are proposed to search for a new type nonlinear optical crystal among the B–O compounds composed of six-membered rings, each having three boron atoms in trigonal and tetrahedral coordination; see $[B_3O_5]^-$ and $[B_3O_6]^{3-}$, pp. 99/116 [8].

Ultrashort tunable pulses of light were generated between 220 and 266 nm with frequency doubling in these crystals of pulses produced by a $K[H_2PO_4]$ parametric oscillator pumped by the second harmonic generation (SHG) of a $YAG:Nd^{3+}$ laser. In the stimulated Raman scattering (SRS) spectrum of this crystal in the Stokes range, the frequency shift of 551 ± 1 cm^{-1} indicates that excitation of the pulsating vibration with frequency ν_p occurs, belonging to symmetry A_1. Besides this frequency, frequencies corresponding to the stretching vibration ν_s of BO_3 of symmetry A_1 between 915 and 921 cm^{-1}, and the bending mode ν_δ (BO_4?) between 1095 and 1109 cm^{-1} are observed. A further bending mode, ν_σ, is between 364 and 370 cm^{-1} of symmetry A_1 (BO_3 or BO_4) [9]. The SHG with λ up to 218 nm from a dye laser using a crystal of $K[B_5O_8]\cdot 4H_2O$ is reported; the influence of power density, wavelength, line width, wave mismatching, and the method of the main radiation focusing of the SHG efficiency is investigated [10].

H_3BO_3 reacts with $(C_2H_5)_3N$ in water at 25 °C to form a borate with the composition $(C_2H_5)_3N\cdot 2.5B_2O_3\cdot 2.5H_2O$, which was characterized by IR spectra and X-ray diffraction analysis as **$[(C_2H_5)_3NH][B_5O_6(OH)_4]$**. The X-ray diffraction pattern resembles that of hexamethylenetetrammonium pentaborate; $D_m = 1.323$ g/cm³ [12]. The addition of ethylenediamine ($C_2H_8N_2$) to aqueous H_3BO_3 gives $C_2H_8N_2\cdot 2.5B_2O_3\cdot 4H_2O$, which is formulated as **$[H_2N\text{–}CH_2CH_2\text{–}NH_3][B_5O_6(OH)_4]\cdot 1.5H_2O$**; $D_m = 1.593$ g/cm³ [13].

References on pp. 278/9

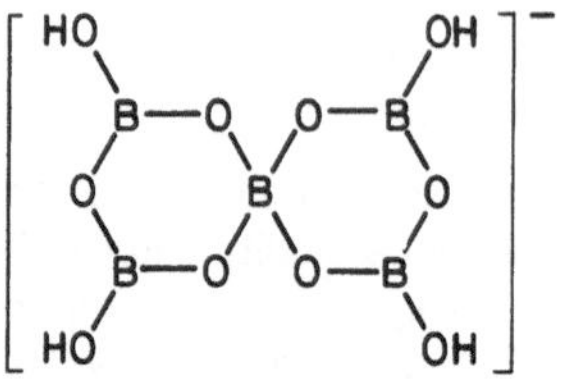

46

The reaction of urotropine (hexamethylenetetramine; $(CH_2)_6N_4$) with boric acid gives crystals with the composition $(CH_2)_6N_4 \cdot 2.5B_2O_3 \cdot 3H_2O$ [14]. This compound crystallizes in the monoclinic space group $P2_1/c-C^5_{2h}$ (No. 14) with a=957.15(2), b=1437.2(1), and c=1155.62(6) pm; β=109.934(4)°; Z=4; D_m=1.623 g/cm³ and D_c=1.636 g/cm³; it was refined to R=4.1% and R_w=6.3% from 1968 independent reflections. In the structure, one nitrogen atom in the hexamethylenetetramine molecule is protonated to give the urotropinium ion. The compound has the structural formula $\mathbf{[(CH_2)_6N_4H][B_5O_6(OH)_4] \cdot 0.5H_2O}$, and the anion has the usual pentaborate structure with the anion **46**. The coordination of the central boron atom, B(1), is tetrahedral, each set of the atoms forming the six-membered rings lie in one plane. The central boron atom deviates from these planes by 11 pm or 24 pm, respectively. The B(1)–O bonds in the flatter ring are shorter (at 145.2(4) pm) than those in the other ring with 148.1(4) pm. The water molecule is disordered between two positions with a side-occupancy factor of 0.5 [15].

3.6.6.2 The Ion $[B_5O_6(OH)_6]^{3-}$

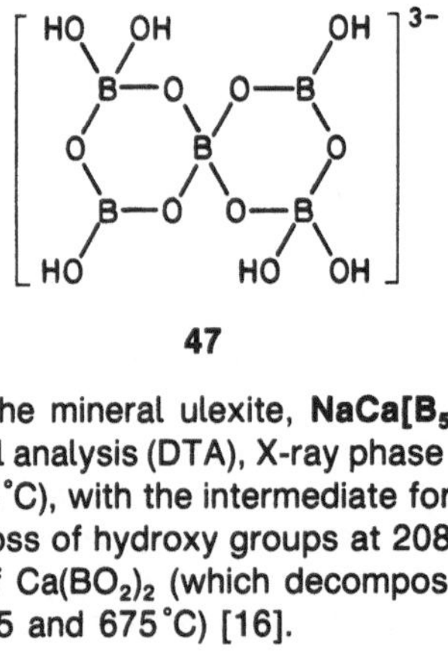

47

The thermal dehydration of the mineral ulexite, $\mathbf{NaCa[B_5O_6(OH)_6] \cdot 5H_2O}$, proceeds, as determined by differential thermal analysis (DTA), X-ray phase analysis, IR spectral and tensimetric data (between 60 and 208 °C), with the intermediate formation of the trihydrate (60 °C) and the monohydrate (120 °C). Loss of hydroxy groups at 208 °C leads to the degradation of the crystal with the formation of $Ca(BO_2)_2$ (which decomposes at 850 °C) and of $Na[B_3O_5]$ (which decomposes between 605 and 675 °C) [16].

References for 3.6.6:

[1] Yuan, Durong (Rengong Jingti **17** [1988] 99/102 from C.A. **111** [1989] No. 205790).
[2] Yuan, Durong; Liu, Xiling (Rengong Jingti **17** No. 1 [1988] 31/6 from C.A. **111** [1989] No. 222366).
[3] Chaudhuri, M. K.; Das, B. (J. Chem. Soc. Dalton Trans. **1988** 243/4).
[4] Ananikyan, T. A.; Nalbandyan, A. G.; Nalbandyan, O. G. (J. Cryst. Growth **73** [1986] 505/9).
[5] Nikogosyan, D. N.; Gurzadyan, G. G. (Kvantovaya Electron. [Moscow] **14** [1987] 1529/41 from C.A. **107** [1987] No. 143793).

[6] Arutyunyan, A. G.; Buniatyan, G. R.; Melkonyan, A. A.; Mesropyan, A. V.; Paityan, G. A. (Nelinein Optich. Vzaimodeistviya Erevan **1987** 135/44 from C.A. **108** [1988] No. 46517).
[7] Arutyunyan, A. G.; Gurzadyan, G. G.; Ispiryan, R. K. (Kvantovaya Electron. [Moscow] **16** [1989] 2493/5 from C.A. **112** [1990] No. 87752).
[8] Wu, Y.-Ch.; Chen, Ch.-T. (Wuli Xuebao **35** [1986] 1/6 from C.A. **104** [1986] No. 159063).
[9] Petrosyan, K. B.; Pokhsraryan, K. M. (Phys. Status Solidi B **127** [1985] K105-K108, 1 plate; Izv. Akad. Nauk Arm. SSR Fiz. **20** [1985] 39/42 from C.A. **103** [1985] No. 45512).
[10] Alekseev, V. A.; Mikhalina, T. I.; Nikiforov, V. G.; Trinchuk, B. F.; Shulenin, A. V. (Zh. Prikl. Spektrosk. **46** [1987] 844/6 from C.A. **107** [1987] No. 86704).

[11] Myers, R. E.; Dickens, E. D.; Licursi, E.; Evans, R. E. (J. Fire Sci. **3** [1985] 432/49 from C.A. **104** [1986] No. 187898).
[12] Skvortsov, V. G.; Tsekhanskii, R. S.; Molodkin, A. K.; Sukova, L. M.; Petrova, O. V.; Dolganev, V. P. (Zh. Neorg. Khim. **32** [1987] 491/4; Russ. J. Inorg. Chem. **32** [1987] 272/4).
[13] Tsekhanskii, R. S.; Skvortsov, V. G.; Molodkin, A. K.; Sukova, L. M.; Rodionov, N. S. (Zh. Neorg. Khim. **30** [1985] 1609/12; Russ. J. Inorg. Chem. **30** [1985/86] 917/9).
[14] Skvortsov, V. G.; Tsekhanskii, R. S.; Molodkin, A. K.; Petrova, O. V.; Akimov, V. M. (Zh. Neorg. Khim. **30** [1985] 277/9; Russ. J. Inorg. Chem. **30** [1985] 158/9).
[15] Batsanov, A. S.; Petrova, O. V.; Struchkov, Yu. T.; Akimov, V. M.; Molodkin, A. K.; Skvortsov, V. G. (Zh. Neorg. Khim. **31** [1986] 1120/6; Russ. J. Inorg. Chem. **31** [1986] 637/40).
[16] Saiko, I. G.; Kononova, G. N.; Burlya'ev, V. V.; Tavrovskaya, A. Ya. (Zh. Neorg. Khim. **30** [1985] 1149/53; Russ. J. Inorg. Chem. **30** [1985] 649/52).

3.6.7 The Hexaborate Ion $[B_6O_7(OH)_6]^{2-}$

For hexaborates, see also Section 3.6.1, p. 259.

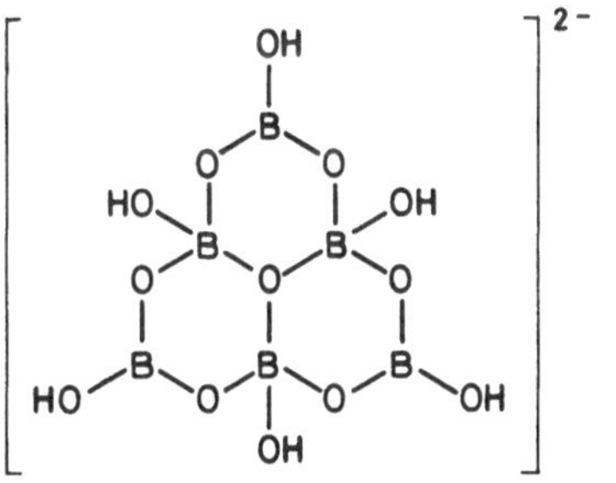

48

The thermal decomposition of synthetic aksaite, $\mathbf{Mg[B_6O_7(OH)_6]\cdot 2H_2O}$, containing the anion **48**, was studied by high-temperature X-ray diffraction analysis and thermogravimetry. Between 160 and 200 °C, one equivalent of water is lost to form the first unstable intermediate monohydrate. The next three equivalents are lost between 250 and 320 °C to give amorphous $\mathbf{Mg[B_6O_7(OH)_6]}$ as a second unstable intermediate; up to 600 °C stable and amorphous $Mg[B_6O_{10}]$ is formed, which decomposes near 700 °C to yield crystalline $Mg[B_4O_7]$ and amorphous B_2O_3 [1].

$\mathbf{\{Mg(H_2O)_4[B_6O_7(OH)_6]\}_2\cdot H_2O}$, or $MgO\cdot 3B_2O_3\cdot 7.5H_2O$, reacts in aqueous solutions of potassium borate to give the compound $K_2O\cdot MgO\cdot 3B_2O_3\cdot 9H_2O$ [2].

References on p. 280

For the formation of hydrated calcium hexaborates in H_3BO_3-containing systems, see also Section 3.5.7.4, p. 203.

The monoclinic modification of **$\{Ni(H_2O)_4[B_6O_7(OH)_6]\}\cdot H_2O$** has the space group $P2_1/b-C^5_{2h}$ (No. 14) with a = 893.6(2), b = 2175.5(6), and c = 719.4(2) pm; β = 99.96(2)°; Z = 4; R = 2.0%. The crystal structure is formed from neutral asymmetric structural units $Ni(H_2O)_4[B_6O_7(OH)_6]$, which are stabilized by intramolecular hydrogen bonds of 291.0 pm to the molecules of crystallization water. The octahedral environment of the nickel atom consists of two OH apices of the boron tetrahedra of **48** and of four H_2O molecules [3].

References for 3.6.7:

[1] Abdulla'ev, G. K.; Aga'ev, A. M. (Zh. Neorg. Khim. **30** [1985] 330/4; Russ. J. Inorg. Chem. **30** [1985] 184/6).
[2] Gode, H.; Majore, I. (Latv. PSR Zinat. Akad. Vestis Kim. Ser. **1988** 276/8).
[3] Silina, E.; Zavodnik, V. E.; Ozolins, G.; Tetere, I. (Latv. PSR Zinat. Akad. Vestis Kim. Ser. **1988** 404/10).

3.6.8 Additional Borates

For salts and complexes derived from hydroxycarbonic acids and boric acid, see Section 3.5.7.5, pp. 218/9. For the anions $[B_8O_{12}(OH)_4]^{4-}$ and $[B_{12}O_{18}(OH)_6]^{6-}$, see Section 3.6.5.2, p. 275.

The structure of **$Tl_4[B_7O_{10}(OH)_3\cdot OBO(OH)]\cdot H_2O$** or $Tl_2[B_4O_7]\cdot 1.5\,H_2O$ (see Section 3.6.5.2, p. 275) contains a unit composed of three B_3O_3 rings which are linked together by boron atoms; each ring is formed by two BO_4 tetrahedra and one BO_3 triangle. The corresponding fully hydrated polyanion is $[B_7O_9(OH)_7]^{4-}$; the chain is constructed of units linked by $BO_2(OH)$ triangles. The shorthand notation of this borate may be $7{:}\infty_1[(3\Delta+4T)+\Delta]$ or, as was proposed in an earlier publication [1], the compound is $\{Tl_2[B_4O_6(OH)_2]\cdot 2H_2O\}_n$, which corresponds to $Tl_2[B_4O_7]\cdot 3H_2O$ with the shorthand notation $4{:}\infty_2[2\Delta+2T]$ [2].

According to the Gmelin System of Last Position, the following compounds containing Pt, Si, U, Cu, W, or V normally would not be discussed in a "Boron" volume. In the following, only a representative selection of these acids and salts is given.

The compound **$K_3[Pt\{B_7O_{11}(OH)_5\}(OH)]\cdot 3H_2O$** crystallizes in the orthorhombic space group $Pnma-D^{16}_{2h}$ (No. 62) with a = 1928.3(3), b = 1148.0(2), and c = 810.6(2) pm; Z = 4; D_c = 2.66 g/cm³; R = 2.7% for 1458 reflections. The structure contains isolated mononuclear $[Pt\{B_7O_{11}(OH)_5\}(OH)]$ complexes which are held together by potassium ions and water molecules. The borate complex acts as a pentadentate ligand which encloses the Pt nucleus on one side. It is the first example of a true heptaborate which forms a new type of molecular structure with the shorthand notation $7{:}[5\Delta+2T]$ [3].

The mineral howlite, **$Ca_4\{[Si_2B_4O_{10}(OH)_6][B_3O_4(OH)_2]_2\}$**, crystallizes in the space group $P2_1/c-C^5_{2h}$ (No. 14) with a = 1282.0(3), b = 935.1(1), and c = 860.8(2) pm; β = 104.84(2)°; refined to R = 2.3% from 3085 observed reflections. The positions of the five dissimilar hydrogen atoms have been located by difference-Fourier synthesis. The principal structural units are $[Si_2B_4O_{10}(OH)_6]^{6-}$ and $[B_3O_4(OH)_2]^-$ polyanions infinitely connected along the c axis. The polyanions comprise $Si_2B_4O_{12}(OH)_6$ fragments attached to one another to form corrugated walls parallel to the (100) plane, and $B_3O_6(OH)_2$ fragments chained together in the same direction. Single CaO_8 polyhedral chains strengthen the corrugated walls and double CaO_8 chains help to connect adjacent polyanionic chains along the a axis. All OH groups participate in hydrogen bonds of various strengths [4].

$K_6[UO_2\{B_{16}O_{24}(OH)_8\}]\cdot 12H_2O$ is monoclinic, space group $P2_1/n-C^5_{2h}$ (No. 15) with a = 1202.4(6), b = 2645(1), and c = 1254.3(4) pm; β = 94.74(3)° at 190 K; Z = 4; D_c = 2.36 g/cm³; final R = 6.1% and R_w = 6.3% for 6158 reflections. The structure contains isolated mononuclear $[(UO_2)\{B_{16}O_{24}(OH)_8\}]^{6-}$ complex anions [5].

New types of polyanions in oxoborates of transition metal ions such as $[(UO_2)\{B_{16}O_{24}(OH)_8\}]^{6-}$, $[Cu_2\{B_{16}O_{24}(OH)_{10}\}]^{6-}$, $[Cu_2\{B_{18}O_{28}(OH)_8\}]^{6-}$, or $[Cu_4O\{B_{20}O_{32}(OH)_8\}]^{6-}$ were described [6]. **$Tl_4Cu[Cu_2\{B_{18}O_{28}(OH)_8\}]\cdot 10H_2O$** is orthorhombic, space group $Amm2-C^{14}_{2v}$ (No. 38) with a = 1283.6(8), b = 2493.3(9), and c = 748.2(3) pm; Z = 2; D_c = 2.728 g/cm³; R = 6.7% and R_w = 8.2%. The structure contains isolated borate complexes $[Cu_2\{B_{18}O_{28}(OH)_8\}]^{6-}$. Each borate anion complex is ring-shaped with rigid tetraborate subunits and encloses two Cu^{2+} ions, thereby forming a chelate complex [7].

The synthetic compound **$HNa_{10}[B_{18}O_{30}(OH)_5]$**, or $9B_2O_3\cdot 5Na_2O\cdot 3H_2O$, crystallizes in the triclinic space group $P\bar{1}-C^1_i$ (No. 2) with a = 932.59(5), b = 1185.1(1), and c = 669.07(7) pm; α = 74.96(1)°, β = 81.40(1)°, and γ = 83.25(1)°; Z = 2; D_c = 2.337 g/cm³; R = 4.5% for 2872 observed reflections. The structure of the ion $[B_{18}O_{30}(OH)_5]^{11-}$ consists of two different (pentaborate and tetraborate) groups linked together to form tubes running along the c axis. The three-dimensional connection is provided by hydrogen bonds and by the (NaO_n) polyhedra which form dense sheets parallel to the (021) plane. The 3+1 coordination around the B(5) atom is discussed [8].

An icosaborate ion in hydrated potassium and sodium copperpolyborates, **$HK_5[Cu_4O\{B_{20}O_{32}(OH)_8\}]\cdot 33H_2O$** or in **$HNa_5[Cu_4O\{B_{20}O_{32}(OH)_8\}]\cdot 32H_2O$**, was found. The potassium compound is triclinic, space group $P\bar{1}-C^1_i$ (No. 2) with a = 1627.4(7), b = 1799.6(9), and c = 1365.2(5) pm; α = 96.37(5)°, β = 110.93(5)°, and γ = 111.36(5)°; Z = 2; D_m = 1.96 g/cm³ (flotation) and D_c = 1.91 g/cm³; R = 15.5% for 5022 observed reflections. The sodium compound is tetragonal, space group $I\bar{4}-S^2_4$ (No. 82) with a = 1866 and c = 1042 pm; Z = 2; D_m = 1.63 g/cm³ and D_c = 1.67 g/cm³. The structure of the icosaborate ion is the largest known isolated borate; it consists of a porphyrine-like ring, and encloses four Cu ions in a nearly planar Cu_4O moiety thus forming a chelate-like complex. While the anionic structure is clear, the positions of the alkali metal ions and the water molecules are not yet known, although an exact analysis and thermoanalytic determinations confirm the formula [9].

The reaction conditions for the formation of hexagonal **$H_5[BW_{12}O_{40}]\cdot 22H_2O$** from H_3BO_3 and tungsten compounds, are a high pH value, a slight excess of H_3BO_3, and low temperatures, those for the formation of tetragonal **$H_5[BW_{12}O_{40}]\cdot 30H_2O$** are low pH, a large excess of H_3BO_3, and high temperatures. Both of the pure acids have been investigated by thermal analysis, UV, IR, and Raman spectroscopy [10]. **$H_5[BW_{12}O_{40}]\cdot 21H_2O$** was also prepared by electrolysis of solutions of $Na_2[WO_4]\cdot 2H_2O$ and H_3BO_3 as the anode liquor with a Pt anode and an Ni cathode at a current density of 0.04 A/cm² and 35 °C for 12 h, extraction with H_2SO_4 and ether, and vacuum evaporation; the tetragonal acid **$H_5[BW_{12}O_{40}]\cdot 7H_2O$** was also prepared [11]. The compound **$K_8H[BW_{11}O_{39}]\cdot 13H_2O$** is cubic, space group $P\bar{4}3m-T^1_d$ (No. 215) with a = 1071.0(1) pm; Z = 1; D_m = 4.39 g/cm³, D_c = 4.42 g/cm³, and R = 4.5%. The $[BW_{11}O_{39}]^{9-}$ anion has an α-Keggin-type structure, with the boron atom located in the center of an oxygen tetrahedron [12].

Cathodes for water electrolysis in acidic media were activated in a 5×10^{-5} M **$H_5[BW_{12}O_{40}]$** solution [13]. Cyclic ethers were prepared by cyclization of diols over anhydrous $H_5[BW_{12}O_{40}]$; for example, tetrahydrofuran was obtained from butane-1,4-diol at 120 °C [14].

Tungsten-183 NMR spectra of the dark brown six-electron reduced Keggin anion α-**$[BW_{12}O_{40}H_6]^{5-}$** show W^{IV} valences located in one of the four groups of the edge-shared WO_6 octahedra that comprise the Keggin structure [15]. Keggin-structured reduced dark brown

References on p. 283

$[BW_9^{VI}O_{37}W_3^{IV}(OH_2)_3]^{5-}$ contains H_2O molecules. Oxygen-17 NMR spectroscopy in aqueous solution show C_{3v} structure. In nonaqueous solutions, the terminal H_2O molecules are lost, and suitable substrates, e.g., $(CH_3)_2SO$, $(C_6H_5)_2SO$, or $(C_6H_5)_3AsO$, are deoxydized, producing $(CH_3)_2S$, $(C_6H_5)_2S$, or $(C_6H_5)_3As$, respectively, in moderate to high yields and regenerating the oxidized W^{VI} heteropoly anion. Stoichiometric and perhaps catalytic oxygen atom transfer possibilities are considered [16]. The Keggin-type acid $H_5[BW_{12}O_{40}]$ and the $[NH_4]^+$ or Ag^+ salts thereof were prepared and were used for the dehydration of different alcohols [17].

$[(n\text{-}C_4H_9)_4N]_5[BW_{12}O_{40}]$, required for an electrochromic film derived by cathodic deposition at −1.0 V versus Ag, was prepared by mixing $[(n\text{-}C_4H_9)_4N]Br$ with **$K_5[BW_{12}O_{40}]$** in water and recrystallization from CH_3CN [18]. The charge-transfer absorption spectrum of **$[(n\text{-}C_4H_9)_4N]_4K[\alpha\text{-}BW_{12}O_{40}]$** (in acetonitrile solution) shows a maximum at 39100 cm^{-1} with an intensity of 45200 $L \cdot mol^{-1} \cdot cm^{-1}$ [19]. Photoexcitation of the oxygen-to-tungsten charge-transfer band of the Keggin **$[BW_{12}O_{40}]^{5-}$** ion in the presence of CH_3OH (between pH 2 and 9.4) gave **$[BW_{12}O_{40}]^{6-}$**, H_2, and HCHO; the electron-transfer integral between adjacent tungsten sites for the unpaired electron was estimated by ESR measurements to be 0.43 eV [20]. Fe^{III} dopant in the Keggin compound **$[(CH_3)_4N]_5[BW_{12}O_{40}] \cdot nH_2O$** occupies slightly distorted tetrahedral positions with a line-width of 2.2 mT, as was shown by an ESR spectrum of an Fe^{III}-doped polycrystalline probe [21].

The following salts were prepared and characterized by elemental analysis, electronic and vibrational spectra, cyclic voltammetry, and ^{11}B and ^{119}Sn NMR spectroscopy: **$\alpha\text{-}K_7\text{-}[BW_{11}O_{39}Sn^{II}] \cdot 8H_2O$** with a voltammetric reduction potential of −1.05 V, **$\alpha\text{-}K_6[BW_{11}O_{39}\text{-}Sn^{IV}OH] \cdot xH_2O$** and **$\alpha\text{-}[(CH_3)_4N]_6[BW_{11}O_{39}Sn^{IV}OH] \cdot xH_2O$** with −0.82 and −1.09 V versus saturated calomel electrode [22]. **$\alpha\text{-}K_7[BW_{11}O_{40}V^{IV}]$** was prepared either from **$K_9[BW_{11}O_{39}]$**, $VO[SO_4]$, and KCl in an acetate buffer (pH 5), or from $Na_2[WO_4] \cdot 2H_2O$, H_3BO_3, $VO[SO_4]$, and KCl in acetic acid (pH 6.3) [23]. **$[BW_{11}O_{39}Mn^{II}(OH_2)]^{7-}$** ions were transferred into benzene or toluene using $[(n\text{-}C_7H_{15})_4N]Br$ and were rapidly oxidized to the Mn^{III} compound above −35 °C with an irreversible color change (absorption at 520 nm) [24].

H_3BO_3 reacts with $U(OC(O)CH_3)_4$, $Na_2[WO_4]$, and KCl in acetic acid to give **$K_{14}[(BW_{11}O_{39})_2U] \cdot 25H_2O$**, which was characterized by IR and UV/VIS spectra and thermal analysis. The dehydration begins at 40 °C and occurs in two stages: between 40 and 170 °C with the loss of 16 zeolitic water molecules, and between 170 and 420 °C with the loss of the remaining 9 zeolitic water molecules; at 573 K, the compound decomposes; the central uranium atom has a coordination number of six [25].

H_3BO_3 reacts with $Th[NO_3]_4$, $Na_2[WO_4]$, and KCl in acetic acid to give **$Na_{14}[(BW_{11}O_{39})_2Th] \cdot 38H_2O$**, which was characterized by IR and UV/VIS spectra and X-ray diffraction analysis (face-centered cubic structure with a = 1840.7 pm) [29].

The ^{17}O and ^{183}W NMR spectra of the **$[BW_{12}O_{40}Co]^{5-}$** and **$[BW_{12}O_{40}Co]^{5-}$** Keggin anions were compared with those of the anion $[BW_{12}O_{40}]^{5-}$ [26]. **$K_7[BW_{11}O_{39}Cu^{II}(OH_2)] \cdot nH_2O$** has been synthesized; in acid solution, it revealed no evidence of oxidation to Cu^{III} derivatives [27].

In the system H_3BO_3–$Na_3[VO_4]$–$Cs[NO_3]$–H_2O there exists an equilibrium solid phase **$B_2O_3 \cdot 2Cs_2O \cdot 6V_2O_5 \cdot 8.3H_2O$**, which is formulated (on the basis of infrared spectral data and X-ray phase analysis) as **$Cs_2[HBV_6O_{18} \cdot 2.15H_2O] \cdot 1.5H_2O$** [28].

References for 3.6.8:

[1] Woller, K.-H.; Heller, G. (Z. Kristallogr. **156** [1981] 151/7).
[2] Touboul, M. (Phosphorus Sulfur **28** [1986] 145/9).
[3] Behm, H. (Acta Crystallogr. C **44** [1988] 1348/51).
[4] Griffen, D. T. (Am. Mineral. **73** [1988] 1138/44).
[5] Behm, H. (Acta Crystallogr. C **41** [1985] 642/5).

[6] Behm, H.; Beurskens, P. T. (Z. Kristallogr. **174** [1986] 10/1).
[7] Behm, H.; Smykalla, C. (Z. Kristallogr. **183** [1988] 63/70).
[8] Bonazzi, P.; Menchetti, S.; Sabelli, C.; Trosti-Ferroni, R. (Z. Kristallogr. **180** [1987] 51/62).
[9] Heller, G.; Pickardt, J. (Z. Naturforsch. **40b** [1985] 462/6).
[10] Fu, Guoyi; Wang, Enbo; Liu, Jinfu; Zheng, R.-L. (Huaxue Xuebao **43** [1985] 949/54 from C.A. **104** [1986] No. 160860).

[11] Peng, J.; Wang, Enbo; Zheng, R.-L. (Kexue Tongbao [Foreign Lang. Ed.] **31** [1986] 526/32 from C.A. **105** [1986] No. 142003).
[12] Fu, Guoyi; Wang, Enbo; Liu, Jinfu; Lin, Yonghua; Jin, Songchun; Shi, Endong (Yingyong Huaxue **6** No. 5 [1989] 16/20 from C.A. **112** [1990] No. 227067).
[13] Nadjo, L.; Keita, B. (Fr. Demande 2573779 [1984/86] 1/13 from C.A. **106** [1987] No. 75016).
[14] Tonomura, S.; Aoshima, A.; Fukui, H. (Jpn. Kokai Tokkyo Koho 86-126080 [1984/86] 1/6 from C.A. **106** [1987] No. 67144).
[15] Piepgrass, K.; Pope, M. T. (J. Am. Chem. Soc. **109** [1987] 1586/7).
[16] Piepgrass, K.; Pope, M. T. (J. Am. Chem. Soc. **111** [1988] 753/4).
[17] Yang, G.-Y.; Cheng, T.-C.; Lin, J.; Liu, R.-Sh. (Proc. Natl. Sci. Counc. Repub. China A **11** No. 1 [1989] 41/9 from C.A. **108** [1988] No. 94036).
[18] Yamase, T.; Sasaki, Y.; Motowaki, T. (Inorg. Chim. Acta **121** [1986] L19-L22).
[19] Nomiya, K.; Sugie, Y.; Animoto, K.; Miwa, M. (Polyhedron **6** [1987] 519/24).
[20] Yamase, T.; Watanabe, R. (J. Chem. Soc. Dalton Trans. **1986** 1669/75).

[21] Weiner, H.; Lunk, H. J.; Stösser, R.; Lück, R. (Z. Anorg. Allg. Chem. **572** [1989] 164/74).
[22] Chorghade, G. S.; Pope, M. T. (J. Am. Chem. Soc. **109** [1987] 5134/8).
[23] Leparulo-Loftus, M. A.; Pope, M. T. (Inorg. Chem. **26** [1987] 2112/20).
[24] Katsoulis, D. E.; Pope, M. T. (J. Chem. Soc. Dalton Trans. **1989** 1483/9).
[25] Rusu, M.; Botar, A. V. (Stud. Univ. Babes-Bolyai Chem. **31** [1986] 84/7 from C.A. **107** [1987] No. 108052).
[26] Kazanskii, L. P.; Fedotov, M. A. (Koord. Khim. **14** [1988] 939/42).
[27] Kelly, C. M. (INTIS-AFIT/CI-Nr-86-178-T [**1986**] 1/101 from C.A. **106** [1987] No. 222843).
[28] Nakhodovna, A. P.; Listratenko, I. V.; Makarova, R. A. (Ukr. Khim. Zh. [Russ. Ed.] **53** [1987] 899/902 from C.A. **109** [1988] No. 44032).
[29] Marcu, G. H.; Botar, A. V. (Inorg. Synth. **23** [1988] 186/91).

3.7 Peroxoboric Acid and Peroxoborates

The names perboric acid and perborate instead of peroxoboric acid and peroxoborate (as in Chemical Abstracts) are not used in this volume because of the possible confusion with tetrahydroxodi-(μ-peroxo)diboric(2−) acid, which is also called perboric acid; for IUPAC recommendations, see [31].

For the formation of peroxoborates in H_2O_2-containing solutions, see Section 3.5.7.4, p. 202, and for spectral data of peroxoborate solutions, see Section 3.7.2, p. 288.

References on pp. 287/8

3.7.1 Preparation

An acidic solution of H_3BO_3 in H_2O_2 at various concentrations did not form any peroxoboron compound, as ^{11}B NMR spectra show; in alkaline solution, the ions $[B(OH)_4]^-$, $[B(OOH)_4]^-$, $[B_2(O_2)_2(OH)_4]^{2-}$, and $[B_2(O_2)_2(OOH)_4]^{2-}$ are present [20].

Analysis of the data from a titration of H_3BO_3-H_2O_2 solutions with standard bases led to the conclusion that the ions $\mathbf{[B(OH)_3(OOH)]^-}$ and $\mathbf{[B(OH)_2(OOH)_2]^-}$ predominate between pH 7 and 9. The equilibria are given by the equations: $H_3BO_3 + H_2O_2 \rightleftharpoons H^+ + [B(OH)_3(OOH)]^-$ (1) with the equilibrium constant $K_1 = 2.0 \times 10^{-8}$; and $[B(OH)_3(OOH)]^- + H_2O_2 \rightleftharpoons [B(OH)_2(OOH)_2]^- + H_2O$ (2) with the equilibrium constant $K_2 = 2.0\ mol^{-1}$. Data of pH mixing experiments at low pH values could not fit in terms of the equations (1) and (2) alone. Rather, addition of a third reaction, $H_3BO_3 + H_2O_2 \rightleftharpoons B(OH)_2(OOH) + H_2O$ (3), with the equilibrium constant $K_3 = 0.01\ mol^{-1}$ accounts for the observed pH change. This reaction is a substitution process in which the product is the trigonal monoperoxoboric acid. A distribution diagram for the system calculated from the determined equilibrium constants is presented in **Fig. 3-23** [29].

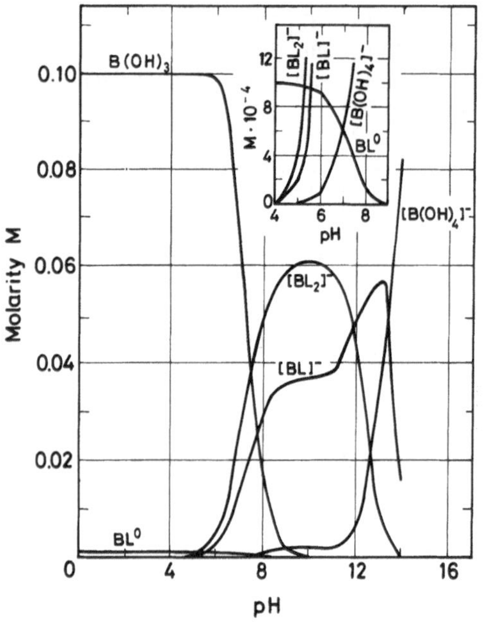

Fig. 3-23. Distribution diagram of 0.10 M $B(OH)_3$/1.00 M H_2O_2. ($BL^0 = B(OH)_2(OOH)$; $[BL]^- = [B(OH)_3(OOH)]^-$; $[BL_2]^- = [B(OH)_2(OOH)_2]^-$) [29].

This distribution diagram can be used to determine appropriate pH conditions for carrying out ^{11}B NMR experiments. At pH 6, only one major boron-containing species, H_3BO_3, is present. At pH 8.08, three peaks are expected. The spectrum at 25.5°C exhibits two sharp peaks and one broad peak, and at 55.5°C a sharpening of the broad peak is observed [29].

Technological and ecological problems in the chemical production of $\mathbf{Na[BO_3] \cdot nH_2O}$ [1], and the requirements for the quality of H_2O_2 used in this synthesis [2] were discussed.

The preparation of $Na[BO_3] \cdot nH_2O$ was investigated by electrolyzing a solution of borax and soda ash employing a Pt anode and a stainless steel cathode. The effect of concentrations of the reactants, temperature, and anodic charge density on the current efficiency of the sodium peroxoborate formation was studied [3]. The preparation of an alkaline solution of H_2O_2 by cathodic reduction of oxygen in a trickle-bed reactor was attempted as a first step in the preparation of sodium peroxoborate. The effect of variation of charge density, liquid flow rate, and concentration of NaOH on the current efficiency of formation of H_2O_2 was studied. Later, attempts were made to prepare $Na[BO_3]\cdot nH_2O$ by adding borax to the electrolyte [4].

The effects of anodic charge density, pH, temperature, carbonate concentration, Na^+, K^+, F^-, or $[Cr_2O_7]^{2-}$ on the current efficiency in the electrochemical production of $Na[BO_3]\cdot nH_2O$ were investigated. The current efficiency is maximal at 400 mA/cm^2, and increases with increasing $[CO_3]^{2-}$ concentration and decreasing temperature of the electrolyte solution. In addition to these factors, the replacement of Na^+ with K^+ and the addition of F^- and $[Cr_2O_7]^{2-}$ to the solution also increases the current efficiency. The current efficiency is 81% in $K_2[CO_3]/H_3BO_3$ solution at 400 mA/cm^2 between pH 9.8 and 11.2 [5].

The electrosynthesis of sodium peroxoborate is selectively catalyzed by platinized titanium film anodes [6]. The anodic characteristics of the composite coatings of, e.g., film-like Ti/SnO_2-Pt anodes were studied under conditions of oxygen evolution and electrosynthesis of sodium peroxoborate. A diaphragm cell was used for polarization measurements in 1.5 M $Na_2[CO_3]$ (or 1.3 M $Na_2[CO_3]$ + 0.2 M $Na[HCO_3]$) + 0.23 M H_3BO_3 solutions at pH 10.5 and at 20 °C. The anodes have a sufficiently high corrosion resistance and an acceptable current efficiency for peroxoborate formation [7]. In the electrosynthesis of sodium peroxoborate, anodes of Ru alloys with Nd, Er, and Fe-group elements (Fe, Co, or Ni) were used to study the kinetic characteristics of the anodic process [8].

The effect of carbonate-borate electrolytes on the corrosion resistance of an In-Bi intermetallic compound with Pb or Sn was investigated. Polarization measurements were made to determine the current efficiency with respect to the flow of active oxygen and to study the kinetics of its catalytic decomposition. The electrolytes contained sodium and potassium carbonates with different initial borates (meta-, ortho-, tetra-, penta-), but with the same content of carbonate ions and boron. One of the solutions simulates the electrolyte used in the electrosynthesis of sodium peroxoborate [9].

In order to optimize the conditions of the electrosynthesis of sodium peroxoborate over a broad range of experimental conditions, the influences of the concentration of the bath components, the pH of the bath, the presence and nature of the diaphragm and of separate electrode compartments, as well as of the nature of the borate introduced, on the current yield of active oxygen were examined. A method of polarization measurements under potentiostatic conditions was used [10]. The anodic oxidation of carbonate and tetraborate was studied on Pt, Pd, Pt-Hf, Pt-Zr, Pt-Pd-Hf, and Pt-Pd-Zr electrodes in alkaline solutions as a function of potential and depolarizer concentration. The formation of peroxoborate and peroxodicarbonate was found. The possibility of oxygen formation is pointed out. The thermal decomposition of the formed peroxo salts is discussed, and its dependence on the electrode material is described [11]. The polarization curves for stationary and rotating electrodes in potassium carbonate and tetraborate solutions, together with the results of bulk electrolysis, suggest the formation of peroxo compounds during electrolysis on Pt alloys with Zr, Hf, or Y as electrode materials [12].

The influence of the concentrations of the basic components of the electrolyte, the solution pH, the temperature, and the electrolyte circulation on the improvement of the process for obtaining sodium peroxoborate by an electrochemical method under conditions of combining electrolysis and crystallization was studied. A solution of 160 to 220 g/L $Na_2[CO_3]$ and 6 to 14 g/L $Na[B(OH)_4]$ was subjected to electrolysis on a Ti-Pt anode in a diaphragm-free electrolyzer of filter-press type at an anodic charge density of 360 mA/cm^2. Then the electrolyte, enriched with active oxygen, was placed into a crystallizing apparatus to form $Na[BO_3]\cdot nH_2O$ [15].

The electrolysis of borate-carbonate electrolytes was studied by ^{11}B NMR spectroscopy. The formation of peroxoborate, borate-carbonate complexes, peroxoboron complexes, or di-

 References on pp. 287/8

and polymeric forms of borates was demonstrated. The correlation between NMR results and potentiostatic curves is discussed. The dependence of the impedance data and the peroxide-oxygen current yield on the boron content was also described [16]. The effects of the rate of the catalytic decomposition of peroxo compounds on the quantitative characteristics of the electrochemical synthesis of sodium peroxoborate was studied. An equation describing the yield of sodium peroxoborate as a function of the peroxo compound concentration was derived. The relationship between the constant of the catalytic decomposition, kinetics of accumulation, and efficiency of formation was discussed [17].

$Na[BO_3]\cdot 4H_2O$ is dried at ≤70°C and 0.03 to 0.13 bar giving **$Na[BO_3]\cdot H_2O$** with good abrasion resistance and a satisfactory dissolution rate [13]. A process for pelletizing powdered sodium peroxoborate gives a granulated product, which satisfies consumer requirements in the manufacturing of synthetic detergents. The highest yield of the 0.25 to 2.0 mm sodium peroxoborate fraction, namely 60 to 72%, was obtained when a 1.5 to 2.5% solution of poly(vinylalcohol) was used as the binder with a feedstock having a moisture content of 12 to 16% [14].

A micropilot-scale turbulent-flow crystallizer was geometrically similar to the industrial-scale crystallizer and had a 5 L nominal capacity. It consisted of three sections: feed, reaction-crystallization, and vacuum. The two feeds were a 20% solution of $NaBO_2\cdot 3H_2O$ and a 40% H_2O_2 solution. The product was $Na[BO_3]\cdot 4H_2O$. Results from kinetic analyses of the data for determining the growth rate are in good agreement with a kinetic expression derived from fluidized-bed experiments [30].

The phase diagram of $NaBO_2$–H_2O_2–H_2O at 0°C shows three solid phases: $Na[BO_3]\cdot 4H_2O$, $Na[BO_3]\cdot 2H_2O$, and $NaBO_2\cdot 4H_2O$, with compositions at the two eutectic points of $NaBO_2$= 3.1 wt%, H_2O_2=5.1 wt%; $NaBO_2$=19.4 wt%, and H_2O_2=1.0 wt%, respectively; at 25°C, the solid phases are $Na[BO_3]\cdot 4H_2O$, $NaBO_2\cdot 4H_2O$, and a solid solution. Optimal conditions are found for the isolation of $Na[BO_3]\cdot 4H_2O$ [18].

The preparation of alkali and alkaline earth metal peroxoborates was studied by high-resolution ^{11}B NMR spectroscopy with regard to cation composition and the pH of the solution [19]. The ^{11}B NMR studies of aqueous $Na[B(OH)_4]$ in the presence of various amounts of H_2O_2 indicate the formation of the ions $[B_2(O_2)_2(OH)_4]^{2-}$, $[B_2(O_2)_2(OOH)_4]^{2-}$, and $[B(OOH)_4]^-$. $[NH_4]_2$-$[B_4O_5(OH)_4]$ solutions in the presence of H_2O_2 form the ion $[B(OOH)_4]^-$ [20]. The mechanisms for the peroxoborate formation in aqueous H_2O_2 solutions containing $Na[B_5O_6(OH)_4]\cdot 3H_2O$, $Na_2[B_4O_5(OH)_4]\cdot 8H_2O$, $Mg[B_3O_3(OH)_4]_2\cdot 3.5H_2O$, $Mg[B_3O_3(OH)_5]\cdot 5H_2O$, $Na_3[B_3O_3F_6]$, $K_2[B_3O_3F_4(OH)]$, $Na[HO\text{–}BF_3]$, or $K[HO\text{–}BF_3]$ was studied by ^{11}B and ^{19}F NMR spectroscopy. The transformation to peroxoborates in acidic or alkaline solutions occurs only in the presence of ligands such as F^- or $[OH]^-$, which convert three-coordinate into four-coordinate boron in the borate structures. Optimal conditions for $[B(OH)_3(OOH)]^-$ formation occur between pH 10 and 12, while higher peroxoborates are formed between pH 4 and 8. The degree of condensation of the peroxoborates depends on the initial borate structures [21].

In the system H_3BO_3–MOH–MF–H_2O_2–H_2O with M=Li, Na, K, or NH_4, ^{11}B and ^{19}F NMR studies indicated the formation of the anions $[BF_4]^-$, $[BF_3(OOH)]^-$, $[BF_2(OOH)_2]^-$, $[BF(OOH)_3]^-$, $[B(OOH)_4]^-$, $[BF_3(OH)]^-$, $[B_2F_4(O_2)_2]^{2-}$, and $[B_2F_2(O_2)_2(OOH)_2]^{2-}$. An increase in the H_2O_2 concentration increases the amounts of $[BF(OOH)_3]^-$ and $[B(OOH)_4]^-$. The change in composition of the products was also studied with the increase of F^- concentration and an increase of time [22]. An additional ^{11}B and ^{19}F NMR study of solutions of hydroxotrifluoroborates, $[BF_3(OH)]^-$, in CH_3COOH and $CH_3C(O)OOH$ shows the existence of the ions $[HOO\text{–}BF_3]^-$, $[BF_2(OOH)_2]^-$, $[CH_3C(O)O\text{–}BF_3]^-$, and $[CH_3C(O)OO\text{–}BF_3]^-$. Moreover, a linear strong-field shift of the signal of the $[BF_3(OH)]^-$ ion and a decrease in the ^{11}B-^{19}F NMR spin-spin-coupling constant was observed at low pH values [23]. For conditions for the formation of fluorine containing peroxoborates like $K[HOO-BF_3]$ [24, 25], $K_2[B(O_2)F_3]\cdot4H_2O$, $Na_2[B(O_2)F_3]\cdot4H_2O$, and $[NH_4]_2[B_2(O_2)_3F_2]$ in the presence of hydrofluoric acid, see Section 3.5.7.4, p. 202 [26].

Alkali metal peroxoborates were prepared from B_2O_3, MOH (M = K, Na, or Li), and H_2O_2, and were characterized by ESR, IR, ^1H NMR and ^{11}B NMR spectra, and thermal analyses. Typically, at pH 1, $[B(OOH)_4]^-$ was formed, but at pH 2, polymerization occurred with the formation of μ-peroxo(hydroperoxo)borates and μ-peroxo(hydroperoxo)peroxoborates. For lithium compounds, highly condensed anions were formed, whereas otherwise the compounds $Na_2[B_2(O_2)_2(OH)(OOH)_3]\cdot5H_2O$, $Na_3[B_3(O_2)_4(OOH)_4]\cdot2H_2O$, $Na_4[B_4(O_2)_6(OOH)_4]\cdot2H_2O$, and $K_2[B_2(O_2)_2(OH)(OOH)_3]$ were found [27].

The thermally stable compounds **$K[BO_3]$** and **$K[BO_3]\cdot H_2O$** were obtained by the conversion of KBO_2 with H_2O_2 at ≤20°C and with a 30 to 200 mol% excess of $K_2[SO_4]$ or KCl. Thus, $K[BO_3]\cdot H_2O$ with the density 0.94 kg/L and an active oxygen content of 13.7% was obtained by continuous addition of KBO_2 solution (KOH = 253.4 g/L, B_2O_3 = 216.6 g/L, and KCl = 151 g/L) and 54% H_2O_2 containing 250 g/L KCl to a mixture of 1.5 L KBO_2 solution, 280 mL 54% H_2O_2, and 70 g KCl. The mixture was stirred at a temperature between 2 and 4°C with continuous removal of the reaction product in order to maintain an H_2O_2 concentration of 96 g/L and a B_2O_3 concentration of 113 g/L in the mother liquor. Fluidized-bed drying of the crystalline precipitate was performed at 30°C [28].

References for 3.7 and 3.7.1:

[1] Futoryanskii, A. Ya.; Rassokhina, V. N.; Shubin, A. S. (Tr. Ural. Neorg. Khim. Inst. No. 57 **1984** 36/9 from C.A. **105** [1986] No. 63123).
[2] Futoryanskii, A. Ya.; Rassokhina, V. N. (Prvo Potreblenie Perekisi Vodoroda. Mater. Vses. Koordinats. Soveshch. L. **1987** 31/3 from C.A. **110** [1989] No. 78683).
[3] Raghavendran, N. S.; Narasimham, K. C. (Bull. Electrochem. **4** [1988] 263/6 from C.A. **108** [1988] No. 212505).
[4] Raghavendran, N. S.; Basha, C. A.; Narasimham, K. C.; Vasu, K. I. (Bull. Electrochem. **4** [1988] 355/7 from C.A. **109** [1988] No. 218044).
[5] Imer, F.; Koseoglu, A.; Bayrak, U. (Doga Turk Kim. Derg. **13** [1989] 215/23 from C.A. **113** [1990] No. 13717).
[6] Kiselev, E. Yu.; Kondrikov, N. B.; Eliseenko, L. G.; Logvinova, V. B. (Izv. Vyssh. Uchebn. Zaved. Khim. Khim. Technol. **30** No. 6 [1987] 68/71 from C.A. **107** [1987] No. 143558).
[7] Kiselev, E. Yu.; Kondrikov, N. B.; Shub, D. M. (Zh. Prikl. Khim. **61** [1988] 1027/31 from C.A. **109** [1988] No. 82010).
[8] Toroptseva, N. T.; Raevskaya, M. V. (Zh. Prikl. Khim. **58** [1985] 795/9; J. Appl. Chem. [USSR] **58** [1985] 720/3 from C.A. **103** [1985] No. 44767).
[9] Toroptseva, N. T. (Zh. Prikl. Khim. **59** [1986] 1111/2; J. Appl. Chem. [USSR] **59** [1986] 1032/3 from C.A. **105** [1986] No. 160777).
[10] Toroptseva, N. T.; Khomutov, N. E.; Niftullaeva, T. A. (Zh. Prikl. Khim. **58** [1985] 2020/7; J. Appl. Chem. [USSR] **58** [1985] 1859/65 from C.A. **103** [1985] No. 202675).

[11] Toroptseva, N. T.; Vaseva, A. Yu. (Zh. Prikl. Khim. **61** [1988] 402/4; J. Appl. Chem. [USSR] **61** [1988] 363/4 from C.A. **109** [1988] No. 29002).
[12] Toroptseva, N. T.; Raevskaya, M. V.; Konobas, Yu. I.; Sokolovskaya, E. M. (Vestn. Mosk. Univ. Khim. **28** [1987] 263/8 from C.A. **107** [1987] No. 163878).
[13] Dugua, J.; Cuer, J. P. (Eur. Appl. 155894 [1984/85] 1/4 from C.A. **104** [1985] No. 111831).
[14] Ovchinnikova, K. N.; Zapol'skii, S. V.; Arkhipkina, V. D.; Fedyushkin, B. F. (Khim. Promst. [Moscow] **1987** No. 6, pp. 344/5 from C.A. **107** [1987] No. 136850).

[15] Kondrikov, N. B.; Berdyugina, V. P.; Chernyshov, B. N.; Kalenik, V. M. (Zh. Prikl. Khim. **62** [1989] 2683/7; J. Appl. Chem. [USSR] **62** [1989] 2486/9 from C.A. **112** [1990] No. 127719).
[16] Il'yin, I. E.; Kondrikov, N. B.; Brovkina, O. V.; Chernyshov, B. N. (Latv. PSR Zinat. Akad. Vestis Kim. Ser. **1989** 275/80).
[17] Il'yin, I. E. (Zh. Prikl. Khim. **62** [1989] 48/53; J. Appl Chem. [USSR] **62** [1989] 42/6 from C.A. **111** [1989] No. 86136).
[18] Xie, Gaoyang; Ma, Shiming; Feng, Qi (Fudan Xuebao Ziran Kexuebao **27** [1988] 21/6 from C.A. **109** [1988] No. 238200).
[19] Shchetinina, G. P.; Brovkina, O. V.; Chernyshov, B. N.; Ippolitov, E. G. (Zh. Neorg. Khim. **32** [1987] 18/24; Russ. J. Inorg. Chem. **32** [1987] 9/12).
[20] Chernyshov, B. N.; Shchetinina, G. P.; Brovkina, O. V.; Ippolitov, E. G. (Koord. Khim. **11** [1985] 31/5).

[21] Chernyshov, B. N.; Brovkina, O. V. (Latv. PSR Zinat. Akad. Vestis Kim. Ser. **1988** 422/8).
[22] Ippolitov, E. G.; Chernyshov, B. N.; Shchetinina, G. P.; Brovkina, O. V.; Martynyuk, Yu. L.; Gorin, Yu. V. (Ukr. Khim. Zh. [Russ. Ed.] **52** [1986]; Sov. Prog. Chem. **52** [1986] 818/23).
[23] Shchetinina, G. P.; Brovkina, O. V. (Russ. J. Inorg. Chem. **30** [1985] 1226/8).
[24] Shchetinina, G. P.; Perschin, V. L.; Chernyshov, B. N. (Zh. Strukt. Khim. **28** [1987] 25/30).
[25] Shchetinina, G. P.; Brovkina, O. V. ; Chernyshov, B. N.; Ippolitov, E. G. (Zh. Neorg. Khim. **30** [1985] 819/21; Russ. J. Inorg. Chem. **30** [1985] 460/1).
[26] Chandhuri, M. K.; Das, B. (Inorg. Chem. **24** [1985] 2580/2).
[27] Martynyuk, Yu. L.; Zakharova, S. A.; Chernyshov, B. N.; Kavun, V. Ya.; Gorin, Yu. V. (Ukr. Khim. Zh. [Russ. Ed.] **54** [1988] 6/10 from C.A. **108** [1988] No. 230914).
[28] Doetsch, W.; Siegel, R. (Ger. Offen. 3347595 [1983/85] 1/10 from C.A. **103** [1985] No. 107091).
[29] Pizer, R.; Tihal, C. (Inorg. Chem. **26** [1987] 3639/42).
[30] Chianese, A.; Condo, A. (ICP **16** No. 9 [1988] 181/8 from C.A. **110** [1989] No. 98107).

[31] IUPAC Nomenclature of Inorganic Chemistry (Recommendations 1990), Blackwell, Oxford 1990, pp. 124/9.

3.7.2 Properties

For properties of peroxoborate containing solutions, see also Section 3.7.1, p. 283. For the system $NaBO_2$–H_2O_2–H_2O, see p. 286.

Kinetic and NMR studies of the decomposition of different sodium peroxoborates and of $K[BF_3(OOH)]$ as a function of pH indicate that the first stage is the partial decomposition of the bridging peroxo ligands. Conditions and results of the subsequent steps are described [1]; cf. also Section 3.7.3, p. 290. The composition and state of functional terminal proton-containing groups in simple lithium peroxoborates and in high-peroxide lithium peroxoborates ($Li_3[B_3(O_2)_4(OH)_{4-x}(OOH)_x]$) were studied by ^1H, ^7Li, and ^{11}B NMR spectroscopy. The presence of labile and nonlabile two-spin proton pairs, stabilized by a system of hydrogen bonds, was established, demonstrating the stability of the hydroperoxo compounds [2].

The ^{11}B NMR spectra of solutions of $[B(OH)_4]^-$ or H_3BO_3 (0.01 to 4 mol/L) containing H_2O_2 (0.01 to 34 mol/L) were studied between pH 4 and 14; ions identified were $[B(OH)_4]^-$, $[B(OH)_3(OOH)]^-$, $[B(OH)_2(OOH)_2]^-$, $[B(OH)(OOH)_3]^-$, $[B(OOH)_4]^-$; $[B_2(O_2)_2(OH)_4]^{2-}$, $[B_2(O_2)_2$-$(OH)_2(OOH)_2]^{2-}$, and $[B_2(O_2)_2(OOH)_4]^{2-}$. Raman spectra were recorded for concentrated solutions (1 to 4 mol/L $[B(OH)_4]^-$ and 1 to 12 mol/L H_2O_2), and partial assignments were proposed

[3]; cf. also Section 3.7.1, p. 284. Raman and infrared spectra were also given for solid $Na_2[B_2(O_2)_2(OH)_4]\cdot6H_2O$ (normal, ^{10}B, ^{11}B, and D-labeled), $Na_2[B_2(O_2)_2(OH)_4]\cdot4H_2O$, $Na_2[B_2(O_2)_2(OH)_4]$, $Li_2[B_2(O_2)_2(OH)_4]$, and $M_2[B_2(O_2)_2(OH)_2(OOH)_2]$ with M = K, Rb, or Cs; the vibrational modes were assigned [4].

Although the concentration of peroxoboric acid at pH 6 is too small to be detected by ^{11}B NMR spectroscopy under any of the experimental conditions, it is possible to determine its dissociation constant from the relationship $K_a\cdot[B(OH)_2(OOH)]=K_1/K_3=2\times10^{-6}$ [5]. This means that peroxoboric acid is a stronger acid than orthoboric acid, a result in agreement with earlier work in this field [11]. One possible explanation for the difference in acidity is that the B–O bond in R_2BOH has more multiple-bond character than that in R_2BOOH, due to the electron-withdrawing nature of the second oxygen atom in the OOH group [5].

For properties of peroxoborate containing solutions, for the ^{11}B NMR spectra of solutions of $[B(OH)_4]^-$ or H_3BO_3 mixed with H_2O_2, and for the reaction of H_3BO_3 with H_2O_2 at pH 7 to 9 [5], see also Section 3.7.1, p. 283.

The primary nucleation kinetics of sodium peroxoborate in aqueous solutions show the great influence of the sodium borate content [6]. The growth and dissolution rates of sodium perborate crystals in aqueous solutions were measured by means of a fluidized-bed apparatus. Between 18 and 28 °C in aqueous solutions, the effects of the sodium borate concentration (up to ca. 5%), of the crystal size (0.27 to 0.72 mm), and of the temperature (18 to 28 °C) on the growth rate of sodium peroxoborate were investigated [7].

Growth and dissolution rates of compressed polycrystalline disks of sodium peroxoborate were measured in aqueous solutions with and without sodium borate as co-solute. When w and w_{eq} are the solute concentrations of sodium peroxoborate in the bulk solution and at equilibrium, respectively, the growth rate with sodium metaborate is $R_G=1.34\times10^3(w-w_{eq})$, and without sodium metaborate the growth rate is $R_G=1.01\times10^3(w-w_{eq})$, both of first order with respect to the concentration gradients. The overall growth process is diffusion-controlled. From the dissolution experiments in pure aqueous solutions, the diffusion coefficient of sodium peroxoborate was first determined and then compared with the values measured by interferometry or calculated by a prediction method. In addition, the overall mass transfer coefficients for the dissolution processes are in good agreement with the rate constant for the bulk diffusion step in crystal growth [8].

Kinetic analyses of the data for determining the growth rate of $Na[BO_3]\cdot4H_2O$, developed with a model that accounts for the classification of the crystals, furnished results that are in good agreement with a kinetic expression that was derived previously from fluidized-bed experiments [9].

The primary crystal nucleation of sodium peroxoborate from an aqueous solution at 20 °C was decreased and the average crystal size was increased by the addition of 80 ppm of an anionic surfactant. The crystal shape was prismatic at the lowest supersaturation and almost spherical at the highest, in contrast to the dendritic surface encountered in the absence of the surfactant [10].

References for 3.7.2:

[1] Chernyshov, B. N.; Vaseva, A. Yu.; Brovkina, O. V.; Ippolitov, E. G. (Dokl. Akad. Nauk SSSR **306** [1989] 1140/4; Dokl. Chem. Proc. Acad. Sci. USSR **306** [1989] 185/8).

[2] Chernyshov, B. N.; Kavun, V. Ya.; Zakharova, S. A.; Martynyuk, Yu. L.; Brovkina, O. V. (Zh. Neorg. Khim. **34** [1989] 1987/91; Russ. J. Inorg. Chem. **34** [1989] 1131/4).

[3] Flanagan, J.; Griffith, W. P.; Powell, R. D.; West, A. P. (J. Chem. Soc. Dalton Trans. **1989** 1651/5).

[4] Flanagan, J.; Griffith, W. P.; Powell, R. D.; West, A. P. (Spectrochim. Acta A **45** [1989] 951/5).
[5] Pizer, R.; Tihal, C. (Inorg. Chem. **26** [1987] 3639/42).
[6] Chianese, A.; Contaldi, A.; Mazzarotta, B. (J. Cryst. Growth **78** [1986] 279/80).
[7] Chianese, A.; Condo, A.; Mazzarotta, B. (J. Cryst. Growth **97** [1989] 375/86).
[8] Chianese, A. (J. Cryst. Growth **91** [1988] 39/49).
[9] Chianese, A.; Condo, A. (ICP **16** No. 9 [1988] 181/8 from C.A. **110** [1989] No. 98107).
[10] Chianese, A.; Condo, A.; Di Berardino, F.; Mazzarotta, B. (Process Technol. Proc. **6** [1989] 261/4 from C.A. **111** [1989] No. 117333).

[11] Menzel, H. Z. (Z. Phys. Chem. **105** [1923] 402/41).

3.7.3 Detergents and Activation of Peroxoborates

The effect of some surfactants on the decomposition of peroxoborates and bleaching of cotton fabric was described [1]. Sodium or potassium peroxoborate solutions decompose differently in the presence of perfume waste products, sodium tripolyphosphate, carboxymethylcellulose, sodium metasilicate, optical whiteners, or calcined soda ash between 289 and 348 K; the solution had been formed by anodic oxidation on Pt electrodes [2].

The decomposition kinetics of sodium peroxoborate was determined in aqueous solution between 30 and 65 °C in the presence of acetylsalicylic acid, glyceryl triacetate, sorbitol hexaacetate, or glucosyl pentaacetate as activators; the latter was the most efficient activator [3]. Changes in the active oxygen concentration in aqueous solutions of sodium peroxoborate in the presence of activators can be described by second order reaction equations. The activation energies of this decomposition are 71.1 kJ/mol in the presence of ethylenediaminetetraacetate (EDTA: the most effective activator), 72.6 kJ/mol in the presence of diacetylurea, and 73.5 kJ/mol in the presence of acetylurea and tetraacetylglycoluril (TAGU), respectively [4]. The decomposition kinetics of organophosphorus esters in aqueous sodium peroxoborate solution was also studied [5].

The decrease in sodium peroxoborate stability in powdered laundry detergents during storage was of a linear nature, with the rate of sodium peroxoborate decomposition depending mainly on the detergent composition and not on the initial sodium peroxoborate content or the formulation technology. For the majority of the detergents, the decrease in active oxygen content ranged from 2.19 to 15.84% after 12 months. There was no direct relationship between the sodium peroxoborate content of the detergents and their ability to remove red wine stains [6].

For use as a sodium peroxoborate activator in laundry formulations, C_8- to C_{16}-alkyl- and alkenylsuccinic anhydrides are considerably less effective than the standard peroxyacetic acid-releasing compounds tetraacetylglycoluril, EDTA, or glucosyl pentaacetate [7].

The use of nitrilotriesters as activators for peroxoborate or alkaline H_2O_2 bleaches in laundry detergent compositions, especially at low temperatures, is described. Thus, wine- or tea-stained cloths were laundered in 1 L water at 30 °C with a composition containing 4 g of surfactant, 0.75 g of $Na[BO_3]\cdot 4H_2O$, 0.1 g of $N(COOC_2H_5)_3$, and 0.15 g of $Na_2[SO_4]$, giving 66.87 or 72.4% stain removal, respectively, as compared to 66.03 or 71.34% removal for a composition containing 0.1 g EDTA as bleach activator [8]. The use of glucosyl pentaacetate as an activator for peroxoborate bleaching agents in detergents for washing machines improved the bleaching capacity and hygienic properties of textiles laundered at low temperatures [9]. For the glucosylpentaacetate and EDTA determination in such bleach solutions, see [10]. 1,5-Diacetylhexahydro-*s*-triazine-2,4-dione is also an activator for sodium peroxoborate laundry bleach [11].

The use of organic phosphonates as builders in laundry detergents containing peroxoborates, and the synergism with other additives was discussed [12]. The bleaching efficiency of peroxoborate solutions was investigated by means of developed photographical layers. It was shown that peroxoborate requires an activator for Ag bleaching similar to persulfates. Copper salts and benzoquinone are suitable for the acceleration of bleaching, but thio compounds are less effective. The advantages and disadvantages of peroxoborate were compared with persulfate bleachbaths [13].

Alkali metal (Li, Na, K, Rb, or Cs) peroxoborate structures and radical-generating properties were discussed; newly reported radical formation in thermally treated $Li[BO_3]\cdot H_2O$, detected by ESR, was interpreted in terms of radical transfer mechanisms involving a symmetric or homolytic cleavage of anion rings [15].

Singlet oxygen does not play an important role in bleaching with $Na[BO_3]\cdot H_2O$, as was shown by kinetic investigations on its decay by the bleaching of 1,3-diphenylisobenzofuran and trapping with tetrapotassium 2,3,8,9-rubrenetetracarboxylate; the addition of β-carotene did not have a significant effect [16].

The mechanism and kinetics of potassium peroxoborate decomposition were studied between 20 and 350°C, with KBO_2 being the final product [17]. The thermal decomposition of magnesium peroxoborate, $Mg[BO_3]_2\cdot 2H_2O$, involves a sequence of reactions [18]. Low cost, readily synthesized iron-free zeolites decompose sodium peroxoborate to an extent less than or equal to that of commercial detergent grade zeolite A [14].

References for 3.7.3:

[1] Vaseva, A. Yu.; Volkov, V. A.; Toroptseva, N. T.; Sokhadze, L. A. (Protsessy Apparaty Predpriyatii Byt. Obsluzh. Naseleniya M. **1986** 5/10 from C.A. **108** [1988] No. 223523).

[2] Vaseva, A. Yu.; Toroptseva, N. T. (Zh. Fiz. Khim. **59** [1985] 1836/9; Russ. J. Phys. Chem. **59** [1985/86] 1090/9).

[3] Bunina, N. A.; Kalinina, N. V.; Nechesnyuk, G. P.; Kruchinin, V. A. (Zh. Prikl. Khim. **60** [1987] 2091/4; J. Appl. Chem. [USSR] **60** [1987] 1932/5 from C.A. **108** [1988] No. 58367).

[4] Bunina, N. A.; Nechesnyuk, G. P.; Slapygina, O. L.; Kruchinin, V. A. (Zh. Prikl. Khim. **62** [1989] 2392/3; J. Appl. Chem. [USSR] **62** [1989] 2223/4 from C.A. **112** [1990] No. 79980).

[5] Konley, R. A.; Lee, G. C.; Winterle, J. S. (J. Org. Chem. **50** [1985] 40/4).

[6] Bolinski, L. (Przemysl Chem. **68** [1989] 274/6 from C.A. **112** [1990] No. 38715).

[7] Smidrkal, J.; Simunek, J. (Abh. Akad. Wiss. DDR Math. Naturwiss. Tech. [1986/87] 575/81 from C.A. **107** [1987] No. 178679).

[8] BP International, Ltd. (Res. Discl. No. 292 [1988] 617 from C.A. **109** [1988] No. 192625).

[9] Franzolin, G.; Colombo, P. (Riv. Ital. Sostanze Grasse **64** No. 3 [1987] 107/12 from C.A. **109** [1988] No. 172579).

[10] Sedea, L.; Franzolin, G. (Tenside Surfactants Deterg. **26** [1989] 211/4 from C.A. **111** [1989] No. 117331).

[11] Schmidt, H. (Wiss. Z. T. H. Carl Schorlemmer Leuna-Merseburg **28** [1986] 579/85 from C.A. **106** [1987] No. 198152).

[12] Paladini, M.; Schnorbus, G. (Seifen Öle Fette Wachse **114** [1988] 756/60 from C.A. **110** [1989] No. 175531).

[13] Hänsel, R.; Böttcher, H. (J. Signalaufzeichnungsmater. **17** [1989] 233/43 from C.A. **111** [1989] No. 123587).
[14] Burriesci, N.; Crisafulli, M.; Giordano, N.; Antonucci, P. L. (Zeolites **6** [1986] 119/24 from C.A. **104** [1986] No. 151721).
[15] Heller, G.; Pawel, A. (Tenside Surfactants Deterg. **23** [1986] 73/5).
[16] Koberstein, E.; Kurzke, H. (Tenside Surfactants Deterg. **24** [1987] 210/2).
[17] Nagaishi, T.; Tanisawa, Y.; Matsumoto, M.; Yoshinaga, S. (Kogyo Kagaku **45** No. 2 [1984] 94/7 from C.A. **103** [1985] No. 12072).
[18] Nagaishi, T.; Inoue, M.; Matsumoto, M.; Yoshinaga, S. (J. Therm. Anal. **31** [1986] 523/9).

3.7.4 Oxidations with Peroxoborates

For oxidation processes with peroxoborates, see also Section 3.7.2, p. 288, and Section 3.7.3, p. 290.

Sodium peroxoborate/acetic acid anhydride in CH_2Cl_2 oxidizes alkenes to oxiranes [1]. Sodium peroxoborate is the best of all oxidizing agents for the purification of acetic acid anhydride, resulting in a product purity of 98% and a $K[MnO_4]$ persistence time of 90 s [2].

N,N-Dialkylhydrazones are efficiently cleaved to their parent ketones under mild conditions at pH 7 with sodium peroxoborate [3]. The oxidation of L-ascorbic acid (H_2A) in perchloric acid medium by sodium peroxoborate is of first order dependence each in H_2A and the peroxoborate [4]. Styrenes were oxidized by sodium peroxoborate in acetic acid to give benzyl alcohols [5]. Sodium peroxoborate in acetic acid/trifluoroacetic acid is an effective reagent for the oxidation of anilines to nitroarenes and of sulfides to either sulfoxides or sulfones [6]. The optimal conditions (94% yield) for the oxidation of di-*n*-butyl sulfide with sodium peroxoborate to give di-*n*-butyl sulfoxide, $(n\text{-}C_4H_9)_2SO$, are three hours reaction time, a temperature of 45°C, and a mole ratio of di-*n*-butyl sulfide:sodium peroxoborate of 1:1.2 [7].

References for 3.7.4:

[1] Xie, Gaoyang; Xu, Linxiao; Hu, Jun; Ma, Shiming; Hou, Wei; Tao, Fenggang (Tetrahedron Lett. **29** [1988] 2967/8).
[2] Rastogi, M. K.; Bisarya, S. C. (Res. Ind. **33** [1988] 25/7 from C.A. **109** [1988] No. 230243).
[3] Enders, D.; Blushan, V. (Z. Naturforsch. **42b** [1987] 1595/6).
[4] Reddy, K. N.; Rajanna, K. C.; Saiprakash, P. K. (Orient. J. Chem. **4** [1988] 159/62 from C.A. **110** [1989] No. 173648).
[5] Gupton, J. T.; Duranceau, S. J.; Miller, J. F.; Kosiba, M. L. (Synth. Commun. **18** [1988] 937/47).
[6] McKillop, A.; Tarbin, J. A. (Tetrahedron **43** [1987] 1753/8).
[7] Yang, Wuzi; Liang, Shiyi; Jin, Suwen (Huadong Huagong Xueyuan Xuebao **13** [1987] 760/3 from C.A. **109** [1988] No. 56979).

3.7.5 Analytical

Capillary isotachophoresis with conductivity detection was used for the determination of peroxoborate (besides sulfate, phosphate, and nitrilotriesters) in detergent compositions, using the external standard calibration method with a minimum of sample pretreatment. Only a short analysis time is necessary [1].

An indirect polarographic microdetermination of H_2O_2 and some inorganic peroxides is described. The method is based on the reaction of the peroxo compounds with an acidified iodide solution to form an equivalent amount of iodine which is extracted into chloroform and then reduced to iodide. The iodide is oxidized to iodate which is measured polarographically. By this method, it is possible to determine 9.6 μg $Na[BO_3]\cdot H_2O$ with good precision and accuracy [2].

Molybdenum(VI) can be determined in trace amounts in water by its catalytic effect on the reaction: $[BO_3]^- + 3I^- + 2H^+ \rightarrow [BO_2]^- + [I_3]^- + H_2O$ in an acidic medium. The indicator reaction is transformed to a Landolt reaction by the addition of L-ascorbic acid which reduces I_2 to $2I^-$; the effect of the sodium peroxoborate concentration on the reciprocal of the induction period was given [3].

References for 3.7.5:

[1] Krestholt, H.; Jeuring, H. J. (Tenside Surfactants Deterg. **25** No. 3 [1988] 162/5).
[2] Rahim, S. A.; Hassan, F. A. (Microchem. J. **37** [1988] 377/81 from C.A. **109** [1988] No. 162537).
[3] Kataoka, M.; Tahara, S.; Ohzeki, K. (Fresenius Z. Anal. Chem. **321** [1985] 146/9).

3.8 Borate Minerals

The following Table 3/26 lists additions to the tables of minerals, given in "Boron Compounds" 1st Suppl. Vol. 1, 1980, pp. 242/50; 2nd Suppl. Vol. 1, 1983, pp. 300/3; and 3rd Suppl. Vol. 2, 1987, pp. 182/4. This table should be used in conjunction with the earlier ones.

Table 3/26
Borate Minerals: New Structural Investigations (RE means rare earth element).

mineral	composition or structural formula	Ref.
azoproite	$Mg_2(Fe^{III},Ti,Mg)[O_2\vert(BO_3)]$	[1]
bonaccordite	$Ni_2Fe^{III}[O_2\vert(BO_3)]$	[1]
danburite	$Ca_{0.95}Al_{0.02}Fe_{0.02}Mg_{0.01}[B_2Si_2O_8]\cdot 0.01H_2O$	[2]
datolite	$Ca[BSiO_4(OH)]$	[3]
diomignite	$Li_2[B_4O_7]$	[4]
fluoborite	$Mg_3[(F, OH)_3\vert(BO_3)]$	[1]
gaudefroyite	$Ca_4Mn_3[O_3\vert(CO_3)\vert(BO_3)_3]$	[1]
henmilite	$Ca_2Cu[(OH)_4\vert\{B(OH)_4\}_2]$	[5]
hilgardite-1Tc	$Ca_2[B_5O_9(OH)]\cdot H_2O$	[24]
hilgardite-3Tc	$Ca_2[B_5O_9(OH)]\cdot H_2O$ (previously parahilgardite)	[24]
hilgardite-4M	$Ca_2[B_5O_9(OH)]\cdot H_2O$	[24, 25]
homilite	$Ca_{2.00}(Fe_{0.90},Mn_{0.03})[B_2Si_2O_{9.86}(OH)_{0.14}]$	[6]

References on p. 295

Table 3/26 (continued)

mineral	composition or structural formula	Ref.
howlite	$Ca_4\{[Si_2B_4O_{10}(OH)_6][B_3O_4(OH)_2]_2\}$	[7]
hulsite	$(Fe^{II}, Mg)_2(Fe^{III}, Sn)[O_2\|(BO_3)]$	[1]
hydroboracite	$CaMg[B_3O_4(OH)_3]_2 \cdot 3H_2O$	[8, 9]
inderite	$Mg[B_3O_3(OH)_5] \cdot 5H_2O$	[10, 11]
jeremejevite	$Al_6[(BO_3)_5F_3]$	[12]
jimboite	$Mn_3[BO_3]_2$	[1]
kotoite	$Mg_3[BO_3]_2$	[1]
kurchatovite	$CaMg[B_2O_5]$	[1]
kurgantaite	$CaSr[(B_5O_9)Cl] \cdot H_2O$	[13]
leucosphenite	$Na_4Ba[TiSi_5(BO_3)O_{12}]_2$	[26]
ludwigite	$Mg_{2.11}Al_{0.31}Fe_{0.53}Ti_{0.05}Sb_{0.01}[O_2\|(BO_3)]$	[14 to 16]
lüneburgite	$Mg_3[B_2O(OH)_4\|(PO_4)_2] \cdot 6H_2O$	[17]
moydite	$(Y, RE)[B(OH)_4\|(CO_3)]$	[18]
nordenskioldine	$CaSn[BO_3]_2$	[1]
orthopinakiolite	$Mg_2Mn^{III}[O_2\|(BO_3)]$	[1]
painite	$CaZr[BAl_9O_{18}]$	[1]
pinakiolite	$Mg_2Mn^{III}[O_2\|(BO_3)]$	[1]
sakhaite	$Ca_3Mg[(BO_3)_2\|CO_3] \cdot H_2O$	[1]
suanite	$Mg_2[B_2O_5]$	[19]
sussexite	$Mn^{II}_2(OH)[B_2O_4(OH)]$	[1]
szaibelyite	$Mg_2(OH)[B_2O_4(OH)]$	[14, 20]
takéuchite	$Mg_{1.59}Mn_{1.20}Fe_{0.19}[O_2\|(BO_3)]$	[21]
Na-Al tourmaline	$(\square_{0.06}Na_{0.94})Al_3Al_6[(Si_{4.22}Al_{1.78})(BO_3)_3O_{19.2}(OH)_{2.8}]$	[22]
	$(\square_{0.24}Na_{0.76})Al_3Al_6[(Si_{5.82}Al_{0.18})(BO_3)_3O_{20.6}(OH)_{1.4}]$	[22]
tyretskite-1Tc	$Ca_2[B_5O_9(OH)] \cdot H_2O$	[24]
ulexite	$NaCa[B_5O_6(OH)_6] \cdot 5H_2O$	[23]
vonsenite	$Fe^{II}_2Fe^{III}[O_2\|(BO_3)]$	[1]
warwickite	$(Mg, Ti)_2[O\|(BO_3)]$	[1]
wigthmanite	$Mg_5[(OH)_5\|O\|(BO_3)] \cdot 2H_2O$	[1]

References for 3.8:

[1] Hawthrone, F. C. (Can. Mineral. **24** [1986] 625/42).
[2] Sugiyama, K.; Takéuchi, Y. (Z. Kristallogr. **173** [1985] 293/304).
[3] Yesinowski, J. P.; Eckert, H.; Rossman, G. R. (J. Am. Chem. **110** [1988] 1367/75).
[4] London, D.; Zolensky, M. E.; Roedder, E. (Can. Mineral. **25** [1987] 173/80).
[5] Nakai, I.; Okada, H.; Masutomi, K.; Koyama, E.; Nagashima, K. (Am. Mineral. **71** [1986] 1234/6).
[6] Miyawaki, R.; Nakai, I.; Nagashima, K. (Acta Crystallogr. C **41** [1985] 13/5).
[7] Griffen, D. T. (Am. Mineral. **73** [1988] 1138/44).
[8] Fischer, W.; Thomas, R.; Loos, G. (Chem. Erde **48** [1988] 33/8 from C.A. **110** [1989] No. 26763).
[9] Gurevich, V. M.; Gorbunov, V. E.; Aksemova, T. D.; Gavrichev, K. S.; Khodanovskii, I. L. (Zh. Fiz. Khim. **62** [1988] 3110/3; Russ. J. Phys. Chem. **62** [1988/89] 1625/6).
[10] Gode, H.; Majore, I.; Sokolova, E. V.; Yamnova, N. A. (Latv. PSR Zinat. Akad. Vestis Kim. Ser. **1986** 532/3).

[11] Gode, H.; Bernane, A. (Latv. PSR Zinat. Akad. Vestis Kim. Ser. **1989** 273/4).
[12] Sokolova, E. V.; Egorov-Tismenko, Yu. K.; Kargal'tsev, S. V.; Klyakhin, V. A.; Urnsov, V. S. (Vestn. Mosk. Univ. Geol. **1987** No. 3, pp. 82/4 from C.A. **108** [1988] No. 67644).
[13] Rumanova, I. M.; Razmanova, Z. P. (Kristallokhim. Rentgenogr. Mineralov L. **1987** 106/12 from C.A. **108** [1988] No. 41194).
[14] Lisitsyn, A. E.; Rudnev, V. V.; Gaft, A. L.; Dobrovol'skaya, N. V.; Dara, O. M.; Tkacheva, T. V. (Zap. Vses. Mineral. Obshch. **114** [1985] 62/73 from C.A. **103** [1985] No. 9130).
[15] Cooper, J. J.; Tilley, R. J. D. (J. Solid State Chem. **58** [1985] 375/82; J. Solid State Chem. **63** [1986] 129/38).
[16] Norrestam, R.; Dahl, S.; Bovin, J.-O. (Z. Kristallogr. **187** [1989] 201/11).
[17] Wei, Dongyan (Yanshi Kuangwu Ji Ceshi **4** [1985] 313/8 from C.A. **104** [1986] No. 189866).
[18] Grice, J. D.; Ercit, T. C. (Can. Mineral. **24** [1986] 675/8).
[19] Lisitsyn, A. E.; Rudnev, V. V.; Yurkina, K. V. (Mineral. Zh. **7** No. 5 [1985] 32/40 from C.A. **104** [1986] No. 92226).
[20] Kwak, T. A. P.; Nicholson, M. (Mineral. Mag. **52** No. 368 [1988] 713/6 from C.A. **110** [1989] No. 11175).

[21] Norrestam, R.; Bovin, J.-O. (Z. Kristallogr. **181** [1987] 135/49).
[22] Rosenberg, P. E.; Foit, F.; Ekambaran, V. (Am. Mineral. **71** [1986] 971/6).
[23] Saiko, I. G.; Kononova, G. N.; Burlya'ev, V. V.; Tavrorskaya, A. Ya. (Zh. Neorg. Khim. **30** [1985] 1149/53; Russ. J. Inorg. Chem. **30** [1985] 649/52).
[24] Ghose, S. (Am. Mineral. **70** [1985] 636/7).
[25] Rachlin, A. L.; Mandarino, J. A.; Murowchik, B. L.; Ramik, R. A.; Dunn, P. J.; Back, M. E. (Can. Mineral. **24** [1986] 689/93).
[26] Peterson, O. V. (Riv. Min. Ital. No. 1 [1989] pp. 39/48 from C.A. **113** [1990] No. 9528).

Physical Constants and Conversion Factors

Avogadro constant N_A (or L)	$=6.02214\times10^{23}$ mol^{-1}	Planck constant	$h=6.62608\times10^{-34}$ J·s
Faraday constant	$F=9.64853\times10^{4}$ C/mol	elementary charge	$e=1.60218\times10^{-19}$ C
molar gas constant	$R=8.31451$ J·mol^{-1}·K^{-1}	electron mass	$m_e=9.10939\times10^{-31}$ kg
molar volume (ideal gas) (273.15 K, 101 325 Pa)	$V_m=2.24141\times10^{1}$ L/mol	proton mass	$m_p=1.67262\times10^{-27}$ kg

1 kg = 2.205 pounds
1 m = 3.937×10^{1} inches = 3.281 feet
1 m³ = 2.642×10^{2} gallons (U.S.)
1 m³ = 2.200×10^{2} gallons (Imperial)

Force	N	dyn	kp
1 N	1	10^{5}	1.019716×10^{-1}
1 dyn	10^{-5}	1	1.019716×10^{-6}
1 kp	9.80665	9.80665×10^{5}	1

Pressure	Pa	bar	kp/m²	at	atm	Torr	lb/in²
1 Pa = 1 N/m²	1	10^{-5}	1.019716×10^{-1}	1.019716×10^{-5}	9.86923×10^{-6}	7.50062×10^{-3}	1.450378×10^{-4}
1 bar = 10^{6} dyn/cm²	10^{5}	1	1.019716×10^{4}	1.019716	9.86923×10^{-1}	7.50062×10^{2}	1.450378×10^{1}
1 kp/m² = 1 mm H_2O	9.80665	9.80665×10^{-5}	1	10^{-4}	9.67841×10^{-5}	7.35559×10^{-2}	1.422335×10^{-3}
1 at (technical)	9.80665×10^{4}	9.80665×10^{-1}	10^{4}	1	9.67841×10^{-1}	7.35559×10^{2}	1.422335×10^{1}
1 atm = 760 Torr	1.01325×10^{5}	1.01325	1.033227×10^{4}	1.033227	1	7.60×10^{2}	1.469595×10^{1}
1 Torr = 1 mm Hg	1.333224×10^{2}	1.333224×10^{-3}	1.359510×10^{1}	1.359510×10^{-3}	1.315789×10^{-3}	1	1.933678×10^{-2}
1 lb/in² = 1 psi	6.89476×10^{3}	6.89476×10^{-2}	7.03069×10^{2}	7.03069×10^{-2}	6.80460×10^{-2}	5.17149×10^{1}	1

Work, Energy, Heat	J	kW·h	kcal	Btu	eV
1 J = 1 W·s = 1 N·m = 10^7 erg	1	2.778×10^{-7}	2.39006×10^{-4}	9.4781×10^{-4}	6.242×10^{18}
1 kW·h	3.6×10^{6}	1	8.604×10^{2}	3.41214×10^{3}	2.247×10^{25}
1 kcal	4.1840×10^{3}	1.1622×10^{-3}	1	3.96566	2.6117×10^{22}
1 Btu (British thermal unit)	1.05506×10^{3}	2.93071×10^{-4}	2.5164×10^{-1}	1	6.5858×10^{21}
1 eV	1.602×10^{-19}	4.450×10^{-26}	3.8289×10^{-23}	1.51840×10^{-22}	1

1 cm^{-1} ≙ 1.239842×10^{-4} eV
2 rydberg = 1 hartree = 27.2114 eV
1 Hz ≙ 4.135669×10^{-15} eV
1 eV ≙ 96.485 kJ/mol

Power	kW	hp	$kp\cdot m\cdot s^{-1}$	kcal/s
1 kW = 10^3 J/s	1	1.35962	1.01972×10^{2}	2.39006×10^{-1}
1 hp (horsepower, metric)	7.3550×10^{-1}	1	7.5×10^{1}	1.7579×10^{-1}
1 $kp\cdot m\cdot s^{-1}$	9.80665×10^{-3}	1.333×10^{-2}	1	2.34384×10^{-3}
1 kcal/s	4.1840	5.6886	4.26650×10^{2}	1

References:

Mills, I. (Ed.), International Union of Pure and Applied Chemistry, Quantities, Units and Symbols in Physical Chemistry, Blackwell Scientific Publications, Oxford 1988.
The International System of Units (SI), National Bureau of Standards Spec. Publ. 330 [1972].
Landolt-Börnstein, 6th Ed., Vol. II, Pt. 1, 1971, pp. 1/14.
ISO Standards Handbook 2, Units of Measurement, 2nd Ed., Geneva 1982.
Cohen, E. R., Taylor, B. N., Codata Bulletin No. 63, Pergamon, Oxford 1986.

Key to the Gmelin System of Elements and Compounds

System Number	Symbol	Element
1		Noble Gases
2	H	Hydrogen
3	O	Oxygen
4	N	Nitrogen
5	F	Fluorine
6	**Cl**	**Chlorine**
7	Br	Bromine
8	I	Iodine
8a	At	Astatine
9	S	Sulfur
10	Se	Selenium
11	Te	Tellurium
12	Po	Polonium
13	B	Boron
14	C	Carbon
15	Si	Silicon
16	P	Phosphorus
17	As	Arsenic
18	Sb	Antimony
19	Bi	Bismuth
20	Li	Lithium
21	Na	Sodium
22	K	Potassium
23	NH_4	Ammonium
24	Rb	Rubidium
25	Cs	Caesium
25a	Fr	Francium
26	Be	Beryllium
27	Mg	Magnesium
28	Ca	Calcium
29	Sr	Strontium
30	Ba	Barium
31	Ra	Radium
32	**Zn**	**Zinc**
33	Cd	Cadmium
34	Hg	Mercury
35	Al	Aluminium
36	Ga	Gallium

System Number	Symbol	Element
37	In	Indium
38	Tl	Thallium
39	Sc, Y La—Lu	Rare Earth Elements
40	Ac	Actinium
41	Ti	Titanium
42	Zr	Zirconium
43	Hf	Hafnium
44	Th	Thorium
45	Ge	Germanium
46	Sn	Tin
47	Pb	Lead
48	V	Vanadium
49	Nb	Niobium
50	Ta	Tantalum
51	Pa	Protactinium
52	**Cr**	**Chromium**
53	Mo	Molybdenum
54	W	Tungsten
55	U	Uranium
56	Mn	Manganese
57	Ni	Nickel
58	Co	Cobalt
59	Fe	Iron
60	Cu	Copper
61	Ag	Silver
62	Au	Gold
63	Ru	Ruthenium
64	Rh	Rhodium
65	Pd	Palladium
66	Os	Osmium
67	Ir	Iridium
68	Pt	Platinum
69	Tc	Technetium[1]
70	Re	Rhenium
71	Np, Pu...	Transuranium Elements

Material presented under each Gmelin System Number includes all information concerning the element(s) listed for that number plus the compounds with elements of lower System Number.

For example, zinc (System Number 32) as well as all zinc compounds with elements numbered from 1 to 31 are classified under number 32.

---

[1] A Gmelin volume titled "Masurium" was published with this System Number in 1941.

A Periodic Table of the Elements with the Gmelin System Numbers is given on the Inside Front Cover